Green Stormwater Infrastructure Fundamentals and Design

# Green Stormwater Infrastructure Fundamentals and Design

*Allen P. Davis, William F. Hunt, and Robert G. Traver*

This edition first published 2022


*Registered Office*
John Wiley & Sons, Inc., 111 River Street, Hoboken, NJ 07030, USA

*Editorial Office*
111 River Street, Hoboken, NJ 07030, USA

For details of our global editorial offices, customer services, and more information about Wiley products visit us at www.wiley.com.

Wiley also publishes its books in a variety of electronic formats and by print-on-demand. Some content that appears in standard print versions of this book may not be available in other formats.



A catalogue record for this book is available from the Library of Congress

Hardback ISBN: 9781118590195; ePDF ISBN: 9781119339779; ePub ISBN: 9781119338024; oBook ISBN: 9781119339786

Cover image: Courtesy of Allen P. Davis
Cover design by Wiley

Set in 9.5/12.5pt STIXTwoText by Integra Software Services Pvt. Ltd., Pondicherry, India

SKY10035589_080122

*To our families and to creating a better earth for our children and grandchildren.*

# Contents

# Preface

The intention of the authors is to present the fundamentals of green urban stormwater infrastructure from an engineering design and performance analysis perspective. This book is intended to be used as a textbook in senior-undergraduate and first-year graduate courses in water resources/environmental engineering. It is also envisioned to be a reference for practicing engineers and other water/environment professionals. The book focuses on novel stormwater control measures (SCMs) and related technologies for the reductions of detrimental impacts from urban stormwater. Stormwater challenges have risen in importance as clean water focus has shifted from point to nonpoint source pollution as a source of water impairments. Stormwater also becomes part of the "one water" focus on long-term sustainable urban water. Many novel SCMs are nature-based and are considered as part of a "green infrastructure" approach that includes bioretention, vegetated swales, vegetated filter strips, green roofs, pervious pavements, water harvesting, and wetlands.

It is expected that users of this book would have had a course in engineering hydraulics/hydrology and some exposure to environmental engineering treatment processes and water quality. It is also complementary to graduate surface water hydrology and traditional water and wastewater treatment engineering. While written with an engineering focus, nonengineers such as landscape architects, planners, and environmental scientists should find the text useful. Specific attempts have been made to integrate both English (US customary) and metric units throughout the book.

The initial chapters provide background information on urban hydrology, water quality, and stormwater generation and characteristics. The preponderance of the book focuses on stormwater control and improvement via a suite of different green infrastructure technologies and techniques. Within this context, background information on engineering unit processes for affecting the water balance and improving water quality are presented. The evolving challenge of setting and meeting stormwater control metrics is discussed. The latter chapters provide specific details on categories of SCMs; topics such as selection, design, performance, and maintenance are presented in detail. SCM selection, treatment trains, and climate change are included as a final chapter. This text provides a baseline as this topic is a rapidly changing field.

## About the Authors

**Allen P. Davis**, Ph.D., P.E., D. WRE, F.EWRI, F.ASCE, is the Charles A. Irish Sr. Chair in Civil Engineering and Professor in the Department of Civil and Environmental Engineering and Affiliate Professor in Plant Science and Landscape Architecture at the University of Maryland, College Park, MD. He is the Editor of the *ASCE Journal of Sustainable Water in the Built Environment.*

**William F. Hunt III**, Ph.D., P.E., D. WRE, is a William Neal Reynolds Distinguished University Professor and Extension Specialist in the Department of Biological and Agricultural Engineering at the North Carolina State University, Raleigh, NC. He is the leader of the Stormwater Engineering Group at NC State.

**Robert G. Traver**, Ph.D., P.E., D. WRE, F.EWRI, F.ASCE, is a Professor in the Department of Civil and Environmental Engineering at Villanova University, Villanova, PA, and former Edward A. Daylor Chair in Civil Engineering. He is the Director of the Villanova Center for Resilient Water Systems and the Villanova Stormwater Partnership.

Collectively, the authors have 90 years of research, education, and outreach experience encompassing the topics covered in this book. They have built, maintained, and monitored hundreds of SCM research practices and have authored over 300 refereed journal articles, including several together. They have presented research results all over the world, hosted international conferences, while also helping address state and local water challenges. The authors love each other, the field in which they work, and the people with whom they partner.

# Acknowledgments

The authors are grateful for the contributions of many, many colleagues in the various research projects that have led to many subjects of this text. These include students, post-doctoral researchers, and faculty colleagues. We also thank the various agencies that supported, and continue to support and promote, green stormwater infrastructure research.

# About the Companion Website

This book is accompanied by a companion website which includes a number of resources created by author for instructors that you will find helpful.

www.wiley.com/go/davis/greenstormwater

The Instructor website includes answers to the end-of-chapter problems

Please note that the resources in instructor website are password protected and can only be accessed by instructors who register with the site.

# 1

# Introduction to Urban Stormwater and Green Stormwater Infrastructure

## 1.1 Population and Urban Infrastructure

Human population continues to increase in most areas of the world, including developed countries such as the United States. Two of the basic needs of humans are shelter and community. As we have progressed over the millennia, the ideas of shelter and community have evolved, first from simple villages to larger cities. More recently, these populations are shifting, generally from rural and inland areas to the coasts, while residents of inner cities are migrating to less dense suburban development. Frequently, the result is the consumption of pristine and agricultural land at rates disproportionately greater than population growth. As part of the development process, natural vegetation is replaced by lawn or pavement, soils are disrupted and compacted, pipes replace natural water courses, and the native topography is smoothed. Even in areas of urban redevelopment, frequently the impervious footprint increases as the living infrastructure becomes larger (Boorstein 2005; Hekl and Dymond 2016; MacGillis 2006).

Our past and current land development practices rely heavily on the use of impervious area infrastructure (materials that cover the ground and do not let water infiltrate down into the ground as it would in an undeveloped area) and piped systems. Large-area rooftops for homes and garages, highways, sidewalks, wide driveways, and generous patios are all desired attributes of increasingly affluent (sub)urban areas. Commercial and institutional properties provide for similar large impervious infrastructure and ample (if not excessive) parking. This urban network has replaced lands that were once undeveloped, such as forest, meadow, or open plains.

Rain that falls on developed areas is transported via impervious conveyance systems rapidly away from the original surface contact point, typically being discharged into the nearest waterway. This impervious area, coupled with a drainage system that accelerates the movement of runoff, vastly alters the water balance in the urban system. A variety of problems, including flooding, stream damage, loss of aquatic habitat, and significant downstream water body degradation, are the result. The amount of urbanization and related impervious area created has, and continues to, expand in many areas as demonstrated in Figure 1.1 for the greater Las Vegas area.

*Green Stormwater Infrastructure Fundamentals and Design*, First Edition. Allen P. Davis, William F. Hunt, and Robert G. Traver.

Companion Website: www.wiley.com/go/davis/greenstormwater

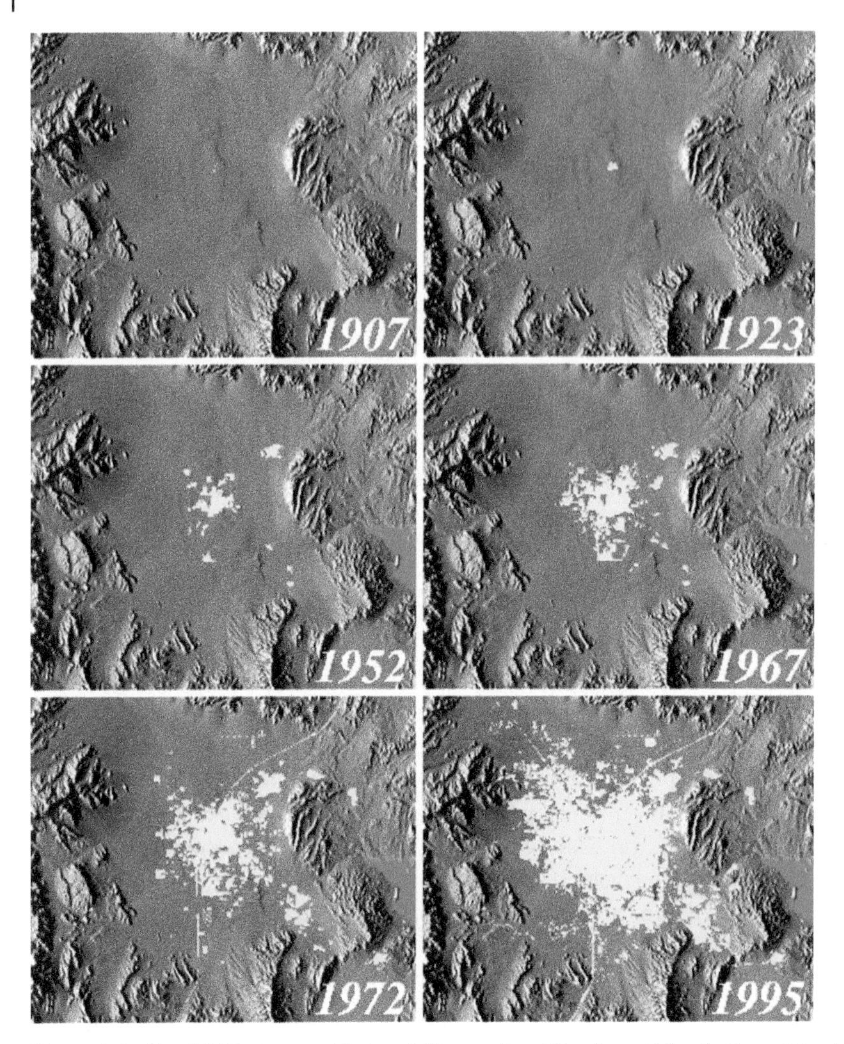


Figure 1.1 Spatial Patterns and Rates of Change Resulting from Urbanization of the Las Vegas Areas. (Credit: US Geological Survey).

## 1.2 Impacts of Urbanization

Our cities, towns, and villages, and the transportation networks that connect them, all rely on impervious infrastructure. Rooftops, roadways, sidewalks, driveways, parking lots, basketball and tennis courts, and patios all direct rainfall rapidly to their periphery, eliminating the natural runoff reduction and filtration of the vegetated systems that have been replaced.

Figure 1.2 shows the water balance around areas with different levels of urban development. In the undeveloped lands (humid regions), about half of the annual incoming water via rainfall infiltrates, supplying both shallow and deep groundwater. Another large fraction of this volume is evaporated from the soil and vegetation and transpired through the leaves of the vegetation, the combined processes known as

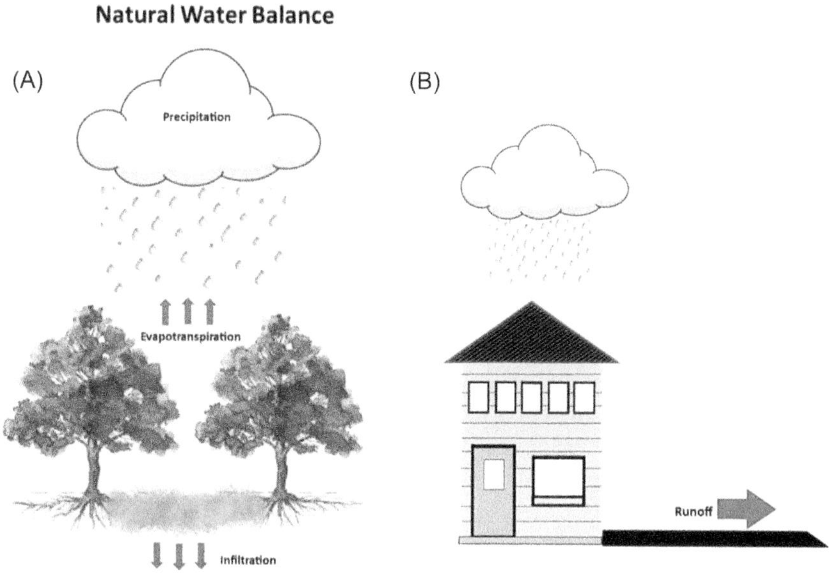

Figure 1.2 Water Balances for Different Land Use Conditions: (A) Natural Water Balance Showing Primary Water Pathways of Evapotranspiration and Infiltration and (B) Urban Water Balance Includes Runoff from Impervious Surfaces.

evapotranspiration (ET). This leaves only a small fraction of the incoming rainfall to become surface runoff.

As the amount of development increases within an area, so does the amount of impervious area. The vegetated land area available for the infiltration and ET of runoff becomes increasingly small. In highly urbanized areas, the water balance changes drastically, as shown in Figure 1.2B. Infiltration and ET are now greatly reduced. The bulk of the incoming rainfall now is converted to surface runoff, which must be responsibly managed so as to not to create public safety and health concerns, and to protect our waterways and water bodies from environmental problems.

Environmental impacts of land development are well known and additional details on these impacts continue to be forthcoming (Booth 2005). The increased volume and flows of stormwater runoff from urbanized areas, coupled with impaired water quality and increased temperature, amplify the magnitude and increase the probability of flooding, decrease stream baseflow, degrade downstream river channels, adversely affect the quality of receiving waters, and impact stream ecology (e.g., Walsh et al. 2005; Wang et al. 2003). High sustained flow rates (not just peaks) are associated with accelerated stream bank erosion and gully formation (Figure 1.3). Elimination of stream baseflow in headwater areas by eliminating rainfall infiltration can greatly impact downstream ecology and ecological processes (Sweeney et al. 2004). Loss of biological nutrient cycling processes in small streams will adversely impact water quality in downstream areas (Peterson et al. 2001).

While certainly flooding occurs with or without urbanization, the changes to the land increase the frequency and magnitude of such events, magnifying the impact to the local waterways. Figure 1.4 shows the great increase in amplitude in flow rate from

Figure 1.3 Stream Impacts from Uncontrolled Stormwater. (Photo by Authors).

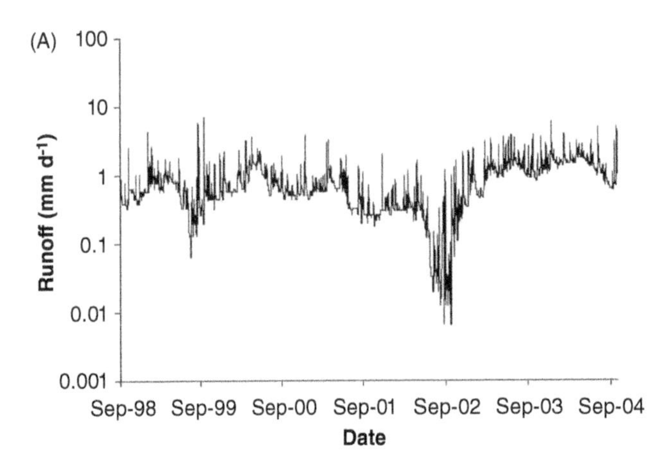


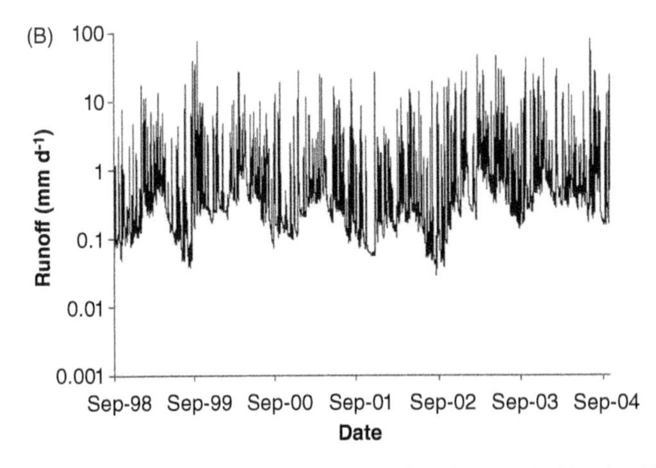


Figure 1.4 Continuous Flow Measured from Streams in Maryland (Normalized by Drainage Area): (A) Forested Stream and (B) Urban Stream. (Shields et al. 2008).

a highly impervious watershed as contrasted to that of a lower impervious watershed (note the log scale). In Figure 1.4A, the flow of a forested stream is given (in units of mm/day, which represents the stream flow divided by the stream catchment area). The flow averages about 1 mm/day, with limited excursions to about 10 mm/day during high flow events and as low as 0.01 mm/day during a very dry period.

Contrast these data to Figure 1.4B, which shows the same data for a highly developed catchment area. The flows are much more erratic and vary significantly throughout the study. Both high and low flows are frequent as the stream responds rapidly to rainfall that falls on the catchment areas.

In the watershed, impervious surface without adequate drainage leads to pooled water during large rain events. This pooled water is dangerous to vehicle travel and pedestrians and can cause flooding of buildings in the urban area. Figure 1.5 shows nuisance flooding in a residential area of New Bern N.C. Note the depth of water as the vehicles pass each other.

Figure 1.5 Nuisance Flooding New Bern NC. (Photo by Authors).

Figures 1.6 and 1.7 show other effects of excess water related to high impervious area. Figure 1.6 clearly shows the accelerated erosion of a drainage swale threatening the stability of the adjacent house. Figure 1.7 shows a flood on the larger Perkiomen Creek in Pennsylvania. While not visible on the photo, cars on the bridge could not move because the bridge approaches were under water. In addition to obvious flood hazards, standing water can lead to other health concerns.

Increased imperviousness from urbanization leads to high flows that also change the river channels through erosion and deposition. Figure 1.8 shows incisions and bank erosion from high flows in streams in Maryland. Over time, soil is washed from tree root structures, the trees become unstable and will fall into the stream.

The relationship between impervious cover and stream biotic health has been documented by many researchers. Figure 1.9 shows declines of macroinvertebrate indicator

**Figure 1.6** Severely Eroded Neighborhood Swale. (Photo by Authors).

**Figure 1.7** Significant Flooding of the Perkiomen Creek, PA. Note Heavy Sediment Load Carried by the River. (Photo by Authors).

taxa in streams in Maryland as a function of the impervious cover in the watershed (King et al. 2011). The dramatic increase in the decline of the taxa demonstrates changes in the physical and chemical conditions of the stream ecosystems. As the fraction of impervious area increases, various alterations to the stream characteristics result, making it a less-favorable habitat for many diverse aquatic species, and indicating poor stream health. This change occurs dramatically, from only about 0.5% to 2% impervious cover.

Figure 1.8 Incised Streams in Maryland, Resulting from Erosive Flows: (A) Small Stream and (B) Large Stream. (Photos by Authors).

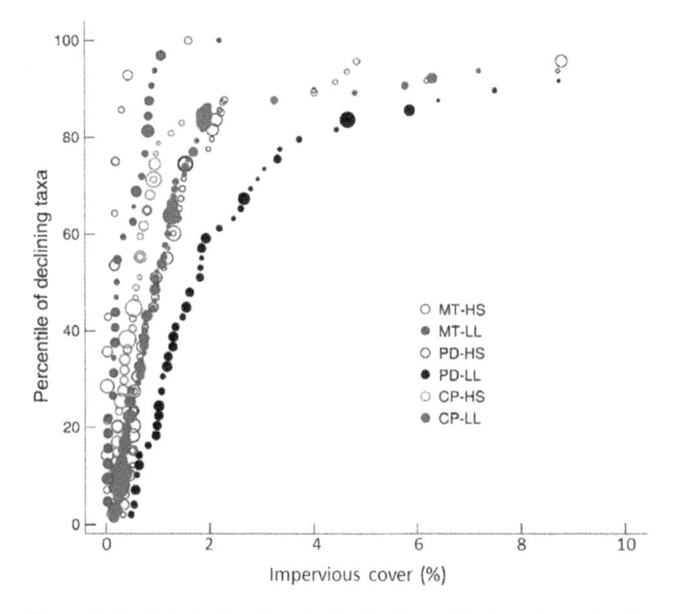


Figure 1.9 Data Indicating the Reduction of Various Organism Populations with Increasing Watershed Urbanization (Impervious Coverage) (King et al. 2011). MT, PD, and CP Represent Mountain, Piedmont, and Coastal Plain Geology, Respectively. HS Represents High Slope Small Watersheds; LL Represents Low Slope Large Watersheds.

## 1.3 The US Regulatory Environment

The governing legislation driving urban stormwater management in the United States is the Clean Water Act (CWA). The CWA was promulgated in the early 1970s to address water pollution in waters of the United States, with a goal to "restore and maintain the chemical, physical, and biological integrity of the nation's waters." Initially enforcement of the CWA focused on discharges of wastewater (sometimes untreated) from municipal wastewater treatment plants and from various industries. This enforcement led to the development of the National Pollutant Discharge Elimination System (NPDES) program. NPDES programs are managed by the states and establish a permitting process for any entity that discharges to the nation's waters. NPDES permits for industry and wastewater treatment plants commonly specify limits for several water-quality parameters. The limits will depend on the industry and the water body into which the discharge occurs.

In 1987, the Water Quality Act, a modification to the CWA, required stormwater discharges to operate under the NPDES system. This includes municipal, construction, and industrial stormwater; agricultural runoff was removed so that it is not regulated under the CWA.

Regulation of municipal separate storm sewer systems (MS4s) was implemented in two phases. In the first, implemented in 1990, large jurisdictions (cities and counties), defined as those with population of 100,000 or more, were issued NPDES permits for their stormwater. Phase I covers about 750 municipalities in the United States (www.epa.gov). Figure 1.10 displays a timeline of stormwater regulatory actions and milestones.

Early CWA regulatory actions primarily focused on point source impacts and have been successful at reducing their impact significantly. Point sources are direct (treated) wastewater discharges from municipal wastewater treatment plants and from industries. As a result of this regulatory structure, the majority of the US water body impairment sources shifted from point to non-point sources (Figure 1.11). Non-point sources

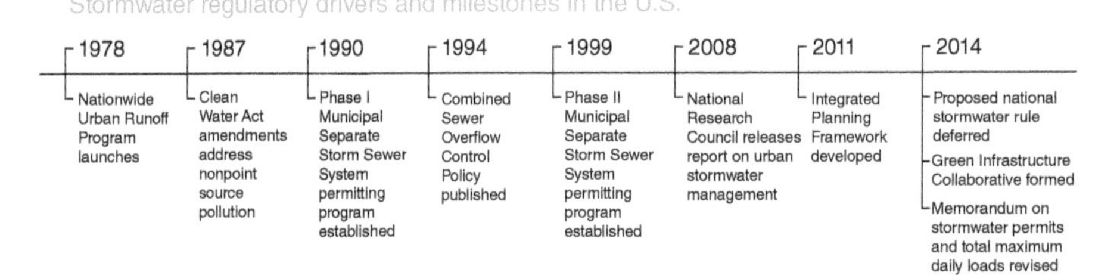


Figure 1.10 Stormwater Regulatory Drivers and Milestones in the United States (with Permission, Water Environment Federation, WEF 2015).

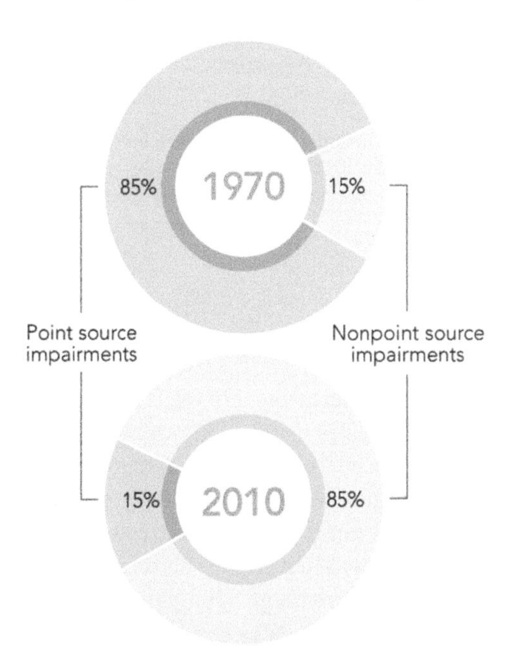


Figure 1.11 Shift of Balance of Impairment Sources from Point to Non-Point after Initial Enforcement of the Clean Water Act (with Permission, Water Environment Federation, WEF 2015).

are primarily stormwater from urban, highway, industrial, construction, and agricultural land uses.

Recognizing the need to address non-point sources, NPDES Phase II was implemented in 1999 targeting smaller urbanizing areas. Phase II covers approximately 6700 jurisdictions (www.epa.gov) and requires programs to reduce pollutant discharge to the "maximum extent practicable" (MEP), protect water quality, and meet the water-quality requirements of the CWA.

In all but five cases, the authority for NPDES permitting and enforcement for Phases I and II has been delegated to each respective state. MS4 NPDES permits are generally issued in 5-year cycles. Stormwater NPDES permits have focused on implementing "Best Management Practices" and public education for stormwater control, targeting runoff from diffuse surfaces. These best management practices (BMPs) can be structural stormwater control measures (SCMs) or nonstructural practices, such as street sweeping, both of which are discussed in later chapters. Recently, especially in areas in which surface water quality has remained poor, NPDES permits are becoming increasingly stringent for both Phases I and II communities.

Twenty-seven industrial sectors are included under the industrial stormwater program. The US EPA has created a multi-sector general permit (MSGP) that specifies benchmark monitoring for most of these sectors. The benchmark monitoring is used as a measure of the effectiveness of stormwater management at the site. Construction permits cover construction activities and focus on land disturbances. The general permits have identical provisions for all facilities under the same sector. For large facilities with unique challenges, an individual NPDES permit can be issued. Frequently, an individual permit would cover all water discharges at a facility: stormwater and process wastewater.

Late in 2000, the CWA was amended to address combined sewer overflows (CSOs). Many older cities combined street drainage and sewage collection and conveyance in the same piping system; originally these networks discharged directly to local water bodies (Figure 1.12). Over time these pipe networks were redirected to wastewater treatment plants. However, with these combined systems, larger stormwater events (0.5 in. (1.2 cm) and up) can overload the pipe and treatment systems, causing

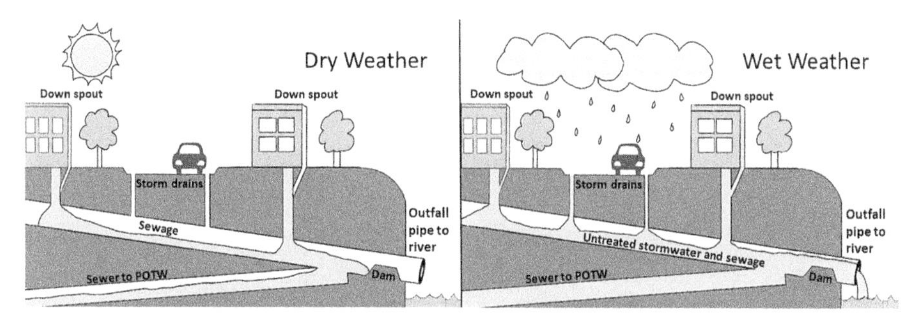


Figure 1.12 A Combined Sewer System. During Dry Weather (and Small Storms), All Wastewater and Stormwater Flows are Handled by the Publicly Owned Treatment Works (POTW). During Large Storms, the Relief Structure Allows Some of the Combined Stormwater and Wastewater to Be Discharged Untreated to an Adjacent Water Body.

discharge of untreated stormwater and sewage, an event known as a CSO. The new legislation requires cities with combined sewers to develop long-term control plans to reduce the impacts of CSOs, to bring them into compliance with the CWA. Figure 1.13 shows CSO locations in the New York City area.

Another section of the CWA that impacts stormwater is the *Total Maximum Daily Load* (TMDL). Water bodies of the United States are designated for specific uses, usually by the respective states. These uses can include drinking, swimming, fishing, and so on. Under Section 303(d) of the CWA, water bodies that cannot meet their designated use, because of poor water quality, are labeled as *impaired*. The impairment is attributed to a specific water-quality parameter, such as bacteria, nutrients, or sediment.

When a water body is classified as impaired, the CWA requires the establishment of a TMDL. A TMDL is set for a water body based on estimates of the pollutant load (mass) that the water body can adequately manage yet still meet its designated uses. In

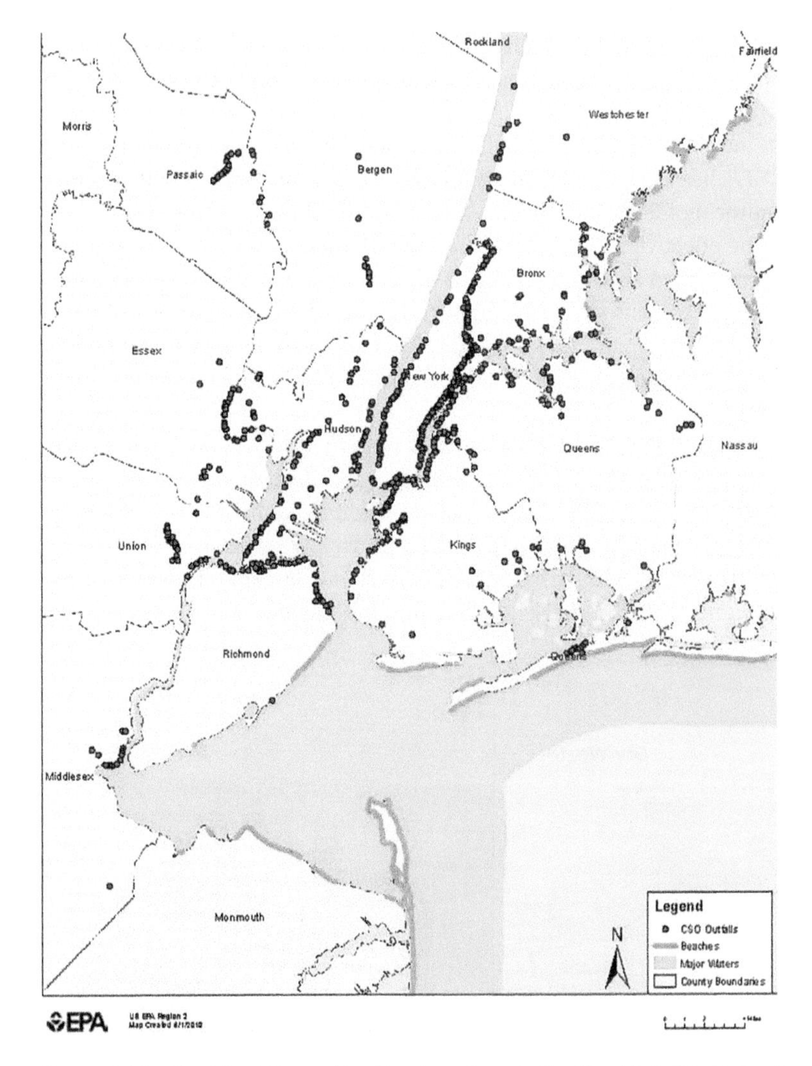


Figure 1.13 Combined Sewer Overflow Locations in the New York City Metro Area. (Credit: U.S. EPA 2011a).

an impaired water body, the overall pollutant load to a water body exceeds the TMDL. In this case, specific reductions to the various water discharge sectors will be required, a so-called pollution diet plan to eliminate the water body state of impairment and return the water quality to the designated use condition. These sectors include municipal and industrial wastewater discharges, agricultural runoff, and urban runoff. Increasingly, TMDL concerns are being written into MS4 NPDES permits. The result can be very stringent requirements for the management and control of urban stormwater.

In addition, many states have developed their own regulations to address stormwater impacts. Most of these state requirements started as flood control criteria and focused on peak runoff flow rates from the site during extreme events. Pennsylvania, for example, passed its stormwater management act in 1978 in response to Hurricane Agnes. While the language of the act addressed increase of runoff from developing areas, the act was interpreted as requiring that the peak flow leaving the project site be maintained at preconstruction levels for extreme events. Later this requirement evolved into reducing peak flow after construction to less than preconstruction in an effort to consider the downstream watershed (Traver and Chadderton 1983).

As the focus of stormwater management has shifted over the past decade to addressing smaller storms, many states and municipalities added volume control to their stormwater regulations. While it is argued that volume or peak rates are not addressed under the CWA, it is not possible to address environmental quality without it (NRC 2009). Table 1.1 compares stormwater volumetric requirements for a few states for comparison.

Table 1.1 Volumetric Retention Standards for Discharges from New Development (Compiled from U.S. EPA 2011b).

| State or locality (date enacted) | Size threshold | Standard |
|---|---|---|
| Vermont (2003, draft 2010) | 1 acre | Capture 90% of the annual storm events |
| New Hampshire (2009) | 1 acre/100,000 ft$^2$ outside MS4 | Infiltrate, evapotranspire or capture first 1.0 in. from 24-h storm |
| Wisconsin (2010) | 1 acre | Infiltrate runoff to achieve 60–90% of predevelopment volume based on impervious cover level |
| West Virginia (2009) | 1 acre | Keep and manage on site 1 in. rainfall from 24-h storm preceded by 48 h of no rain |
| Montana (2009) | 1 acre | Infiltrate, evapotranspire, or capture for reuse runoff from first 0.5 in. of rain |
| Portland, OR (1990) | 500 ft$^2$ of impervious cover | Infiltrate 10-year, 24-h storm |
| Anchorage, AK (2009) | 10,000 ft$^2$ | Keep and manage the runoff generated from the first 0.52 in. of rainfall from a 24-h event preceded by 48 h of no measurable precipitation |

## 1.4 Urban Stormwater Management

As stated earlier, without the ability to infiltrate, rain that falls on impervious surfaces will collect and travel quickly over these surfaces, moving polluted waters to our stream systems and causing erosion and sediment deposition. In a highly developed area, without a place to go, this water will pool, creating a flooding hazard, and increase flooding in area streams.

### 1.4.1 Flood Control

The first generation of stormwater management was developed to reduce flooding hazards. Storm drains and storm sewer networks were installed to collect runoff from impervious areas. These drains were directed into the nearest stream or river so that rainfall that fell on the impervious area could be conveyed away as quickly as possible. In many older cities, the sanitary sewer system (for conveyance of wastewater to treatment plants) was already in place. In some situations, the urban flooding challenge was addressed by piping the stormwater into the sanitary sewer networks, creating combined sewers. These engineering projects addressed the urban flooding problem but created others.

During heavy rainfall events, these drainage systems put a tremendous water burden on the repository of the flow, either the stream outfall or the sanitary conveyance and treatment network. This increased flow comes quickly, with high volumes and velocities. The streamflows are increased dramatically, resulting in erosion of the streambed, scour, and stream flooding. Loss of aquatic habitat occurs, including beneficial stream processes, such as nitrogen processing. These problems associated with stormwater discharges have been termed *urban stream syndrome* (Barco et al. 2008; Walsh et al. 2005). In many cases, due to the perceived need for space and to prevent erosion, entire streams were replaced with concrete channels and ditches (Figure 1.14).

**Figure 1.14** Hardened Urban Stream, Crow Branch in Laurel, MD. (Photo by Authors).

During heavy rains in combined sewer areas, very large volumes of water are dumped to the sanitary sewer system. This runoff volume can be too much for the sewer network and wastewater treatment plant to handle. As a result, relief areas are constructed into the sewer system so that if the flows become too large, they will overflow into the nearby streams and rivers. The result of this relief is that during large rainfall events, runoff, mixed with raw sewage, is directly discharged, untreated into the environment. This condition, obviously, creates major public health and environmental problems and is a violation of the CWA. CSOs can occur many times per year in some cities.

### 1.4.2 Peak Flow Control

Recognizing that direct connections to the nearby streams were causing environmental damage to the streams and surroundings, efforts were subsequently made to incorporate some degree of runoff storage to reduce extreme event peak flows into the newer stormwater systems that were being installed. Generally, this consisted of some type of dry or wet pond that was placed between the new impervious infrastructure and the receiving stream. This pond would fill during the rain event and was managed with weirs so that it would restrict the outflow to preconstruction levels. Figure 1.15 shows an early 1980s Pennsylvania wet pond, designed to hold peak flows at preconstruction levels for the 24 hour 2–100-year design storms (Chapter 6). The ponds were designed to be deep to prevent growth of vegetation.

The ponds addressed the peak flow problem directly at the point of design, but still the challenge of high erosive flows remained, which was commonly exacerbated by the combination effect of multiple individually designed storage facilities

Figure 1.15 Stormwater Management Retention Pond Circa 1980s. (Photo by Authors).

within a watershed (Emerson et al. 2005; Traver and Chadderton 1983). While arguably effective at the property line for extreme events, the increased volume and extended increased velocities exacerbated the erosive discharge for the stream. McCuen and Moglen (1988) stated, "Both theory and experience indicate that, while detention basins designed to control peak discharge are effective in controlling the peak rates, the basins are ineffective in controlling the degradation of erodible channels downstream of the basin."

### 1.4.3 Watershed Approach to Peak Flow

Recognizing that timing of release from detention basins could actually increase flood peaks (Emerson et al. 2005; McCuen and Moglen 1988), many regions in the 1980s started to require downstream analysis for extreme events to ensure that the cumulative increased peak flow effects from detention basins did not increase river flows downstream of the developed properties. Termed *Release Rates*, this analysis was codified based on watershed modeling of extreme design storms, often requiring that outflows from individual extreme events be reduced below preconstruction levels to avoid unintentional downstream peak flow increases due to the extended outflow of runoff. For example, it is common to require that the 2-, 10-, and 100-year storm to be reduced to a fraction of the preconstruction peak level, often as much as 50%, resulting in much larger regulatory structures.

A few areas, early on moved away from individual storm analysis, instead using a continuous simulation approach to look at the annual impact. As the mechanisms for stream erosion and sedimentation are related to both flow rate and duration of these flows, Western Washington requires a continuous simulation analysis that demonstrates that the postconstruction flow durations are held for selected extreme events ranging from 50% of the 2-year storm to that of the 50-year storm (Ecology 2005).

### 1.4.4 Water-Quality Control

In the 1990s, it was recognized that more and more, urban runoff was a significant contributor to quality problems in receiving waters. Regulations promulgated in the 1970s and 1980s placed severe restrictions on discharges from point sources, that is, industrial and municipal wastewater treatment plants. As the water quality from industrial discharges improved, and more urban infrastructure was installed, pollutant loads from non-point sources, such as urban runoff, were becoming a significant contributor of the overall pollutant burden of many water bodies (Amandes and Bedient 1980).

In response, water-quality requirements were added to stormwater regulations. In many jurisdictions, this led to the definition of a *water quality volume*. The water-quality volume is a runoff volume defined by the regulatory agency that must be captured and treated. This volume is found as a rainfall or runoff depth, over the drainage area (or some fraction of the drainage area). It is assumed that the majority of the pollutant load is present in this initial runoff volume (Chapter 3), and that it is a significant fraction of the annual runoff.

## 1.5 Climate Change and Stationarity

Most hydrologic design is based on the concept of stationarity. Stationarity assumes that events of the future can be predicted by understanding events from the past; that is the population distribution of events does not change. Standard hydrologic design has always assumed that rainfall frequencies are constant over the long term. This allows us to design infrastructure based on probabilities for rainfall, floods, and so on. Return periods used for design are based on historical data sets that are assumed to match future events.

Nonetheless, global climate is changing now. Overall global temperatures are increasing (Melillo et al. 2014). This impacts the hydrologic cycle, and accordingly, stormwater, in many ways. Additionally, changes at the regional and local level can be very different from global trends. The most recent data and predictions indicate that generally areas will become wetter and exposed to more intense rainfall during wet seasons and dryer during dry seasons. As precipitation is a key driver of SCM performance, changes to precipitation volume, intensity, and frequencies will drive our stormwater management approaches. Much of the United States is expecting more frequent and higher intensity events, with periods of increased drought. Regulatory goals and design concepts will need to be rethought as precipitation patterns change. As will be discussed in future chapters, green stormwater infrastructure (GSI) can be more flexible and resilient than traditional curb, gutter, and piping systems ("gray infrastructure"), as there may be some dampening of the effect due its functional dependence on natural processes.

## 1.6 Green Stormwater Infrastructure

As stormwater management criteria expanded, better ways to address the urban runoff challenge were developed and, a number of topics began to emerge and coalesce. Philosophies were introduced, such as implementing ways to manage stormwater directly at the source, rather than downstream after it has been combined with flow from large areas. New performance metrics were discussed, focusing on having the land behave hydrologically similar to that when it was undeveloped, a goal of restoring the watershed to "pre-development hydrology." These ideas led to interest in incorporating green space into the urban areas and making these green spaces functional with respect to hydrology and water-quality management. This philosophy has gradually matured under several different concept titles. An example of this concept is shown in Figure 1.16, with a water balance diagrammed in Figure 1.17. These titles include low-impact development, sustainable urban drainage, and GSI.

The ideals of GSI are to mitigate the deleterious effects of urban stormwater using natural processes such as vegetation and soils at or near where the rain falls. The water balance in urban areas is modified so that less surface runoff is created and more rainfall is allowed to infiltrate and evapotranspire. Water quality is improved by various natural processes, including sedimentation, filtration, adsorption, and biological processes. Overall, the water balance more closely mimics the preconstruction conditions.

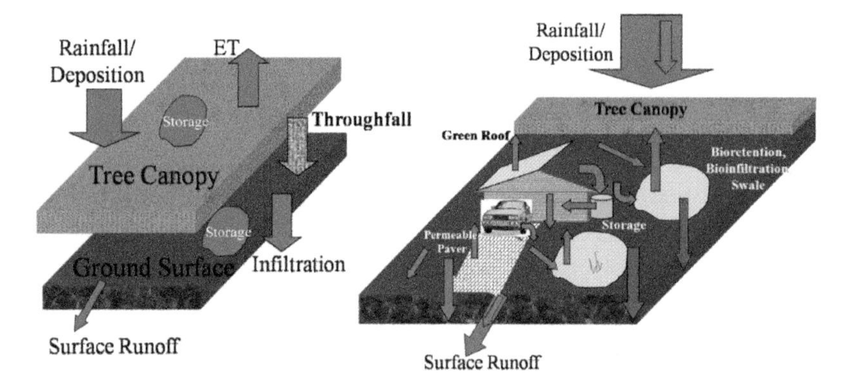


Figure 1.16 Land Development Using the Various Concepts of Green Stormwater Infrastructure to Mimic Pre-Development Hydrologic and Water Quality Conditions.

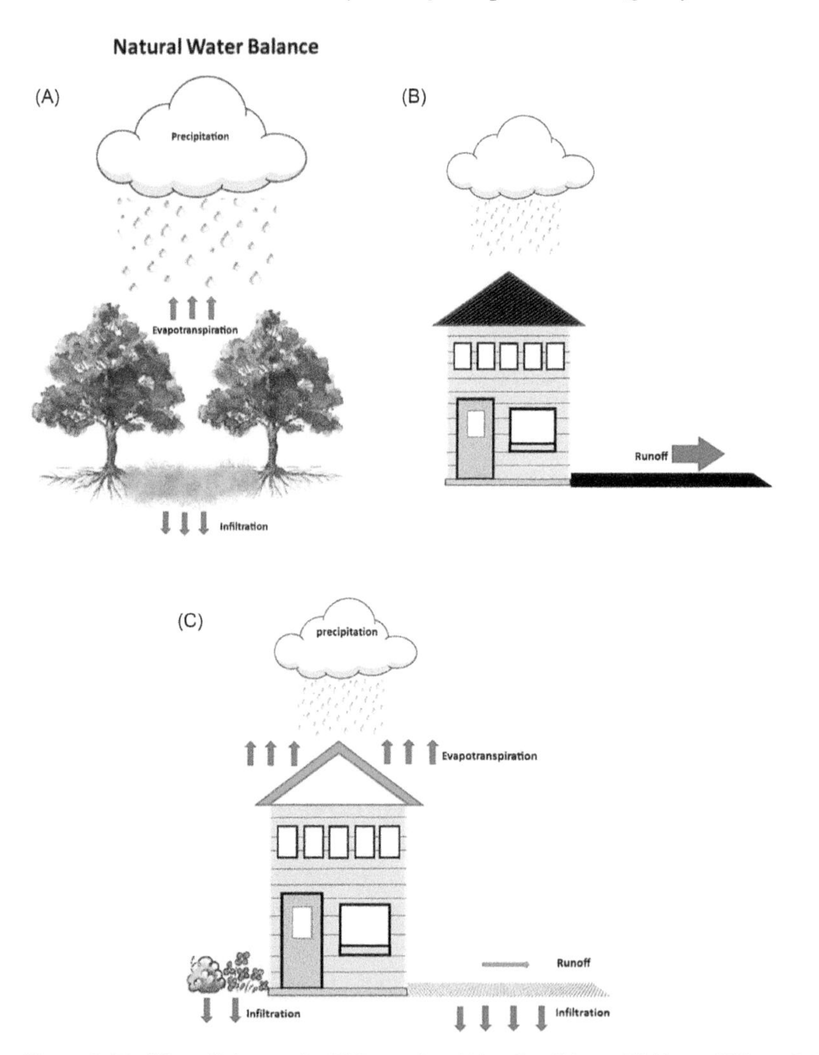


Figure 1.17 Water Balances for Different Land Use Conditions: (A) Natural Water Balance Showing Primary Water Pathways of Evapotranspiration and Infiltration; (B) Urban Water Balance Includes Runoff from Impervious Surfaces; and (C) Green Infrastructure Water Balance Promotes Evapotranspiration and Infiltration in the Built Environment from Green Roofs, Permeable Pavements, and Other Stormwater Control Measures.

## 1.7 Stormwater Control Measures

A number of techniques and processes are employed to reduce the impacts of urban stormwater runoff. Collectively, these are known as SCMs. These processes have been historically designated as stormwater BMPs. This designation is still prevalent but is not as precise and specific as SCM (NRC-National Research Council 2009).

Common green infrastructure SCMs include vegetated technologies such as vegetated swales and filter strips, rain gardens and bioretention, green roofs, and wetlands. Other SCMs that reduce runoff and are also considered part of green stormwater infrastructure include water-harvesting technologies, infiltration basins, and permeable pavements. Green infrastructure SCMs attempt to beneficially affect the urban water balance, and to reduce the pollutant load, to counteract the problems created by urban development, as noted in Figure 1.17. These SCMs provide storage and promote infiltration and ET of rain and runoff, reducing volumes, flows, and velocities. In most SCMs, specific design and operational characteristics will promote the inclusion of environmental unit processes that will improve (or in some cases of poor design, worsen) water quality in the runoff. (Water-quality improvement may be less of an issue in CSO watersheds.) Selection and sizing of SCMs depend on many parameters, including catchment area size and land use, hydrologic and water-quality goals, soil and geologic conditions, and available land space for the SCM. More and more, knowledge is available to tailor specific SCM selection and designs to area needs, and even reversing development impacts through retrofitted older paved areas. SCMs are engineered technologies and techniques that will follow specific hydrologic and water-quality improvement process rules. This selection process and design parameters are covered in the later chapters of this book.

## 1.8 Stormwater Infrastructure and Equity

It is recognized that stormwater infrastructure historically has not been equitably distributed throughout the built landscape. It is well documented that people of color and other minorities, through various policies and initiatives, have been forced to live in areas that are more prone to negative environmental factors, including flooding, poor infrastructure, and air pollution.

Incorporation and selection of green stormwater infrastructure in any neighborhood, but specifically in underserved areas, must be carefully done, with consideration of the past, present, and future. This is necessary to balance equity and to address previous poor and racist decisions and policies. GSI implementation must balance the needs and recognize the history of the neighborhood. Sudden large investments in infrastructure, including GSI, will alter, maybe drastically, the characteristics and personality of the neighbor. While hopefully being beneficial, a large infrastructure investment can impact housing prices and related cost of living issues in the neighborhood. This can lead to gentrification in established neighborhoods that have developed over many years. Engineers and other stormwater professions should work with the communities throughout the entire project to understand the needs and constraints of the community as the GSI projects are implemented.

## References

Amandes, C. and Bedient, P. (1980). Stormwater detention in developing watersheds. *Journal of the Environmental Engineering Division-ASCE* 106 (2): 403–419.

Barco, J., Hogue, T., Curto, V., and Rademacher, L. (2008). Linking hydrology and stream geochemistry in urban fringe watersheds. *Journal of Hydrology (Amsterdam)* 360 (1–4): 31–47.

Boorstein, M. (2005). Giant houses rankle residents of college heights. The Washington Post, Page C05. July 10, 2005.

Booth, D.B. (2005). Challenges and prospects for restoring urban streams: A perspective from the Pacific Northwest of North America. *Journal of the North American Benthological Society* 24 (3): 724–737.

Ecology- Washington State Department of Ecology Water Quality Program (2005) *Stormwater Management in Western* Washington *Volume III Hydrologic Analysis and Flow Control Design/BMPs*, Publication No. 05- 10-31(A revision of Publication No. 99-13).

Emerson, C.H., Welty, C., and Traver, R.G. (2005). Watershed-scale evaluation of a system of storm water detention basins. *Journal of Hydrologic Engineering* 10 (3): 237–242.

Hekl, J.A. and Dymond, R.L. (2016). Runoff impacts and LID techniques for mansionization-based stormwater effects in Fairfax County, Virginia. *Journal of Sustainable Water in the Built Environment* 2 (4): 05016001.

King, R.S., Baker, M.E., Kazyak, P.F., and Weller, D.E. (2011). How novel is too novel? Stream community thresholds at exceptionally low levels of catchment urbanization. *Ecological Applications* 21 (5): 1659–1678.

MacGillis, A. (2006). The growing problem of floods. The Washington Post Page T22. December 14, 2006.

McCuen, R. and Moglen, G. (1988). Multicriterion stormwater management methods. *Journal of Water Resources Planning and Management-ASCE* 114 (4): 414–431.

Melillo, J.M., Richmond, T. (T.C.), and Yohe, G.W. (eds.) (2014). *Climate Change Impacts in the United States: The Third National Climate Assessment*, 841 pp. U.S. Global Change Research Program. doi:10.7930/J0Z31WJ2.

NRC-National Research Council (2009). *Urban Stormwater Management in the United States*. Washington, DC: The National Academies Press.

Peterson, B.J., Wollheim, W.M., Mulholland, P.J., Webster, J.R., Meyer, J.L., Tank, J.L., Martí, E., Bowden, W.B., Valett, H.M., Hershey, A.E., McDowell, W.H., Dodds, W.K., Hamilton, S.K., Gregory, S., and Morrall, D.D. (2001). Control of nitrogen export from watersheds by headwater streams. *Science* 292: 86–90.

Shields, C.A., Band, L.E., Law, N., Groffman, P.M., Kaushal, S.S., Savvas, K., and Fisher, G.T. (2008). Streamflow distribution of non-point source nitrogen export from urban-rural catchments in the Chesapeake Bay watershed. *Water Resources Research* 44 (9): W09416.

Sweeney, B.W., Bott, T.L., Jackson, J.K., Kaplan, L.A., Newbold, J.D., Standley, L.J., Hession, W.C., and Horwitz, R.J. (2004). Riparian deforestation, stream narrowing, and loss of stream ecosystem services. *Proceedings of the National Academy of Sciences* 101 (39): 14132–14137.

Traver, R.G., and R.A. Chadderton (1983). The downstream effects of storm water detention basins. *International Symposium on Urban Hydrology, Hydraulics and Sediment Control*, University of Kentucky, July 1983.

U.S. EPA (2011a). *Keeping Raw Sewage & Contaminated Stormwater Out of the Public's Water*. New York: US EPA Region 2.

U.S. EPA (2011b). *Summary of State Stormwater Standards*. US EPA, Office of Water, Office of Wastewater Management, Water Permits Division.

Walsh, C., Roy, A., Feminella, J., Cottingham, P., Groffman, P., and Morgan, R. (2005). The urban stream syndrome: Current knowledge and the search for a cure. *Journal of the North American Benthological Society* 24 (3): 706–723.

Wang, L.Z., Lyons, J., and Kanehl, P. (2003). Impacts of urban land cover on trout streams in Wisconsin and Minnesota. *Transactions of the American Fisheries Society* 132 (5): 825–839.

WEF-Water Environment Federation (2015). *Rainfall to Results: The Future of Stormwater*. Alexandria, VA.

## Problems

1.1 Does your state or local jurisdiction have a stormwater manual? Try to find it on the web. What year was it created? What stormwater control measures does it promote and describe?

1.2 How many 303(d) impaired waters are listed in your state? What is the greatest cause of the impairment?

1.3 Find the river, lake, or reservoir closest to your home or school. Is it listed as 303(d) impaired? If so, describe the impairment. If not, find the nearest river with an impairment. Does it have an approved TMDL?

# 2

# Precipitation: The Stormwater Driver

## 2.1 Introduction

Effective implementation of green stormwater infrastructure (GSI) requires an understanding of the hydrological processes that occur within landscapes that have been altered by human activities and the resulting impact to the water cycle. To mitigate these impacts, the GSI professional must be able to evaluate the hydrologic processes for both the preconstruction and postconstruction conditions, select a strategy and level of green infrastructure mitigation, and track the precipitation hydraulically from the impervious surface to, and through, the green infrastructure stormwater control measure (GI SCM). This chapter introduces the reader to urban hydrology concepts and rainfall characterization. Chapter 6 further develops this topic, where hydrologic and hydraulic processes are discussed in detail. It is presumed that the reader has a fundamental understanding of fluid mechanics and hydrology.

## 2.2 The Urban Hydrologic Cycle

The hydrologic cycle comprises the movement of water from the clouds, to rainfall, to runoff and infiltration into the soil, entering our streams and groundwater sources, frequently to the ocean, where evaporation and/or transpiration transfer the water back to the clouds again. As the system is powered by energy from the sun, it is continuous. Figure 2.1 describes the "natural" hydrologic cycle of many temperate regions, which demonstrates that on an annual basis, the majority of the rainfall returns to the atmosphere through evapotranspiration (ET), with a fraction becoming surface runoff. The remaining water soaks into the ground, either replenishing the groundwater storage or becoming the baseflow of our streams and rivers.

As described in Chapter 1, urban development interrupts and short-circuits the natural hydrologic cycle. Ubiquitous impervious area in developed regions prevents the rainfall from entering the ground and accelerates the speed that the runoff enters our streams and rivers. Note that as shown in Figure 2.2, the urbanized watershed transfers water that previously was destined for groundwater, baseflow, or ET to surface runoff. Thus, the urban hydrology focus then is on these transport pathways and their effects.

*Green Stormwater Infrastructure Fundamentals and Design*, First Edition. Allen P. Davis, William F. Hunt, and Robert G. Traver.

Companion Website: www.wiley.com/go/davis/greenstormwater

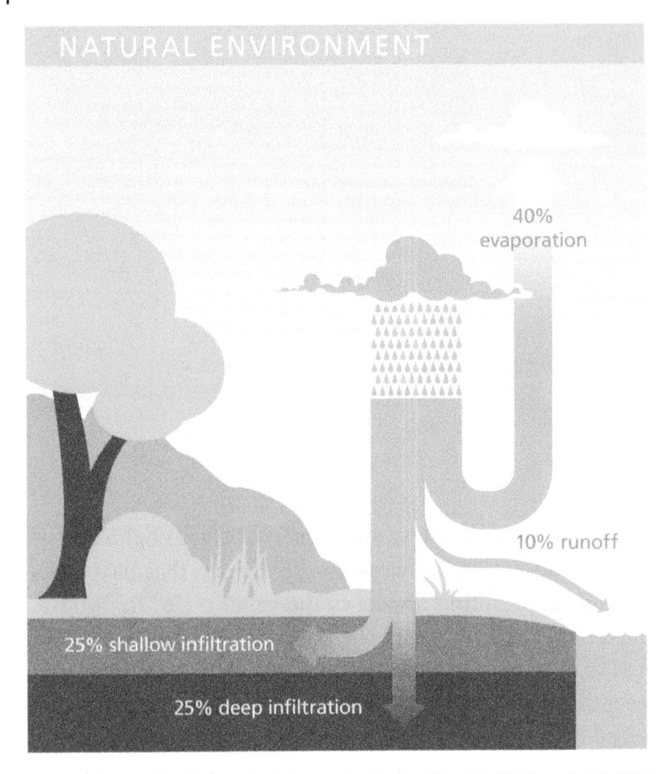


**Figure 2.1** The Natural Hydrologic Cycle. (Credit: Philadelphia Water Department).

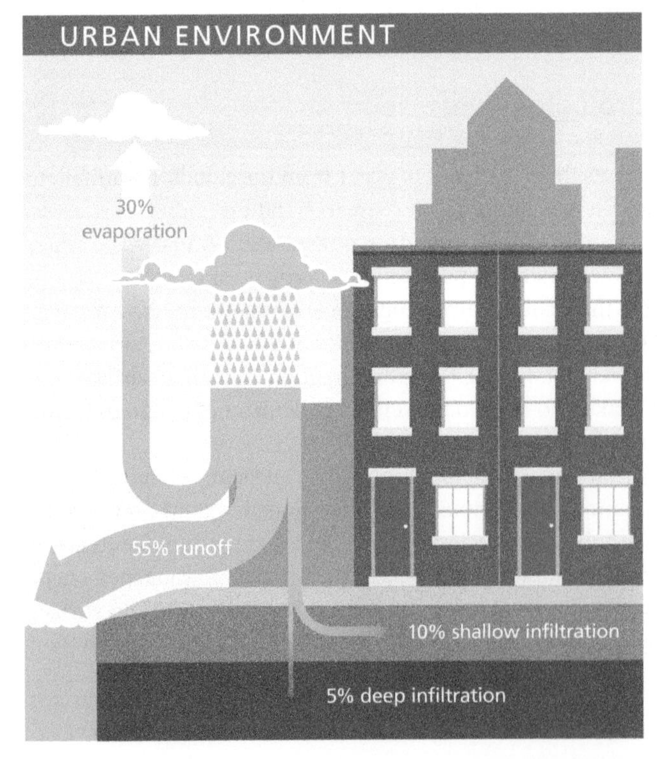


**Figure 2.2** The Urban Hydrologic Cycle. (Credit: Philadelphia Water Department).

Stormwater runoff from urbanized areas is characterized by rapidly changing and widely fluctuating flows and pollutant concentrations, which are heavily tied to the contributing impervious area and flow conveyance pathways. Runoff from pavement and compacted soils can be extremely flashy due to the lack of soil retention and speed of the runoff traveling over various impervious surfaces or in shallow concentrated flow. Drainage area characterization must take into account not only the land form but the stormwater conveyance and collection systems, which often may behave differently depending on the size of the storm event being considered and the season. As the magnitude and pattern of the hydrologic cycle is different across the United States and throughout the world, so are the design goals and the regulatory structures. Design goals may be related to nuisance flooding, water quality and stream geomorphology, combined sewer overflows, or more extreme flooding events. Some regions deal with extreme precipitation events separately from the more frequent storms both from a regulatory and design approach perspective. The differences are further complicated due to the historic development of the regulatory structure, but in all scenarios the fundamental hydrologic processes remain.

Urban stormwater hydrologic goals are normally related to an occurrence probability, a specific time duration, and the regulatory process. These goals can include extreme event peak flow rates from a specified return period or historic storm, an annual view of velocity changes of rate and duration based on geomorphic impacts, and daily or seasonal volumes. Other goals can include a hybrid time-related volume focus, for example, capturing the first flush of rainfall (commonly somewhere between 1 and 3 cm; also known as the water quality volume), or a combination of volume and annual duration of flow rates. These principles are discussed in Chapter 6. In any event, to meet the design intent, the GSI professional must be able to evaluate the hydrologic characteristics for both the preconstruction and postconstruction conditions in order to determine what mitigation approach is required and to predict the performance of the GSI and GI SCMs.

Depending on the approach, the specific time focus for GI evaluation may extenuate or negate individual components of the design. For example, when considering a peak flow during a storm event, ET may be considered insignificant. However, if that focus is lengthened, then ET becomes a key component when determining the rate of recovery of storage capability prior to the next storm, or when looking at the annual performance of a SCM designed for smaller individual storms. In any event, the intent of GSI is to reverse the effects of urbanization, inserting "green" processes within the gray infrastructure to approach the natural hydrologic cycle as a goal.

## 2.3 Precipitation

Understanding GSI starts with a fundamental characterization of rainfall. For every region and climate throughout the world, the rainfall volumes and patterns influence the GSI strategy approach to be used. For example, the weather in Seattle, Washington, is dominated by frequent smaller storm events for most of the year. Therefore, a small runoff volume can be an effective target, which is ideal for green infrastructure mitigation. However, the challenge for Seattle and other municipalities with similar climates is how to implement GSI that can endure the annual multi-month dry season. A very different challenge occurs in Austin, Texas, where the majority of rainfall falls during relatively few infrequent, but large storm events.

Conversely the rainfall on the east coast of the United States is more uniformly distributed, but with higher precipitation influenced by occasional tropical storms leading to large erosive channel forming events. Higher temperatures in the lower latitudes (which impact plant selection and water needs) and snow and ice (and snow removal) in the northern regions are factors that must be part of GI design considerations. An effective GI strategy requires understanding the climate seasonality and future-predicted climate change trends.

## 2.4 Precipitation Depths

Precipitation depth is characterized based on a specified time frame of interest, such as a year, season, individual storm event, or peak rate within that storm event. Statistical analysis is used to relate the rainfall depth to a time interval and frequency of occurrence, to develop target storm characteristics commonly known as *design storms*. A more sophisticated approach would be to use an annual or multiyear climate record as further discussed in Chapter 6. A statistical comparison of precipitation data, based on historical records of precipitation, is presented in Table 2.1 for four US cities. Differences in rainfall characteristics become clear when comparing the number of rainy days (≥0.1 in.; ≥0.25 cm) to the number of days with large rain events (≥1.0 in.; ≥2.54 cm).

From the GSI perspective, of interest is how many times per year a set rainfall event occurs. From this information must come the storage volume a SCM must contain to hold the runoff from these events. For example, in Washington, DC, the great majority of rainfall events are less than 1 in. (2.54 cm). Therefore, most runoff volumes will be produced from storms 1 in. or less, and the 1-in. runoff volume is exceeded on average less than 10 times a year (Table 2.1).

The long-term rain gage data available from the US National Oceanographic and Atmospheric Administration (NOAA) is a valuable resource for these types of analyses. Consider the 53-year rainfall record from the Philadelphia (PA) Airport rain gage (www.ncdc.noaa.gov/cdo-web). Figure 2.3 shows the rainfall daily volumes (depths) sorted from smallest to largest in a cumulative distribution curve (Figure 2.3, percent storm). This curve gives the percentage of total rainfall depth represented by the corresponding depth, and smaller depths.

**Table 2.1** Comparison of Annual Precipitation Data for Four Locations within the United States 1980–2010 (Data from NOAA National Center for Environmental Information)

| | Washington DC | Houston TX | Minneapolis MN | Seattle WA |
|---|---|---|---|---|
| Average precipitation (in.) (cm) | 39.7 (100) | 49.8 (126) | 30.6 (78) | 37.5 (95) |
| Average snowfall (in.) (cm) | 15.4 (39) | 0.1 (0.25) | 54.4 (138) | 6.8 (17) |
| Average number of precip. events ≥ 0.1 in. (≥0.25 cm) | 70.1 | 64.2 | 61.8 | 91.0 |
| Average number of precip. days ≥ 1.0 in. (≥2.54 cm) | 9.4 | 15.1 | 6.0 | 4.6 |

*Capture depth* then adds to the storm depth the additional depth from removing the storm depth from larger events (Figure 2.3, percent capture). For example, Figure 2.3 shows that roughly 45% of the average rainfall falls in storms of one inch (2.54 cm) or less. However, when the first inch of larger storms is included, the resulting capture depth percentage rises to 80%. Other uses of this daily volume data set relate to understanding of how often a SCM would exceed capacity (Chapter 6). For the region of this daily data set, seventeen events a year on average exceed 1 in. (2.5 cm), 4.3 events exceed 2 in. (5 cm), and 1.2 events exceed 3 in. (7.5 cm). Understanding of the regional climate in this manner is an important consideration for GSI SCM selection and design.

As many GSI approaches rely on the use of vegetation, it is also necessary to understand the monthly or seasonal average rainfall, as plant health is an important design consideration. Figure 2.4 shows average monthly rainfalls for several US cites. Note the

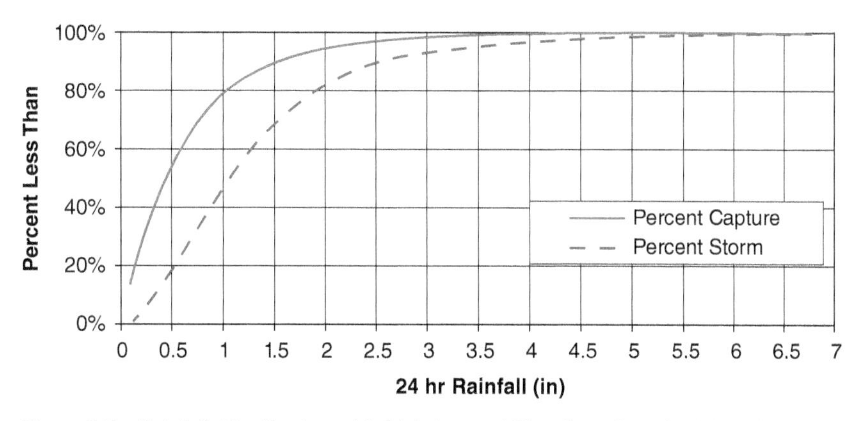


Figure 2.3 Rainfall Distribution with 24-h Interval Time Data from NOAA National Center for Environmental Information–Philadelphia (1948–2011).

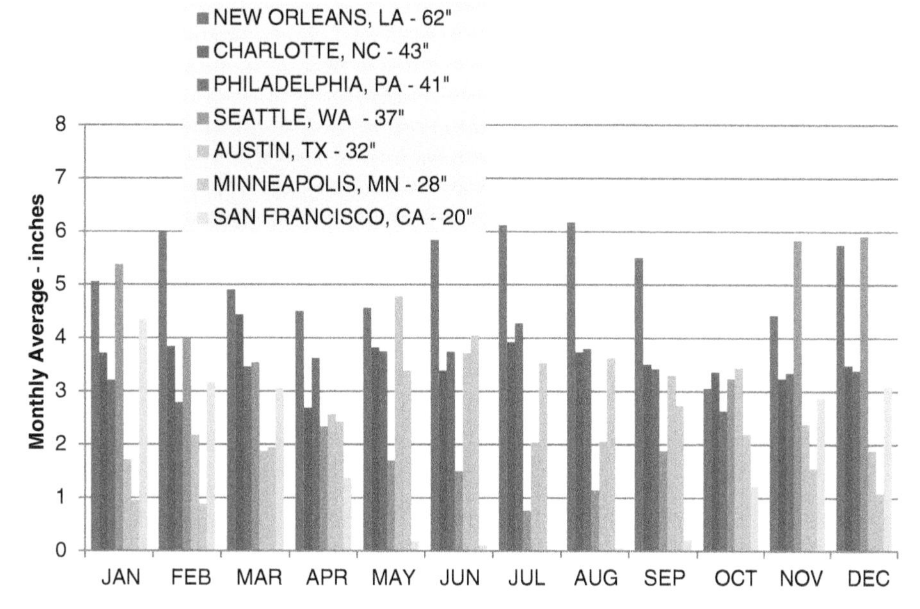


Figure 2.4 Average Monthly Precipitation for Selected Cities–1961–1990. (Data from NOAA National Center for Environmental Information).

wide variation in rainfalls in June through August. When coupled with high temperatures, vegetated SCMs such as green roofs or bioretention facilities in San Francisco or Seattle may require irrigation systems that are not needed in other cities.

## 2.5 Rainfall Patterns

Patterns of rainfall vary greatly over the course of a year. The 2014 yearly storm event summary from Villanova University's green roof (Figure 2.5) highlights the larger events and displays as well the many periods with no rain. The daily measured quantity of liquid (or liquid equivalent in the case of frozen precipitation) precipitation is shown over the course of 2014, with the median non-zero quantity (thick gray line)

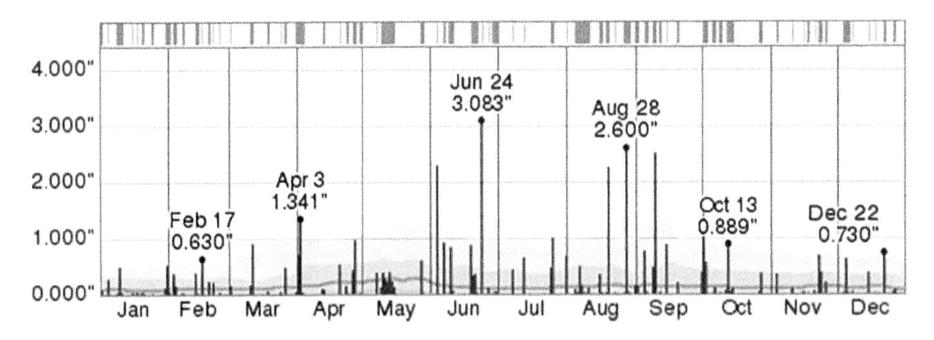


Figure 2.5 2014 Daily Precipitation Variation from Villanova University's Green Roof.

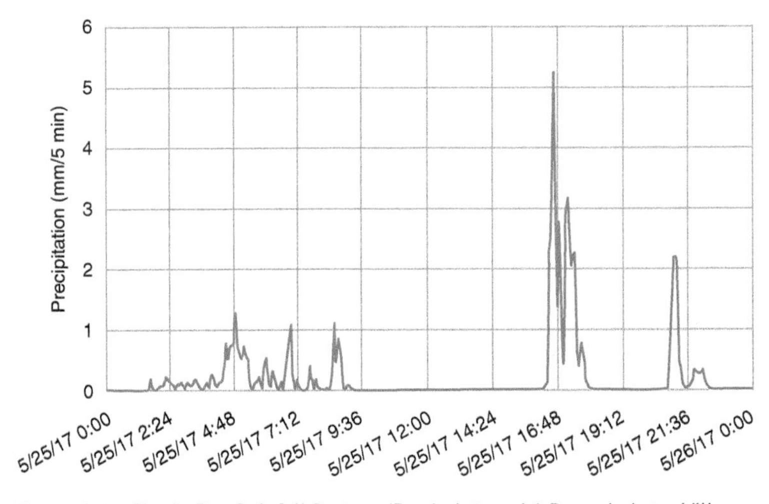


Figure 2.6 Single Day Rainfall Pattern (5-min Intervals) Recorded at a Villanova University Research Site in Philadelphia.

and 10th, 25th, 75th, and 90th non-zero percentiles (shaded areas). The bar at the top of the graph is dark if any precipitation was measured that day and white otherwise. The question of what makes up an individual storm is always a challenge, as can be seen by several clusters of events within smaller time periods (few minutes/few hours), as well as areas of well-defined separation.

Rainfall patterns also vary during individual storms as depicted in Figure 2.6, which shows the rainfall as recorded by a rain gage at a Villanova research site in the city of Philadelphia. Even within a small time duration, the rainfall pattern is rarely uniform. Note that for this storm, the rainfall has several clusters of precipitation throughout the 24-h time period.

## 2.6 Inter-event Interval

As stated earlier, understanding of the frequency and time between events is critical in understanding the ability of GI SCMs to recover this treatment capacity between storm events. During this period, both infiltration and ET are occurring, removing water from the SCM and restoring volumetric capacity for future storms. The size of the storms is also of interest, as a second storm that exceeds the restored capacity from the proceeding event would be of concern. Generally known as drawdown time, current regulations routinely adopted criteria from the detention era of design, requiring a 24–48-h period of surface flow drawdown. While region specific, more recent studies have found that the probability of back-to-back storms large enough to cause overflow is small, frequently less than 0.1 average occurrences per year (Wadzuk et al. 2017); thus extending the drawdown period would be limited more by nuisance than reduced performance.

## 2.7 Extreme Event Precipitation

The intensity, duration, and frequency (*i–d–f*) of a rainfall event are used to characterize an event. Extreme rainfall events are commonly defined through statically based *i–d–f* relationships. Historical rainfall data then can be developed into a rainfall pattern or design storm if needed for stormwater management design. The statistics are generally based upon an extended data record of the largest storm event in each year. The extreme events are related to a recurrence interval, or average return period, $T_R$ (years), defined as the inverse of the probability, $p$.

$$T_R = \frac{1}{p} \tag{2.1}$$

The probability is the chance of annual occurrence for an event of a defined magnitude. The return period is a way of expressing the *average* yearly time interval expected for an event of this size or greater. For example, a 100-year storm has a 1% chance of being equaled or exceeded every year. Similarly, the median extreme event is the two-year storm (50% chance of being exceeded in any particular year).

The probability, $z$, of an event happening with a series of years, $n$, is given by

$$z = 1 - \left(1 - \frac{1}{T_R}\right)^n \tag{2.2}$$

**Example 2.1** Find the risk of a 100-year storm occurring over the life of a 30-year mortgage

*Solution* Using Eq. 2.2 with $T_R = 100$ and $n = 30$ gives

$$z = 1 - \left(1 - \frac{1}{100}\right)^{30} = 0.26$$

The 100-year storm would have a 26% chance of occurring at least once over the life of a 30-year mortgage; it would have a 74% chance of not occurring.

Often the maximum rainfall intensity is used to design stormwater and SCM conveyance requirements, but the overall rainfall depth (volume) would be of more interest in setting stormwater volume control criteria. The maximum rainfall intensity depends on the duration and frequency of the event selected. The intensity of a 100-year return period with a 10-min duration is much greater than the same return period lasting an hour. However, the smaller duration event would produce a much smaller rainfall depth over the duration. The design intent then drives the selection of the duration and frequency. *i–d–f* curves are generally developed for regions from available historical rainfall data, as shown in Figure 2.7 for eastern New Jersey.

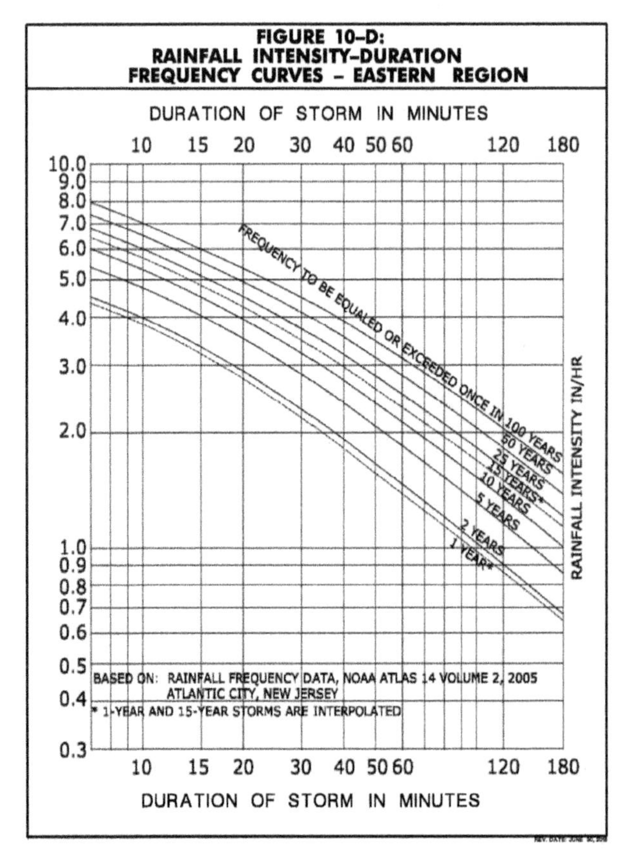


Figure 2.7 Intensity–Duration–Frequency (i–d–f) Curves for Eastern New Jersey (with Permission, New Jersey Department of Transportation, NJDOT 2015).

**Example 2.2** Contrast the rainfall intensity and depth for a 15- and 30-min duration, 10-year storm in Eastern New Jersey

*Solution* Using Figure 2.7, at 30 min on the *x*-axis, move upward to meet the 10-year curve. The intersection occurs at about 3.2 in./h on the *y*-axis. For the 15-min axis, the rate is 4.5 in./h. The depths would be 1.6 and 1.125 in., respectively. Note that both axes are log scale.

## 2.8 Introducing the Rainfall–Runoff Relationship

While the rainfall drives the GSI design, to be useful for design, the rainfall must be converted to the runoff generated. An overview is provided here, with details given in Chapter 6.

The first step in this process is to delineate the watershed, as shown in Figure 2.8. To delineate is to identify all areas that contribute runoff to the point of interest, usually a SCM or other drainage point. Simply stated "water flows downhill" is the criterion, but delineation must include the stormwater collection system as well. In Figure 2.8, the

Figure 2.8 Contributing Drainage Area (Light Blue) Entering Philadelphia Water Department Rain Gardens Developed from LIDAR/GIS Study. (with Permission, Humaira Jahangiri and Virginia Smith, Villanova University).

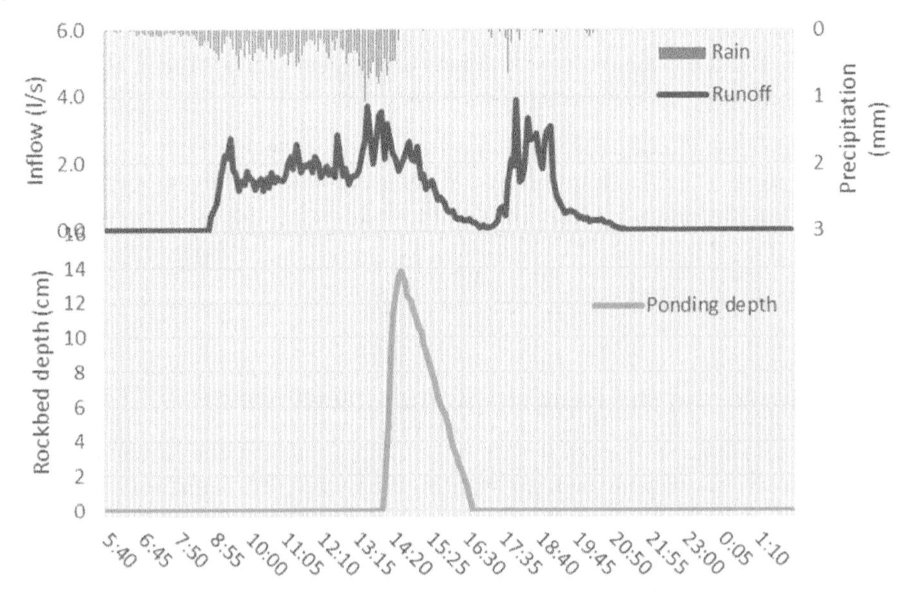


**Figure 2.9** Rainfall, Runoff, and Ponding Depth Measured at an Infiltration Trench. (Authors' Research Group).

light area delineates the area contributing to the bioinfiltration raingardens in the center. The areas outside the light area drain elsewhere, as defined by the flow paths that separate different drainage areas. Rain that falls inside the light area will flow to the SCM; that which falls outside will flow elsewhere. Accurate topographic information is important here; for small drainage areas common to GSI design, a few centimeter difference in topography can make a large difference in delineating a drainage area, especially when defining parking areas, which tend to be fairly flat. Note that when delineating a watershed, the transport of runoff water through any drainage pipes must be included.

The second step in this process is converting the rainfall into runoff and transporting it over the watershed. This is done using some type of water balance model, as described in detail in Chapter 6. This process considers stormwater volume losses through infiltration, depression storage, and the travel time over the watershed. This process can be used to assess the performance of the GI SCM, resulting in a hydrologic pattern as shown in Figure 2.9 for a large rainfall event. The rainfall intensity is shown across the top of the figure, with the scale on the right. The inflow (center line) and ponding depth (lower line) of the infiltration trench SCM are given for a large rainfall event. The plot itself can be used to identify stormwater metrics to include flow peaks, volumes, and depths. Further analysis then may look at the soil–water relationships that control runoff storage in the bioinfiltration facility, and the infiltration and ET mechanisms following the storm event which will allow for the recovery of the storage.

## 2.9 Precipitation and Water Quality

In general, the quality of rainwater is good. The process of evaporation in the hydrologic cycle is cleansing, where the only water is vaporized, leaving most contaminants behind. In the atmosphere, the moisture can absorb air pollutants, which can affect the water quality. The pollutants of greatest concern are usually nitrogen oxides. These compounds are formed from the conversion of nitrogen gas in the atmosphere during fossil fuel combustion processes, especially from automobiles. Some sulfur species were problematic a few years ago as sulfur emissions from combustion created sulfuric acid in atmospheric moisture, and "acid rain," but recent air pollution regulations have drastically reduced this concern.

Once the rainfall strikes various surfaces on the earth, it picks up and mobilizes various pollutants, including particulate matter, nutrients, various toxic chemicals, microorganisms, and salts. At this point, it becomes polluted runoff. The degree of pollution will depend on many factors discussed in this chapter, including rainfall intensity and depth, time between storm events, and characteristics of the drainage area. All these issues are discussed in more detail in Chapter 5.

## 2.10 Climate Change

As precipitation is a key driver of stormwater GSI analysis, design, and performance, changes to precipitation depth, intensity and frequencies will alter all of these GSI considerations. Much of the United States is predicted to receive more frequent and higher intensity rainfall events, with periods of increased drought. Regulatory goals and design concepts will need to be rethought as precipitation patterns change. As will be discussed in future chapters, GSI is frequently more flexible than gray stormwater infrastructure in managing highly variable stormwater conditions, as there may be some dampening of the stormwater control effects. More details are provided in subsequent chapters.

## References

NJDOT (2015). *Roadway Design Manual*. Department of Transportation, The State of New Jersey.

Wadzuk, B.M., Lewellyn, C., Lee, R., and Traver, R.G. (2017). Green infrastructure recovery: Analysis of the influence of back-to-back rainfall events. *Journal of Sustainable Water in the Built Environment* 3 (1): 04017001.

## Problems

2.1 For your region, find out what your yearly rainfall is, and how much goes to baseflow, streamflow, and evapotranspiration.

2.2 Compare and contrast your state's hydrologic stormwater criteria to one state with similar weather conditions, and one that is different.

2.3 Find the probability of a 10-year storm occurring over the 25-year design life of a structure.

2.4 Find the return period of the storm that should have a 10% probability of occurring over a 30-year time frame.

2.5 A 50-year storm event would have a 10% chance of occurring over what time frame?

2.6 Find the rainfall intensity for a 60-min duration, 25-year storm in Eastern New Jersey.

**2.7** Find the return period for a 6.0-in./h rainfall intensity for a 15-min duration in Eastern New Jersey.

**2.8** Find the rainfall duration for a 2.0-in./h intensity, 5-year storm in Eastern New Jersey.

**2.9** Determine the probability for a 3.7-in./h rainfall intensity for a 30-min duration in Eastern New Jersey occurring over a 10-year time period.

**2.10** Determine the probability for a 2.1-in./h rainfall intensity for a 60-min duration in Eastern New Jersey occurring over a 10-year time period.

# 3

# Water Quality

## 3.1 Introduction

While achieving "good" (or "excellent") water quality is an easy metric to set as an environmental goal, in practice it is difficult to define. The definition of "good" water quality depends on the designated use of the water. Additionally, physical, chemical, and biological properties must be considered in concert when defining water quality. These properties define conditions that are conducive for various types of desired aquatic life to thrive. While an individual pollutant may be of great importance, it must be understood that in many cases pollutants that can be toxic or otherwise lead to undesirable water characteristics may be present at low concentrations in rivers that are in "good" condition.

## 3.2 Designated Water Uses

The Clean Water Act requires that the 50 states (and the District of Columbia and authorized Native American Tribes) set designated uses of water bodies within their boundaries. These uses can include recreation, public water supply, habitat for various fish, shellfish, and wildlife species, agriculture, industrial, and navigation. Clearly these uses will require different levels of water quality. Once the state sets the designated use, water quality parameters and other criteria are used to determine if the use can be met. If the water body does not meet its intended use, then a plan must be put in place to achieve that use. Parameters that may be evaluated to consider designated use may include:

- conditions that are detrimental to human health through drinking or contact
- conditions that render fish or shellfish unhealthy for human consumption
- agricultural irrigation quality
- aquatic life toxicity from acute or chronic toxicity impacts

An example of a water body that does not meet its intended use is the Chesapeake Bay, one of the largest estuaries in the world. The Chesapeake Bay is in Maryland, and Virginia, but with a watershed that covers parts of four other states and the District of Columbia. Monitoring data and observations on the estuary continue "to show that the

*Green Stormwater Infrastructure Fundamentals and Design*, First Edition. Allen P. Davis, William F. Hunt, and Robert G. Traver.

Companion Website: www.wiley.com/go/davis/greenstormwater

Bay has poor water quality, degraded habitats and low populations of many species of fish and shellfish." Based on this classification, a total maximum daily load (TMDL, as described in Chapter 1) has been completed on the Bay and pollutant load allocations have been established. The TMDL requires the contributing states within the watershed to reduce sediment, nitrogen, and phosphorus contributions. Other examples include a TMDL for trash on the Los Angeles River and sediment on the Minnesota River Basin.

## 3.3 Water-Quality Parameters and Measures

Good water quality means having acceptable levels of desired constituents, such as dissolved oxygen (DO) and low levels of undesired contaminants. Other parameters, such as water temperature and pH, must also be within acceptable ranges. Pollutants can enter natural waters from natural or anthropogenic sources. These pollutants can impact the DO levels in the water, directly or indirectly, cause turbidity or toxicity, or create some other water quality problem.

Most water quality parameters are defined in terms of the *concentration*, the amount or mass of a material in the water per unit volume. The material may be a desirable constituent, such as DO, or a pollutant, such as a metal like lead. The typical units of concentration are mg/L. However, in doing chemical calculations, such as with pH, or when calculating stoichiometric relationships, calculations must be done using moles, a measure of the number of molecules, with concentrations given in mol/L, designated as M. Moles and milligrams are related through the molecular mass of the substance.

**Example 3.1** The concentration of dissolved oxygen in a highway runoff is 4.4 mg/L. What is this concentration in molar units?

*Solution* From a periodic table or table of molecular masses, the molecular mass of oxygen is 16. Since oxygen gas and dissolved oxygen is of the form $O_2$, the mass of $O_2$ is 32 g/mol. This is equal to 32,000 mg/mol. Therefore:

$$\text{Dissolved oxygen} = \frac{4.4 \text{ mg/L}}{32{,}000 \text{ mg/mol}} = 1.38 \times 10^{-4} M$$

## 3.4 Temperature

Most aquatic life thrives in cooler water temperatures. High temperatures can stress aquatic life, such as fish. At temperatures greater than about 90°F (32°C), some fish species are unable to reproduce. Water temperature will also impact the saturated DO concentration, with high temperatures reducing available oxygen, as discussed later.

High temperatures can be problematic in stormwater runoff as rain comes into contact with hot asphalt in the summer. This thermal energy is transferred from the pavement to the water, which can then be carried to a water body.

## 3.5 pH

The pH is a measure of the concentration of hydrogen ions, $H^+$, in water. Many chemical and biological reactions are affected by hydrogen ions. Also, extremely high concentrations of $H^+$ can be corrosive and toxic.

Hydrogen ions exist naturally in water from its autodissociation, also producing hydroxide ions, $OH^-$:

$$H_2O \leftrightarrow H^+ + OH^- \tag{3.1}$$

The equilibrium constant for this reaction is given by

$$K_w = \left[H^+\right]\left[OH^-\right] \tag{3.2}$$

The value of $K_w$ is $10^{-14}$ at 25°C.

Because the concentration of $H^+$ can vary over many orders of magnitude, pH is generally used as a measurement parameter. Therefore, pH is defined as

$$pH = -\log\left[H^+\right] \tag{3.3}$$

When $[H^+]$ is high, pH is low, indicating acidic conditions.

Under ideal conditions, a pure water has a pH of 7, which is neutral. By Equation 3.3, pH 7 corresponds to $[H^+]$ equal to $10^{-7}$ M. Continuing using Equation 3.2, the concentration of $OH^-$ in a pure water is also $10^{-7}$ M. However, even in nature most waters are not pure. Contact with carbon dioxide in the environment creates a small amount of carbonic acid, a weak acid, in the water, which can reduce the pH as low as 5.6.

Generally, a pH between 6.5 and 8.5 is required for ambient waters.

**Example 3.2** Find the values of $[H^+]$ and $[OH^-]$ at pH 8.5.

*Solution* Using Equation 3.3, at pH 8.5:

$$H^+ = 10^{-pH} = 10^{-8.5}\,M$$

From Equation 3.2:

$$\left[OH^-\right] = \frac{K_w}{\left[H^+\right]} = \frac{10^{-14}}{\left[10^{-8.5}\right]} = 10^{-5.5}\,M$$

## 3.6 Dissolved Oxygen

Oxygen is necessary for the survival of most aquatic fauna. Animals such as fish, crabs, oysters, and clams need oxygen dissolved in the water. This oxygen is used in respiration in the same manner as terrestrial animals, to breathe air to get oxygen. Most aquatic animals have limited mobility, and when DO levels fall too low, they die.

Generally, the accepted threshold for DO is 5 mg/L. Below this value many fish and aquatic organisms are stressed, may not reproduce, and struggle to survive. While some species can survive at lower DO concentrations, many desirable fish, such as trout and bass, generally cannot. The condition of very low DO is known as *anoxia*.

**Table 3.1** Saturated Dissolved Oxygen Levels in Fresh Water.

| Temp. (°C) | DO (mg/L) | Temp. (°C) | DO (mg/L) | Temp. (°C) | DO (mg/L) |
|---|---|---|---|---|---|
| 1 | 14.2 | 11 | 11.1 | 21 | 9.0 |
| 2 | 13.8 | 12 | 10.8 | 22 | 8.8 |
| 3 | 13.5 | 13 | 10.6 | 23 | 8.7 |
| 4 | 13.1 | 14 | 10.4 | 24 | 8.5 |
| 5 | 12.8 | 15 | 10.2 | 25 | 8.4 |
| 6 | 12.5 | 16 | 10.0 | 26 | 8.2 |
| 7 | 12.2 | 17 | 9.7 | 27 | 8.1 |
| 8 | 11.9 | 18 | 9.5 | 28 | 7.9 |
| 9 | 11.6 | 19 | 9.4 | 29 | 7.8 |
| 10 | 11.3 | 20 | 9.2 | 30 | 7.6 |

The DO concentration in water is limited by saturation. Under equilibrium conditions, the maximum DO concentration is fixed, as given by Henry's Law. This saturated value is dependent on temperature. At higher temperatures, the saturated DO value is low, with DO increasing at lower temperatures. Table 3.1 gives values for saturated DO concentrations as a function of water temperature.

The values in Table 3.1 are saturated values, that is, the most oxygen that water can hold. If something consumes the oxygen, the DO concentration can be much less. Note from Table 3.1 that at the highest water temperatures, the difference between the saturated DO concentration and the 5 mg/L limit where fish are stressed is small. Any process or contamination that can reduce DO concentrations by only a few mg/L can result in summer fish kills when temperatures are high.

The addition of organic matter (often readily biodegradable) can lead to reductions in DO in waters. This organic matter can be waste material discharged by industry or municipal treatment plants, or the result of decaying substances, such as vegetation or algae. In either case, natural bacteria in the water will consume the organic matter in a respiration process.

$$\text{Biodegradable organic matter} + O_2 \xrightarrow{\text{microbes}} CO_2 + H_2O + \text{cells} \tag{3.4}$$

As indicated in Equation 3.4, the biodegradation process will consume oxygen. If concentrations of biodegradable organic matter are too high, the organic matter respiration process will utilize too much of the available DO, resulting in levels too low to support fish and shellfish.

As an example, Figure 3.1 shows measured and model-predicted DO concentrations at the Miller Island monitoring site in the Kalmath River in Oregon (Sullivan et al. 2012). During warmer months, the average DO concentration decreases at all points and the number of DO samples that fall below a threshold criterion, such as 4 mg/L, increase. A TMDL has been established for this river to limit the addition of biodegradable organic matter and nutrients that cause algae growth.

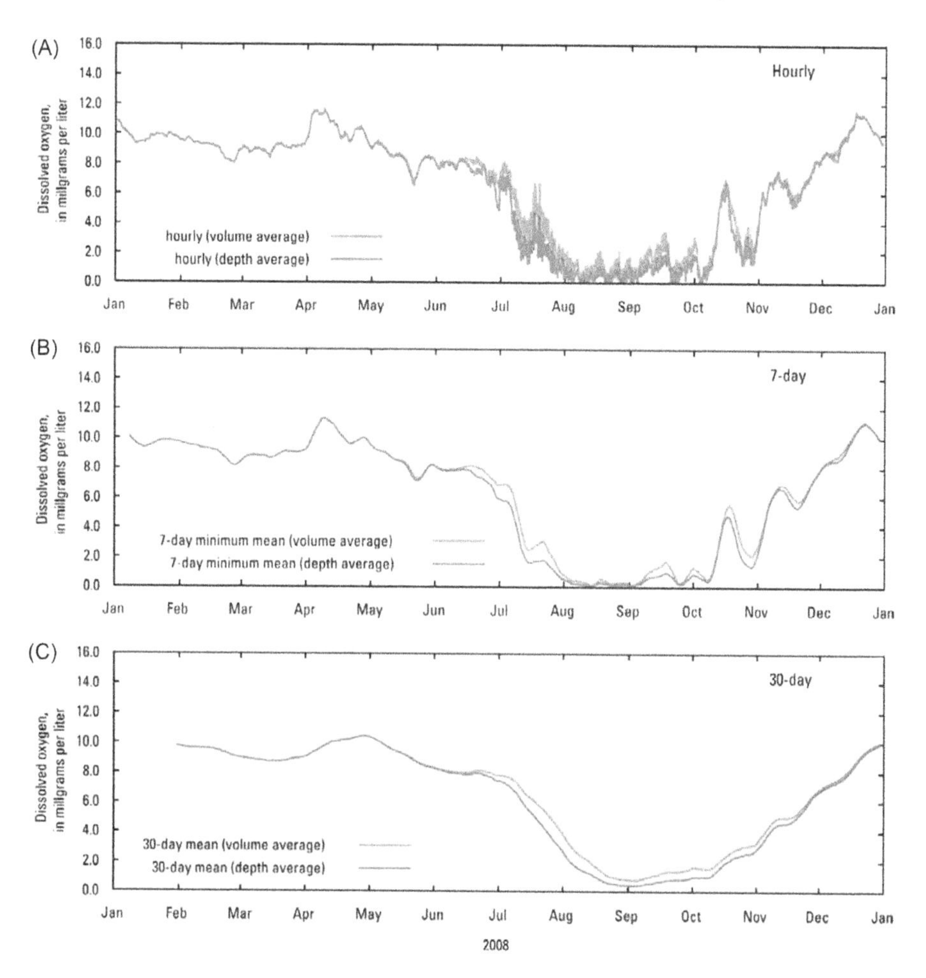


**Figure 3.1** Low Summer Dissolved Oxygen Concentrations in 2008 at a Point in the Klamath River in Oregon, Leading to Establishment of TMDL. (A) Hourly Readings, (B) 7-day Averages, and (C) 30-day Averages (Sullivan et al. 2012). Credit: US Geological Survey.

## 3.7 Turbidity and Particulate Matter

Particulate matter in water results naturally from soil components, microorganisms, and particulate organic materials, such as parts of leaves and other vegetation. In the built environment, other sources include wear and attrition of asphalt, building materials, such as brick and concrete, and car components, including brake pads and tires.

Particulate matter can cause water quality problems as it can make the water murky and prevent the penetration of light to lower depths where it is needed by aquatic vegetation. High sediment loads can deposit on water bottoms and choke off vegetation and aquatic benthic organisms. Particulate organic matter can contribute to oxygen demand and to excess nitrogen and phosphorus in waters.

Turbidity is a measure of light blockage through water; the blockage results from the presence of small suspended particles in the water. Turbidity is easily measured by focusing a light source through a water sample and determining the amount of light that passes through; the scattered light is proportional to the turbidity and is calculated by difference. The process of measuring light blockage is known as *nephalometry* and the units of turbidity are *nephalometric turbidity units, NTU*. Natural waters are generally limited to turbidity of <50 NTU to protect aquatic life.

Two methods are commonly used for measurement of particulate matter concentrations in stormwaters. The first is the *total suspended solids* (TSS). TSS is a measurement common to the water/wastewater fields in which a sample of water is pipetted from the original sample and then filtered. The filter is dried and the collected particulate matter is weighed. The TSS concentration is equal to the mass of filtered particulate matter divided by the volume of water sampled.

*Suspended sediment concentration* (SSC) is used to measure particulate matter in streams. In this case, all of the water in the collected sample is filtered. The filter is dried and the SSC calculated as mass per volume. The SSC is commonly greater than the TSS in stormwater samples because the SSC will normally include larger particles that would not be collected in the TSS pipetting procedure.

There is some disagreement as to which method, TSS or SSC, is the best to use for measuring urban stormwater particulate matter concentrations. SSC will capture the larger particles that the TSS method will not. However, these larger particles are generally, chemically, and biologically benign. They should also settle quickly under quiescent conditions. The proper test may depend on the application and need for the particulate matter measurement.

TSS and SSC both measure the total sum of particulate matter in the water sample. However, the particulate matter will be present at many different sizes, from the very small clay sizes (<2 μm) to the quite coarse (>2 mm). Frequently, knowledge of the particle size distribution (PSD) is of interest in determining detrimental impacts to water bodies and certainly when evaluating SCM treatment efficiencies. As discussed in Chapter 8, larger particles are much easier to remove than smaller particles.

A PSD can be characterized graphically by cumulatively examining the particle sizes. Figure 3.2 shows a PSD curve. The values on the *x*-axis represent the particle

**Example 3.3** A 200-mL runoff sample is used in a TSS test. The mass of the filter before the test is 0.140 g. After the filtration and drying, the filter mass is 0.155 g. Find the TSS concentration.

***Solution*** The particulate matter mass is the difference between the filter masses:

$$\text{Particulate mass} = 0.155\ \text{g} - 0.140\ \text{g} = 0.015\ \text{g} = 15\ \text{mg}$$

Concentration is the mass divided by the water sample volume:

$$\text{TSS} = \frac{15\ \text{mg}}{0.2\text{L}} = 75\ \text{mg/L}$$

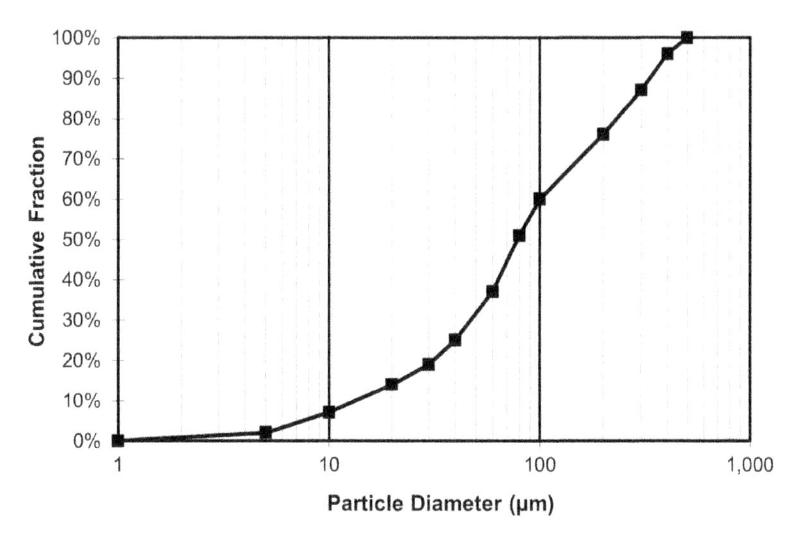


Figure 3.2 Example Particle Size Distribution for Particulate Matter in a Stormwater Sample.

size (usually based on a sieve size, but possibly through use of some other technique). The *y*-axis values represent the cumulative mass of the particle distribution. Numerically, the distribution can be represented by values such as $d_{10}$, $d_{50}$, and $d_{90}$. Ten percent of the total particulate matter mass is less than the $d_{10}$. Fifty percent of the mass is less than the $d_{50}$ and similar for $d_{90}$.

**Example 3.4** Find the values of as $d_{10}$, $d_{50}$, and $d_{90}$ from the particle size distribution presented in Figure 3.2.

*Solution* The $d_{10}$ is found at the 10% point. At a cumulative mass of 10%, the corresponding diameter is 13 μm. Using a similar procedure at 50% and 90%, $d_{50}$ = 80 μm and $d_{90}$ = 330 μm.

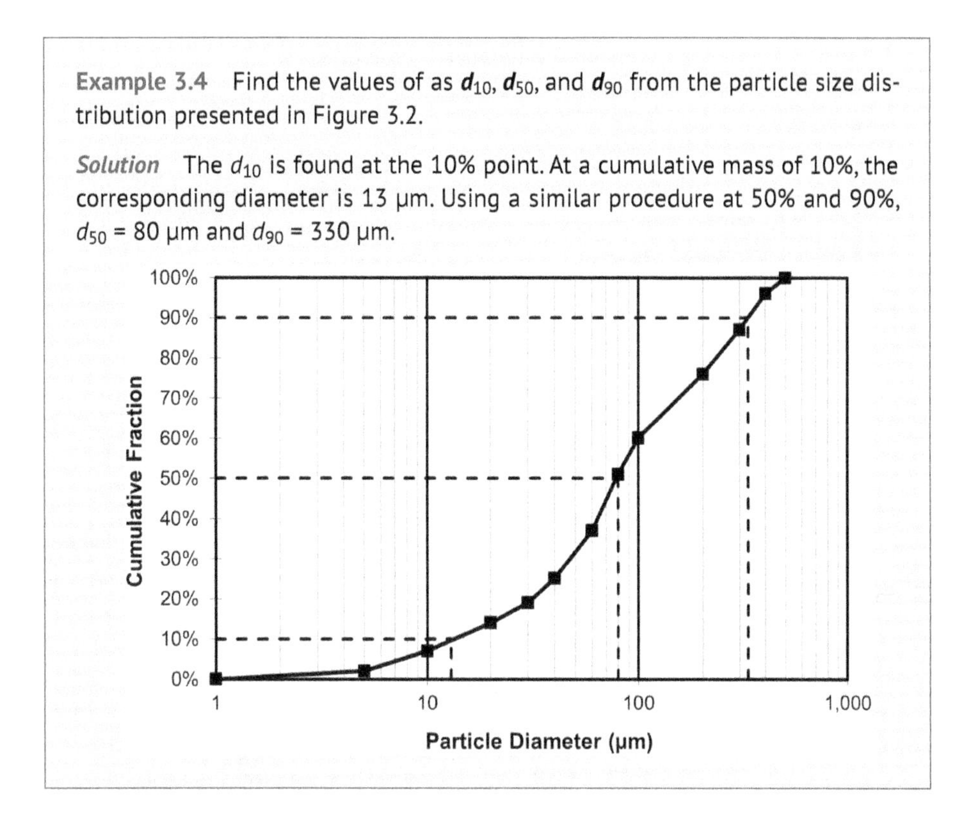

## 3.8 Biodegradable Organic Matter or "Oxygen Demand"

The addition of biodegradable organic matter to a water will lead to lowered DO concentrations and possibly death to aquatic organisms. This organic matter is degraded via Equation 3.4.

The oxygen consumption characteristics of the water discharge can be measured in several ways. The traditional measurement is the *biochemical oxygen demand* (BOD). The BOD test is a standard measurement developed many years ago to test the strength of treated and untreated municipal sewage.

In the BOD test, a 300-mL water sample is collected, or a smaller sample is diluted to 300 mL and placed in a brown bottle. A microbial inoculum may be added, but this is usually unnecessary. The DO concentration in this initial sample is measured, $DO_i$. The sample is placed in the dark at 20°C and held there for 5 days. During this time, the organic matter is biodegraded, consuming the oxygen, as in Equation 3.4. At the end of 5 days, the DO is again measured, $DO_f$. The difference in DO, with consideration of the dilution, will give the BOD, as

$$BOD_5 = \left(DO_i - DO_f\right)DF \tag{3.5}$$

where DF is the dilution factor, equal to the total sample (300 mL), divided by the volume of water being tested. To be a valid test, both $DO_f$, and the difference between $DO_i$ and $DO_f$ must be greater than 1 mg/L.

In some cases, the value is reported as BOD, in other as $BOD_5$. The "5" indicates the standard 5-day BOD test. Both BOD and $BOD_5$ indicate the same measurement. The units of BOD are in mg/L of oxygen and represent the concentration of oxygen that would be consumed during the natural biodegradation process.

**Example 3.5** 75 mL of a water sample is used in a 300-mL BOD test. The initial DO measurement is 6.2 mg/L and the 5-day measurement is 2.5 mg/L. Find the $BOD_5$ of the water sample.

*Solution* The dilution factor is 300/75 = 4.0. Using Equation 3.5, the $BOD_5$ can be found:

$$BOD_5 = (6.2\text{ mg/L} - 2.5\text{ mg/L})(4.0) = 14.8\text{ mg/L}$$

Because of the long time required for the BOD test, other tests are also used to measure the oxygen demand of a water. The *chemical oxygen demand* (COD), uses a strong oxidizing chemical, chromic acid at high temperature, to oxidize the organic matter to $CO_2$. The amount of chromic acid used during the oxidation process is measured and converted to equivalent units of mg/L of oxygen consumed.

The COD test is faster than the BOD test. However, the results are not equivalent. The COD test will measure organic compounds that are not readily biodegradable. As a result, nearly always, the value of COD for a water will be greater than the $BOD_5$.

## 3.9 Nitrogen

The presence of high concentrations of nutrients in water will trigger a series of events that lead to low DO, anoxia, and mortality of aquatic life. The principal nutrients of concern are nitrogen (N) and phosphorus (P). High levels of N and P in waters will stimulate excess algae growth, utilizing carbon from the atmosphere. Once nutrient levels become limiting, the algae will die. The dying and decaying algae are a source of organic matter to the water; this organic matter is readily biodegraded by ambient microorganisms, as designated in Equation 3.4. Rapidly, the DO levels decline, resulting in anoxia, fish kills, and loss of other aquatic organisms. This process, known as *eutrophication*, is prevalent in many areas where wastewater and runoff discharges occur, transporting nutrients into receiving waters.

The biogeochemistry of nitrogen is complex. The largest global pool of nitrogen is in the atmosphere, as nitrogen gas, $N_2$. Nitrogen is transported throughout the aquatic and terrestrial ecosystems as a number of different nitrogen species, as shown in Figure 3.3. The development of technologies for converting nitrogen gas into ammonia and nitrate greatly benefited mankind through the creation of inexpensive fertilizers for food production but has also fueled major increases in nitrogen in natural water environments. Automobiles also create oxidized nitrogen species during the combustion process, combining oxygen and nitrogen in the air.

In water systems, nitrogen is present in three general forms: oxidized N ($NO_x$), consisting of nitrate ($NO_3^-$) and nitrite ($NO_2^-$), ammonia N ($NH_3$–N), and organic N, with total N being the sum of these three forms. Organic nitrogen can be dissolved (*dissolved organic nitrogen* [DON]) or may be particulate (*particulate organic nitrogen* [PON]).

$$NO_x = NO_3^- + NO_2^- \tag{3.6}$$

$$TN = NO_x + (NH_3 - N) + \text{Organic N}. \tag{3.7a}$$

$$TN = NO_3^- + NO_2^- + (NH_3 - N) + DON + PON. \tag{3.7b}$$

Each of these forms has very different fate and transport characteristics in the stormwater and green infrastructure environments.

### 3.9.1 Nitrate

Nitrate is the most oxidized form of nitrogen, with oxidation state of +5. The chemical form is $NO_3^-$ and does not change with pH. Chemically, is it fairly inert under typical environmental conditions. However, it is biologically active and is the most bioavailable form of nitrogen. It is the form common in many fertilizers. Nitrate can be converted to $N_2$ via anoxic denitrification, as discussed in Chapter 8.

As an inert anion, nitrate is very mobile in the aquatic and terrestrials environments. Plant uptake and biological conversions are relatively slow, on the order of hours to days. Nitrate will not adsorb to soil or other natural media and will leach to and through the subsurface. Control of nitrate in water systems is a challenge.

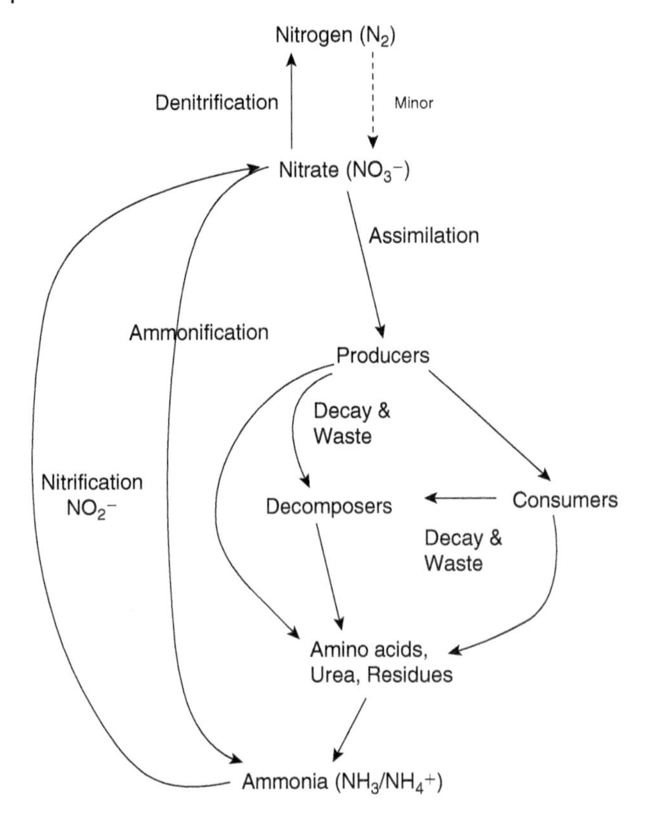


Figure 3.3 The Nitrogen Cycle in the Environment.

### 3.9.2 Nitrite

Nitrite, $NO_2^-$, is a relatively unstable nitrogen species in the aquatic environment. Generally nitrite is converted into nitrate chemically or biologically within minutes to hours. As a result, concentrations of nitrite tend to be small compared to the other nitrogen species. Nitrite is a short-lived intermediate species in the nitrification process, where ammonium is biologically transformed to nitrate. Frequently, nitrite can be omitted from the total nitrogen calculation without much loss of accuracy.

### 3.9.3 Ammonium

Ammonium is the most reduced inorganic nitrogen form, with an oxidation state of −3. Ammonium is a weak acid, with chemical form of $NH_4^+$. It can dissociate to ammonia.

$$NH_4^+ \leftrightarrow NH_3 + H^+. \tag{3.8}$$

The p$K_a$ for ammonium is about 9.0. Therefore, at pH below 9.0 (typical of most natural waters), $NH_4^+$ is the dominant species. Ammonia, $NH_3$, is a gas and is quite toxic to fish.

Ammonium, as a cation, can adsorb/ion exchange to soils and similar media. Ammonium is formed from the breakdown of organic compounds that contain N. Ammonium can also be transformed to nitrate via biological nitrification, as described in Chapter 8.

### 3.9.4 Organic Nitrogen

Nitrogen is present in biological materials as proteins and amino sugars. The amino acids are the basic building blocks. Organic N can exist in many forms and has many different physicochemical characteristics. Vegetation and similar natural materials, as well as their breakdown products, will contain organic N as both PON and DON.

Large organic molecules that contain N can adsorb and partition into organic media. Smaller organic N compounds, such as amino acids, may be more mobile in natural environments. Organic N compounds will biodegrade over time, breaking down into smaller and smaller molecules. Eventually, the N will be converted to an inorganic species, most likely ammonium. Organic N represents a very wide range of compounds and its properties cannot be readily generalized.

### 3.9.5 Nitrogen Measurements

Inorganic forms of nitrogen in water are relatively straightforward to measure directly. Wet chemistry methods are available for measurement of nitrate, nitrite, and ammonium. These species can also be measured via ion chromatography. Because organic N is a collection of various species, the analytical procedure usually involves a digestion to convert all organic N, and possibly other N forms, to a common form for measurement. An alkline persulfate digestion process is available that converts all forms of N to nitrate for analysis, giving a value for TN.

*Total Kjeldahl nitrogen* (TKN) is a digestion process that converts all organic N to ammonium, and the ammonium concentration is measured. Therefore:

$$\text{TKN} = \text{Organic N} + [\text{NH}_3 - \text{N}] \tag{3.9}$$

An independent measurement of ammonium will allow the determination of organic N from TKN by difference.

The units used for N analysis and evaluation are mg/L "as N." This terminology includes just the N in the various species and allows simple summations to be used in expressions such as Equations 3.7 and 3.9. It is a designation only and calculates the concentrations of the N species using only the molecular weight of N, 14. Using this designation, the various species of N may change through reactions and conversions, but the total N should remain constant. Commonly, the ammonium concentration as N may be presented as $NH_3$–N or $NH_4$–N, as above. Similarly, nitrate or oxidized N may be designated as $NO_3$–N or $NO_x$–N.

**Example 3.6** The concentration of ammonium in a water sample is 6.8 mg/L. What is the ammonium concentration in mg/L as N?

*Solution* The molecular weight of ammonium is 14 + 4(1) = 18. The concentration is converted to "as N" using the ratio of the molecular weight of N to that of the original compound.

$$NH_4 - N = 6.8\ \text{mg/L}\left(\frac{14}{18}\right) = 5.3\ \text{mg/L as N}$$

**Example 3.7** Results from the water quality analysis of a runoff sample give TKN = 2.2 mg/L, nitrate = 1.1 mg/L as N, nitrite = 0.2 mg/L as N, and ammonium = 0.8 mg/L as N. Find the organic N and Total N concentrations.

*Solution* Since all concentrations are "as N," they can be simply added or subtracted. Using a modification of Equation 3.9, the organic N can be calculated.

$$\text{Organic N} = \text{TKN} - (NH_3 - N) = 2.2\ \text{mg/L} - 0.8\ \text{mg/L} = 1.4\ \text{mg/L}$$

Total N is found using Equation 3.7:

$$\text{TN} = NO_x + (NH_3 - N) + \text{Organic N} = NO_3^- + NO_2^- + \text{TKN}$$

$$\text{TN} = 1.1\ \text{mg/L} + 0.2\ \text{mg/L} + 2.2\ \text{mg/L} = 3.5\ \text{mg/L}$$

## 3.10 Phosphorus

Phosphorus speciation in the environment can also be very complex. Similar to nitrogen, P can be present in organic compounds of biological origin. It can attach to particulate matter and be present as various dissolved species. A diagram of the P cycle is shown in Figure 3.4. Total phosphorus, TP, is the sum of all species:

$$\text{TP} = \text{PP} + \text{DP} \tag{3.10}$$

with PP and DP corresponding to particulate and dissolved phosphorus. These two phosphorus forms are separated via a filtration process.

$$\text{DP} = \text{Organic P} + \text{Inorganic P} \tag{3.11}$$

Both organic and inorganic P can attach to particulate matter and be part of PP. Also, large particles of P-containing organic matter can be considered as part of the PP fraction.

Inorganic P is measured as *soluble reactive P* or using a method for *orthophosphorus*; these measurements are generally considered as equivalent. Orthophosphorus represents the total $H_xPO_4^{x-3}$ species. $H_3PO_4$ can lose all three protons, depending on pH.

$$H_3PO_4 \leftrightarrow H_2PO_4^- + H^+ \quad pK_a = 2.1 \tag{3.12a}$$

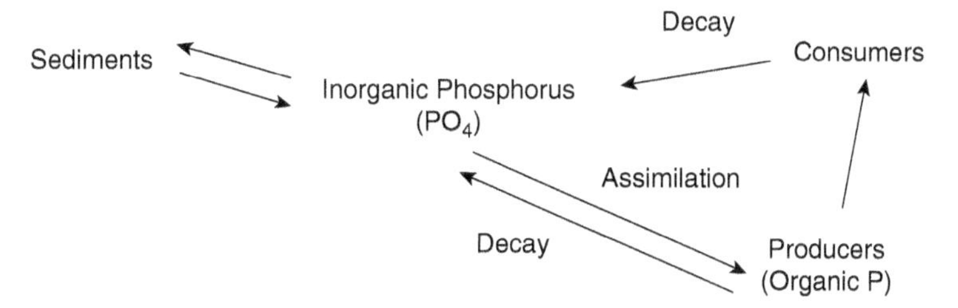


Figure 3.4 The Phosphorus Cycle in the Environment.

$$H_2PO_4^- \leftrightarrow HPO_4^{2-} + H^+ \; pK_a = 7.2 \tag{3.12b}$$

$$HPO_4^{2-} \leftrightarrow PO_4^{3-} + H^+ \; pK_a = 12 \tag{3.12c}$$

Runoff and natural water pH values are expected to be between 5 and 8; therefore, the dominant orthophosphate species are $H_2PO_4^-$ and $HPO_4^{2-}$.

Similar to N, the units for representing phosphorus are "as P." This measurement counts only the phosphorus in the species and allows total P and various other summations to be used without changing units to account for changes in speciation.

Ortho-P is the form most readily available to organisms. However, organic P can break down, releasing ortho-P. Also, as conditions change in a water or sediment, P of various forms can be released from the particulate phase.

**Example 3.8** Results from the water quality analysis of a runoff sample give TP = 1.26 mg/L, DP = 0.31 mg/L, and ortho-P = 0.19 mg/L. Fill in the table below showing the phosphorus speciation breakdown.

<table>
<tr><td colspan="3">TP =</td></tr>
<tr><td rowspan="2">PP =</td><td colspan="2">DP =</td></tr>
<tr><td>Soluble organic P =</td><td>Ortho-P =</td></tr>
</table>

*Solution* TP, DP, and Ortho-P are all given. From Equation 3.10, PP can be calculated:

$$PP = TP - DP = 1.26 \text{ mg/L} - 0.31 \text{ mg/L} = 0.95 \text{ mg/L}$$

Similarly, SOP is found using Equation 3.11:

$$\text{Organic P} = DP - \text{Inorganic P} = 0.31 \text{ mg/L} - 0.19 \text{ mg/L} = 0.12 \text{ mg/L}$$

<table>
<tr><td colspan="3">TP = 1.26 mg/L</td></tr>
<tr><td rowspan="2">PP = 0.95 mg/L</td><td colspan="2">DP = 0.31 mg/L</td></tr>
<tr><td>Soluble organic P = 0.12 mg/L</td><td>Ortho-P = 0.19 mg/L</td></tr>
</table>

## 3.11 Heavy Metals

A number of common heavy metals are utilized throughout the urban environment; many of these metals can be toxic to humans and aquatic life. The most common are lead (Pb), copper (Cu), and zinc (Zn). Other metals or metalloids, including cadmium, chromium, nickel, arsenic, and mercury, can be found in runoff, but generally at concentrations near or below 1 μg/L, and little information is available on their occurrence and treatment through SCMs.

Cu and Zn exhibit significant aquatic toxicity, especially Cu, which can be toxic to some fish at only a few μg/L. Pb has a greater human toxicity, but less so to aquatic life. If metals are captured in an SCM, they cannot be degraded or transformed. They will continue to accumulate.

Metals can be found in water and runoff as part of the particulate fraction and in dissolved form. Therefore, similar to phosphorus, total metals (TM) can be speciated into a particulate fraction (PM) and dissolved fraction (DM).

$$TM = PM + DM \tag{3.13}$$

The chemistry of dissolved metals can be complex. The basic form of dissolved metals is the divalent cation, that is, $Cu^{2+}$, $Pb^{2+}$, and $Zn^{2+}$. However, changes in water pH and the presence of other compounds can result in complexation of the dissolved metal, producing species that may be cationic, nonionic, and even anionic. Examples can include $ZnOH^+$ (aq), $PbCO_3^0$ (aq), and Cu-$OM^-$ (where OM may be some organic matter molecule). These complexed metals may behave differently than that expected for the basic divalent cation.

## 3.12 Hydrocarbons and Other Organic Pollutants

A number of organic pollutants may be present in urban stormwater. Hydrocarbons can be present in fuels or other components of vehicle fluids. Large organic molecules may be released from tars and asphalts. Pesticide use in green areas may lead to leaching to runoff. Many of these organic compounds can be toxic to aquatic life.

### 3.12.1 Hydrocarbons

By definition, hydrocarbons consist only of hydrogen and carbon. They are common components of vehicle fuels and other fluids, such as lubricating oils. Simple hydrocarbons may be linear chains, such as octane (eight carbons).

```
  H  H  H  H  H  H  H H
H-C—C—C—C—C—C—C-C-H
  H  H  H  H  H  H  H H
```

Other common hydrocarbon compounds may include benzene, toluene, and xylene, which are aromatic hydrocarbons, that is, they contain an aromatic ring.

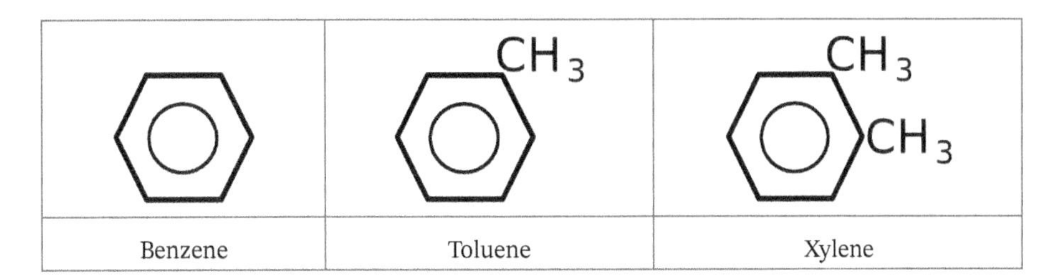


Heavier components of oils, tars, or asphalts may contain *polycyclic aromatic hydrocarbons*. These compounds have multiple aromatic rings; the simplest is naphthalene. Larger ones can contain many rings.

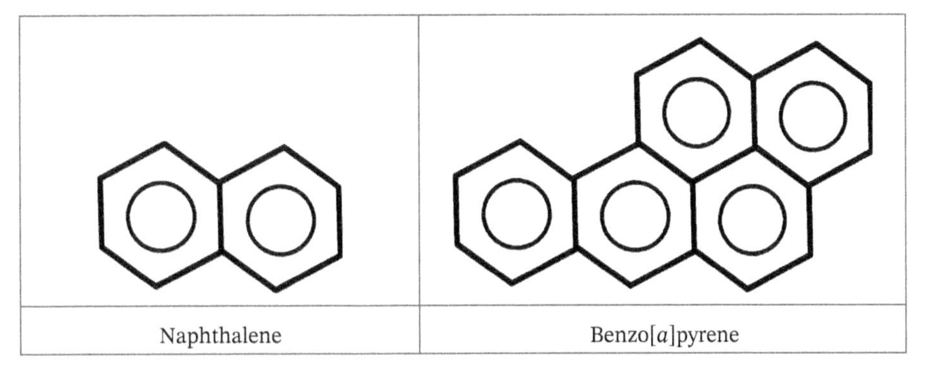


Hydrocarbons will tend to partition to organic matter, especially in soil or similar media. The larger the hydrocarbon, generally, the stronger the partitioning. Smaller hydrocarbons can be biodegraded in the environment. Larger ones tend to be more persistent.

Hydrocarbon concentrations can be measured commonly via some form of gas or liquid chromatography. Standard methods for bulk hydrocarbon measurements include that of "Oil & Grease" or "Total Petroleum Hydrocarbons." These related methods use an organic solvent or solid to extract all of the hydrocarbons from the water. The solvent is evaporated or the hydrocarbons are extracted from the solid and the hydrocarbons are measured gravimetrically.

### 3.12.2 Pesticides and Other Organic Chemicals

Pesticides are used in urban areas on and near lawns and other vegetated areas to control unwanted weeds and pests. Excess pesticide can run off of grassed areas. Also, pesticides that are applied to control weeds in cracks and openings in sidewalks and driveways may be mobilized in runoff. Other organic compounds, such as dioxins, can be present in the urban environment, resulting from combustion byproducts and/or some legacy source.

Polychlorinated biphenyls (PCBs) are a group of highly persistent organic chemicals. A PCB can contain from 1 to 10 chlorines at various places on the biphenyl molecule, resulting in the possibility of 209 different chemical forms (known as congeners). PCBs were used in a number of industrial applications for many years. In the United States, the production of PCBs has been banned since 1979, but legacy sources exist.

PCBs are found in urban stormwater, almost entirely affiliated with particulate matter (Hwang and Foster 2008; Zgheib et al. 2011). However, the sources of the PCBs are not clear.

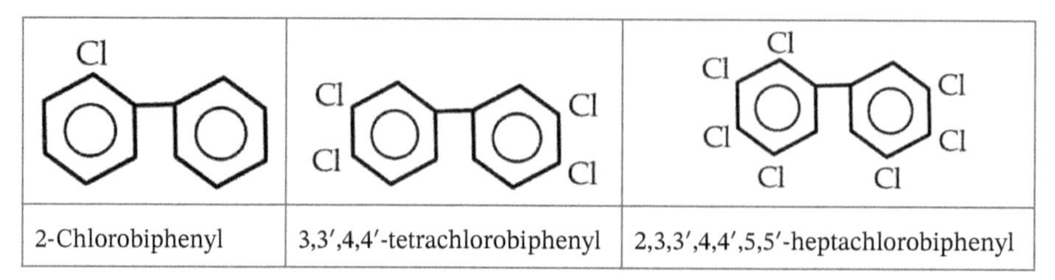


| 2-Chlorobiphenyl | 3,3′,4,4′-tetrachlorobiphenyl | 2,3,3′,4,4′,5,5′-heptachlorobiphenyl |
|---|---|---|

These organic compounds will have different fate and transport characteristics depending on the physicochemical and biological properties of the compound. Large hydrophobic organics will adsorb strongly to various media and will have limited mobility. Soluble organic compounds may be quite mobile in the environment.

## 3.13 Pathogens

Pathogens are disease-causing organisms. These organisms may include bacteria, viruses, and protozoa. Common microbial pathogens include those that cause waterborne diseases such as salmonella, cholera, hepatitis, and giardiosis.

Not all microorganisms are pathogens. The environment contains many different microbes that perform numerous important ecological functions. This includes important roles in the cycling of many elements. Microorganisms assist in the breakdown of organic matter and the cycling of carbon and nitrogen. A few grams of soil can contain millions of microorganisms.

Pathogenic microorganisms thrive in the guts of warm-blooded animals, including humans. These microorganisms are shed with the waste from these animals. In the urban environment, a number of opportunities exist for excrement to be deposited and mobilized. Indiscriminant disposal of diapers and old clothing can spread human waste. Uncollected pet waste occurs in residential areas. Many wild animals still make their homes in suburban and urban areas. These can include many bird species, squirrels, rabbits, deer, and others. The prevalence of impervious areas in the urban environment allows these wastes to be mobilized in stormwater, rather than be assimilated naturally into the soil.

Hundreds of different pathogens may be found in the urban environment. Analyzing and quantifying these different microorganisms can be technically very difficult and costly. As a result, the concept of the *indicator organism* is used. An indicator organism is a specific species or type of microorganism that is measured to indicate the presence of pathogenic contamination. Historically, the *coliform* organism has been used as the indicator of pathogens. Coliforms are found in the gut of warm-blooded animals; not all coliforms are pathogenic. Fairly simple tests are available for measuring coliform concentrations in water, such as the most probable number test. The presence of coliform bacteria indicates waste contamination and the likelihood that other pathogens are present at high levels. Conversely, low coliform levels suggest low pathogen concentrations. Coliform limits are still commonly used to determine if waters are safe for swimming at beach areas. Various species of *Escherichia coli* and *enterococcus* bacteria are also used as indicator organisms.

Continuous advances in biotechnology are allowing easier measurement of actual pathogen species, such as salmonella, and provide more specific information on pathogen concentrations, fates, transport, exposures, and risks.

## 3.14 Dissolved Solids and Conductivity

Dissolved solids are salts that will remain after water is evaporated from a filtered sample. Dissolved solids cannot be seen in the water. They result from soluble minerals that have been dissolved by the water. The analytical procedure involves first filtering the water sample to remove particulate matter, as in the TSS measurement discussed in Section 3.7. In the remaining sample, a known volume of water is dried at 104°C. The salts that remain are weighed. The concentration, known as *total dissolved solids (TDS)*, is given as the mass of salt divided by the volume of water.

The TDS is an operationally defined water quality parameter and the makeup of the dissolved solids cannot be inferred from the measurement. TDS will likely be dominated by sodium chloride and other salts that may contain calcium, potassium, magnesium, sulfate, and carbonate. TDS may also contain heavy metals, but these concentrations are likely much lower than the compounds listed above.

Generally, natural fresh waters, such as streams, rivers, and lakes, will have TDS in the range of about 200–400 mg/L. Higher TDS waters are considered brackish. Seawater TDS is relatively constant at approximately 35,000 mg/L, mostly from sodium chloride.

**Example 3.9** After a snow event and highway salting, a runoff sample is collected for TDS measurement. Eighty-five milliliters of runoff is filtered to remove suspended solids. The filtered sample is placed in a dish that weighs 85.337 g. After drying, the dish weighs 85.590 g. Find the TDS concentration and comment on the relative value.

*Solution* The dissolved solid mass is the difference between the masses in the dish:

$$\text{Dissolved solid mass} = 85.590\text{ g} - 85.337\text{ g} = 0.253\text{ g} = 253\text{ mg}$$

Concentration is the mass divided by the water sample volume:

$$\text{TDS} = \frac{253\text{ mg}}{0.085\text{ L}} = 2980\text{ mg/L}$$

This value is greater than a typical fresh water, indicating a high salt level from the highway salting.

Specific compounds that make up the TDS may be measured directly. These can include sodium, chloride, calcium, sulfate, magnesium, carbonate, and potassium. Other dissolved compounds, such as heavy metals, nitrogen, and phosphorus species, will be captured in the TDS measurement.

The TDS concentration is also proportional to the water electrical conductivity (EC). The conductivity is a measure of the ability of an electrical current to pass through the water. The higher the concentration of dissolved solids, the higher the EC will be. EC is measured in numerically equivalent units of micromhos per cm (μmho/cm) or microsiemens per cm (μS/cm). Different ions provide different degrees of conductivity, so EC and TDS do not have a predictive relationship. However, for the same ions, they should be linearly related. The TDS (in mg/L) is approximately equal to the EC (μS/cm), multiplied by a factor of 0.55–0.9, with 0.64 being commonly used. EC is measured using a probe and meter; thus, it is an easy measurement to make. Fresh water conductivity generally ranges from 150 to 500 μS/cm.

## 3.15 Trash

Trash is a visual water pollutant. Trash is carried in stormwater and with the wind. It consists of cans, bottles, wrappers, plastic bags, newspapers, and many other materials. Additionally, trash can cause clogging of stormwater inlets, pipes, and SCMs. Trash measurements in water cannot be directly made, but accumulations over time can be. Maryland and California both have water bodies that are regulated for trash as a pollutant.

## References


Hwang, H.-M. and Foster, G.D. (2008). Polychlorinated biphenyls in stormwater runoff entering the tidal Anacostia River, Washington, DC, through small urban catchments and combined sewer outfalls. *Journal of Environmental Science and Health Part a-Toxic/ Hazardous Substances & Environmental Engineering* 43 (6): 567–575.

Sullivan, A.B., Rounds, S.A., Deas, M.L., and Sogutlugil, I.E. (2012). Dissolved oxygen analysis, TMDL model comparison, and particulate matter shunting—Preliminary results from three model scenarios for the Klamath River upstream of Keno Dam, Oregon: U.S. Geological Survey Open-File Report 2012-1101, 30 p.

Zgheib, S., Moilleron, R., Saad, M., and Chebbo, G. (2011). Partition of pollution between dissolved and particulate phases: What about emerging substances in urban stormwater catchments? *Water Research* 45 (2): 913–925.


## Problems

3.1 The concentration of dissolved lead in a water sample is listed as $10^{-7}$ M. What is the concentration in mg/L?

3.2 Dissolved zinc is found in water at $5 \times 10^{-6}$ M. What is this concentration in mg/L?

3.3 An analysis of a runoff sample shows toluene present at 0.5 mg/L. What is this concentration in M?

3.4 The nitrate concentration in runoff from a lawn is $2 \times 10^{-4}$ M. What is this concentration in mg/L of nitrate and in mg/L nitrate as N?

3.5 The value of $[H^+]$ is $6.3 \times 10^{-9}$ M. What is the $[OH^-]$? What is the pH?

3.6 The pH of a runoff sample is 6.2. What are the values of $[H^+]$ and $[OH^-]$?

3.7 The measured dissolved oxygen content in a small stream is 7.8 mg/L at 12°C. What is the oxygen deficit for the stream?

3.8 During a $BOD_5$ test for a runoff sample, the initial dissolved oxygen concentration was 7.2 mg/L and the 5-day DO was 1.4. What is the $BOD_5$?

3.9 A 150-mL stormwater sample is used in a 300-mL $BOD_5$ test. The initial DO is 6.8 mg/L and the final DO is 2.1 mg/L. What is the $BOD_5$ of the stormwater sample?

3.10 The $BOD_5$ of a water is expected to be around 60 mg/L. What would be a good volume of water to add to the BOD test bottle to obtain a valid $BOD_5$ measurement?

3.11 A 200-mL sample of runoff is filtered through a filter weighing 4.251 g. After filtering and drying, the filter weighs 4.295 g. What is the TSS of the runoff?

3.12 A 200-mL sample of runoff is filtered through a filter weighing 4.102 g. After filtering and drying, the filter weighs 4.242 g. What is the TSS of the runoff?

3.13 A 200-mL sample of runoff is filtered through a filter weighing 5.102 g. The TSS in the runoff is 110 mg/L. What is the weight of the filter after drying?

3.14 Nitrogen analysis of a water gives TKN = 3.8 mg/L, nitrite = 0.2 mg/L as N, and nitrate = 1.8 mg/L as N. What is the total N concentration?

3.15 Nitrogen analysis of a water gives TKN = 3.2 mg/L, nitrite = 0.1 mg/L as nitrite, and nitrate = 2.2 mg/L as nitrate. What is the total N concentration?

3.16 Nitrogen analysis of a water gives TKN = 3.6 mg/L, ammonia-N = 1.2 mg/L, nitrite = 0.1 mg/L as N, and nitrate = 2.6 mg/L as N. What is the total N concentration?

3.17 Nitrogen analysis of a water gives TKN = 2.8 mg/L and ammonia-N = 0.9 mg/L. What is the organic N concentration?

3.18 Results from the water quality analysis of a runoff sample gives total phosphorus = 1.77 mg/L, dissolved P = 0.52 mg/L, and ortho-P = 0.24 mg/L. Fill in the table below showing the phosphorus speciation breakdown.

<table>
<tr><td colspan="3">TP =</td></tr>
<tr><td rowspan="2">PP =</td><td colspan="2">DP =</td></tr>
<tr><td>Soluble organic P =</td><td>Ortho-P =</td></tr>
</table>

3.19 Phosphorus analysis of a water gives total phosphorus = 1.77 mg/L and dissolved P = 0.52 mg/L. What fraction of TP is present as particulate P?

3.20 Copper analysis of a water gives total Cu = 0.092 mg/L and dissolved Cu = 0.052 mg/L. What fraction of Total Cu is present as particulate Cu?

3.21 A 500-mL runoff sample is filtered and then the water is evaporated. The resulting solids weigh 248 mg. What is the TDS of the runoff?

3.22 A 500-mL stream sample is filtered and then placed in a dish that weighs 32.054 g. After the water is evaporated, the dish weighs 32.398 g. What is the TDS of the stream water?

**3.23** A 500-mL stormwater sample is evaporated, leaving 385 mg of solids. The TSS of the stream was measured to be 212 mg/L. What is the TDS of the stormwater?

**3.24** After a snow/ice event, a roadway runoff sample was collected. A 200-mL sample is filtered and then the water is evaporated. The resulting solids weigh 4048 mg. What is the TDS of the runoff?

# 4

# Ecosystem Services

## 4.1 What Are Ecosystem Services?

Ecosystem services refer to any of the benefits that ecosystems—both natural and semi-natural—provide to human society (MEA 2005). These services include food and raw material provision, air and water purification, biodiversity maintenance, and aesthetic and other cultural benefits. Ecosystem services are a product of the structure (e.g., plant and animal community composition) and processes (e.g., nutrient cycling and decomposition) characteristic of an ecosystem and can be ascribed economic, social, and ecological values. Ideally, the inherent value of these services can then be used to guide management and policy decisions regarding the use and/or preservation of ecosystems (Figure 4.1).

The concept of ecosystem services was introduced in the early 1970s. Since then, investigations of the dependency of human health upon properly functioning ecosystems have surged, as have attempts to assign monetary value to the services provided by ecosystems. In one of the most widely cited ecosystem service valuation studies, Costanza et al. (1997) estimate the value of the services provided by earth's ecosystems to be at least \$33 trillion per year. (For comparison, the global gross national product at the time of this study was \$18 trillion.)

A summary of ecosystem services offered through green stormwater infrastructure is presented in Figure 4.2. All systems are highly effective in stormwater runoff reduction, but many benefits beyond stormwater management are listed. These include improved air quality and energy reduction, along with various aspects of improved community livability.

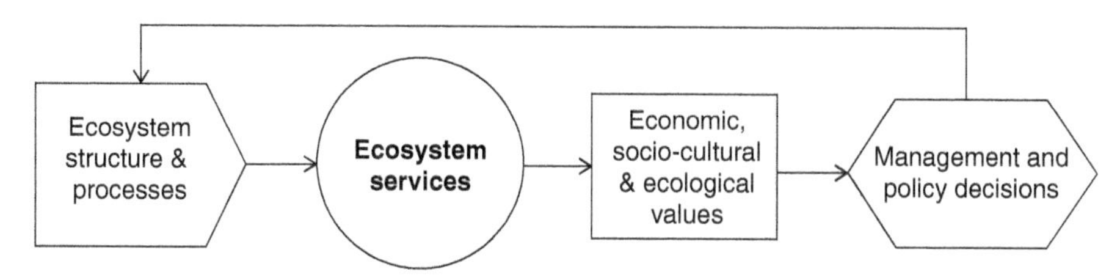


Figure 4.1 Relationship between Ecosystem Services, Ecosystem Structure and Processes, and Their Value in Decision-Making. (Adapted from de Groot et al. 2002).

*Green Stormwater Infrastructure Fundamentals and Design*, First Edition. Allen P. Davis, William F. Hunt, and Robert G. Traver.

Companion Website: www.wiley.com/go/davis/greenstormwater

Much of the initial evaluation of ecosystem services by green infrastructure has been through wetlands. Of the types of ecosystems included in Costanza et al. (1997), the services provided by wetlands were among the most valuable with a global average value of $6,000 per acre ($14,800/ha). A brief summary of the services provided by naturally-occurring and created wetland ecosystems, as well as other SCMs associated with GI, is provided in Table 4.1, along with the economic value of these services as estimated by Costanza et al. (1997), where applicable.

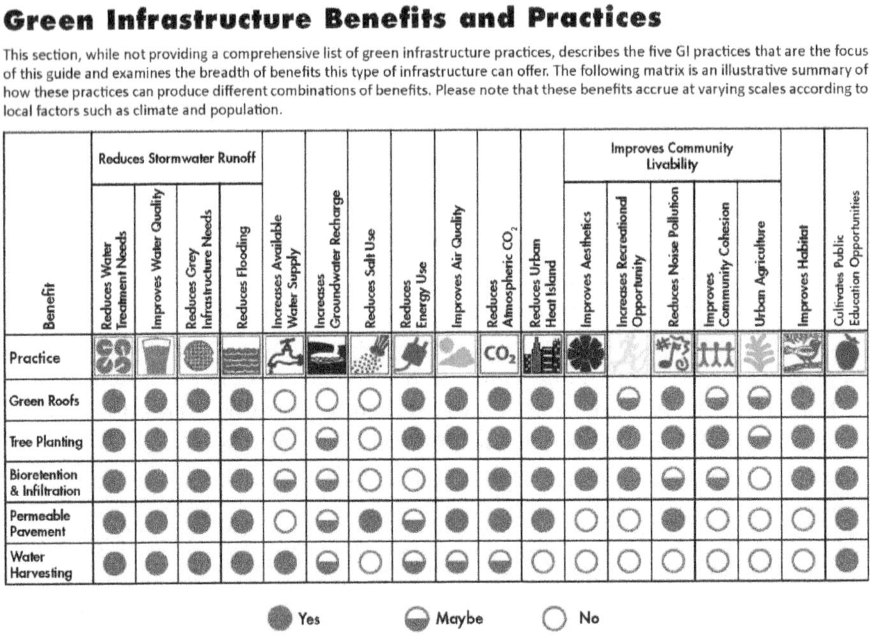

## Green Infrastructure Benefits and Practices

This section, while not providing a comprehensive list of green infrastructure practices, describes the five GI practices that are the focus of this guide and examines the breadth of benefits this type of infrastructure can offer. The following matrix is an illustrative summary of how these practices can produce different combinations of benefits. Please note that these benefits accrue at varying scales according to local factors such as climate and population.

| | Reduces Stormwater Runoff | | | | | | | | | | | Improves Community Livability | | | | | | |
|---|---|---|---|---|---|---|---|---|---|---|---|---|---|---|---|---|---|---|
| Benefit | Reduces Water Treatment Needs | Improves Water Quality | Reduces Grey Infrastructure Needs | Reduces Flooding | Increases Available Water Supply | Increases Groundwater Recharge | Reduces Salt Use | Reduces Energy Use | Improves Air Quality | Reduces Atmospheric $CO_2$ | Reduces Urban Heat Island | Improves Aesthetics | Increases Recreational Opportunity | Reduces Noise Pollution | Improves Community Cohesion | Urban Agriculture | Improves Habitat | Cultivates Public Education Opportunities |
| Practice | | | | | | | | | | $CO_2$ | | | | | | | | |
| Green Roofs | Yes | Yes | Yes | Yes | No | No | No | Yes | Yes | Yes | Yes | Yes | Maybe | Yes | Maybe | Maybe | Yes | Yes |
| Tree Planting | Yes | Yes | Yes | Yes | No | Maybe | No | Yes | Yes | Yes | Yes | Yes | Yes | Yes | Yes | Maybe | Yes | Yes |
| Bioretention & Infiltration | Yes | Yes | Yes | Yes | Maybe | Maybe | No | No | Yes | Yes | Yes | Yes | Yes | Maybe | Maybe | No | Yes | Yes |
| Permeable Pavement | Yes | Yes | Yes | Yes | No | Maybe | Yes | Maybe | Yes | Yes | Yes | No | No | Yes | No | No | No | Yes |
| Water Harvesting | Yes | Yes | Yes | Yes | Yes | Maybe | No | Maybe | Maybe | Maybe | No | No | No | No | No | No | No | Yes |

Yes Maybe No

Figure 4.2 Ecosystem Services Offered by Green Infrastructure Stormwater Systems. (Credit: Center for Neighborhood Technology 2010).

Table 4.1 Ecosystem Services Provided by Natural and/or Created Wetlands. Adapted from Costanza et al. (1997) and de Groot (2006).

| Service | Examples of goods and services derived | Estimated value (1994 US$/ac/year)[1] |
|---|---|---|
| Regulation services | | |
| Water quality | | |
| Erosion control and sediment retention | Sediment filtration and storage capabilities of SCMs prevent downstream migration of sediment and improve downstream water quality | NA |
| Waste treatment | Excess nutrient, organic, and metals loadings reduced to improve water quality through microbial degradation and/or sorption. Shading and heat capacity of water also reduce runoff temperature | 1,690 |

| Service | Examples of goods and services derived | Estimated value (1994 US$/ac/year)[1] |
|---|---|---|
| Nutrient cycling | Nitrogen and phosphorus concentrations reduced to improve water quality through adsorption and denitrification and biological uptake | NA |
| Hydrologic regulation | SCMs moderate the rate, volume, and frequency of surface runoff to provide flood and storm surge protection | 1,860 |
| Climate regulation | | |
| Greenhouse gas regulation | Maintenance of air quality and $CO_2/CH_4$ balance (through C sequestration); regulation of gases also influences climate effects | 54 |
| Microclimate regulation | Maintenance of a favorable climate (temperature, precipitation, etc.) for human habitation, health, and cultivation | NA |
| Soil formation | Land surface can be built through the accumulation of organic material in some SCMs | NA |
| Habitat services | | |
| Refugia | Maintenance of biological and genetic diversity through provision of suitable habitat for resident or migratory plant and animal species. Includes the maintenance of populations of commercially harvested species and biological pest control services. This diversity forms basis of many other ecosystem services | 123 |
| Production services | | |
| Food production | Production of fish, game, fruits for small-scale hunting/gathering or aquaculture. Some concern over toxicity, however | 104 |
| Raw materials | Production of trees, peat, and other biomass appropriate for lumber, fuel, or fodder | 43 |
| Information services | | |
| Recreation | Larger SCMs provide opportunities for hiking, bird watching, hunting, or other recreational uses | 232 |
| Cultural | Provides opportunities for noncommercial uses, including the use of SCMs for school excursions/education and for scientific research. Aesthetic, artistic, and spiritual values are also included | 357 |

[1] Value estimates for each service reprinted by permission from Springer Nature: Nature Costanza et al. (1997).
A listing of NA for individual services indicates that a formal valuation of this service had not yet been conducted.

## 4.2 Ecosystem Services and Stormwater Management

As a natural area is developed, the ability of the landscape to provide ecosystem services is diminished. This is particularly evident in urban areas, which are characterized by reduced flood and climate regulation ability, poor air and water quality, and a loss of native biodiversity. Ecologically engineered stormwater control measures, however, provide an opportunity to restore the ability of the landscape to provide some of these services. Because many of the services provided by these engineered systems have tangible economic value, developers and municipalities alike can benefit by selecting green infrastructure practices based on the suite of services they provide. Additionally, space limitations in urbanizing areas magnify the need to design SCMs that provide flows of multiple services—such as carbon sequestration, biodiversity, and recreation and education opportunities—in addition to runoff quantity and quality management.

## 4.3 Stormwater Wetlands and Ecosystem Services

Among stormwater control measures, constructed stormwater wetlands are considered to have the most potential to provide a great quantity of high-quality ecosystem services (Moore and Hunt 2012). Indeed, naturally occurring and created wetland ecosystems were the most valuable terrestrial ecosystem service providers included in the economic review of Costanza et al. (1997). Information on stormwater wetlands performance and design is given in Chapter 20.

## 4.4 Regulation Services

### 4.4.1 Water Treatment

Water treatment services are perhaps the most widely recognized service provided by SCMs, and indeed this is the purpose for which most SCMs are typically designed. As listed in Table 4.1, water-quality-related services include waste/pollutant treatment, nutrient cycling, and erosion control/sediment retention. SCMs remove and transform pollutants through a combination of physical, chemical, and biological processes; these complimentary processes include sedimentation, filtration, adsorption, chemical precipitation, microbial transformation, and biological uptake (Chapter 8).

### 4.4.2 Hydrologic Regulation

Hydrologic regulation services include regulation of the peak flow rate, volume, and frequency of surface runoff from the urban landscape. Individual and series of SCMs regulate the flashy hydrology of urban areas. This is typically accomplished through peak rate control and runoff storage and “conversion” to infiltration, evapotranspiration (ET), and seepage (shallow interflow cognate).

### 4.4.3 Climate Regulation

#### 4.4.3.1 Microclimate

SCMs can play an important role in climate regulation at a local scale and may contribute to climate regulation on a global scale as well. Climate regulation at the local scale is of particular interest in urban areas, where urban heat island effects may raise the temperature by as much as 3°C (5°F). Although this service has not been adequately quantified, the potential cooling effects of green infrastructure have been acknowledged (Bolund and Hunhammar 1999). The primary mechanism through which SCMs may regulate the urban microclimate is ET. ET consumes a large amount of heat energy via the water heat of vaporization, thus helping to regulate temperatures during the summer.

#### 4.4.3.2 Global Climate and Carbon Sequestration

Most vegetated SCMs (like constructed wetlands, bioretention, and swales) are also recognized for their role in regulating carbon dioxide ($CO_2$) and methane ($CH_4$), two greenhouse gases implicated as main drivers of global climate change. Vegetation removes $CO_2$ from the atmosphere and stores it in above- and below-ground biomass. When this vegetation dies, the saturated conditions typical of an anaerobic, reducing wetland environment force organic matter decomposition to proceed at a relatively slow rate, thus promoting a buildup of carbon in the soil. Through the ongoing processes of carbon accumulation and subsequent burial, some naturally occurring wetlands hold massive soil carbon stores, representing the largest component of the earth's terrestrial biological carbon pool, though they occupy less than 8% of the earth's surface (Mitsh and Gosselink 2007). These functions could be heavily valued in the future with the coming of a carbon market and provide another means for developers and others to further value SCMs economically.

Saturated soil conditions also promote the generation of $CH_4$, a potent greenhouse gas. $CH_4$ is produced as anaerobic bacteria degrade organic matter, particularly after supplies of more energetically favorable electron acceptors such as nitrate, manganese, iron, and sulfate have been exhausted. Although the rate at which methanogenic bacteria decompose organic matter is slow, significant quantities of methane can be evolved; when rice paddies are included, naturally occurring wetlands are estimated to account for about 30% of global methane emissions (Mitsh and Gosselink 2007). This will be a factor to consider in future SCM design.

### 4.4.4 Air Quality Regulation

Air quality is a concern in urban and urbanizing areas, especially where transportation and other activities contribute to air pollution. The effect of green infrastructure, particularly tree-based practices, on air quality in urban areas is receiving increased attention. The main process by which vegetation improves air quality is through the physical filtration of pollutants from the air (Sæbø et al. 2012), though local climatic changes caused by vegetation may also impact air quality. For example, model simulations by Taha (1996) indicated that the cooling effect produced by increasing tree cover by 2% in the greater Los Angeles area would slow photochemical reactions and ozone production such that ambient air quality standards for ozone during peak smog conditions would be exceeded 14% less frequently. The air quality benefits of wetlands have not been quantified; however, a survey of urban land uses indicated that urban wetlands have the potential to provide this service (Bolund and Hunhammar 1999). The potential for SCMs to improve air quality by filtering particulates will depend on the types of vegetation and the size of the practice. Larger applications of green infrastructure (namely, constructed stormwater wetlands) are likely to have a more pronounced impact on air quality.

## 4.5 Habitat Services

Some types of green infrastructure provide habitat for a wide variety of plant and animal species, including fish, birds, insects, amphibians, and aquatic invertebrates. Stormwater wetlands and green roofs have been recognized for the habitat services they provide when designed properly and have demonstrated the ability to support diverse bird (Brenneisen 2006; Duffield 1986), aquatic macroinvertebrate, and vegetative communities (Jenkins and Greenway 2007) (Figure 4.3). However, because SCMs also accumulate contaminants from the urban landscape, some have questioned the value of the wildlife habitat these ecosystems provide. Sparling et al. (2004) investigated the effects of contaminant exposure on red-winged blackbirds nesting in

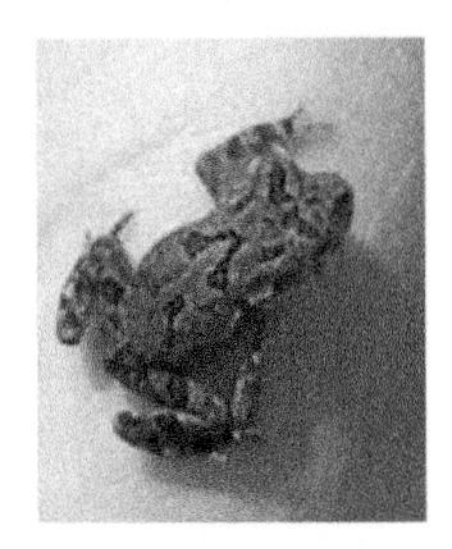

Figure 4.3 Stormwater Wetlands Can Support a Diverse Vegetative Community, as Well as Animals and Insects Such as Dragonflies (above) Frogs (Right) and Green Herons (Far Right). (Credit: Authors' Research Group).

stormwater wetlands near Washington, DC. They found that the hatching success of stormwater wetland blackbird populations compared favorably to national averages, although zinc concentrations were elevated in the tissues of birds inhabiting wetlands in industrial areas. The authors concluded that the benefits of the habitat provided by stormwater wetlands likely outweighed the negative impacts of contaminant accumulation in wildlife and that the habitat provided by certain SCMs may be especially valuable in urban areas where such habitat is scarce. However, the long-term effects of stormwater contaminant exposure on other biota are yet to be explored.

One of the major challenges to SCM habitat in cooler climate areas is the impact of roadway deicing salts on biota. Numerous studies on stormwater management ponds, related SCMs, and controlled laboratory studies have demonstrated negative impacts of high salinity on animals commonly found in these SCMs, typically amphibians (Brand et al. 2010; Mayer et al. 2008; Snodgrass et al. 2008).

Kazemi et al. (2009a, 2009b) found a statistically significant difference in invertebrate biodiversity between bioretention facilities and lawn-type greenspace, with an average of 22 invertebrate species in the bioretention cells vis-à-vis 5 species in lawn-type greenspace. In their study, the highest biodiversity was found in sites with a greater depth of leaf/plant litter, the highest number of plant taxa, and a higher amount of mid-stratum (i.e., not trees or groundcover) vegetation. Bioretention cells with complex and varied vegetation therefore have the potential to provide more invertebrate habitat than monoculture facilities with low-growing plant species.

Habitat provisioning services are crucial as the biological and genetic diversity maintained within an SCM forms the basis for most other ecosystem services provided by green infrastructure (de Groot et al. 2002). For instance, a diverse plant community may contribute to improved water treatment services (Line et al. 2008) and increase the resilience of the SCM to environmental stressors such as pollutant pulses, extreme climatic events, or disease (Hansson et al. 2005). The extent to which SCMs provide habitat for terrestrial and aquatic organisms depends largely upon their size, location within the urban landscape, degree of water saturation, and their connectivity with other natural ecosystems.

## 4.6 Production Services

Many of the plants and animals present in SCMs can potentially provide beneficial consumptive uses to people, as food and raw materials. The fishing industry relies heavily on the provisioning services of naturally occurring wetlands (upon which, to some extent, several green infrastructure SCMs are patterned); in 1998, the harvest of wetland-dependent saltwater fish and shellfish totaled nearly $950 million (Mitsh and Gosselink 2007). Tree-based SCMs may produce valued timber such as bald cypress, tupelo, and oaks. Nonetheless, due to the urban context in which SCMs are located and the perceptions of local citizens, the practicality of harvest may be limited. Several SCMs support productive herbaceous vegetative communities that hold potential as sources of energy, fiber, and other commodities, though this potential has not been widely explored in the United States (Mitsh and Gosselink 2007). Many plants are also edible and include species found abundantly in SCMs such as duck potato (*Sagitaria latifolia*) and blackberries. However, contaminant accumulation in SCMs may limit their value for food production services (Deng et al. 2004) and, until further research is conducted, it is not advisable to consume plants, particularly root tissues, in direct contact with SCMs. The ornamental value of some SCM species, such as water lilies (*Nymphaea odorata*) from constructed stormwater wetlands, is yet to be exploited as well.

Interest in rooftop farming continues to magnify. This is a confluence of expanding the many benefits of urban agriculture, with the possibility of using a green or living roof for stormwater management (Proksch 2011). Rooftop farming helps to address the challenge of obtaining fresh, locally grown food in highly urban areas. Rainfall falling on the rooftop is relatively clean compared to runoff in most urban areas, so exposure to toxic compounds is mostly minimized and edible fruits and vegetables may be obtained. Nonetheless, the nutrient needs of the crop must be balanced with the likely wash-off of excess nutrients from the roof media, creating a runoff problem.

## 4.7 Information Services

Information services contribute to human well-being by providing a place for recreation, education, and aesthetic experiences as well as opportunities for reflection, spiritual enrichment, and even artistic inspiration (de Groot 2006). Some SCMs, like stormwater wetlands, are particularly well suited to provide these services as they are located in close proximity to residential areas and schools and are often easily accessible. Increasingly, SCMs are being integrated into the urban landscape to provide recreational and aesthetic amenities to the surrounding community (Figure 4.4). For instance, walking trails, boardwalks, and wildlife viewing areas can be maintained around and through larger installations of green infrastructure to provide hiking and bird-watching opportunities. Educational signs can also be placed around SCMs to inform the public of the regulating, habitat, and provisioning services provided by stormwater wetlands and other vegetated SCMs.

Communicating the value of SCMs as recreational and aesthetic amenities can help improve the overall public perception of these stormwater control measures (Adams et al. 1984). The value of this group of services can also translate to economic benefits, particularly for developers. The US EPA found that homebuyers were willing to pay up to $18,000 more for lots adjacent to aesthetically pleasing SCMs (US EPA 1995).

Many elementary, middle, and secondary schools, as well as colleges and universities, are using green infrastructure as part of hands-on science activities. Bioretention and other SCMs can be installed on the school grounds for runoff management and can be integrated into various aspects of the curriculum. While effectively managing runoff from the school building and parking areas, the SCM can be followed throughout the year, lending itself to learning about math (surface area and data analysis), physics (water flow and temperature), chemistry (water quality and treatment), biology (microbiology, vegetation, and nutrient cycles), and other topics. A sense of ownership is instilled in the students, leading to a more environmentally informed population.

Figure 4.4 Incorporating Walking Trails, Picnic Areas, and Educational Signs (Such as These (Clockwise) in Los Angeles, CA, Charlotte, NC, El Cerrito, CA, and Minneapolis, MN) Enhance the Value of the Information Services Provided by Green Infrastructure Stormwater Control Measures. (Photos by Authors).

## 4.8 Designing SCMs for Ecosystem Services

Ultimately, the design of any SCM will depend on many parameters, including the objectives of the project and local site constraints. Fortunately, many of the services described above are not mutually exclusive, giving the designer the opportunity to design a system to provide multiple ecosystem services. Since the biological and genetic diversity of an ecosystem underpins many of the services it generates, promoting vegetative diversity through design and construction practices is encouraged.

An evaluation of characteristics of a wetland/wetpond and its associated ecosystem services has been published (Natarajan and Davis 2016). In addition to the expected services related to stormwater hydrologic and water-quality benefits, habitat value and cultural uses are discussed. Important factors include the availability of water (as a source for plants and animals), vegetation cover and type, wildlife supported, and location in the watershed (surrounding land use cover).

## References

Adams, L.W., Dove, L.E., and Leedy, D.L. (1984). Public attitudes toward urban wetlands for stormwater control and wildlife enhancement. *Wildlife Society Bulletin* 12 (3): 299–303.

Bolund, P. and Hunhammar, S. (1999). Ecosystem services in urban areas. *Ecological Economics* 29 (2): 293–301.

Brand, A.B., Snodgrass, J.W., Gallagher, M.T., Casey, R.E., and Van Meter, R. (2010). Lethal and sublethal effects of embryonic and larval exposure of *Hyla versicolor* to stormwater pond sediments. *Archives of Environmental Contamination and Toxicology* 58 (2): 325–331.

Brenneisen, S. (2006). Space for urban wildlife: Designing green roofs as habitats in Switzerland. *Urban Habitats* 4 (1): 27–36.

Center for Neighborhood Technology and American Rivers. (2010). *The Value of Green Infrastructure: A Guide to Recognizing Its Economic, Environmental and Social Benefits*. Chicago, IL: American Rivers.

Costanza, R., d'Arge, R., de Groot, R., Farber, S., Grasso, M., Hannon, B., Limburg, K., Naeem, S., O'Neill, R., Paruelo, J., Raskin, R., Sutton, P., and van Den Belt, M. (1997). The value of the world's ecosystem services and natural capital. *Nature* 387: 253–260. 15 May 1997.

de Groot, R.S. (2006). Function-analysis and valuation as a tool to assess land use conflicts in planning for sustainable, multi-functional landscapes. *Landscape and Urban Planning* 75 (3–4): 175–186.

de Groot, R.S., Wilson, M.A., and Boumans, R.M. (2002). A typology for the classification, description and valuation of ecosystem functions, goods and services. *Ecological Economics* 41: 393–408.

Deng, H., Ye, Z.H., and Wong, M.H. (2004). Accumulation of lead, zinc, copper and cadmium by 12 wetland plant species thriving in metal-contaminated sites in China. *Environmental Pollution* 132 (1): 29–40.

Duffield, J.M. (1986). Waterbird use of an urban stormwater wetland system in Central California, USA. *Colonial Waterbirds* 9 (2): 227–235.

Hansson, L.A., Bronmark, C., Anders Nilsson, P., and Abjornsson, K. (2005). Conflicting demands on wetland ecosystem services: Nutrient retention, biodiversity or both? *Freshwater Biology* 50: 705–714.

Jenkins, G.A. and Greenway, M. (2007). Restoration of a constructed stormwater wetland to improve its ecological and hydrological performance. *Water Science & Technology* 56 (11): 109–116.

Kazemi, F., Beecham, S., Gibbs, J., and Clay, R. (2009a). Factors affecting terrestrial invertebrate diversity in bioretention basins in an Australian urban environment. *Landscape and Urban Planning* 92: 304–313.

Kazemi, F., Beecham, S., and Gibbs, J. (2009b). Streetscale biodiversity basins in Melbourne and their effect on local biodiversity. *Ecological Engineering* 35: 1454–1465.

Line, D.E., Jennings, G.D., Shaffer, M.B., Calabria, J., and Hunt, W.F. (2008). Evaluating the effectiveness of two stormwater wetlands in North Carolina. *Transactions of the ASABE* 51 (2): 521–528.

Mayer, T., Rochfort, Q., Borgmann, U., and Snodgrass, W. (2008). Geochemistry and toxicity of sediment porewater in a salt-impacted urban stormwater detention pond. *Environmental Pollution* 156 (1): 143–151.

Millennium Ecosystem Assessment. (2005). *Ecosystem services and human well-being: Wetlands & water: Synthesis*. World Resources Institute, Washington, DC.

Mitsh, W.J. and Gosselink, J.G. (2007). *Wetlands*. Hoboken, NJ: John Wiley and Sons, Inc.

Moore, T.L.C. and Hunt, W.F. (2012). Ecosystem service provision by stormwater wetlands and ponds – A means for evaluation? *Water Research* 46 (20): 6811–6823.

Natarajan, P. and Davis, A.P. (2016). Ecological assessment of a transitioned stormwater infiltration basin. *Ecological Engineering* 90: 261–267.

Proksch, G. (2011). Urban rooftops as productive resources: Rooftop farming versus conventional green roofs. In: *ARCC2011 - Considering Research: Reflecting upon Current Themes in Architecture Research*, pp. 497–509.

Sæbø, A., Popek, R., Nawrot, B., Hanslin, H.M., Gawronska, H., and Gawronski, S.W. (2012). Plant species differences in particulate matter accumulation on leaf surfaces. *Science of the Total Environment* 427–428: 347–354.

Snodgrass, J.W., Casey, R.E., Joseph, D., and Simon, J.A. (2008). Microcosm investigations of stormwater pond sediment toxicity to embryonic and larval amphibians: Variation in sensitivity among species. *Environmental Pollution* 154 (2): 291–297.

Sparling, D.W., Eisemann, J.D., and Kuenzel, W. (2004). Contaminant exposure and effects in red-winged blackbirds inhabiting stormwater retention ponds. *Environmental Management* 33 (5): 719–729.

Taha, H. (1996). Modeling the impacts of increased urban vegetation on the ozone air quality in the South Coast air basin. *Atmospheric Environment* 30: 3423–3430.

U.S. EPA (1995). *Economic Benefits of Runoff Controls*. Washington, DC: Office of Wetlands, Oceans and Watersheds. EPA 841-5-95-002.

## Problems

4.1 Find a vegetated SCM at near your school or home. Make a list of the ecosystem services supplied by this SCM.

4.2 Find a vegetated SCM at near your school or home. Photo-document any wildlife found in the SCM. Document time and weather. Visit the site several times and note the changes.

4.3 Compare the ecosystem services likely provided by the following SCMs: bioretention, constructed stormwater wetlands, rainwater harvesting, sand filters, swales, and retention ponds.

# 5

# Stormwater Quality

## 5.1 Introduction

Urban runoff contains a variety of pollutants that are washed off impervious, and to a lesser extent, the pervious surfaces of urban areas. Pollutants include particulate matter, toxic compounds, nutrients, and pathogens, with details on the properties and characteristics of these pollutants discussed in Chapter 3. These pollutants accumulate on the watershed between storm events from both natural and anthropogenic sources. Rainfall/Runoff will mobile these pollutants due to the energy of impingement, high scour velocity, and solubilization. Great variability exists in concentrations found and expected in urban runoff, both within and among storm events, as well as in runoff from different land uses and climates. Just as the runoff flow can be characterized by a hydrograph as a dynamic response with time, a curve showing variations in pollutant concentrations during a storm event is known as the *pollutograph* (Figure 5.1). Water-quality challenges related to pollution prevention, treatment, and impacts must consider the high variability of runoff concentrations during a storm event. More details on the dynamic nature of stormwater quality are presented later in this chapter.

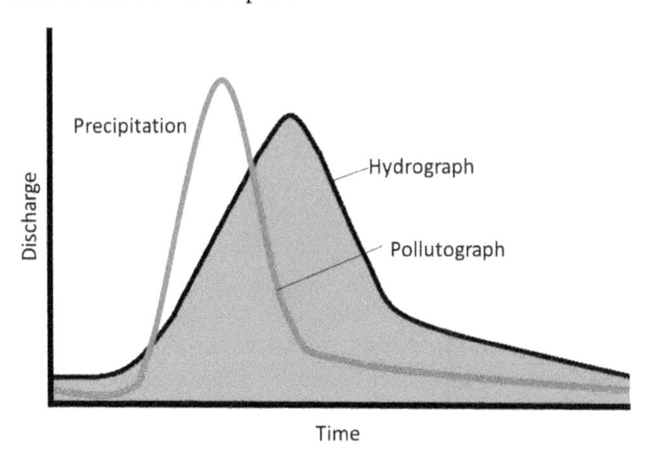


Figure 5.1 Diagram Showing the Hydrograph and Pollutograph for Runoff over an Urban Drainage Area.

*Green Stormwater Infrastructure Fundamentals and Design*, First Edition. Allen P. Davis, William F. Hunt, and Robert G. Traver.

Companion Website: www.wiley.com/go/davis/greenstormwater

## 5.2 Event Mean Concentrations

Because the concentration of a water-quality parameter is not constant throughout a storm event, a volume-weighted average concentration is commonly used to describe a representative concentration for a particular storm event. This concentration is known as the *event mean concentration* or simply the *EMC*. Since it is volume weighted, the EMC represents the concentration that would exist if all the runoff were collected into a large container and completely mixed. It is therefore the ratio of the total pollutant mass, $M$, to the total runoff volume, $V$.

$$\text{EMC} = \frac{M}{V} \tag{5.1}$$

Total runoff volume is measured as the integral of flow, $Q$, with time; similarly, pollutant mass is the integral of concentration, $C$, and flow with time, throughout the duration of the stormwater event, $t_d$. From a practical perspective, these integrals are determined via summations from discrete measurements, or using specialized sampling equipment, as to be discussed in Section 5.4.

$$\text{EMC} = \frac{M}{V} = \frac{\int_0^{t_d} QCdt}{\int_0^{t_d} Cdt} \approx \frac{\sum_0^{t_d} QC\Delta t}{\sum_0^{t_d} C\Delta t} \tag{5.2}$$

**Example 5.1** The table below gives concentrations of TSS in runoff from a campus parking lot. Find the total TSS mass and EMC for this storm event.

| Time of day | Flow rate (L/s) | TSS (mg/L) |
|---|---|---|
| 10:00 | 0 | – |
| 11:00 | 2.8 | 175 |
| 12:00 | 4.9 | 45 |
| 13:00 | 5.7 | 75 |
| 14:00 | 20 | 280 |
| 15:00 | 15 | 145 |
| 16:00 | 5.4 | 115 |
| 17:00 | 1.2 | 21 |
| 18:00 | 0 | |

**Solution** The runoff volume and TSS mass both must be determined via integration from the data set. The total runoff volume is found as the denominator of Equation 5.2. The trapezoidal rule is used for the integration, so the average flow rate for each time interval, $Q_{ave}$, is employed in the calculation.

| Time of day | Flow rate (L/s) | $Q_{ave}$ (L/s) | $\Delta t$ (s) | $Q\Delta t$ (L) |
|---|---|---|---|---|
| 10:00 | 0 | | | |
| 11:00 | 2.8 | 1.4 | 3600 | 5040 |
| 12:00 | 4.9 | 3.85 | 3600 | 13,860 |

| Time of day | Flow rate (L/s) | $Q_{ave}$ (L/s) | $\Delta t$ (s) | $Q\Delta t$ (L) |
|---|---|---|---|---|
| 13:00 | 5.7 | 5.3 | 3600 | 19,080 |
| 14:00 | 20 | 12.85 | 3600 | 46,260 |
| 15:00 | 15 | 17.5 | 3600 | 63,000 |
| 16:00 | 5.4 | 10.2 | 3600 | 36,720 |
| 17:00 | 1.2 | 3.3 | 3600 | 11,880 |
| 18:00 | 0 | 0.6 | 3600 | 2160 |
| | | | | $\Sigma$ = 198,000 L |

For the mass, the numerator of Equation 5.2 is used. $Q$ and $C$ are multiplied before they are averaged.

| Time of day | Flow rate (L/s) | TSS (mg/L) | $QC_{ave}$ (mg/s) | $\Delta t$ (s) | $QC\Delta t$ (mg) |
|---|---|---|---|---|---|
| 10:00 | 0 | – | | | |
| 11:00 | 2.8 | 175 | 245 | 3600 | $0.882 \times 10^6$ |
| 12:00 | 4.9 | 45 | 355 | 3600 | $1.279 \times 10^6$ |
| 13:00 | 5.7 | 75 | 324 | 3600 | $1.166 \times 10^6$ |
| 14:00 | 20 | 280 | 3014 | 3600 | $10.85 \times 10^6$ |
| 15:00 | 15 | 145 | 3888 | 3600 | $14.00 \times 10^6$ |
| 16:00 | 5.4 | 115 | 1398 | 3600 | $5.03 \times 10^6$ |
| 17:00 | 1.2 | 21 | 323 | 3600 | $1.163 \times 10^6$ |
| 18:00 | 0 | – | 126 | 3600 | $0.045 \times 10^6$ |
| | | | | | $\Sigma = 34.4 \times 10^6$ mg |

Therefore, from Equation 5.2, $\text{EMC} = \dfrac{M}{V} = \dfrac{34.4 \times 10^6 \text{ mg}}{198{,}000 \text{ L}} = 174 \text{ mg/L}$

A glance at the TSS concentrations in the original data set suggests that this value is reasonable.

The pollutograph based on these data is shown below. Note that the concentration of TSS is not highest at the beginning of the storm but shows up later.

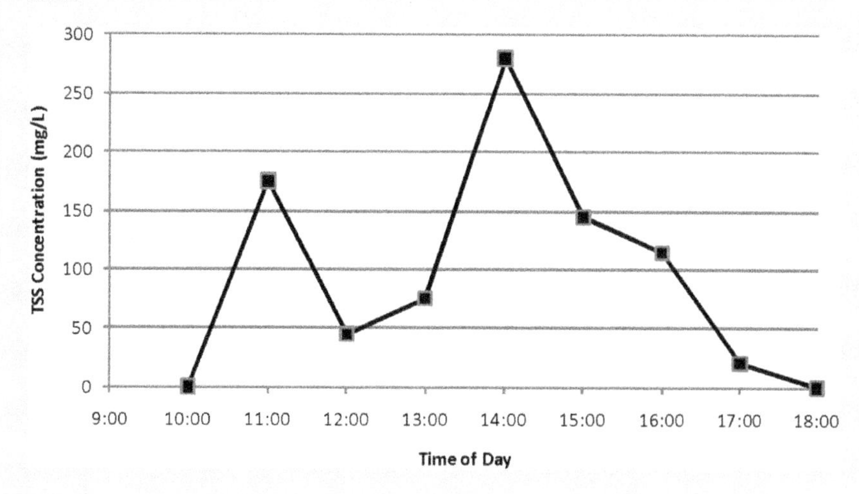

## 5.3 Urban Runoff Pollutant Concentrations

Urban runoff water-quality data have been collected by many groups over the years. The best resource for stormwater concentrations in the United States is the National Stormwater Quality Database (NSQD, https://bmpdatabase.org/national-stormwater-quality-database). This database contains water-quality data that have been collected by various permitted municipalities on their stormwater discharges. Datasets for some of the more common water-quality parameters contain thousands of measurements, allowing a reasonable characterization of urban stormwater runoff throughout the US. Stormwater quality data are generally similar for most developed countries. Concentrations in some countries, for example, China, may be different (e.g., Li et al. 2015; Wu et al. 2015). The NSQD was developed by Dr. Robert Pitt and others at the University of Alabama. It has now being maintained with the International Stormwater BMP Database. Summaries are periodically produced and posted at the Database site. Data can be evaluated based on several characteristic parameters, including land use and geographic locations. Over 9000 values currently exist in the Database. Some common water-quality parameters have thousands of measurements; much fewer measurements are available for less-common parameters. A portion of these water-quality data, compiled using all data in the database, are presented in Table 5.1.

Median and mean values in Table 5.1 indicate that concentrations of several of the water-quality parameters are of environmental concern. Specifically, TSS, oil and grease, total phosphorus, fecal coliform, and several of the metals exceed concentrations that may considered as acceptable for discharge into sensitive natural waters. Note also the large coefficients of variations (COV), which (excluding pH, dissolved oxygen)

**Table 5.1** Summary of Urban Runoff Event Mean Concentrations from the National Stormwater Quality Database (https://bmpdatabase.org/national-stormwater-quality-database).

| | | All data | | | |
|---|---|---|---|---|---|
| **Water-quality parameter** | **Units** | **Median** | **Mean** | **Coefficient of variation** | **Number of data points** |
| pH | – | 7.30 | 7.31 | 0.11 | 3179 |
| TSS | mg/L | 64 | 143 | 2.2 | 7586 |
| Turbidity | NTU | 18.7 | 39 | 2.5 | 935 |
| TDS | mg/L | 80 | 144 | 3.1 | 4097 |
| Conductivity | μS/cm@25°C | 104 | 231 | 2.0 | 1516 |
| Chloride | mg/L | 7.5 | 26 | 2.5 | 733 |
| Oil and grease | mg/L | 4.7 | 21 | 4.6 | 1539 |
| Total petroleum hydrocarbons | mg/L | 2.5 | 3.9 | 1.1 | 192 |
| $BOD_5$ | mg/L | 8.8 | 14.3 | 1.4 | 4914 |
| COD | mg/L | 53 | 79 | 1.1 | 5018 |
| Dissolved $O_2$ | mg/L | 8.2 | 7.9 | 0.23 | 192 |
| TN* | mg/L as N | 2.2 | 2.9 | 1.3 | 2259 |

*(Continued)*

Table 5.1 (Continued)

| Water-quality parameter | Units | All data | | | |
|---|---|---|---|---|---|
| | | Median | Mean | Coefficient of variation | Number of data points |
| $NH_3$ | mg/L as N | 0.43 | 0.77 | 1.5 | 2127 |
| TKN | mg/L as N | 1.4 | 2.0 | 1.6 | 6795 |
| Organic N | mg/L as N | 1.8 | 2.6 | 1.1 | 66 |
| $NO_3$ | mg/L as N | 0.62 | 1.0 | 1.4 | 947 |
| $NO_2$ | mg/L as N | 0.10 | 0.17 | 2.2 | 714 |
| TP | mg/L as P | 0.25 | 0.40 | 1.7 | 7729 |
| DP (filtered) | mg/L as P | 0.14 | 0.22 | 2.0 | 3408 |
| Ortho phosphate | mg/L as P | 0.13 | 0.22 | 1.6 | 626 |
| Fecal coliform | Colonies/100 mL | 4800 | 60,470 | 4.7 | 1987 |
| Fecal streptococcus | Colonies/100 mL | 18,800 | 73,200 | 3.3 | 1233 |
| Antimony, Sb–T | μg/L | 3.0 | 11.1 | 1.6 | 80 |
| Arsenic, As–T | μg/L | 3.0 | 5.9 | 2.5 | 796 |
| Barium, B–T | μg/L | 33 | 55 | 1.9 | 384 |
| Cadmium, Cd–T | μg/L | 1.0[#] | 3.5 | 3.4 | 1594 |
| Chromium, Cr–T | μg/L | 7.0 | 12.1 | 1.8 | 1294 |
| Copper, Cu–T | μg/L | 17 | 33 | 3.6 | 5179 |
| Cu–D | μg/L | 10.4 | 16 | 1.2 | 727 |
| Iron, Fe–T | μg/L | 605 | 2719 | 4.8 | 522 |
| Fe–D | μg/L | 240 | 602 | 2.2 | 224 |
| Lead, Pb–T[$] | μg/L | 14 | 34 | 1.9 | 3688 |
| Pb–D[$] | μg/L | 4.7 | 12 | 1.8 | 266 |
| Mercury, Hg–T | μg/L | 0.20[#] | 2.9 | 6.0 | 117 |
| Nickel, Ni–T | μg/L | 8.0 | 14 | 1.3 | 1059 |
| Selenium, Se–T | μg/L | 5.0[#] | 5.9 | 1.2 | 100 |
| Silver, Ag–T | μg/L | 1.0[#] | 5.6 | 3.0 | 167 |
| Zinc, Zn–T | μg/L | 100 | 199 | 3.0 | 6285 |
| Zn–D | μg/L | 92 | 238 | 3.3 | 736 |
| Cyanide–T | μg/L | 10[#] | 27 | 2.1 | 114 |
| 2-Chloroethylvinyl-ether | μg/L | 2.1 | 3.4 | 0.77 | 362 |
| Chloroform | μg/L | 14 | 75 | 2.1 | 20 |
| 1,2-Dichloroethane | μg/L | 0.82 | 1.5 | 2.4 | 52 |
| Methylene chloride | μg/L | 11 | 12 | 0.77 | 90 |
| Toluene | μg/L | 1.3[#] | 1.5 | 1.4 | 28 |

*T = Total; d = dissolved or filtered sample.
[#]At typical limit of detection.
[$]After 1984 (phase out of lead in gasoline).

Table 5.2 Summary of Urban Runoff Concentrations from Other Sources.

| Water-quality parameter | Units | Range | Median | Number of data points | Reference |
|---|---|---|---|---|---|
| Hg-T | ng/L | 5.0–35.2 | 14.1 | 10 | Eckley and Branfireun (2008) |
| Hg-D | ng/L | 0.6–9.9 | 5.3 | 10 | Eckley and Branfireun (2008) |
| Ethylbenzene | μg/L | <0.5–1 | <0.5 | 14 | Zgheib et al. (2012) |
| Toluene | μg/L | <0.5–1 | <0.5 | 14 | Zgheib et al. (2012) |
| Xylenes | μg/L | <0.5–1 | <0.5 | 14 | Zgheib et al. (2012) |
| Trichloroethylene | μg/L | <0.02–1.3 | <0.02 | 14 | Zgheib et al. (2012) |
| Total PAH | ng/L | 23.6–357 | 281 | 62 | Lau et al. (2009) |
| Σ16 PAH | ng/L | 677–6477 | 1327 | 16 | Zgheib et al. (2012) |
| Pyrene | ng/L | 19–3254 | 177 | 16 | Zgheib et al. (2012) |
| Fluoranthene | ng/L | 23–945 | 134 | 16 | Zgheib et al. (2012) |
| PCB 28, 52, 101, 118, 138, 153, 180 | ng/L | <10–727 | 259 | 16 | Zgheib et al. (2012) |
| Total PCBs | ng/L | 9.8–211 | NA | NA | Hwang and Foster (2008) |

range from 0.77 to 6.0. These coefficients indicate the very wide range of concentrations measured and that pollutant concentrations can be much higher than those listed in Table 5.1 during specific individual storm events. Note also that these are EMCs for storm events. Individual concentrations during storms will be higher at some times.

As an example of the variability, 7586 data points are available for TSS. While the median and mean are 64 and 143 mg/L, respectively, the 10th percentile is 13 mg/L and the 90th percentile is 313 mg/L, so variation over an order of magnitude is expected, as indicated by the COV. This variation complicates the design of stormwater control measures, renders quantitative evaluation of performance difficult (especially comparisons among SCMs), and makes discharge and accumulated load calculations challenging. These challenges will be discussed in various sections of this text.

Because different studies measure different parameters and not all parameters are measured in each study, frequently, species mass balances may not work. For example, total nitrogen is the sum of organic + ammonium + nitrate + nitrite nitrogen. However, this equality is not met in Table 5.1 data set. Data like this should be used with caution.

Recent research-focused studies have been completed to determine concentrations of other pollutants of interest in urban runoff. A few of these studies are summarized in Table 5.2. A recent compilation of urban stormwater quality is available in Pamuru et al. (2022).

### 5.3.1 Particulate Matter and Particle Size Distributions

The median TSS concentration, as a measure of particulate matter, given in Table 5.1 is 59 mg/L, with a COV of 1.8, suggesting common values ranging from 33 to 106 mg/L. (This simple analysis assumes that the pollutant concentrations are normally

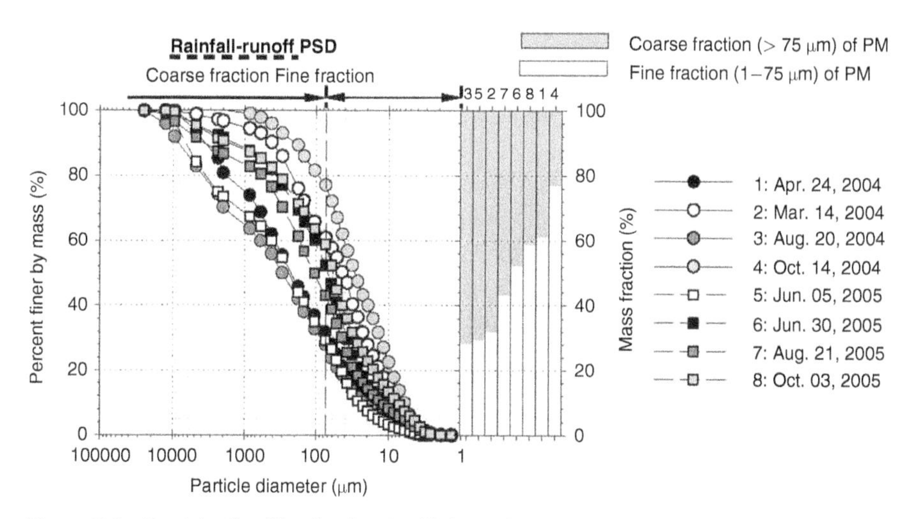


Figure 5.2 Particle Size Distribution for Highway Runoff in Baton Rouge LA (Kim and Sansalone 2008).

distributed, but they are not, as discussed later in this chapter.) TSS EMCs of 100 mg/L or more are not uncommon in urban stormwater.

The measurement of particulate matter as TSS provides a mass-based concentration but does not consider the distribution of particle sizes that can exist in urban runoff. Size distributions can vary widely, ranging from very small near-colloidal material to sand and gravel. Different particle sizes will behave differently. Larger particles are easier to removal via sedimentation and filtration processes. However, the smaller particle fraction is generally more chemically active and is commonly responsible for the particulate fraction of pollutants that affiliate with particles, such as metals, hydrocarbons, and phosphorus.

The use of Stokes Law (Chapter 8) requires information on particle size to evaluate sedimentation removal and the filtration equation (Chapter 8) similarly requires this information. For effective design and evaluation of the treatment performance, particle size distribution information is important.

Kim and Sansalone (2008) found highway particulate matter to range from 1 μm to 24.5 mm in size. The $d_{50}$ for eight runoff events ranged from 29 to 300 μm, with a mean value of 136 μm; the data are shown in Figure 5.2. Particle fractions can be characterized as a coarse fraction for sizes greater than 75 μm. A settleable fraction is defined for particulate matter sizes 25–75 μm. Suspended particulate matter ranges from 0.45 to 25 μm.

### 5.3.2 Nitrogen and Nitrogen Speciation

Table 5.1 suggests a total nitrogen concentration of 2.0 mg/L ($TN = TKN + NO_2 + NO_3$). As noted in Chapter 3, nitrogen in water can exist in a variety of forms, primarily dissolved but also particulate. A study investigating N characteristics in runoff from a parking/roadway area noted the complexity of the N distribution. With an overall TN concentration of 1.62 mg/L, the majority of the N was particulate in form, at 0.93 mg/L

(Li and Davis 2014). The dissolved N was dominated by nitrate (0.28 mg N/L), followed by ammonium (0.15 mg N/L) and nitrite (0.02 mg N/L). Work by Jani et al. (2020) noted similar but somewhat different results, with the average speciation of DON (47%) > PON (22%) > $NO_x$–N (17%) > $NH_4$–N (14%) for a mean TN concentration of 2.5 mg/L. The organic N fraction appears to be biogenic and is composed of vegetative matter, pollen, and like material. This diverse mix of N forms suggests that a single type of treatment may not be effective for removal of N.

Recent work has provided more detailed information on stormwater organic nitrogen characteristics and its complexity. Stormwater from a residential area in Florida had DON concentrations averaging 1.1 mg/L (Lusk and Toor 2016). The DON consisted of more than 1200 different compounds covering a wide variety of chemical characteristics. Compounds were identified with lipid-, protein-, amino sugar-, and lignin-like properties. Many appear to have biogenic origin with about 10% being readily biodegradable.

### 5.3.3 Phosphorus and Phosphorus Speciation

The median total phosphorus concentration is 0.27 mg/L (Table 5.1). About half or more of the total phosphorus is operationally defined as particulate phosphorus, attached to particulate matter, with the remainder being dissolved. Liu and Davis (2014) noted the particulate fraction to be 77% in roadway/parking runoff; similar values of 69% and 74% were reported by Berretta and Sansalone (2011a). Table 5.1 suggests 52% particulate phosphorus based on median values. The distribution of particulate-bound phosphorus among particulate sizes for various storms is shown in Figure 5.3, not dominated by a single size fraction. Of the dissolved phosphorus, 13% was inorganic phosphorus ($H_2PO_4^-/HPO_4^{2-}$) and 10% was organic phosphorus (Berretta and Sansalone 2011b; Liu and Davis 2014).

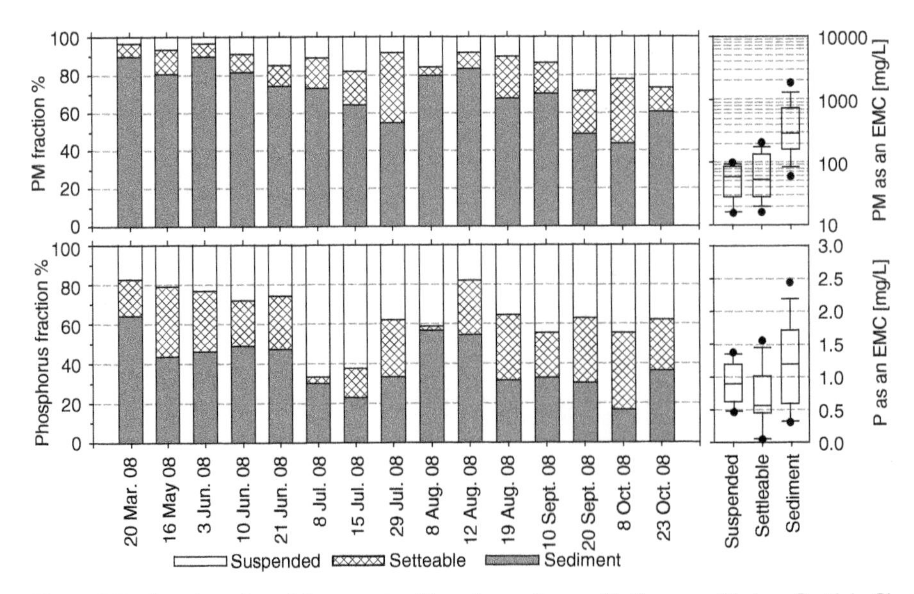


Figure 5.3 Fractionation of Stormwater Phosphorus (Lower Plot) among Various Particle Size Fractions (Berretta and Sansalone 2011a).

### 5.3.4 Heavy Metals Concentrations and Speciation

Heavy metal concentrations listed in Table 5.1 range from the highest concentration of Zn at 73–116 μg/L (all data and residential only) to a low for Hg at 0.20 μg/L. The ranking follows: Zn > Pb ≈ Cu > Ni ≈ Cr > Cd > Hg.

Heavy metals are found partially in dissolved form and affiliated with particulate matter. Sansalone et al. (2010) found particulate-bound fractions for Cu, Pb, Zn, Cr, Cd, and As to be 89%, 87%, 98%, 85%, 73%, and 95%, respectively. In general, the metals were distributed throughout the entire particle size gradation evaluated. Metal concentrations on smaller particles were greater than for larger particle for all metal(loid)s (Figure 5.4). Zanders (2005) found that stormwater Cu and Zn were both associated at

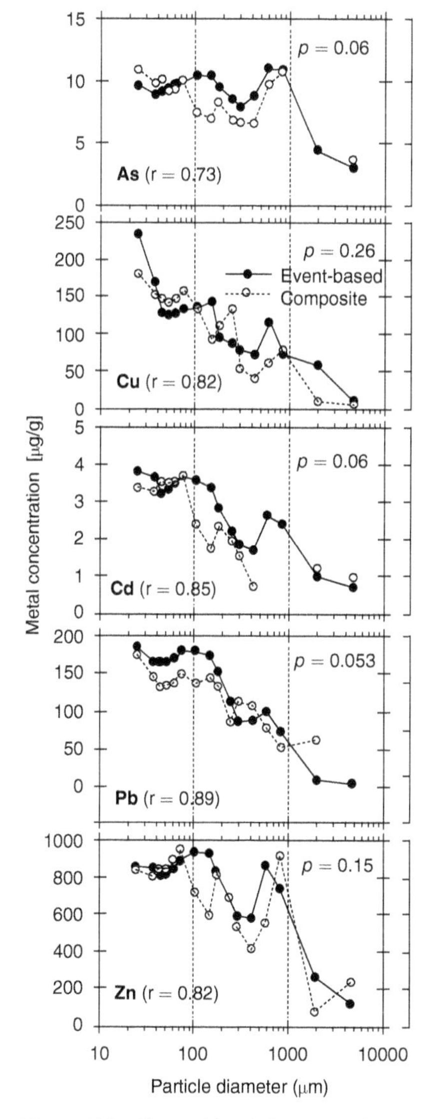


Figure 5.4 Heavy Metal Concentrations in Various Stormwater Particle Size Fractions (Sansalone et al. 2010, with Permission from ASCE).

Table 5.3 Total Metal Concentrations as a Function of Particle-size Fraction (Adapted from Zanders 2005).

| Particle-size fraction (μm) | Total metal concentration (mg/kg) | | | Particle density (kg/m$^3$) |
|---|---|---|---|---|
| | *Cu* | *Zn* | *Pb* | |
| 0–32 | 181 | 2080 | 316 | 2140 |
| 32–63 | 197 | 1695 | 322 | 2150 |
| 63–125 | 212 | 1628 | 334 | 2190 |
| 125–250 | 184 | 1073 | 251 | 2330 |
| 250–500 | 85 | 507 | 193 | 2530 |
| 500–1000 | 26 | 268 | 323 | 2540 |
| 1000–2000 | 21 | 226 | 36 | 2390 |

much higher fractions to particle sizes less than 250 μm, than for sediment particle sizes greater than 250 μm (Table 5.3). The relationship for Pb was much weaker. Smaller particles (<125 μm) also had lower particle densities (<2200 kg/m$^3$) than larger particles (densities ranging from >2300 to >2500 kg/m$^3$), further diminishing the fraction of metal able to be captured along with particulate matter capture. Because the study was conducted in New Zealand, it should be noted that this relatively low particle density may in part be due to lighter weight volcanic rock-based sediment (Zanders 2005).

### 5.3.5 PAH and PCBs

PAH and PCBs are persistent organic pollutants. Because of their hydrophobicity, all but the small PAH and all PCBs are predominantly affiliated with stormwater particulate matter and the concentrations in the dissolved phase are small. These organic pollutants attach strongly to the particulate matter, most likely affiliating with the organic fraction (Hwang and Foster 2006, 2008; Zgheib et al. 2012).

## 5.4 Urban Stormwater Pollutant Sources

Because of the heterogeneity of urban land areas, many sources contribute to the pollutant load in stormwater runoff. Pollutants are deposited onto impervious surfaces, or are transformed into mobile fractions, during dry periods between rainfall events. This deposition and corrosion will depend on factors that may vary throughout the year.

Particulate matter has both natural and anthropogenic origin. Local soils can be deposited onto impervious surfaces by wind or dropped by vehicles after they have traveled in muddy areas. Any type of construction activity can greatly exacerbate the concentrations of particulate matter in runoff, and stormwater control in construction areas requires special protection and provisions.

Particulate matter is also contributed by materials that wear from our built urban infrastructure. Asphalt and concrete pavements gradually wear over time, releasing small pieces of particulate matter. As automobile tires and brakes are used, they also release small particles. The same is true for many building materials. Therefore, the makeup of particulate matter in stormwater can vary widely.

Many studies have indicated that nitrogen and phosphorus in urban runoff are mostly contributed by runoff of fertilizer from lawns and other urban grassed areas. Fertilizers that are applied in excess, or that incidentally end up on sidewalks or roadways, can readily wash off with stormwater. The nitrogen form with the largest concentration, as noted in Table 5.1 and by others (Collins et al. 2010; Jani et al. 2020; Li and Davis 2014), is organic N. This suggests that materials of vegetative origin may be the largest source of N, and possibly P, in urban runoff. Grass clippings, leaf fall, pollen, and other natural organic matter can deposit onto impervious surfaces, to be readily transported during runoff events (Figure 5.5). The degree of these natural contributions may be seasonal. Increasing evidence suggests that leaf litter is a major source of stormwater phosphorus (Bratt et al. 2017; Wang et al. 2020) and isotopic analysis (in Florida) suggests oak detritus and grass clippings as organic N sources (Jani et al. 2020; Lusk et al. 2020). The importance of these biogenic sources suggests that management practices that focus on removal of leaves, grass clippings, and other vegetative sources may provide importont reductions in stormwater nitrogen (Lusk et al. 2020).

Stormwater nitrogen also originates from atmospheric $NO_x$ ($NO + NO_2$) compounds. Oxidized nitrogen species are created during combustion in automobiles:

$$\frac{1}{2}N_2 + \frac{1}{2}O_2 \xrightarrow[\text{temperature}]{\text{combustion}} NO \tag{5.3a}$$

$$\frac{1}{2}N_2 + O_2 \xrightarrow[\text{temperature}]{\text{combustion}} NO_2 \tag{5.3b}$$

The $NO_x$ can be converted in the air to nitrate, to be subsequently deposited on the landscape via wet deposition (Jani et al. 2020).

The origin of microbial pathogens in urban runoff is not entirely clear. It likely that pathogens result from feces of a number of different animals that can include urban wildlife (e.g., deer, squirrels, and rabbits), birds, and pets. Human sources may include carelessly discarded diapers. Dumpsters containing food or other wastes may be pathogen sources in the urban environment. The presence of the impervious surfaces provides a stormwater conduit for these pathogens.

Figure 5.5 Grass Clippings and Leaves along Curb and Gutter as Sources of Stormwater Nutrients (Photos by Authors).

Figure 5.6 Architectural Copper as a Possible Stormwater Copper Source (Photo by Authors).

The origin of oils and grease is clear. Oils are contributed by leaking fluids from vehicles parking and moving over the drainage area.

Heavy metals originate from vehicles and buildings in the urban watershed (Davis et al. 2001). Tires and brakes wear during automobile operation; tires commonly contain high levels of zinc (Councell et al. 2004; Smolders and Degryse 2002) and brakes may contain copper (Hur et al. 2003, 2004) and possibly some lead. Motor oils and other vehicle fluids may also contain metals such as zinc as breakdown inhibitors. Copper and zinc are commonly used in building construction for roofing, flashing, or other uses (Figure 5.6). These metals will slowly corrode over time, allowing metals to wash off during rainfall events.

## 5.5 Pollutant Buildup and Wash Off

### 5.5.1 Pollutographs

Just as the dynamic response of runoff flow as a function of time can be presented as the hydrograph, the changing runoff concentrations of pollutants in a storm event can be presented as the *pollutograph*. Figure 5.7 shows discrete TSS measurements during a storm event that can be used to construct a pollutograph. The product of the pollutograph with the hydrograph presents the pollutant mass flux as a function of time.

### 5.5.2 First Flush

Pollutants will accumulate in the watershed during dry periods. The sources described in the previous section will contribute the numerous pollutants to the watershed at various rates. This suggests that rainfall that occurs after long dry periods will produce runoff that carries a higher pollutant load than runoff from a short antecedent dry period. This concept of buildup and wash off is also the basis for most runoff

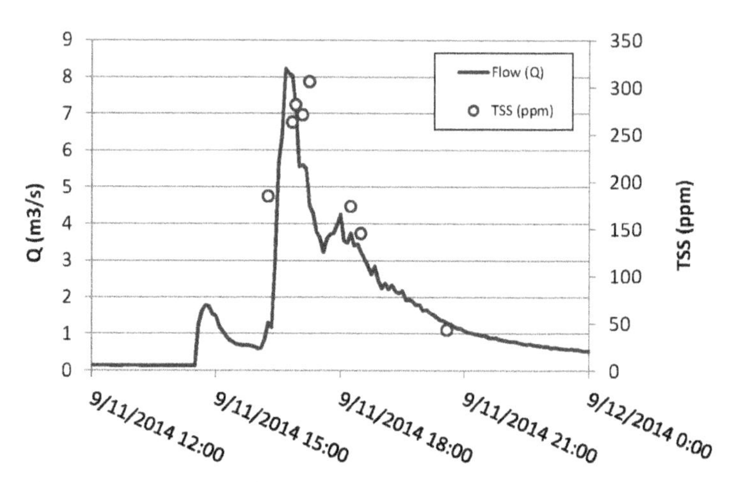


Figure 5.7 Measured TSS Water-Quality Values that Can Be Used to Create a TSS Pollutograph. These Values are Overlain the Hydrograph of the Event (with Permission, Joshua Baird).

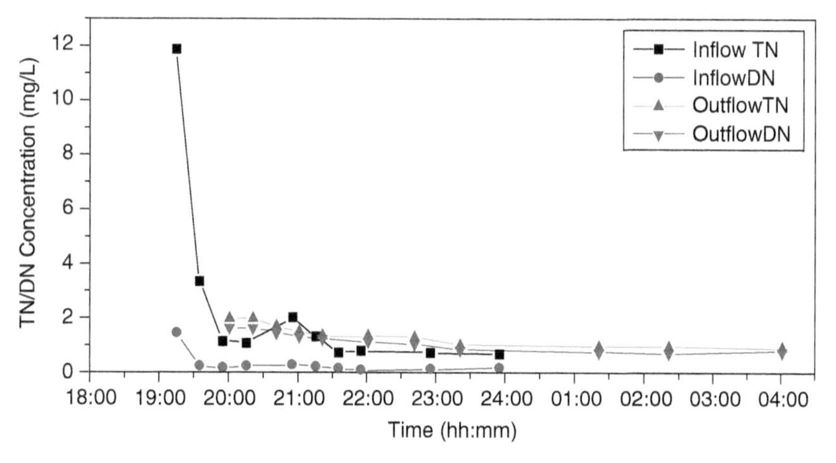


Figure 5.8 Concentrations of Total (Inflow TN) and Dissolved Nitrogen (Outflow TN) Washed from a Roadway and Parking Lot, Showing a Discernible First Flush (Data from Li and Davis 2014).

water-quality models, such as SWMM (Storm Water Management Model), developed by the US EPA (Chapter 7).

Because of the variable buildup and wash-off concepts, the concentrations of pollutants in runoff tend to vary significantly from rainfall event to event, throughout a rainfall event, and for different climates and land uses. The first, initial runoff volume is commonly the most polluted, as the runoff originates from the watershed in its dirtiest state. This phenomenon of higher concentrations in the first portions of the runoff is known as a *first flush*. Subsequent rainfall/runoff over the drainage area is exposed to a cleaner catchment area and the resulting runoff is generally cleaner. Figure 5.8 shows a pollutograph with a clear first flush concentration for both total and dissolved nitrogen. Photos of the runoff collected during the storm shown in Figure 5.8 are presented in Figure 5.9. A distinct difference in the quality of the water can be seen from the differences in color and clarity.

Figure 5.9 Runoff Samples Washed from a Roadway and Parking Lot, Showing a Discernible First Flush, Corresponding to Data in Figure 5.8 (Credit: Authors' Research Group).

While qualitatively easy to describe, a quantitative definition of the first flush is more challenging. Concentration-based methods have been used to describe first flush, where the first (or some measurement near the first) concentration is higher than the EMC or a later concentration.

Nonetheless, most first flush definitions are mass-based and are defined around the ratio of cumulative pollutant mass to the cumulative runoff volume. The mass ratio $m'(t)$ is defined as

$$m'(t) = \frac{m(t)}{M} \tag{5.4}$$

where $m(t)$ is the cumulative pollutant mass at time t and M is the total pollutant mass. Similarly, the volume ratio is given by

$$v'(t) = \frac{v(t)}{V} \tag{5.5}$$

Graphically, Equation 5.4 can be plotted as a function of Equation 5.5. If the curve lies above the 45° line, the pollutant mass is being washed from the catchment at a greater rate than the corresponding volume. The 45° line indicates both mass and volume flushed at the same rate. A line below the 45° occurs when the mass flush is slower than the volume. A curve showing these conditions is presented in Figure 5.10.

Quantitatively, the $m'(t):v'(t)$ can be given as:

$$m'(t) = [v'(t)]^b \tag{5.6}$$

A value of b equal to 1.0 indicates the $m'(t) = v'(t) \cdot b < 1.0$ describes the curve where $m'(t) > v'(t); m'(t) < v'(t)$ is described by $b > 1.0$. The value of b can be found from a data set by taking the log of both sides of Equation 5.5 to linearize it. The slope is equal to $b$.

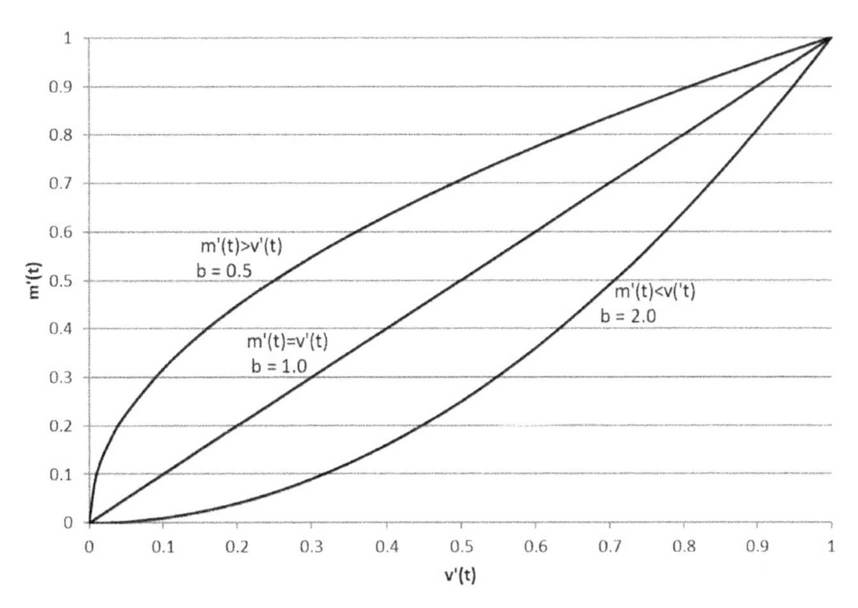


Figure 5.10 Relationships between Runoff Volume and Pollutant Mass Flushing during a Runoff Event. The Relationship Where $m'(t) > v'(t)(b < 1.0)$ Indicates Mass Flushing Faster than the Runoff Volume.

**Example 5.2** Find the value of *b* (Equation 5.6) for the TSS data given in Example 5.1.

*Solution* The runoff volume and TSS mass for each sample interval are presented in Example 5.1. This represents the starting point for this example. $v(t)$ and $m(t)$ are found as the cumulative volume and mass, with $v'(t)$ and $m'(t)$ calculated using Equations 5.4 and 5.5. The log of both are calculated to find *b* from the slope. All data are presented in the following two tables:

Runoff Volume

| Time of day | $Q\Delta t$ (*L*) | $v(t) = \sum Q\Delta t$ (*L*) | $v'(t)$ | Log($v'(t)$) |
|---|---|---|---|---|
| 10:00 | | | | |
| 11:00 | 5040 | 5040 | 0.025 | -1.594 |
| 12:00 | 13,860 | 18900 | 0.095 | -1.020 |
| 13:00 | 19,080 | 37980 | 0.192 | -0.717 |
| 14:00 | 46,260 | 84240 | 0.425 | -0.371 |
| 15:00 | 63,000 | 147,240 | 0.744 | -0.129 |
| 16:00 | 36,720 | 183,960 | 0.929 | -0.032 |
| 17:00 | 11,880 | 195,840 | 0.989 | -0.005 |
| 18:00 | 2160 | 198,000 | 1.000 | 0.000 |
| | $\Sigma$ = 198,000 L = V | | | |

TSS Mass

| Time of day | $Q C \Delta t$ (mg) | $m(t) = \sum Q C \Delta t$ (mg) | $m'(t)$ | $\text{Log}(m'(t))$ |
|---|---|---|---|---|
| 10:00 | | | | |
| 11:00 | $0.882 \times 10^6$ | $0.882 \times 10^6$ | 0.026 | −1.591 |
| 12:00 | $1.279 \times 10^6$ | $2.161 \times 10^6$ | 0.063 | −1.202 |
| 13:00 | $1.166 \times 10^6$ | $3.327 \times 10^6$ | 0.097 | −1.015 |
| 14:00 | $10.85 \times 10^6$ | $14.18 \times 10^6$ | 0.412 | −0.385 |
| 15:00 | $14.00 \times 10^6$ | $28.18 \times 10^6$ | 0.819 | −0.087 |
| 16:00 | $5.03 \times 10^6$ | $33.21 \times 10^6$ | 0.965 | −0.016 |
| 17:00 | $1.163 \times 10^6$ | $34.37 \times 10^6$ | 0.999 | −0.001 |
| 18:00 | $0.045 \times 10^6$ | $34.42 \times 10^6$ | 1.000 | 0.000 |
| | $\Sigma = 34.4 \times 10^6$ mg = M | | | |

Plots of $m'(t)$ versus $v'(t)$, without and with the log transformation, are shown below. The $m'(t)$ versus $v'(t)$ does not show a dominant trend above or below the 45° line. The slope of the log plot is 1.09, indicating that the cumulative mass slightly lags behind the cumulative volume. No first flush is indicated by these data.

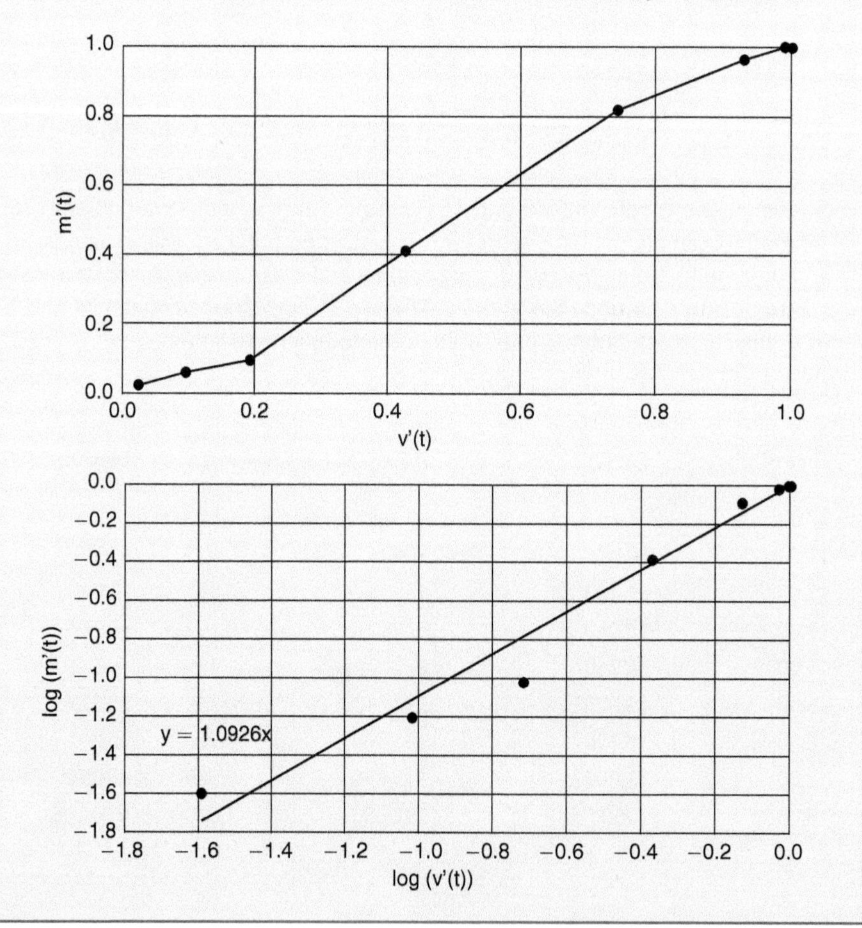

Evaluation of runoff pollutant dynamics from a parking desk (100% impervious) showed first flush for TSS, dissolved copper, dissolved cadmium, nitrate, and chloride (Batroney et al. 2010). Average values of $b$ were, respectively, 0.607, 0.484, 0.744, 0.638, and 0.644. Total nitrogen, total phosphorus, nitrite, and chromium did not demonstrate first flush.

A common quantitative definition for a first flush is that a first flush occurs when at least 50% of the pollutant mass is flushed in the first 25% of the runoff volume (Flint and Davis 2007). Only nitrate met this condition for parking deck stormwater; TSS, dissolved copper, and dissolved cadmium had greater than 40% mass in the first 25% volume (Batroney et al. 2010). Nitrite met this first flush criterion for 33% of the measured events in an urban area in Maryland. This was followed by total phosphorus at 27%, nitrate and TKN at 22%, copper at 21%, and TSS at 17%. Others have used more restrictive definitions, such as at least 80% mass in the first 20% runoff volume, but this highly constrained definition is rarely, if ever, noted in stormwater studies (Batroney et al. 2010; Flint and Davis 2007; Sansalone and Cristina 2004).

First flush events are common for many pollutants but do not always occur (He et al. 2010), especially for bacteria (Hathway and Hunt 2011; He et al. 2010). First flush of particulate matter is common in runoff situations. Higher flow rates generally produce higher particulate matter and TSS concentrations (He et al. 2010; Sansalone and Kim 2008). The initial parts of the hydrograph commonly show the highest particulate matter concentrations and, incrementally, the particulate matter mass is higher than the corresponding incremental runoff volume, although some studies have noted particulate matter at high concentrations during times of high rainfall intensity, possibly later in the storm (McCarthy et al. 2012). First flush of particulate matter is common since higher flow rates are generally found earlier in the event. Regardless of the timing, as the kinetic energy of the rainfall increases during times of high intensity, this energy is able to mobilize the particulate matter from impervious surfaces and the higher flows can also mobilize larger particles in runoff (Sansalone and Kim 2008). Hathway et al. (2012) noted the strongest first flush in TSS, with the strength of first flush for other pollutants following the order: $NH_3$-N, TKN, $NO_x$, TP, ortho-P.

First flush tends to be more common in smaller drainage areas because larger areas present more opportunities for mixing of flow and temporal impacts of rainfall on drainage areas. Generally, more soluble pollutants exhibit first flush more so than suspended matter.

Trends of $v'(t)$ (runoff ratio) and $m'(t)$ (loading ratio) for TKN in a highly urbanized catchment in Maryland are presented in Figure 5.11. Most of the storm events show some indication of a first flush in TKN, but not all.

A first flush corresponds to a mass-limited watershed wash-off model (Sheng et al. 2008). A mass limited situation occurs when adequate runoff volume exists over the drainage area and the runoff flow mobilizes the accumulated pollutant mass. The pollutant concentrations in the runoff and the accumulated mass on the catchment area decrease with the event duration. From a simple modeling perspective, this pollutant decrease follows an exponential decay.

In situations where runoff volume is small, the event is considered as flow-limited. In this case, the volume is inadequate to wash of large amount of the accumulated pollutant mass. The pollutant concentration in the runoff tends to be more constant.

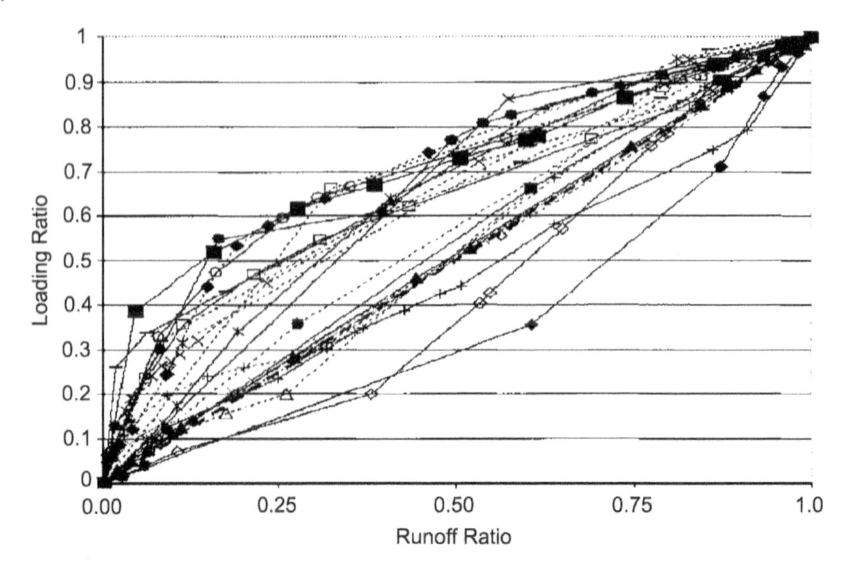


Figure 5.11 Relative TKN Mass Flushed as a Function of Volume Flushed for 23 Storm Events in Maryland (Flint and Davis 2007, with Permission from ASCE).

The consideration of the first flush is very important for the design and operation of SCMs. By treating only the first flush fraction of a runoff event, since the pollutant mass is front loaded in the storm event, and because the majority of the storm events are small in many areas, a significant impact on pollutant loads can be made by treating only an initial fraction of the runoff. For example, based on data from a highly urbanized catchment in Maryland, Flint and Davis (2007) found that the first 12.7 mm (0.5 in.) of runoff contained 81% of the measured TSS mass load, 82% of the $NO_X$ mass, 85% of TKN, and 86% of TP, TPb, TCu, and TZn. Increasing the runoff capture depth to 19.5 mm (0.75 in.), the values for TSS, $NO_2$, $NO_3$, TKN, TP, TPb, TCu, and TZn increased respectively to 90%, 94%, 89%, 89%, 96%, 96%, 94%, and 92%.

In a similar manner, a "slicing" method to define the first flush has been defined by Bach et al. (2010). In this method, the runoff event is divided into depth "slices," such as 2 mm depths. The mean and ranges of pollutant concentrations are determined for each depth slice and compared statistically. If the values for a slice are different, then a first flush may be indicated; if the values are statistically identical, the depth slices are combined. By focusing on a specific depth slice, evaluation of water-quality volumes can be better defined for use in SCMs.

Studies of the toxicity of urban highway runoff in California to various aquatic species have shown that the majority of the toxicity was present in the first flush. The toxicity was attributed to copper, zinc, and in some cases, surfactants in the runoff (Kayhanian et al. 2008).

A seasonal first flush occurs in climates where dry and rainy seasons are common. During a dry season, little rainfall occurs. Significant pollutant loads can accumulate on urban impervious and pervious surfaces. This buildup can occur for several months. As a result, the first rainfall that occurs after this dry period can be extremely polluted; it will be the most polluted event of the year.

## 5.6 Annual Pollutant Loads

While many environmental impacts of runoff are important on an event time scale, others are more appropriately measured as the total pollutant load delivered to a water body. This mass load is calculated as an annual average and usually normalized by the drainage area. Thus the result, as an annual load, is given as lb/ac year or kg/ha year. The methods used for calculating these loadings are discussed in Chapter 9. Table 5.4 gives values of annual loads for TSS, nitrogen, and phosphorus for urban, turf, and forest areas in the Chesapeake Bay watershed.

Pollutant loadings are proportional to mean runoff concentrations. There, using nitrogen and phosphorus data from Tables 5.1, 5.2, and 5.4, loadings for other pollutants can be estimated. These values are presented in Table 5.5.

Table 5.4 Annual Sediment and Nutrient Loads for Urban Areas in the Chesapeake Bay Watershed (Adapted from MDE 2020).

| **Parameter** | **Impervious road** | **Aggregate impervious** | **Turf** | **True forest** |
|---|---|---|---|---|
| TSS (t/ac year) | 10.03 | 4.40 | 1.78 | 0.37 |
| (kg/ha year) | 11.24 | 4.92 | 1.99 | 0.41 |
| Total N (lbs/ac year) | 36.43 | 20.39 | 13.43 | 2.31 |
| (kg/ha year) | 40.82 | 22.84 | 15.04 | 2.59 |
| Total P (lbs/ac year) | 6.89 | 2.55 | 2.10 | 0.32 |
| (kg/ha year) | 7.71 | 2.86 | 2.35 | 0.36 |

Table 5.5 Annual Pollutant Loads for Urban Areas Based on Data from Tables 5.1, 5.2, and 5.4.

| | **Loading** | |
|---|---|---|
| **Parameter** | **(lbs/ac year)** | **(kg/ha year)** |
| Oil and grease | 26 | 29 |
| Cd | $6.0 \times 10^{-3}$ | $6.7 \times 10^{-3}$ |
| Cr | $4.2 \times 10^{-2}$ | $4.7 \times 10^{-2}$ |
| Cu | $9.6 \times 10^{-2}$ | $1.1 \times 10^{-1}$ |
| Pb | $1.0 \times 10^{-1}$ | $1.1 \times 10^{-1}$ |
| Hg | $1.2 \times 10^{-3}$ | $1.3 \times 10^{-3}$ |
| Ni | $4.8 \times 10^{-2}$ | $5.4 \times 10^{-2}$ |
| Zn | $6.9 \times 10^{-1}$ | $7.8 \times 10^{-1}$ |
| Pyrene | $1.1 \times 10^{-3}$ | $1.2 \times 10^{-3}$ |
| Fluoranthene | $8.0 \times 10^{-4}$ | $9.0 \times 10^{-4}$ |

## 5.7 Sampling and Measurements

Quantitative information on flows and water quality for stormwater runoff and SCM treatment requires careful and reliable measurements. Flow rates are typically measured using calibrated flumes and weirs that are placed in flow channels and drainage pipes. Some type of depth measuring device, such as a bubble meter, is placed near the flume or weir. Depth measurements are taken in short increments (every few minutes). Based on calibrations, depth measurements are converted to water flow data.

Samples for water-quality evaluation are commonly taken using automated sampling equipment. These samplers can collect samples at discrete times or based on a ratio of volume sampled per volume of stormwater flow past the measuring device. A composite sample collected using this latter technique will be representative of the overall EMC water quality. Stormwater monitoring protocols, including field equipment, QA/QC, and statistical analysis of sampling data, have been established (e.g., Ecology 2011; Geosyntec Consultants and Wright Water Engineers, Inc. 2009). Once samples are collected, they are taken to an analytical laboratory for analysis and determination of the various water-quality parameters. Strict quality assurance and quality control must be maintained over the entire sampling, transport, measurement, and data-handling processes.

## 5.8 A Note about Stormwater Quality

Defining the quality of urban stormwater is a considerable challenge. Stormwater characteristics are highly dynamic and can vary geographically due to climate variations and diverse activities taking place on the watershed.

Even more, in determinations of water quality, we can only find something if we look for it. Many possible chemicals (and microorganisms) are extremely difficult to detect and to quantify at extremely low levels in which they may be found in stormwater. Recent advances in analytical chemistry have allowed more sophisticated analysis of pollutants of interest in urban stormwater. Recent work has emphasized the complexity of urban stormwater quality. Lusk and Toor (2016) noted that dissolved organic nitrogen may result from many different natural compounds, including those with lipid, protein, amino sugar, and lignin characteristics, as noted previously. Masoner et al. (2019) found over 100 organic chemicals in urban stormwater, including cholesterol, caffeine, nicotine, acetaminophen, bis(2-ethylhexyl) phthalate, various pesticides, and other pharmaceuticals. Niles et al. (2020) recently noted that dozens of water-soluble oxygenated hydrocarbons can be generated from photooxidation of petroleum-based pavement binder. It is reasonable to expect continued identification of exotic pollutants in stormwater.

## References

Bach, P.M., McCarthy, D.T., and Deletic, A. (2010). Redefining the stormwater first flush phenomenon. *Water Research* 44 (8): 2487–2498.

Batroney, T., Wadzuk, B.M., and Traver, R.G. (2010). Parking deck's first flush. *Journal of Hydrologic Engineering* 15 (2): 123–128.

Berretta, C. and Sansalone, J. (2011a). Hydrologic transport and partitioning of phosphorus fractions. *Journal of Hydrology* 403 (1–2): 25–36.

Berretta, C. and Sansalone, J. (2011b). Speciation and transport of phosphorus in source area rainfall — Runoff. *Water, Air, and Soil Pollution* 222 (1–4): 351–365.

Bratt, A.R., Finlay, J.C., Hobbie, S.E., Janke, B.D., Worm, A.C., and Kemmitt, K.L. (2017). Contribution of leaf litter to nutrient export during winter months in an Urban residential watershed. *Environmental Science & Technology* 51 (6): 3138–3147.

Collins, K.A., Lawrence, T.J., Stander, E.K., Jontos, R.J., Kaushal, S.S., Newcomer, T.A., Grimm, N.B., and Cole Ekberg, M.L. (2010). Opportunities and challenges for managing nitrogen in urban stormwater: A review and synthesis. *Ecological Engineering* 36: 1507–1519.

Councell, T.B., Duckenfield, K.U., Landa, E.R., and Callender, E. (2004). Tire-wear particles as a source of zinc to the environment. *Environmental Science & Technology* 38 (15): 4206–4214.

Davis, A.P., Shokouhian, M., and Ni, S.B. (2001). Loading estimates of lead, copper, cadmium, and zinc in urban runoff from specific sources. *Chemosphere* 44 (5): 997–1009.

Eckley, C.S. and Branfireun, B. (2008). Mercury mobilization in Urban stormwater runoff. *Science of the Total Environment* 403 (1–3): 164–177.

Ecology (2011). *Technical Guidance Manual for Evaluating Emerging Stormwater Treatment Technologies*. No. 11-10-061, Washington: Washington State Department of Ecology, Olympia.

Flint, K. and Davis, A.P. (2007). Pollutant mass flushing characteristics of highway stormwater runoff from an ultra urban area. *Journal of Environmental Engineering* 133 (6): 616–626.

Geosyntec Consultants and Wright Water Engineers, Inc (2009). Urban stormwater BMP performance monitoring. https://static1.squarespace.com/static/5f8dbde10268ab224c895ad7/t/604926dae8a36b0ee128f8ac/1615406817379/2009 MonitoringManualSingleFile.pdf (accessed December 16, 2021).

Hathway, J.M. and Hunt, W.F. (2011). Evaluation of first flush for indicator bacteria and total suspended solids in urban stormwater runoff. *Water, Air, & Soil Pollution* 217: 135–147.

Hathway, J.M., Tucker, R.S., Spooner, J.M., and Hunt, W.F. (2012). A traditional analysis of the first flush effect for nutrients in stormwater runoff from two small urban catchments. *Water, Air, & Soil Pollution* 223: 5903–5915.

He, J., Valeo, C., Chu, A., and Newmann, N.F. (2010). Characteristics of suspended solids, microorganisms, and chemical water quality in event-based stormwater runoff from and urban residential area. *Water Environment Research* 82 (12): 2333–2345.

Hur, J., Schlautman, M.A., and Yim, S. (2004). Effects of organic ligands and ph on the leaching of copper from brake wear debris in model environmental solutions. *Journal of Environmental Monitoring* 6: 89–94.

Hur, J., Yim, S., and Schlautman, M.A. (2003). Copper leaching from brake wear debris in standard extraction solutions. *Journal of Environmental Monitoring* 5: 837–843.

Hwang, H.M. and Foster, G.D. (2006). Characterization of polycyclic aromatic hydrocarbons in urban stormwater runoff flowing into the tidal Anacostia River, Washington, DC, USA. *Environmental Pollution* 140 (3): 416–426.

Hwang, H.-M. and Foster, G.D. (2008). Polychlorinated biphenyls in stormwater runoff entering the tidal Anacostia River, Washington, DC, through small urban catchments and combined sewer outfalls. *Journal of Environmental Science and Health Part a-Toxic/Hazardous Substances & Environmental Engineering* 43 (6): 567–575.

Jani, J., Yang, -Y.-Y., Lusk, M.G., and Toor, G.S. (2020). Composition of nitrogen in urban residential stormwater runoff: Concentrations, loads, and source characterization of nitrate and organic nitrogen. *PLOS ONE* Public Library of Science, 15 (2): e0229715.

Kayhanian, M., Stransky, C., Bay, S., Lau, S.-L., and Stenstrom, M.K. (2008). Toxicity of urban highway runoff with respect to storm duration. *Science of the Total Environment* 389: 386–406.

Kim, J.-Y. and Sansalone, J.J. (2008). Event-based size distributions of particulate matter transported during urban rainfall-runoff events. *Water Research* 42: 2756–2768.

Lau, S.-L., Han, Y., Kang, J.-H., Kayhanian, M., and Stenstrom, M.K. (2009). Characteristics of highway stormwater runoff in Los Angeles: Metals and polycyclic hydrocarbons. *Water Environment Research* 81 (3): 308–318.

Li, D., Wan, J., Ma, Y., Wang, Y., Huang, M., and Chen, Y. (2015). Stormwater runoff pollutant loading distributions and their correlation with rainfall and catchment characteristics in a rapidly industrialized city. *Plos One* 10 (3): e0118776.

Li, L. and Davis, A.P. (2014). Urban stormwater nitrogen composition and fate in bioretention systems. *Environmental Science & Technology* 48 (6): 3403–3410.

Liu, J. and Davis, A.P. (2014). Phosphorus speciation and treatment using enhanced phosphorus removal bioretention. *Environmental Science & Technology* 48 (1): 607–614.

Lusk, M.G. and Toor, G.S. (2016). Biodegradability and molecular composition of dissolved organic nitrogen in urban stormwater runoff and outflow water from a stormwater retention pond. *Environmental Science & Technology* 50 (7): 3391–3398.

Lusk, M.G., Toor, G.S., and Inglett, P.W. (2020). Organic nitrogen in residential stormwater runoff: Implications for stormwater management in urban watersheds. *Science of the Total Environment* 707: 135962.

Masoner, J.R., Kolpin, D.W., Cozzarelli, I.M., Barber, L.B., Burden, D.S., Foreman, W.T., Forshay, K.J., Furlong, E.T., Groves, J.F., Hladik, M.L., Hopton, M.E., Jaeschke, J.B., Keefe, S.H., Krabbenhoft, D.P., Lowrance, R., Romanok, K.M., Rus, D.L., Selbig, W.R., Williams, B.H., and Bradley, P.M. (2019). Urban stormwater: An overlooked pathway of extensive mixed contaminants to surface and groundwaters in the United States. *Environmental Science & Technology* 53 (17): 10070–10081.

MDE — Maryland Department of the Environment. (2020). *Wasteload Allocations and Impervious Acres Treated. Guidance for National Pollutant Discharge Elimination System Stormwater Permits*. Baltimore, MD. https://mde.maryland.gov/programs/Water/StormwaterManagementProgram/Documents/2020%20MS4%20Accounting%20Guidance.pdf (accessed December 16, 2021).

McCarthy, D.T., Hathway, J.M., Hunt, W.F., and Deletic, A. (2012). Intra-event variability of Escherichia coli and total suspended solids in urban Stormwater Runoff. *Water Research* 46: 6661–6670.

Niles, S.F., Chacón-Patiño, M.L., Putnam, S.P., Rodgers, R.P., and Marshall, A.G. (2020). Characterization of an Asphalt Binder and Photoproducts by Fourier Transform Ion Cyclotron Resonance Mass Spectrometry Reveals Abundant Water-Soluble Hydrocarbons. *Environmental Science & Technology, American Chemical Society* 54 (14): 8830–8836.

Pamuru, S.T., Forgione, E., Croft, K., Kjellerup, B.V., and Davis, A.P. (2022). Chemical characterization of urban stormwater: Traditional and emerging contaminants. *Science of The Total Environment*, 813: 151887.
Sansalone, J.J. and Cristina, C.M. (2004). First flush concepts for suspended and dissolved solids in small impervious watersheds. *Journal of Environmental Engineering* 130 (11): 1301–1314.
Sansalone, J.J. and Kim, J.-Y. (2008). Transport of particulate matter fractions in urban source area pavement surface runoff. *Journal of Environmental Quality* 37: 1883–1893.
Sansalone, J., Ying, G., and Lin, H. (2010). Distribution of metals for particulate matter transported in source area rainfall-runoff. *Journal of Environmental Engineering-Asce* 136 (2): 172–184.
Sheng, Y., Ying, G., and Sansalone, J. (2008). Differentiation of transport for particulate and dissolved water chemistry load indices in rainfall-runoff from urban source area watersheds. *Journal of Hydrology* 361 (1–2): 144–158.
Smolders, E. and Degryse, F. (2002). Fate and effect of zinc from tire debris in soil. *Environmental Science & Technology* 36 (17): 3706–3710.
Wang, Y., Thompson, A.M., and Selbig, W.R. (2020). Leachable phosphorus from senesced green ash and Norway maple leaves in urban watersheds. *Science of the Total Environment* 743: 140662.
Wu, J., Ren, Y., Wang, X., Wang, X., Chen, L., and Liu, G. (2015). Nitrogen and phosphorus associating with different size suspended solids in roof and road runoff in Beijing, China. *Environmental Science and Pollution Research* 22 (20): 15788–15795.
Zanders, J.M. (2005). Road sediment: Characterization and implications for the performance of vegetated strips for treating road run-off. *Science of the Total Environment* 339 (1): 41–47.
Zgheib, S., Moilleron, R., and Chebbo, G. (2012). Priority pollutants in urban stormwater: Part 1-Case of separate storm sewers. *Water Research* 46 (20): 6683–6692.

## Problems

The following table presents runoff flows and pollutant concentrations measured from a highway during a storm event. When needed, integrate using trapezoids; extrapolate 1 $\Delta t$ at the beginning and end of the storm to 0 for each parameter.

| Time of day | Flow rate (L/s) | TSS (mg/L) | TP (mg/L) | $NO_3$–N (mg/L) | TKN (mg/L) | T–Cu (μg/L) | T–Pb (μg/L) |
|---|---|---|---|---|---|---|---|
| 15:00 | 6.4 | 241 | 1.52 | 0.95 | 3.9 | 169 | 23 |
| 15:20 | 12.7 | 370 | 0.71 | 0.47 | 7.4 | 54 | 33 |
| 15:40 | 2.1 | 148 | 0.23 | 0.70 | 1.4 | 25 | 15 |
| 16:00 | 22.0 | 215 | 0.48 | 0.19 | 4.8 | 53 | 33 |
| 16:20 | 6.1 | 108 | 1.02 | 0.35 | 2.9 | 88 | 17 |
| 16:40 | 1.8 | 59 | 0.22 | 0.52 | 0.98 | 24 | 11 |
| 17:00 | 1.3 | 41 | 0.28 | 0.52 | 0.77 | 17 | 3 |
| 17:20 | 0.82 | 26 | 0.19 | 0.57 | 0.56 | 15 | 3 |

5.1 Plot the hydrograph and pollutograph for TSS.

5.2 Plot the hydrograph and pollutograph for TP.

5.3 Plot the hydrograph and pollutograph for T–Pb.

5.4 Find the EMC and total TSS mass in the runoff from the storm event.

5.5 Find the EMC and total TP mass in the runoff from the storm event.

5.6 Find the EMC and total $NO_3$–N mass in the runoff from the storm event.

5.7 Find the EMC and total TKN mass in the runoff from the storm event.

5.8 Find the EMC and total T–Cu mass in the runoff from the storm event.

5.9 Find the EMC and total T–Pb mass in the runoff from the storm event.

5.10 Plot the $m(t)$ vis-à-vis $v(t)$ relationship for TSS for the storm event. Find the value of the $b$ exponent. Comment on the first flush behavior.

5.11 Plot the $m(t)$ vis-à-vis $v(t)$ relationship for TP for the storm event. Find the value of the $b$ exponent. Comment on the first flush behavior.

5.12 Plot the $m(t)$ vis-à-vis $v(t)$ relationship for $NO_3$–N for the storm event. Find the value of the $b$ exponent. Comment on the first flush behavior.

5.13 Plot the $m(t)$ vis-à-vis $v(t)$ relationship for TKN for the storm event. Find the value of the $b$ exponent. Comment on the first flush behavior.

5.14 Plot the $m(t)$ vis-à-vis $v(t)$ relationship for T–Cu for the storm event. Find the value of the $b$ exponent. Comment on the first flush behavior.

5.15 Plot the $m(t)$ vis-à-vis $v(t)$ relationship for T–Pb for the storm event. Find the value of the $b$ exponent. Comment on the first flush behavior.

5.16 Determine the fraction of total TSS mass that is flushed in the first 20% of runoff volume.

5.17 Determine the fraction of total TP mass that is flushed in the first 20% of runoff volume.

5.18 Determine the fraction of total $NO_3$–N mass that is flushed in the first 20% of runoff volume.

5.19 Determine the fraction of total TKN mass that is flushed in the first 20% of runoff volume.

5.20 Determine the fraction of total T–Cu mass that is flushed in the first 20% of runoff volume.

5.21 Determine the fraction of total T–Pb mass that is flushed in the first 20% of runoff volume.

5.22 The watershed for a small river contains 200 ac of agricultural land, 300 ac of forested land, and 50 ac of developed land. Assuming that the total phosphorus loads for each of these land uses are, respectively, 7.2, 1.5, and 3.5 lb TP/ac/year, what is the total average delivery of TP to the river each year?

5.23 The watershed for a lake contains 450 ha acres of forested land, 200 ha of agricultural land, and 350 ha of developed land. The total nitrogen loads for each of these land uses are 8, 35, and 22 kg TN/ha/year, respectively. Find the total average delivery of TN to the lake each year.

# 6

# Watershed Hydrology

## 6.1 Introduction

Hydrologic unit processes are used for estimation of runoff volumes and flowrates from an individual event, to obtain a broader view of the duration of flooding and erosive velocities, and to approximate event exceedance probabilities. Many of these fundamental processes are applied to estimate the watershed runoff entering a stormwater control measure (SCM), as well to predict the hydrologic and water quality performance of the SCM. These unit processes connect the rainfall and its pattern to the land cover, storage, infiltration, evapotranspiration (ET), and overland flow travel. These hydrologic parameters are used to in conjunction with appropriate stormwater performance metrics to guide the SCM design. While many of the unit processes are complementary, this material has been divided into two chapters, Chapter 6 *Watershed Hydrology* and Chapter 7 *Stormwater Control Measure Hydrologic Processes*.

This hydrologic unit process discussion starts with precipitation, then transitions to hydrologic unit processes methods that are used to develop storm inflows to the SCM. For example, the National Resource Conservation Service of the US Department of Agriculture (NRCS) has developed several of the methods and tools that are often used in the watershed hydrologic analysis; however these NRCS approaches are not always the most appropriate for SCM applications and physically based methods are discussed. For example, for a continuous simulation study, while NRCS curve numbers can be used, a model based on soil physics such as Green and Ampt is preferred for SCM design to more realistically represent climate, soil processes, and vegetation influences on the watershed runoff. The soil physics model also better represents SCM performance, especially for continuous simulation approaches and is presented in Chapter 7.

The primary focus of this chapter is on applications to Green Infrastructure Stormwater Control Measures. These systems typically have smaller, less complex drainage areas than many other hydrologic applications and the watershed runoff section of this chapter is focused on these urban applications. While the SCM unit processes are the same no matter the complexities of the watershed contributing to the SCM, the runoff modeling is not. The modeling of runoff from larger more

*Green Stormwater Infrastructure Fundamentals and Design*, First Edition. Allen P. Davis, William F. Hunt, and Robert G. Traver.

Companion Website: www.wiley.com/go/davis/greenstormwater

complex watersheds where routing is required is outside the scope of this text, and readers are referred to a Surface Hydrology course or textbook. SCM unit processes, including evapotranspiration and infiltration are presented in Chapter 7. It should be noted that these unit properties can also be used for modeling watershed hydrology.

## 6.2 Precipitation

Two approaches are in common use for hydrologic analysis and design. The first is a *design storm* that is developed from the rainfall statistical record and ordered normally with the bulk of the precipitation at the center of the storm. A legacy method from the stormwater detention era, design storms do not consider rainfall variations or seasonal influences, and should be considered a conservative flood peak approach, not an accurate depiction of the overall hydrologic cycle (Table 6.1). The second approach is the use of *continuous simulation*, which includes the rainfall time record of a year, or multiple years, to multiple decades. Temperature and other meteorological data may accompany the rainfall record allowing for inclusion of seasonal and climatic conditions. An average or extreme year or decade, or an enhanced record to simulate climate changes is commonly used to evaluate SCM performance metrics, and also used to consider the hydrologic water balance. Continuous simulation may replace the design storm in the near future as the method of choice to assess performance of SCMs.

### 6.2.1 Design Storms

To simplify the analysis of the impact of a storm event on a watershed or SCM, many regions create a statistically based design storm or hyetograph that is codified in regulation. For design of GI SCMs, a design storm may be selected and routed over the

Table 6.1 Differences between Design Storms and Actual Storms (Based on Figure 4-28 Part 630 *National Engineering Handbook*, Chapter 4 (NRCS 2019)) (Revised by Authors).

| Storm Characteristic | Design Storms | Actual Storms | Continuous Simulation |
|---|---|---|---|
| Storm duration | 24-h duration | Any duration from minutes to days | Same as actual storm. Multiple years, includes all storms, not just extreme events. |
| Temporal rainfall distribution | Smoothly increasing and decreasing rainfall intensity | Irregular rainfall pattern with respect to time, possibly including intervals of no rainfall | Same as actual storm. When used with evapotranspiration data, informs SCM design based on regional precipitation patterns |
| Intensity/duration relationship | Based on intensity/duration data for a single return period such as 25-year | Generally, includes intensity/duration data for different return periods | Allows SCM design to match rainfall intensity with SCM volume removal and storage capacity. |

**Example 6.1** Create a 2-h duration 2-year design storm for Baltimore, MD

*Solution* The 2-year event data for Baltimore is given in the third column of Table 6.2. The table below allows the calculation of the intensity for different durations from these data. In column 3, the 2-year rainfall depth at a duration is subtracted from the previous duration; in column 4, the duration is subtracted from the previous duration. The intensity is found as the depth divided by the duration.

| Duration (min) | 2-Year Rainfall Depth (in.) | Rainfall Depth after Subtracting Lower Duration (in.) | Duration after Subtracting Lower Duration Time (h) | Intensity | |
|---|---|---|---|---|---|
| | | | | (in./h) | (cm/h) |
| 5 | 0.41 | 0.41 | 0.0833 | 4.92 | 12.50 |
| 10 | 0.66 | 0.25 | 0.0833 | 3.00 | 7.62 |
| 15 | 0.83 | 0.17 | 0.0833 | 2.04 | 5.18 |
| 30 | 1.14 | 0.31 | 0.25 | 1.24 | 3.15 |
| 60 | 1.43 | 0.29 | 0.5 | 0.58 | 1.47 |
| 120 | 1.70 | 0.27 | 1 | 0.27 | 0.69 |

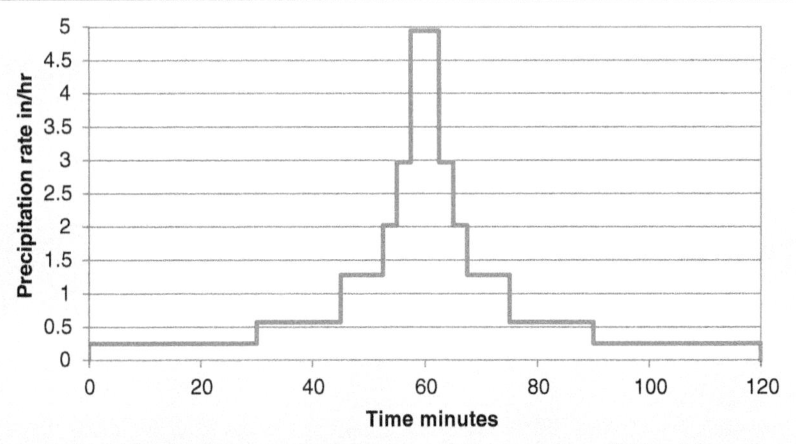


NOAA 14 2-h Design Storm for Baltimore, MD. Based on Data of Table 6.2.

To turn this data set into the design storm, first a 2-h event is centered at 60 min (1 h). The 5-min intensity of 4.92 in./h is extended 2.5 min on each side of the center. The 10-min intensity extends another 2.5 min on each side (for a total of 10 min, 5 on each side). This process continues, producing the design storm diagrammed below.

watershed to produce a hydrograph. This hydrograph, along with other site and regulatory opportunities and constraints, is employed to guide the design of the SCM.

Extreme event *intensity/duration/frequency* (IDF-Chapter 2) rainfall data are available in most developed areas of the United States from NOAA's Hydrometeorological Design Studies Center (https://hdsc.nws.noaa.gov/hdsc/pfds). The term "extreme event" means that the data set is based on a record of the largest events, not the overall rainfall record. Simply by selecting a point on a map, current rainfall data linked to surrounding rain gages is developed for a location. Table 6.2 shows an example for

Table 6.2 NOAA 14 Rainfall Intensities for Baltimore MD (BWI Marshall Airport) Downloaded from https://hdsc.nws.noaa.gov/hdsc/pfds.

| | PDS-based point precipitation frequency estimates with 90% confidence intervals (in inches)[1] | | | | | | | | | |
|---|---|---|---|---|---|---|---|---|---|---|
| Duration | Average recurrence interval (years) | | | | | | | | | |
| | 1 | 2 | 5 | 10 | 25 | 50 | 100 | 200 | 500 | 1000 |
| 5-min | 0.343<br>(0.312–0.378) | 0.411<br>(0.373–0452) | 0.489<br>(0.443–0.538) | 0.546<br>(0.494–0.601) | 0.617<br>(0 555–0.680 ) | 0.670<br>(0.600–0. 739) | 0.722<br>(0.643–0. 798) | 0.771<br>(0. 683– 0.856) | 0.833<br>(0. 730–0.931) | 0.881<br>(0.766–0. 991) |
| 10-min | 0.548<br>(0.499–0.603) | 0.657<br>(0. 596–0.723) | 0.783<br>(0.710–0.862) | 0.873<br>(0. 791–0.961) | 0.984<br>(0.884–1.08) | 1.07<br>(0.955–1.18) | 1.15<br>(1.02–1. 27) | 1.22<br>(1 08–1.36) | 1.32<br>(1.15–1.47) | 1.39<br>(1.21–1 .56) |
| 15-min | 0.686<br>(0.624–0.754) | 0.826<br>(0.750–0. 909) | 0.991<br>(0.898–1.09) | 1.11<br>(1.00–1. 22) | 1.25<br>(1.12–1.37) | 1.35<br>(1.21–1.49) | 1.45<br>(1.29–1.60) | 1.54<br>(1.37–1.71) | 1.66<br>(1.45–1.85) | 1.74<br>(1.51–1.96) |
| 30-min | 0.940<br>(0. 855–1.03) | 1.14<br>(1.03–1 .25) | 1.41<br>(1.28–1.55) | 1.60<br>(1.45–1.76) | 1.85<br>(1.66–2.03) | 2.04<br>(1.82–2.24) | 2.22<br>(1 98–2.46) | 2.40<br>(2. 13–2.67) | 26.4<br>(2.31–2.95) | 2.82<br>(2.45–3.17) |
| 60-min | 1.17<br>(1.07–1.29) | 1.43<br>(1 .30–1 .58) | 1.81<br>(1.64–1.99) | 2.08<br>(1.89–2 .29) | 2.46<br>(2.21–2.71) | 2.76<br>(2.47–3.04) | 3.06<br>(2. 73–3.38) | 3.37<br>(2 .98–3 .74) | 3.79<br>(3.32–4.23) | 4.12<br>(3.58–4.63) |
| 2-h | 1.40<br>(1.27–1.54) | 1.70<br>(1. 55–1.88) | 2.16<br>(1.95–2.37) | 2.51<br>(2.26–2.76) | 3.00<br>(2.69–3.30) | 3.40<br>(3.04–3.74) | 3.82<br>(3 .38–4 .21) | 4.25<br>(3. 74–4.71) | 4.87<br>(4.23–5.44) | 5.37<br>(4.62–6 .03) |
| 3-h | 1.51<br>(1.37–1 .67) | 1.83<br>(1.66–2 .02) | 2.33<br>(2.11–2.57) | 2.71<br>(2.45–2.99) | 3.27<br>(2.93–3.60) | 3.72<br>(3.32–4.11) | 4.20<br>(3. 72–4.65) | 4.71<br>(4.13–5 .23) | 5.45<br>(4.70–6.08) | 6.04<br>(5.15–6.79) |
| 6-h | 1.86<br>(1.70–2 .05) | 2.25<br>(2.05–2. 47) | 2.84<br>(2.58–3.13) | 3.32<br>(3.01–3.66) | 4.05<br>(3.63–4.45) | 4.66<br>(4.14–5.13) | 5.32<br>(4 .68–5 .88) | 6.04<br>(5 .25–6. 69) | 7.09<br>(6.06–7.94) | 7.98<br>(6.73–9.00) |
| 12-h | 2.25<br>(2.04–2.52) | 2.72<br>(2. 46–3.04) | 3.46<br>(3.12–3.87) | 4.09<br>(3.67–4.57) | 5.06<br>(4.49–5.64) | 5.91<br>(5.19–6. 59) | 6.86<br>(5.94–7.67) | 7.92<br>(6 .76–8.88) | 9.53<br>(7.94–10.8) | 10.9<br>(8.93–12.4) |
| 24-h | 2.61<br>(2.39–2. 88) | 3.16<br>(2 .89–3.49) | 4.06<br>(3.71–4.48) | 4.86<br>(4.42–5 .34) | 6.08<br>(5.48–6.64) | 7.16<br>(6.41–7. 80) | 8.38<br>(7. 44–9.10) | 9.77<br>(8. 57–10.6) | 11.9<br>(10.3–12.8) | 13.8<br>(11.7–14.8) |

(*Continued*)

Table 6.2 (Continued)

| PDS-based point precipitation frequency estimates with 90% confidence intervals (in inches)[1] | | | | | | | | | | |
|---|---|---|---|---|---|---|---|---|---|---|
| Duration | Average recurrence interval (years) | | | | | | | | | |
| | 1 | 2 | 5 | 10 | 25 | 50 | 100 | 200 | 500 | 1000 |
| 2-day | 3.03<br>(2.76–3 .33) | 3.66<br>(3. 34–4.04) | 4.70<br>(4.29–5.17) | 5.59<br>(5.08–6 .14) | 6.93<br>(6.25–7.58) | 8.09<br>(7.26–8.84) | 9.39<br>(8. 36–10.2) | 10.8<br>(9. 55–11.8) | 13.0<br>(11 .3–14. 2) | 14.9<br>(12.8–16.2) |
| 3-day | 3.19<br>(2.91–3 .51) | 3.86<br>(3. 53–4.24) | 4.94<br>(4.51–5.43) | 5.87<br>(5.34–6.44) | 7.27<br>(6.56–7.94) | 8.48<br>(7.61–9.25) | 9.83.<br>(8 .76–10 .7) | 11.3<br>(10.0–12 .3) | 13.6<br>(11. 8–14.8) | 15.6<br>(13.3–16.9) |
| 4-day | 3.35<br>(3.06–3.68) | 4.05<br>(3.71–4 .45) | 5.18<br>(4. 73–5.69) | 6.15<br>(5.60–6 .74) | 7.61<br>(6.88–8.31) | 8.87<br>(7.97–9. 67) | 10.3<br>(9.16–11.2) | 11.8<br>(10.5–12.9) | 14.2<br>(12.3–15.4) | 16.2<br>(13.9–17.6) |
| 7-day | 3.89<br>(3.57–4.26) | 4.68<br>(4.30–5.12) | 5.92<br>(5.42–6.47) | 6.98<br>(6.38–7.61) | 8.56<br>(7.77–9.31) | 9.92<br>(8 .95–10. 8) | 11.4<br>(10.2–12.4) | 13.1<br>(11. 6–14.2) | 15.6<br>(13.6–16.9) | 17.7<br>(15.3–19. 2) |
| 10-day | 4.43<br>(4.09–4.82) | 5.33<br>(4 .91–5 .80) | 6.65<br>(6.12–7.23) | 7.75<br>(7.12–8 .43) | 9.36<br>(8.56–10.1) | 10.7<br>(9.74–11.6) | 12.2<br>(11.0–13.2) | 13.7<br>(12. 3–14.9) | 16.0<br>(14.2–17.3) | 17.9<br>(15.8–19.5) |
| 20-day | 5.99<br>(5.58–6.44) | 7.12<br>(6. 64–7.66) | 8.61<br>(8.02–9.25) | 9.81<br>(9.12–10. 5) | 11.5<br>(10.6–12.3) | 12.9<br>(11 .9–13.8) | 14.3<br>(13. 1–15.3) | 15.7<br>(14.4–16.8) | 17.8<br>(16.1–19.0) | 19.4<br>(17.4–20.8) |
| 30-day | 7.39<br>(6.92–7.90) | 8.75<br>(8.19–9.35) | 10.4<br>(9.74–11.1) | 11.7<br>(11.0–12. 5) | 13.6<br>(12. 6–14 .5) | 15.0<br>(14.0–16.0) | 16.5<br>(15. 3–17.6) | 18.1<br>(16. 6–19.3) | 20.1<br>(18.4–21.5) | 21.8<br>(19.8–23.3) |
| 45-day | 9.31<br>(8.76–9.90) | 11.0<br>(10.3–11.7) | 12.8<br>(12.1–13.6) | 14.3<br>(13.4–15.1) | 16.1<br>(15.1–17.1) | 17.6<br>(16.4–18. 7) | 19.0<br>(17.7–20.1) | 20.3<br>(18.9–21.6) | 22.1<br>(20.5–23.6) | 23.4<br>(21.6–25 .0) |
| 60-day | 11.1<br>(10.5–11.7) | 13.0<br>(12.3–13.8) | 15.1<br>(14 .2–15.9) | 16.6<br>(15.7–17.6) | 18.6<br>(17.5–19.7) | 20.1<br>(18.9–21. 2) | 21.5<br>(20. 1–22.7) | 22.8<br>(21. 3–24.2) | 24.5<br>(22.8–26 .0) | 25.7<br>(23.9–27.3) |

[1]Precipitation frequency (PF) estimates in this table are based on frequency analysis of partial duration series (PDS). Numbers in parenthesis are PF estimates at lower and upper bounds of the 90% confidence interval. The probability that precipitation frequency estimates (for a given duration and average recurrence interval) will be greater than the upper bound (or less than the lower bound) is 5%. Estimates at upper bounds are not checked against probable maximum precipitation (PMP) estimates and may be higher than currently valid PMP values.Please refer to NOAA Atlas 14 document for more information.

Baltimore, MD. These data can be used to set a specific runoff volume or develop a rainfall pattern. Note that the 90% confidence interval is also presented in Table 6.2 to provide the user with a better understanding of the precision of the data.

These precipitation volumes are typically transformed into 24-h central peaking design storms. This approach stacks the rainfall from each duration creating the center peak. It conservatively assumes that the 24-h, 12-h, 6-h, 1-h, etc., precipitation depths are included in the same storm distribution.

First, each of the respective depths are divided by their associated duration to obtain the intensity. Next the central peak volume is used to anchor the center of the rainfall pattern. For the remaining times, volumes are adjusted; the 10-min volume includes the 5-min rainfall, the 15-min volume includes the 10-min rainfall, and so on. After the highest intensity is set as the center, the remaining periods are draped symmetrically on either side of this center peak. The resulting rainfall pattern can be used as the basis for a simulation model and SCM design.

The NRCS distribution method is another approach in common use to transform the depth data to a design storm. A 24-h rainfall data set similar to Table 6.2 is used for any location with available rainfall data. This method, developed in the 1960s, applies a simple mass curve to distribute a daily volume in smaller increments over a 24-h period. The United States is divided into four regional patterns for this analysis (Type I, Ia, II, and III) (Figure 6.1).

As shown in Figure 6.1a, the mass curve has a vertical component for the common type II distribution at the 12-h mark, signifying that the majority of rainfall occurs at the center of the storm. Figure 6.1b shows the geographic distributions of the four rainfall distributions. While widely used based upon its simplicity, it should be recognized that over 50 years of data have been collected since this method was first established and analysis indicates that it only coarsely reflects regional rainfall differences and assumes that the larger rainfall intensities occur distributed normally around the peak. More recently NRCS no longer recommends the use of this method for the great majority of the United States and instead advises to create a distribution or design storm using NOAA-14.

New rainfall distributions have been developed based upon the NOAA-14 data for different regions of the United States (Merkel et al. 2015). Figure 6.2 shows the regions

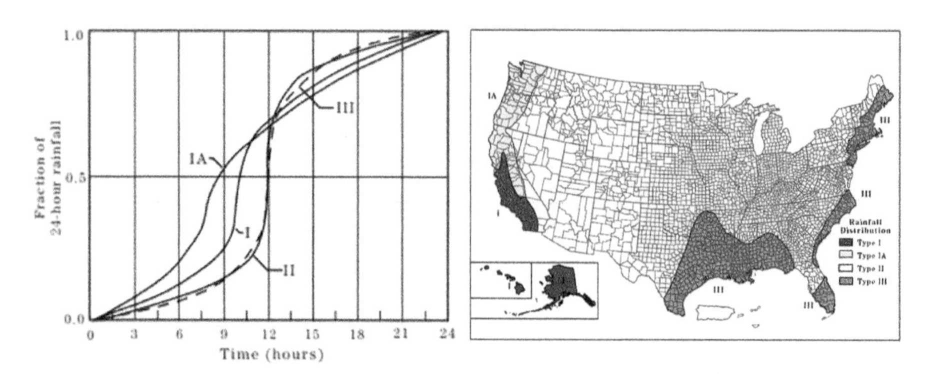


Figure 6.1 A) Plot of Types I, IA, II, and II Synthetic Rainfall Distributions. (Credit: US Department of Agriculture (NRCS 2019)).

labeled A through D to differentiate them from the older Type II distribution. For example, Pennsylvania is now represented by four regions, compared to one in the older methodologies which did not include the influence of the mountain ranges and lake effects on the precipitation patterns. This is significant, as evidenced by the different ratios for the shorter duration shown in Table 6.3.

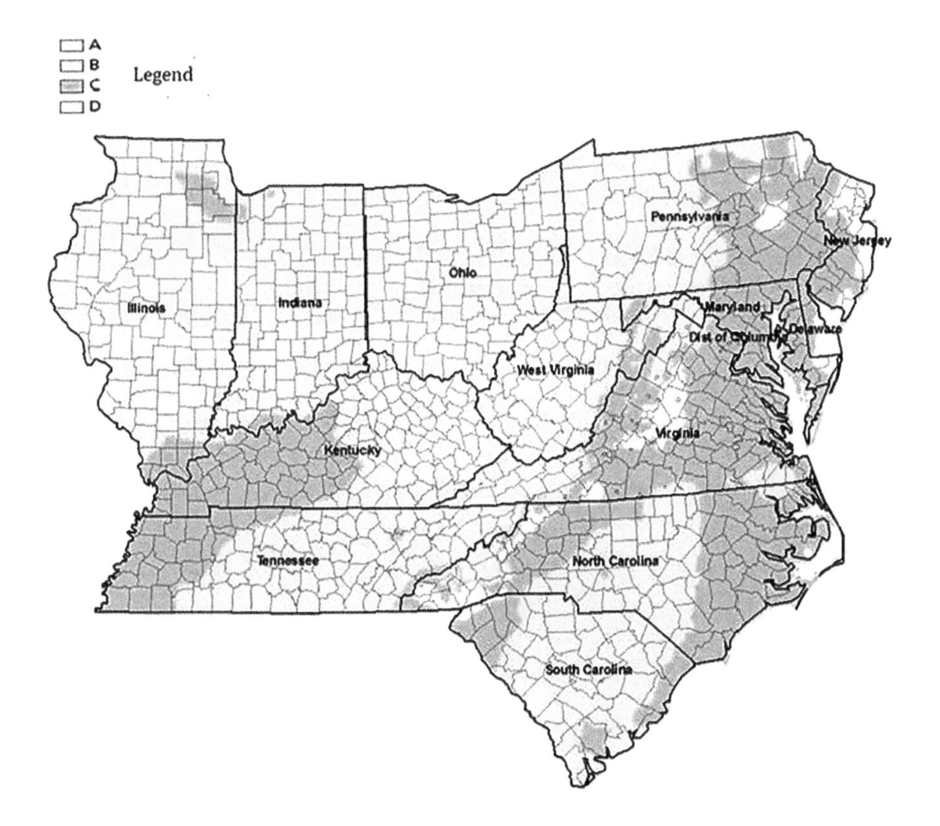


Figure 6.2 NOAA Atlas 14 Volume 2 Region, Rainfall Distribution Regions (NRCS 2019).

Table 6.3 Mean Ratios for Four Rainfall Distribution Regions NOAA Atlas 14, Ohio Valley and Neighboring States (NRCS 2019).

| Duration Ratio | Region A | Region B | Region C | Region D |
|---|---|---|---|---|
| 5 min/24 h | 0.143 | 0.121 | 0.105 | 0.094 |
| 10 min/24 h | 0.219 | 0.189 | 0.166 | 0.149 |
| 15 min/24 h | 0.272 | 0.237 | 0.210 | 0.188 |
| 30 min/24 h | 0.386 | 0.344 | 0.308 | 0.276 |
| 60 min/24 h | 0.502 | 0.453 | 0.409 | 0.366 |
| 120 min/24 h | 0.594 | 0.543 | 0.500 | 0.454 |
| 3 h/24 h | 0.635 | 0.585 | 0.545 | 0.501 |
| 6 h/24 h | 0.749 | 0.705 | 0.672 | 0.636 |
| 12 h/24 h | 0.864 | 0.840 | 0.823 | 0.805 |

**Example 6.2** Create a 2-h duration 2-year design storm for Baltimore, MD using the NRCS Storm Distribution.

***Solution*** The process is similar to Example 6.1, except we need the 2-year 24-h event volume from Table 6.2, which is 3.16 in., and we use the Type C storm which is appropriate for Baltimore (Figure 6.2). The table below allows the calculation of the intensity for different durations from these data. In column 3, the 2-year rainfall depth is found by multiplying the 2-year 24-h rainfall depth (3.16 in) by the Type C distribution ratio. The other columns are found using the same methods as the previous example. Note in the figure the results from both examples are plotted to show the differences.

| Duration (min) | Type C Distribution Ratio (Table 6.3) | 2-Year Rainfall Depth (in.) | Rainfall Depth after Subtracting Lower Duration (in.) | Duration after subtracting Lower Duration Time (h) | Intensity | |
|---|---|---|---|---|---|---|
| | | | | | (in./h) | (cm/h) |
| 5 | 0.105 | 0.33 | 0.33 | 0.0833 | 3.98 | 10.12 |
| 10 | 0.166 | 0.52 | 0.19 | 0.0833 | 2.31 | 5.88 |
| 15 | 0.21 | 0.66 | 0.14 | 0.0833 | 1.67 | 4.24 |
| 30 | 0.308 | 0.97 | 0.31 | 0.25 | 1.24 | 3.15 |
| 60 | 0.409 | 1.29 | 0.32 | 0.5 | 0.64 | 1.62 |
| 120 | 0.5 | 1.58 | 0.29 | 1 | 0.29 | 0.73 |

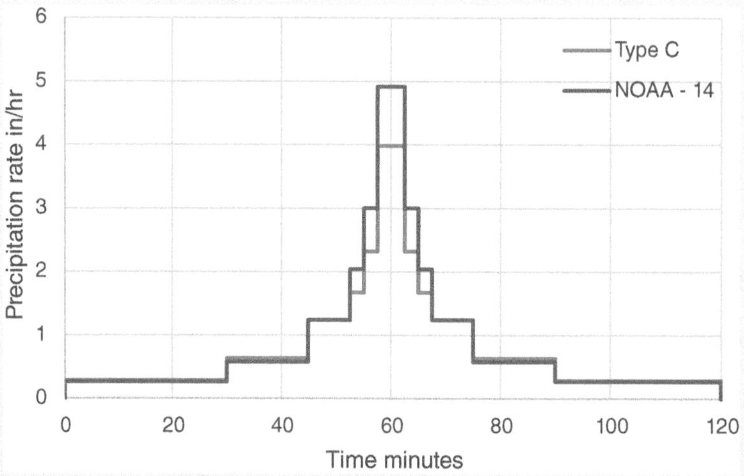

Comparing NOAA 14 and NRCS (Type C) 2-h Design Storms for Baltimore, MD.

Figure 6.3 demonstrates the conservative aspects of this method by comparing the measured rainfall pattern for a 3.76-in. event to the corresponding NOAA Type C Distribution. Note that the peak flow of the design storm does not reflect the pattern of the recorded event. The actual rainfall pattern showed steady rain and multiple smaller peaks over the event duration.

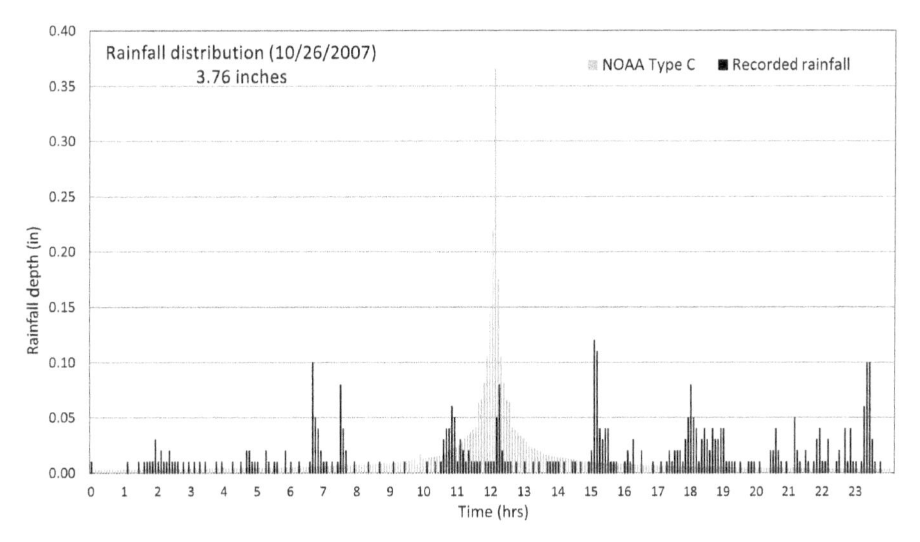


Figure 6.3 Comparison of a 24-h Recorded Storm Event at Villanova University Compared to that of Type C Rainfall Distribution. Rainfall Depths Are Recorded in 5-min Intervals.

### 6.2.2 Continuous Simulation

To incorporate the variability of regional climate, and to encompass seasonal variation in infiltration and evapotranspiration, a continuous simulation approach to rainfall analysis is required. Recorded data from the NOAA National Climatic Data is available for various locations in daily, 15- and 60-min intervals. Other weather information such as air temperature and wind speed is available and matched to the rainfall.

Use of continuous simulation can take many forms. The Western Washington Hydrology Model (Ecology 2020) uses multiple-year NOAA 15-min rainfall data and pan evaporation records to address both water quality volume, and flow peaks and durations in design of GI SCMs. The post construction condition requires both peak rate and flow duration criteria to address stream channel erosion. As another example, the city of Philadelphia as part of their long-term control plan to address combined sewer overflows created a representative year as a design and assessment tool. The 2005 rainfall records were employed as a base, using the annual number of storm events, total rainfall, and the best fit cumulative distribution function of hourly peak rainfall (Philadelphia Water Department 2009). The 2005 record was adjusted statistically to meet these criteria.

The use of continuous simulation as a hydrologic design and analysis approach has grown as the profession understands the need to represent the variabilities of climate patterns and soil/vegetation processes. As our weather data record grows, the use of representative years, decades, and flooding and drought outliers are under discussion for use in design. Recent work by Albright and Schramm (2018) created a methodology to create a reference year and extreme years, which may have promise as replacement for the design storm methodology. Figure 6.4 shows the suggested reference year, and an outlier year that includes the hurricane season. Other uses of continuous

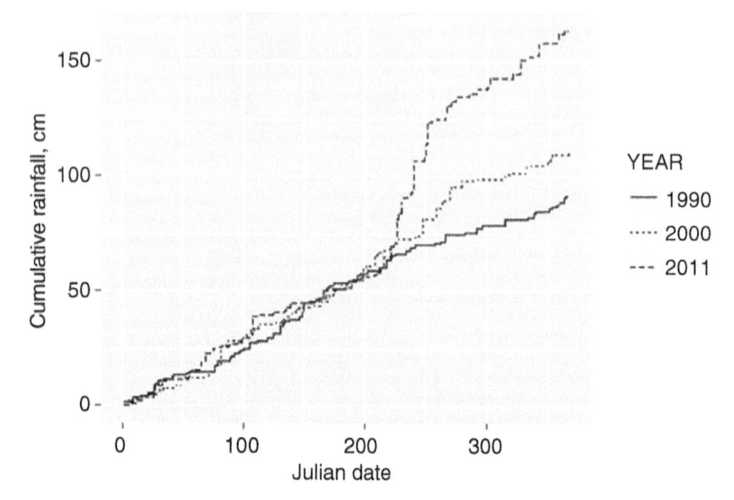


Figure 6.4 Cumulative Plot of an Outlier Year (2011) within the 25-Year Analysis Period for Philadelphia in Comparison with Two Reference Years (1990 and 2000) (Albright and Schramm 2018).

simulation include evaluation of climate change through amplification of current records. Continuous simulation allows for a more dynamic look at GI as a system, and to consider the hydrologic budget.

## 6.3 Watershed Hydrology

After selecting the design rainfall event, the rainfall must be applied over the contributing watershed to represent the losses due to depression storage, infiltration, and evapotranspiration to transform the remaining water to a pattern of runoff over time. This section will review common watershed hydrologic practices for smaller drainage areas often associated with distributed SCMs. Of these practices, the most popular due to its ease of use is the NRCS hydrology approach. Readers interested in modeling more complex watersheds that include reach routing should consult a watershed hydrology text or enroll in a watershed hydrology course. As stated earlier, other methodologies for determining losses are developed in Chapter 7 but are also useful for watershed modeling, especially when conducting continuous simulation modeling.

### 6.3.1 Drainage Area Delineation

The first step as outlined in Chapter 2 is to delineate the drainage area, ensuring that all areas draining to the SCM are accounted for. Figure 6.5 demonstrates this process, showing the outlined drainage area contributing to the bioinfiltration SCM at Villanova University. Note that the pervious and impervious land uses are identified separately, to include their flow paths. For smaller directly connected drainage areas, pervious and impervious land cover act differently when considering both infiltration losses and the flow velocity.

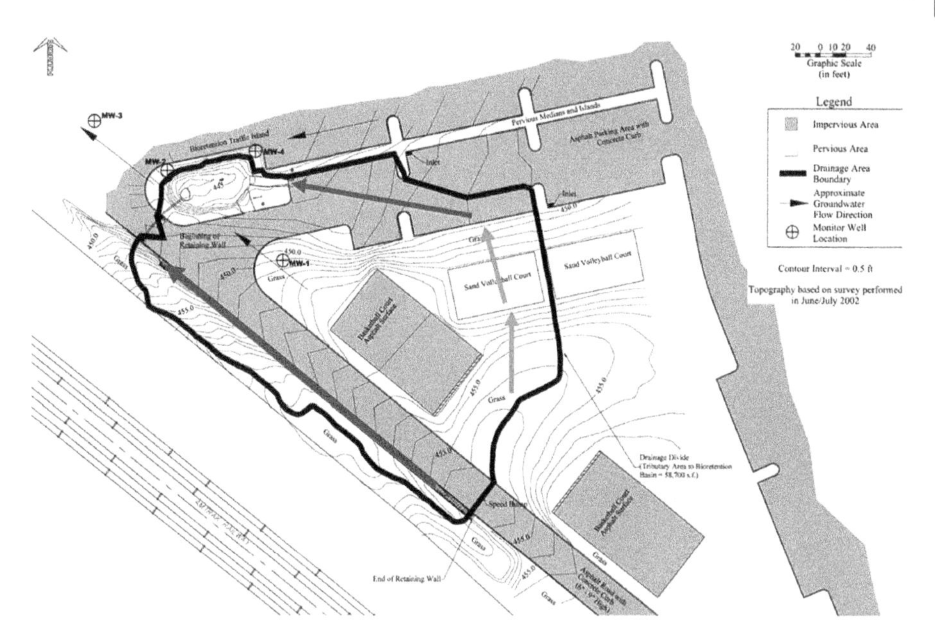


Figure 6.5 Villanova BioInfiltration Drainage Area (Modified from Machusick et al. 2011, with Permission from ASCE).

## 6.3.2 Interception and Depression Storage

Losses due to initial abstractions and infiltration must be estimated as the next step to determine runoff from precipitation. Interception is the capture of rainfall on the vegetation canopy and depression storage is the small volumes of water caught in cracks or voids in the pavement, or in small low points throughout the land surface. Interception and depression storage can be extremely significant when performing a continuous simulation, but not as significant when modeling a design storm due to the short duration and high intensities used.

Tables 6.4 and 6.5 show recommended interception and depression storage values developed for the *Minnesota Stormwater Best Management Practices Manual* from an engineering report (Barr Engineering 2010). For interception, much more information is available through the USDA for both rural and urban forest canopies. The volume of rainfall captured in depressions will depend on the volume of storage available throughout the drainage area. Holes and other depressions will contribute to storage. At a microscopic level, tiny holes, even in impervious surfaces such as asphalt will contain a measurable, albeit small, amount of storage. Flat surfaces will have some degree of storage as the water surface tension will hold the water into place before it moves as runoff, such as water that lands on leaves. For paved areas, the slope of the land and surfaces will determine the amount of storage and initial abstraction. The rainfall captured or held then either infiltrates or evaporates during dry times. The contribution of the urban street trees to urban stormwater initial abstraction is of great interest and is discussed in Chapter 10. Fundamentally, increasing the initial losses throughout the watershed is an important means to reduce runoff. While extremely significant when addressing the long-term hydrologic cycle, both interception and depression storage are much less significant during extreme events.

Table 6.4 Rainfall Interception for Selected Vegetation Types (Minnesota Pollution Control Agency https://www.pca.state.mn.us/water/enhancing-stormwater-management-minnesota).

| Vegetation Type | Interception (in.) | Region Data | Collected Source |
|---|---|---|---|
| Coniferous Trees | 0.11–0.17 | Douglas-fir-western hemlock ecosystem in SW Washington State | Link et al. 2004 |
| Deciduous Trees | 0.09–0.14* | Oak Tree in Davis, CA | Xiao et al. 2000 |
| Meadows – 1-foot | 0.08 | ** | Linsley et al. 1982 |
| Cropland – Corn – 6 feet | 0.03 | ** | Linsley et al. 1982 |
| Cropland – Small Grains – 3 feet | 0.16 | ** | Linsley et al. 1982 |

*Interception values are valid for full-leaf canopy. Xiao found that leaf-off interception was 0.04 in. for pear trees.
** Linsley et al. used Horton's equations for crop interception, which are based on experiments made in Seneca Falls, NY, in 1914.

Table 6.5 Depression Storage for Selected Land Covers (Minnesota Pollution Control Agency https://www.pca.state.mn.us/water/enhancing-stormwater-management-minnesota).

| Land Covers | Depression Storage (in) | Source |
|---|---|---|
| Impervious, 1% slope, flat roofs, parking lots, roads | 0.0625 – 0.125 | Tholin and Kiefer 1960 |
| Impervious, 2.5% slope, sloped roofs | 0.05 | Viessman and Lewis 1996 |
| Turfgrass | 0.25 | Tholin and Keifer 1960 |
| Open fields | 0.40* | Urban Drainage and Flood Control District 2008 |
| Wooded areas | 0.40* | Urban Drainage and Flood Control District 2008 |

*These values include interception losses by vegetation

### 6.3.3 The Simple Method

An advantage of the Simple Method is that it only requires the watershed drainage and impervious areas, and rainfall depth to estimate the volume of runoff. It was designed for small noncomplex drainage areas and developed statistically from rainfall/runoff measurements (Schueler 1987).

$$R_V = 0.05 + 0.9 * IA \tag{6.1a}$$

$$V = 3630 * RD * Rv * A \tag{6.1b}$$

With $R_V$ representing the unitless runoff coefficient, IA the impervious fraction of the drainage area, V the volume of runoff (ft$^3$), RD the Rainfall Depth (in), and A the watershed area (ac). The coefficients in Eq. 6.1a account for abstraction processes; the 3630 in Eq. 6.1b is a units conversion. This method was developed for individual drainage areas.

**Example 6.3** Find the runoff volume resulting from a ¾-in. rainfall over a 2.0-acre drainage area. 1.0 acres have been developed into single family homes (35% impervious) and 0.5 acres into multifamily attached residences (65% impervious). The other 0.5 acres have remained undeveloped.

*Solution* With the rainfall depth and area given, first the weighted IA must be determined since multiple land uses are present. The table below shows the weighting process, with the area of each land use divided by the total area.

| | Total Area (acres) | Impervious Fraction | Impervious Area (acres) |
|---|---|---|---|
| Single Family | 1.0 | 0.35 | 0.35 |
| Multifamily Attached | 0.5 | 0.65 | 0.325 |
| Undeveloped | 0.5 | 0 | 0 |
| *Total* | *2.0* | - | *0.675* |

Therefore, the impervious area fraction is 0.675/2.0 = 0.338. Using Eqs. 6.1a and 6.1b:

$$R_V = 0.05 + 0.9 * IA = 0.05 + 0.9 * 0.338 = 0.354$$

$$V = 3630 * RD * Rv * A = 3630 * 0.75 in * 0.354 * 2 ac = 1926 ft^3$$

### 6.3.4 NRCS Curve Number Method

In the 1950s, NRCS developed a coupled approach combining a loss, time of concentration, and unit hydrograph method to estimating both the volume and pattern of runoff off the land (NEH 2004). After selecting the design rainfall, this method requires the user to determine the soil grouping, and land cover. Soils characteristics will control hydrologic runoff creation and transport.

#### 6.3.4.1 NRCS Soil Groups

The NRCS has associated various soil textures to four soil groupings. The hydrologic soil groups are based on the lowest hydraulic conductivity and depth to impermeable layer and are used to simplify the estimation of infiltration during rainfall events (USDA 1986). Further discussion on the soils, soil textures, and soil groupings is provided in Chapter 7.

- Group A soils are primarily sand, loamy sand, or sandy loam soils and have low runoff potential and high water transmission rates (greater than 0.30 in./h, >0.76 cm/h).
- Group B soils are loam or silt loam. They have moderate infiltration rates when wetted, and moderate transmission rates (0.15 to 0.30 in./h, 0.38 to 0.76 cm/h).
- Group C soils are sandy clay loams. They have low infiltration rates when wetted, and a low rate of water transmission (0.05–0.15 in./h, 0.13–0.38 cm/h).
- Group D soils are clay loams, silty clay loams, sandy clay, silty clay, or clays. They have high runoff potential and very low infiltration rates when wetted, and a very low rate of water transmission (<0.05 in./h, < 0.13 cm/h).

Coarse-textured soils will allow for more rapid infiltration than fine soils. Hydrologic Group A and B soils will infiltrate at a faster rate than Groups C and D.

#### 6.3.4.2 NRCS "Curve Number" Method

The NRCS or "curve number" method is commonly used to predict runoff volumes under different land development scenarios. The Curve Number (CN) method was initially developed in the 1950s based on field studies, supplemented with infiltrometer tests. It is a statistical evaluation, plotting direct runoff as a function of storm rainfall. The NRCS method considers the NRCS hydrologic soil group, the land cover, summation rainfall, and antecedent moisture conditions. Three different antecedent moisture conditions (AMC) are considered in the method: I – Dry, II – Average, and III – Wet. AMC I and AMC III are developed from the average AMC II condition, which is the most common usage for stormwater applications. Initial abstraction is included in the water loss calculation, and defined as the volume of rainfall that needs to fall prior to the start of runoff, including interception, depression storage, and some soil volume infiltration. The curve number itself ranges from 30 to 100, with the latter value corresponding to completely impervious area.

Table 6.6 gives AMC II curve numbers for various urban areas in the corresponding soil groups. The original version of Table 6.6 included area-weighted curve numbers based on the percentage imperviousness for different urban settings. This is no longer recommended, as for smaller rainfalls and for the initial start of all storms the averaging of initial abstractions resulted in predictions of no flow running off the impervious surfaces. Instead it is recommended that users separately compute runoff volumes from pervious and impervious areas. Referring back to Figure 6.5, the pervious and impervious area can be treated as separate subareas, with different flow paths. The impervious area has low initial abstractions and speed of runoff are very different than

Table 6.6 NRCS Runoff Curve Numbers for Urban Areas. (USDA-United States Department of Agriculture 1986, Revised by Authors).

**Fully Developed Urban Areas (Vegetation Established)**

| Cover Description | | Curve Numbers for Hydrologic Soil Group | | | |
|---|---|---|---|---|---|
| | | A | B | C | D |
| Open space (lawns, parks, golf courses, cemeteries, etc.) | Poor condition (grass cover < 50%) | 68 | 79 | 86 | 89 |
| | Fair condition (grass cover 50% to 75%) | 49 | 69 | 79 | 84 |
| | Good condition (grass cover > 75%) | 39 | 61 | 74 | 80 |
| Impervious areas | Paved parking lots, roofs, driveways, etc. (excluding right of way) | 98 | 98 | 98 | 98 |
| Streets and roads | Paved; curbs and storm sewers (excluding right of way) | 98 | 98 | 98 | 98 |
| | Paved; open ditches (including right of way) | 83 | 89 | 92 | 93 |
| | Gravel (including right of way) | 76 | 85 | 89 | 91 |
| | Dirt (including right of way) | 72 | 82 | 87 | 89 |

(Continued)

Table 6.6 (Continued)

**Fully Developed Urban Areas (Vegetation Established)**

| Cover Description | | Curve Numbers for Hydrologic Soil Group | | | |
|---|---|---|---|---|---|
| | | A | B | C | D |
| Western desert urban areas | Natural desert landscaping (pervious area only) | 63 | 77 | 85 | 88 |
| | Artificial desert landscaping (impervious weed barrier, desert shrub with 1- to 2-in. sand or gravel mulch and basin borders) | 96 | 96 | 96 | 96 |
| Newly graded areas | Pervious areas only, no vegetation | 77 | 86 | 91 | 94 |
| Meadow | Continuous grass, protected from grazing and generally mowed for hay | 30 | 58 | 71 | 78 |
| Woods | Poor: Forest litter, small trees, and brush are destroyed by heavy grazing or regular burning | 45 | 66 | 77 | 83 |
| | Woods are grazed but not burned, and some forest litter covers the soil | 36 | 60 | 73 | 79 |
| | Woods are protected from grazing, and litter and brush adequately cover the soil | 30 | 55 | 70 | 77 |

the pervious environment, and they act as very different systems. The one outlier is disconnection of impervious, where runoff can be routed as sheet flow over the pervious CN (Chapter 10).

From the curve number, the potential maximum soil moisture retention after runoff begins, S, is calculated. S has units of inches.

$$S = 1000 / CN - 10 \tag{6.2}$$

The initial abstraction, $I_a$ (in), was statistically generated as a percentage of the maximum soil moisture, and is calculated from:

$$I_a = 0.2S \tag{6.3}$$

The depth of runoff (in inches over the drainage area) is then calculated as:

$$Q = (P - I_a)^2 / (P - I_a + S) \tag{6.4}$$

where P is the summation depth of rainfall (in).

If $I_a$ is greater than P, the NRCS method should predict no runoff from the site. Care must be exercised in application in Eq. 6.4, as a negative result of the (P-$I_a$) subtraction is masked by the square of the term. Note as the Initial Abstraction term includes infiltration prior to the start of runoff, use of depression storage as a substitute violates the statistical basis of the method.

For a drainage area that is made up of multiple land uses, the site should be split into pervious and impervious areas as stated above. A weighted CN can be used for the pervious areas though pervious and impervious areas should not be combined, as the averaging disguises some initial runoff from paved areas due to the difference in

**Example 6.4** Find the runoff volume resulting from a ¾-in. rainfall over a 2-acre drainage area with soil hydrologic group C. One acre has been developed into 1/3-acre residential lots (30% impervious) and 0.5 acres into townhouse residences (65% impervious). The other 0.5 acres have remained undeveloped (good condition open space).

*Solution* The overall runoff volume is calculated by finding the volume from each of the areas. First, the impervious surface is obtained by multiplying the % impervious by the land use.

$$\text{Impervious} = 0.3^*(1\text{ ac}) + 0.65^*(0.5\text{ ac}) = 0.63\text{ acres}$$

$$\text{Pervious} = 2 - 0.63 = 1.37\text{ acres}$$

From Table 6.6, the curve number for the pervious area for C soil in good condition is 74, and that of the impervious area 98.

Therefore, the initial abstraction, potential maximum soil moisture retention, and predicted runoff depth from the two areas are calculated as:

| Impervious | Pervious |
|---|---|
| $S_{impervious}$ = 1000/98 − 10 = 0.20 in. | $S_{pervious}$ = 1000/74 − 10 = 3.51 in. |
| $I_{a\ impervious}$ = 0.2 (0.20) = 0.04 in. | $I_{a\ pervious}$ = 0.2 (3.51) = 0.70 in. |
| $Q_{imp} = (0.75 - 0.04)^2/(0.75 - 0.04 + 0.20) = 0.55$ in. | $Q_p = (0.75 - 0.70)^2/(0.75 - 0.70 + 3.51) = 0.00$ in. |

Multiplying the runoff by the area gives

$$V = 0.55\text{in}(0.63\text{ac}) + 0.00\text{in}(1.37\text{ac}) = 0.35\text{ac-in}$$

$$0.37\text{ac-in}\left(43{,}560\text{ft}^2/\text{ac}\right)\left(1\text{ ft}/12\text{in}\right) = 1258\text{ft}^3$$

Note in this example that no runoff is calculated from the pervious areas, which would be expected, while impervious areas contribute a sizable percentage of the rainfall to runoff.

curve number. While the NRCS curve number method uses the accumulated rainfall and reports the cumulative runoff, the incremental runoff can be determined by extracting the difference between each time step.

One of the drawbacks of the NRCS runoff method is that as a storm progresses, if the total rainfall is large, infiltration is predicted to approach zero instead of the saturated hydraulic conductivity of the soils. Note that in the previous example the losses decrease rapidly; if rainfall continued the losses would reach zero. This leads to overestimation of runoff for larger events, in turn leading to overdesign in GSI applications. In addition, when used in continuous simulation models, a method is needed for the soil moisture to recover capacity during periods without rain. The NRCS is best used for less complex watersheds to determine flows entering an SCM. Note that while there are many publications where modified CNs are advanced to account for SCMs,

**Example 6.5** Find the incremental runoff volume in 20 min increments for the rainfall record given below. The curve number of the watershed is 74.

| Time (min) | 0 | 20 | 40 | 60 | 80 | 100 | 120 |
|---|---|---|---|---|---|---|---|
| Incremental Rainfall (in.) | 0 | 0.4 | 1.0 | 1.3 | 1.0 | 0.50 | 0.40 |

Using the same equations as used in Example 6.4,

$$S = 1000/74 - 10 = 3.51 \text{ in}$$

$$I_a = 0.2(3.51) = 0.70 \text{ in}$$

The next step is to create a summation rainfall row and apply Eq. 6.4 to each cumulative rainfall value to determine the cumulative runoff volume. See table below.

| Time (min) | 0 | 20 | 40 | 60 | 80 | 100 | 120 |
|---|---|---|---|---|---|---|---|
| Inc. Rainfall (in.) | 0.00 | 0.40 | 1.00 | 1.30 | 1.00 | 0.50 | 0.40 |
| Cum. Rainfall – P (in.) | 0.00 | 0.40 | 1.40 | 2.70 | 3.70 | 4.20 | 4.60 |
| Cum. Q (in.) | 0.00 | 0.00 | 0.12 | 0.73 | 1.38 | 1.74 | 2.05 |
| Inc. Q (in.) | 0.00 | 0.00 | 0.12 | 0.61 | 0.66 | 0.36 | 0.30 |
| Inc. Losses (in.) | 0.00 | 0.40 | 0.88 | 0.69 | 0.34 | 0.14 | 0.10 |

Note that at the 20-min mark, since the cumulative rainfall has not exceeded the value of $I_a$, it is set to zero.

At 40 min:

$$Q_{40} = (1.40 - 0.70)^2 / ((1.40 - 0.70) + 3.5)) = 0.12 \text{ in.(cumulative)}$$

To determine the incremental Q values, each cumulative value is subtracted from the previous step. The incremental losses were determined by subtracting the losses from the rainfall.

such as green roofs and raingardens (as discussed in later chapters), this should be considered a simple estimate, as the NRCS model was not developed for these conditions and considerable error may be introduced.

Starting in the late 1980s, discussions have been held to update the curve number method, to include more data sources, newer methodologies, and smaller rainfall events. An initiative was begun in 2015 and completed in 2017 (CNTG 2017), but the work has not yet been adopted. The major change is to the Initial Abstractions formula, where the percentage of initial abstractions is changed to 5%. It should be remembered that this value is statistically derived, so this also changes the land-use curve numbers as well.

$$I_a = 0.05S \tag{6.5}$$

With the change in $I_a$, a change in the curve number is required. In Eq. 6.6, the 20 and 05 subscripts refer to the curve numbers using the original and revised methodology, respectively.

$$CN_{05} = CN_{20} / \left(1.42 - 0.0042 CN_{20}\right) \tag{6.6}$$

In the previous example, the curve number 74 would change to 67, and 98 would change to 97. Readers are advised to watch for these changes.

**Example 6.6** Repeat the problem of ***Example 6.4*** using the revised Initial Abstraction of 5% and discuss the differences in the two runoff values.

*Solution* The pervious and impervious areas do not change. The "old" CN values must be modified using Eq. 6.6.

| **Impervious Area** | **Pervious Area** |
|---|---|
| $CN_{05}$ = 98/(1.42 − 0.0042(98)) = 97 | $CN_{05}$ = 74/(1.42 − 0.0042(74)) = 67 |
| $S_{impervious}$ = 1000/97 − 10 = 0.31 in. | $S_{pervious}$ = 1000/67 − 10 = 4.93 in. |
| $I_{a\ impervious}$ = 0.05 (0.31) = 0.016 in. | $I_{a\ pervious}$ = 0.05 (4.93) = 0.25 in. |
| $Q_{imp}$ = $(0.75 - 0.016)^2/$ (0.75 − 0.016 + 0.31) = 0.52 in. | $Q_p$ = $(0.75 - 0.25)^2/$ (0.75 − 0.25 + 4.93) = 0.05 in. |

Multiplying the runoff by the area gives

$$V = 0.52\ \text{in}\left(0.63\ \text{ac}\right) + 0.05\ \text{in}\left(1.37\ \text{ac}\right) = 0.40\ \text{ac-in}$$

$$0.40\ \text{ac-in}\left(43{,}560\ \text{ft}^2/\text{ac}\right)\left(1\ \text{ft} / 12\ \text{in}\right) = 1438\ \text{ft}^3$$

The runoff from the impervious area is less in the new method, while that from the impervious is greater. The overall predicted runoff volume is higher under the new method by (1438−1258)/1258 = 14%.

### 6.3.5 NRCS "Time of Concentration"

To convert precipitation excess to runoff requires an understanding of the time it takes runoff to reach the outlet or most downstream point of the watershed. These relations are used to build a pattern of runoff that changes based on the time of travel across the watershed. Time of Concentration (TC), is defined by NRCS (2010) as "the time required for runoff to travel from the hydraulically most distant point of the drainage watershed." While many time of concentration relationships exist throughout the literature, the NRCS velocity method is in common use. This method divides the runoff conveyance pathway into separate velocity components, commonly sheet flow, shallow concentrated flow, and open channel flow. The pathway with the longest travel time through the drainage area determines the time of concentration.

Table 6.7 Manning's Roughness Coefficients for Sheet Flow (Flow Depth Generally ⩽ 0.1 ft (3 cm)). (NRCS (2010), Revised by Authors).

| Surface | $n$ |
|---|---|
| Smooth surface (concrete, asphalt, gravel, or bare soil) | 0.011 |
| Fallow (no residue) | 0.05 |
| Cultivated Soils: | |
| Residue cover ≤ 20% | 0.06 |
| Residue cover > 20% | 0.17 |
| Grass: | |
| Short-grass prairie | 0.15 |
| Dense grasses* | 0.24 |
| Bermudagrass | 0.41 |
| Range (natural) | 0.13 |
| Woods: | |
| Light underbrush | 0.40 |
| Dense underbrush | 0.80 |

*Includes species such as weeping lovegrass, bluegrass, buffalo grass, blue grama grass, and native grass mixtures.
**When selecting $n$, consider cover to a height of about 0.11 ft. This is the only part of the plant cover that will obstruct sheet flow.

Sheet flow, as the name implies, is the flow spread over a plane surface. Normally stormwater at the upper reaches of the drainage area is considered to follow sheet flow, occurring before the flow has a chance to concentrate. In pervious areas, sheet flow may include flow through the top portion of the soil. It is generally accepted that sheet flow runoff will transition to shallow concentrated flow after 100 feet (30.5 m). A kinematic wave approach is used for the travel time estimate:

$$Tt_1 = \frac{0.007\left(nL\right)^{0.8}}{\left(P_2\right)^{0.5} S^{0.4}} \tag{6.7}$$

where $Tt_1$ = Travel time (h)
$n$ = Manning's sheet flow roughness factor (Table 6.7)
$L$ = Sheet flow length (ft)
$P_2$ = 2-year, 24 h rainfall depth (in)
$S$ = Slope of land surface (depth/length)

Note that the Manning's $n$ values are specifically derived for sheet flow conditions for depths less than 0.1 ft (30.5 mm).

The second travel segment is termed shallow concentrated flow. This condition follows the sheet flow as the water starts to channelize. This should be viewed as a small swale, gully, or gutter flow where the water depth is less than 0.5 ft (0.15 m). Manning's equation is used to estimate the velocity, which is then divided by the flow path length to determine the Travel Time. Note that in this segment, the Manning roughness values for channel flow, which is introduced and discussed later in this chapter, are used.

$$V = \frac{1.486 R_H^{\frac{2}{3}} S^{0.5}}{n} \tag{6.8}$$

$$Tt_2 = \frac{L}{3600V} \tag{6.9}$$

where $Tt_2$ = Travel Time (hr)
$V$ = velocity (ft /sec)
$n$ = Manning's Channel Roughness
$L$ = Travel Path Length (ft)
$R_H$ = Hydraulic Radius (ft)

For paved areas, the NRCS recommends the use of $R_H = 0.2$ and $n = 0.025$; for unpaved areas $R_H = 0.4$ and $n = 0.050$ are recommended.

The final segment now represents deeper open channel flow. This either includes streams, deeper ditches, or culvert flow. Again, Manning's equation or more complex energy relationships are needed. For decentralized SCMs, this segment may not be appropriate, as the flow may enter the SCM from the shallow concentrated condition. Looking back to Figure 6.5 note the differences in flow paths for the impervious and pervious flows. For the pervious area there is clearly a length of sheet flow, then a change to shallow concentrated flow when the runoff meets the pavement. For the impervious area, the sheet flow is more likely to represent half of the roadway width as it is crowned at the center, and then transitions to shallow concentrated gutter flow.

### 6.3.6 NRCS Unit Hydrograph

Several methods are available for use to transform the incremental runoff volumes to a flow rate and pattern in a stream (storm hydrograph). These include development of individual watershed unit hydrographs based on collected stream flow data, methods that include the hydraulic wave movement (Kinematic Wave) across the landscape, and other statistical approaches. The Kinematic Wave methodology is used within many continuous simulation models such as the USEPA Storm Water Management Model (SWMM), but detailed discussion on this methodology is better suited for a surface hydrology course.

Of the two remaining methods, the statistical approach is the most appropriate as monitored data are rarely available to the designer to develop a data-driven unit hydrograph. Again, multiple statistical approaches are available, but the *NRCS Dimensionless Unit Hydrograph* is the most popular due to its simplicity.

A hydrograph is defined as a temporal pattern of stream flow from a rainfall event exiting a watershed. The Unit Hydrograph (UH) is a pattern related to a unit volume (1 in. or 1 cm) of runoff over the time period of incremental precipitation excess. For most computer models, when the time period of precipitation excess is selected, this sets the UH time period. The unit hydrograph assumes that the rainfall runoff response

**Example 6.7** An urbanized watershed near Baltimore, MD with three flow segments is displayed in the table below. Compute the travel time for each flow segment, and the time of concentration.

| Reach | Description | Length (ft) | Slope |
|---|---|---|---|
| 1 | Overland – wooded (light underbrush) | 100 | 6.0% |
| 2 | Overland – shallow gutter (paved) | 300 | 1.50% |
| 3 | Storm drain $D$ = 3 ft, $n$ = 0.015 | 600 | 0.1% |

*Solution* Each segment is evaluated separately. For Reach 1, since overland flow occurs, Eq. 6.7 is used. The value of Manning's $n$ is taken from Table 6.7, equal to 0.4. The 2-year, 24-h rainfall for Baltimore MD is given in Table 6.2 and is equal to 3.16 in.

$$Tt_1 = \frac{0.007(nL)^{0.8}}{(P_2)^{0.5} S^{0.4}} = \frac{0.007((0.4)(100))^{0.8}}{(3.16)^{0.5}(0.06)^{0.4}}$$

$$= 0.23\text{hr} = 13.9\text{min}.$$

For paved reach 2, $R_H$ is assumed as 0.2 and $n$ = 0.025.

$$V = \frac{1.486 R_H^{\frac{2}{3}} S^{0.5}}{n} = \frac{1.486(0.2)^{\frac{2}{3}}(0.015)^{0.5}}{0.025}$$

$$= 4.97\text{ft/s}$$

$$Tt_2 = \frac{L}{3600V} = 300/(3600 * 4.97) = 0.016\text{hr} = 1.0\text{min}$$

Assuming the culvert is flowing half full, the hydraulic radius is equal to one half the diameter of the pipe, or 1.5 ft.

$$V = \frac{1.486 R_H^{\frac{2}{3}} S^{0.5}}{n} = \frac{1.486(1.5)^{\frac{2}{3}}(0.001)^{0.5}}{0.015}$$

$$= 4.10\text{ft/s}$$

$$Tt_2 = \frac{L}{3600V} = \frac{600}{(3600 * 4.10)} = 0.041hr = 2.4\text{min}$$

$$\text{Total Travel Time} = Tt_1 + Tt_2 + Tt_3 = 13.9 + 1.0 + 2.4 = 17.3\text{ min}$$

Note that though shorter in distance, the sheet flow segment usually has the longest travel time.

is the same for the same volume, and the hydrograph ordinates are proportional to the precipitation excess. In other words, two units of precipitation excess would cause all the runoff ordinates to double.

The NRCS Dimensionless UH is in common use due to its simplicity and packaging with the other NRCS methods of Time of Concentration (TC) and Curve Number. Only two parameters are required, time to peak ($T_p$) and peak flow ($Q_p$), and the equations below.

$$\text{Lag} = 0.6\,\text{TC}(\text{hrs}) \tag{6.10a}$$

$$T_p = D/2 + \text{Lag}(\text{hrs}) \tag{6.10b}$$

$$Q_p = 484\,A/T_p\left(\text{ft}^3/\text{s}\right) \tag{6.10c}$$

where A = Drainage area ($\text{mi}^2$) and D = Time period of the unit hydrograph precipitation excess (hrs). D is usually related to the time increment of rainfall, or the discretization time period selected for the model. The duration should be generally less than or equal to 0.133 times the TC.

The hydrologic peaking factor (484) in Eq. 6.10c was originally part of the development of the NRCS UH. Some areas recommend different values depending on the land use and slope. As shown in Figure 6.6, the relationship is dimensionless, as it is plotted as a ratio of time and flow for each respective peak. Table 6.8 presents the numeric ratios used to create the graph.

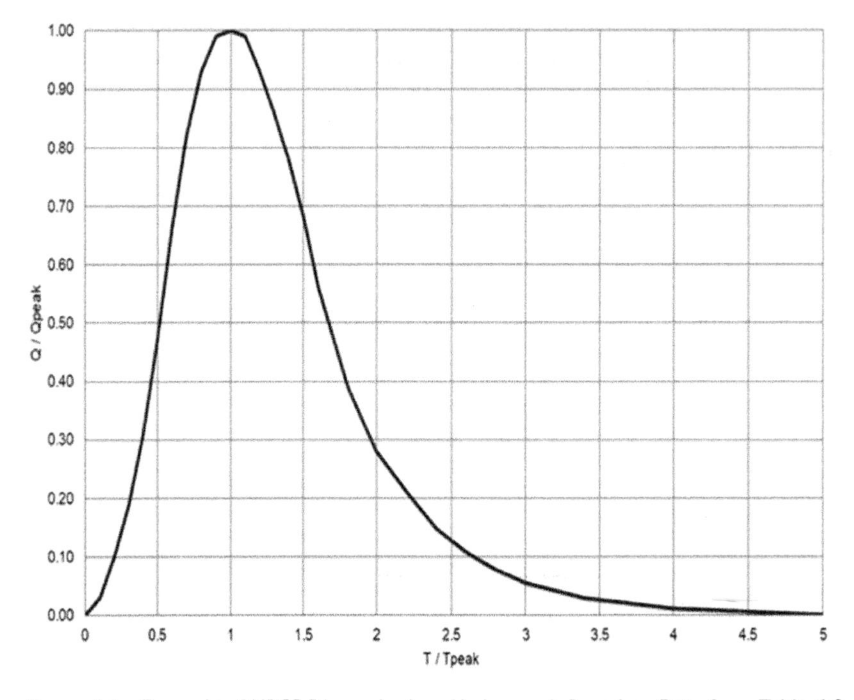


Figure 6.6 Example of NRCS Dimensionless Hydrograph Based on Data from Table 6.8.

Table 6.8 Ratios Used to Create Dimensionless Hydrograph Shown in Figure 6.6.

| **Time Ratios** ($t/T_p$) | **Discharge Ratios** ($Q/Q_p$) | **Time Ratios** ($t/T_p$) | **Discharge Ratios** ($Q/Q_p$) |
|---|---|---|---|
| 0 | 0.000 | 1.7 | 0.460 |
| 0.1 | 0.030 | 1.8 | 0.390 |
| 0.2 | 0.100 | 1.9 | 0.330 |
| 0.3 | 0.190 | 2.0 | 0.280 |
| 0.4 | 0.310 | 2.2 | 0.207 |
| 0.5 | 0.470 | 2.4 | 0.147 |
| 0.6 | 0.660 | 2.6 | 0.107 |
| 0.7 | 0.820 | 2.8 | 0.077 |
| 0.8 | 0.930 | 3.0 | 0.055 |
| 0.9 | 0.990 | 3.2 | 0.040 |
| 1.0 | 1.000 | 3.4 | 0.029 |
| 1.1 | 0.990 | 3.6 | 0.021 |
| 1.2 | 0.930 | 3.8 | 0.015 |
| 1.3 | 0.860 | 4.0 | 0.011 |
| 1.4 | 0.780 | 4.5 | 0.005 |
| 1.5 | 0.680 | 5.0 | 0.000 |
| 1.6 | 0.560 | | |

**Example 6.8** Determine the NRCS Dimensionless Unit Hydrograph for a time period unit hydrograph precipitation excess (D) of 20 mins, a time of concentration (TC) of 2.5 h, and an area of 0.5 square miles.

*Solution* Equations (6.10a) through (6.10c) are used:

$$\text{Lag} = 0.6\ \text{TC} = 0.6(2.5) = 1.5\ \text{h}$$

$$T_p = (20/60)/2 + 1.5 = 1.67\ \text{h}$$

$$Q_p = 484(0.5)/1.67 = 145.2\ \text{ft}^3/\text{s}$$

Note that D is equal to 0.133*TC so the interval is appropriate.

Apply the Dimensionless Unit Hydrograph to the $Q_p$ and $T_p$ by multiplying each by the Time Ratios and Discharge Ratios given in Table 6.8. The final product is presented in the figure below, showing that the peak flow of 145.2 $\text{ft}^3/\text{s}$ occurs at 1.67 h. Note that this procedure is presented in detail in any introductory hydrology text (e.g., McCuen 2017; Viessman and Lewis 2003).

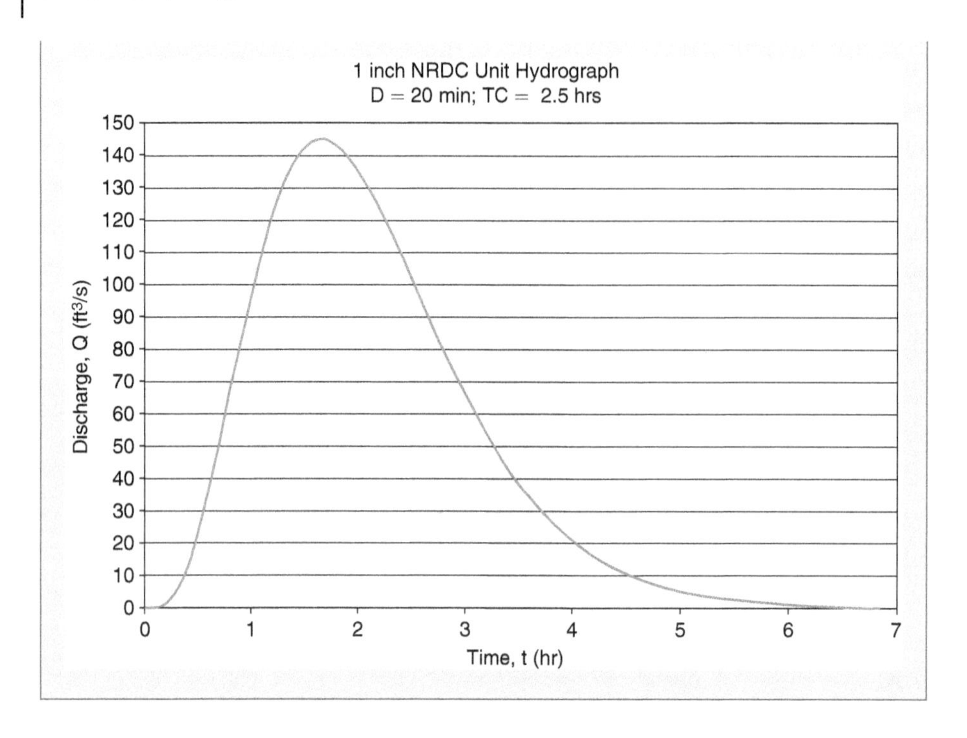


### 6.3.7 Creating the Storm Hydrograph

The final step in developing the storm inflow hydrograph for a subarea draining to an SCM is multiplying the precipitation excess by the unit hydrograph for the same duration. While detailed discussion of this process is beyond the scope of this text, the procedure can be demonstrated. In Example 6.8, the precipitation excess for a 120-min storm in 20-min increments was demonstrated. Given a precipitation excess depth record, each depth is then multiplied by the unit hydrograph of Example 6.8 and lagged the duration of 20 min.

Using the incremental precipitation excess data given in Example 6.5, reproduced below, the hydrograph can be synthesized.

| Time (min) | 0 | 20 | 40 | 60 | 80 | 100 | 120 |
|---|---|---|---|---|---|---|---|
| Precipitation Excess (in) | 0 | 0.40 | 1.0 | 1.3 | 1.0 | 0.50 | 0.40 |

The 20-min curve is the rainfall excess for the 20-min period (0.4 in.) multiplied by the unit hydrograph and moved forward 20 min in time. The 40-min curve, since the rainfall is 1 in., uses the unit hydrograph, moved 40 min to coincide with the beginning of that time period (60 min). The 60-min curve is created by multiplying the UH by 1.3 and lagging by 60 min. This process is repeated for all increments. The storm hydrograph then is found by adding the runoff response from each individual rainfall period, as demonstrated in Figure 6.7. The precipitation excess is shown at the top, along with the individual rainfall excess curves.

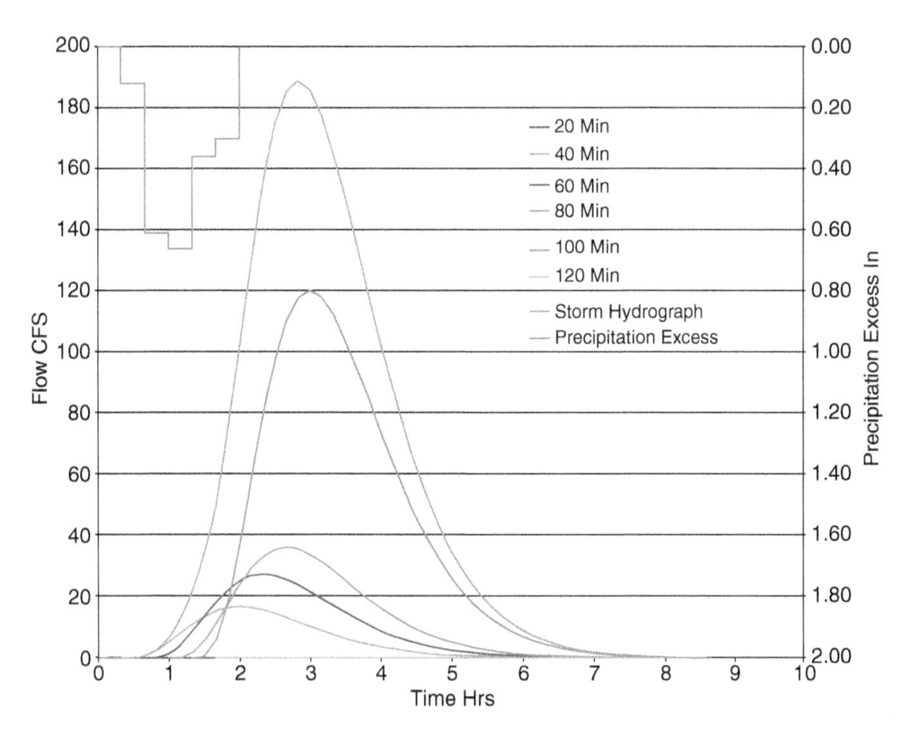


Figure 6.7 Example Summation Storm Hydrograph Using NRDC Unit Hydrograph Method.

This represents an example of a simple watershed response from one drainage area. If for example the watershed is divided into a pervious and impervious fraction, they can be directly summed as they are produced by the same rainfall increment and exit at the same point. For more complex hydrologic analysis that includes systems that would require hydraulic routing, a course or text in hydrology is recommended (e.g., McCuen 2017; Viessman and Lewis 2003).

## 6.4 Peak Flow Methods

Hydrologic tools are necessary to estimate peak runoff flows from developed watersheds. Peak and high flows can be erosive and damaging to receiving streams. Many stormwater regulations require that stormwater control measures reduce the peak flow from the developed drainage area.

### 6.4.1 The Rational Method

The Rational Method is the simplest model used for estimating peak flow from a drainage area. In this form, the Rational method is given as:

$$Q_p = ciA \tag{6.11}$$

where $Q_p$ is the peak runoff flow, c is the Rational method coefficient, i is the rainfall intensity and A is the drainage area. The rational method coefficient basically gives the fraction of rainfall that hits the ground surface that becomes runoff. As such, the value of c depends on the land cover. Natural land cover allows infiltration and the values of c are low. Conversely, impervious surfaces result in large values of c, approaching 1, where the majority of rainfall becomes runoff. Values of c for different land uses are given in Table 6.9.

The rainfall intensity is found from historical rainfall information from the site, such as described in Section 6.2. To select the intensity, the storm return period and storm duration must be specified. In the Rational Method, the storm duration is calculated as the time of concentration in the drainage area.

**Table 6.9** Values of Rational Method c for various land uses.

| Land Use | Rational Method c | Land Use | Rational Method c |
|---|---|---|---|
| Unimproved | 0.10–0.30 | Neighborhood commercial | 0.50–0.70 |
| Urban green areas | 0.10–0.25 | Light industrial | 0.50–0.80 |
| Single family residential | 0.30–0.50 | Multifamily attached residential | 0.60–0.75 |
| Multifamily detached residential | 0.40–0.60 | Heavy industrial | 0.60–0.90 |
| Apartment | 0.50–0.70 | Downtown | 0.70–0.95 |

**Example 6.9** Find the peak runoff flow rate resulting from a 10-year storm over a 20-acre drainage area near Baltimore, MD. Ten acres have been developed into single family homes and 5 acres into multifamily attached residences. The other 5 acres have remained undeveloped. The time of concentration is estimated as 30 min.

*Solution* With the depth and area given, the weighted c must be determined. The table below shows the weighting process. In the absence of additional information, midpoint values for c from Table 6.9 are used.

| | c | Weight | Weighted c |
|---|---|---|---|
| 1.0 acres SF | 0.4 | 10/20 | 0.20 |
| 5 acres MFA | 0.68 | 10/20 | 0.17 |
| 5 acres undeveloped | 0.2 | 5/20 | 0.05 |
| | | | 0.42 |

From Table 6.2, the 10-year, 30 min rainfall depth for Baltimore, MD is 1.60 in. The average intensity is calculated as 1.60 in/0.5 h = 3.20 in./h.

Therefore, $Q_p$ = ciA = 0.42(3.20 in./h)(20 ac) = 27 ac–in./h = 27 ft$^3$/s.

An acre-in./h is equal to ft$^3$/s within less than 1%.

### 6.4.2 The NRCS Unit Hydrograph Method

Peak flow is directly determined in the Dimensionless UH method, as discussed above. Given the time increment of rainfall, the time of concentration, and the drainage area, peak flow ($Q_p$) is given by Eqs. 6.10a through 6.10c.

## 6.5 Watershed and SCM Hydraulics

As noted above, hydraulic considerations are necessary in evaluating watershed hydrology. In addition, the control of water flows into, through, and, especially out of, most GI SCMs is critical to ensuring their proper performance. SCM hydraulics represent a critical part of most designs.

### 6.5.1 Open Channel Flow

Open channels are used for the conveyance of stormwater from one place to another. Steady flow of water in channels that are open to the atmosphere is described by Manning's Equation

$$v = \frac{1.49}{n} R_h^{\frac{2}{3}} s^{\frac{1}{2}} \tag{6.12}$$

where $v$ is the water velocity in the channel in ft/s, $n$ is the Manning roughness coefficient, $R_h$ is the hydraulic radius (in ft), and $s$ is the slope or hydraulic gradient. Eq. 6.12 can be used with SI units without the 1.49 coefficient; in this case, $v$ is in m/s and $R_h$ in m.

The slope is the physical slope of the channel, given as the vertical fall divided by the horizontal distance (therefore dimensionless). The hydraulic radius is defined as the cross-sectional area of the channel ($A$), divided by the wetted perimeter ($P_w$).

$$R_h = \frac{A}{P_w} \tag{6.13}$$

The wetted perimeter is the length of the perimeter in contact with water in the channel. Manning's Equation can be used with a channel of any cross-sectional shape as long as the hydraulic radius can be calculated. Figure 6.8 shows the area and wetted perimeter for rectangular, triangular, and trapezoidal channels.

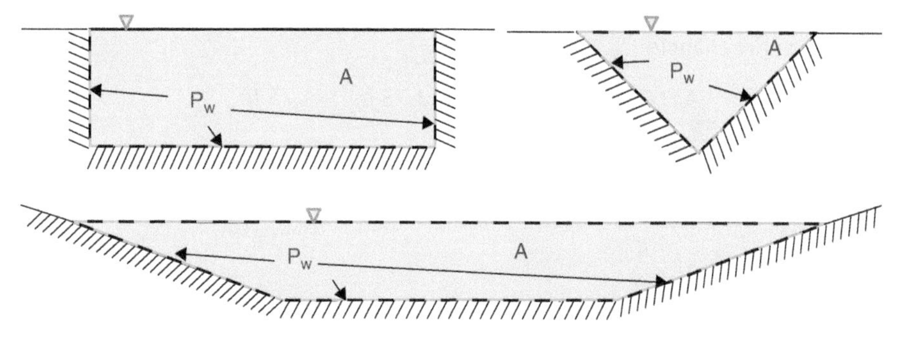


Figure 6.8 Open Channel Geometries Showing Area and Wetted Perimeter. (A) Rectangle. (B) Triangle. (C) Trapezoid.

**Table 6.10** Values of Manning's *n* Roughness Coefficient for Various Surfaces (Linsley et al. 1982).

| Channel Material | *n* | Channel Material | *n* |
|---|---|---|---|
| Ordinary concrete | 0.013 | Natural channels with stones and weeds | 0.035 |
| Smooth earth | 0.018 | Very poor natural channels | 0.060 |
| Firm gravel | 0.023 | Grasses | 0.25–1.0 |
| Natural channels in good condition | 0.025 | | |

Manning's roughness coefficient, *n*, describes the roughness of the channel surface. A rough surface will lead to lower channel velocities than a smooth surface. Table 6.10 provides a list of Manning's *n* values for various surfaces.

The flow rate in an open channel can be found by multiplying the velocity by the cross-sectional area.

$$Q = vA = \frac{1.49}{n} A R_h^{\frac{2}{3}} s^{\frac{1}{2}} \tag{6.14}$$

Because of the complex relationship between flow rate and area and wetted perimeter, a guess and check solution may be necessary for many open channel designs.

**Example 6.10** A triangular channel lined with firm gravel is flowing at 1 ft depth. The channel slope is 0.7% and the angle of the vee is 120°. Find the flow rate through the channel.

***Solution*** The slope is 0.007 and from Table 6.10, *n* is 0.023. Trigonometric analysis can be used to find the area and wetted perimeter.

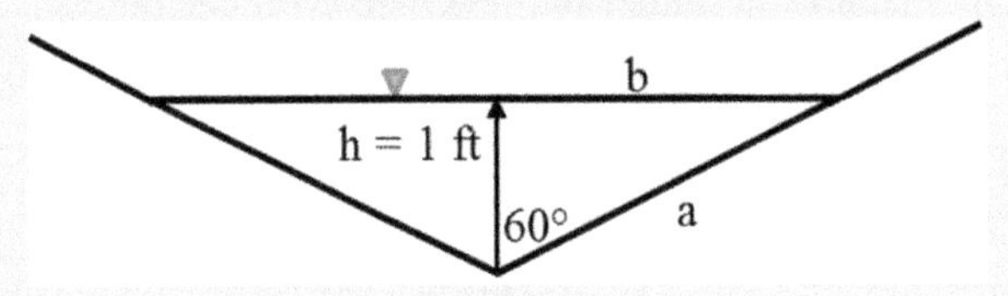


$$\cos 60° = \frac{h}{a} = \frac{1\,ft}{a} = 0.5;\ a = 2\,ft$$

$$\tan 60° = \frac{b}{h} = \frac{b}{1\,ft} = 1.73;\ b = 1.73\,ft$$

For the entire triangle:

$$A = \frac{1}{2}(h)(2b) = \frac{1}{2}(1\,ft)(2(1.73\,ft)) = 1.73\,ft^2$$

$$P_w = 2a = 4\,ft$$

$$R_h = \frac{A}{P_w} = \frac{1.73\,ft^2}{4\,ft} = 0.433\,ft$$

and using Eq. 6.14:

$$Q = \frac{1.49}{0.023}(1.73\,ft^2)(0.433\,ft)^{\frac{2}{3}}(0.007)^{\frac{1}{2}} = 5.37\frac{ft^3}{s}$$

**Example 6.11** A rectangular concrete channel is used to convey runoff down a 0.5% slope to a SCM. The channel width should be 3 times the depth at a design flow of 14 $ft^3/s$. Find the channel dimensions.

*Solution* The slope is 0.005 and from Table 6.10, $n$ is 0.013. For the rectangular channel the wetted perimeter is the sum of the width and 2 times the depth. A width of 3 ft is used as an initial guess, leading to a flow rate of 17.3 $ft^3/s$ – too high. Subsequent guesses lead to the correct results. The table below shows the iterative procedure to solve this problem.

| Depth (ft) | Width (ft) | Area ($ft^2$) | $P_w$ (ft) | $R_h$ (ft) | Q ($ft^3/s$) |
|---|---|---|---|---|---|
| 1 | 3 | 3 | 5 | 0.60 | 17.3 |
| 0.75 | 2.25 | 1.688 | 3.75 | 0.450 | 8.04 |
| 0.917 (11 in.) | 2.75 | 2.52 | 4.58 | 0.55 | 13.7 |

To the nearest inch, the depth is 11 in. at a width of 33 in.

### 6.5.2 Orifices

An orifice is a submerged opening that allows fluid to pass through, therefore the water level must be higher than the top of the opening. The flow rate through an orifice is controlled by the area of the opening and the height of the water above the orifice. An orifice is commonly used in SCMs to control the flow rate of water exiting a storage volume.

The orifice equation is given by:

$$Q = C_d A (2gh)^{\frac{1}{2}} \tag{6.15}$$

where, $Q$ = outflow rate through orifice, $C_d$ = coefficient of discharge (the default value is 0.60 for a circular orifice), $A$ = cross-sectional area of opening, g = gravity, and $h$ = height of water over centerline of the opening (Figure 6.9).

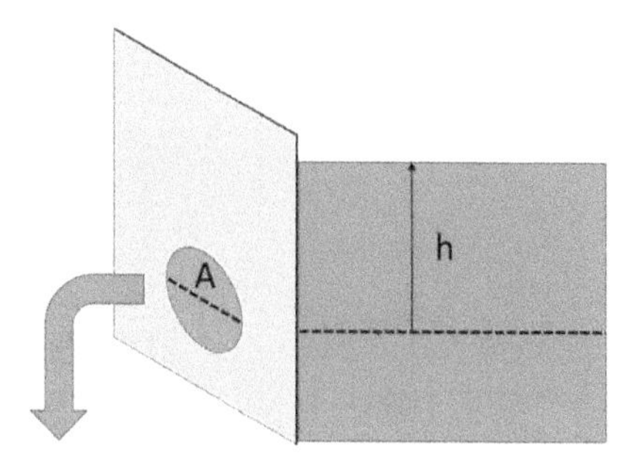


Figure 6.9 Diagram Showing Parameters Describing Orifice Flow.

Since most orifice applications involve the controlled drainage of stored water, the height above the orifice does not remain constant and, accordingly, the flow rate decreases with time. The orifice equation can be integrated to find the time, $t$, required to drain a stored water volume from the initial height above the orifice center line, $h_0$, to the final value, $h_f$, assuming that the cross-sectional area of the storage system, $A_S$, is known.

$$\frac{-dV}{dt} = Q = -A_S \frac{dh}{dt} \tag{6.16a}$$

Substituting Eq. 6.15 and separating variables:

$$-\int_{h_0}^{h_f} h^{-\frac{1}{2}} dh = C_d \frac{A}{A_S} (2g)^{\frac{1}{2}} \int_0^t dt \tag{6.16b}$$

$$t = \frac{2A_S \left( h_0^{\frac{1}{2}} - h_f^{\frac{1}{2}} \right)}{C_d A (2g)^{\frac{1}{2}}} \tag{6.16c}$$

**Example 6.12** A circular storage tank, 8 ft tall with an 8 ft diameter is filled with water. A small hole, 1/8 in. diameter is drilled at the bottom edge of the tank. Find the time for the tank to drain to the top of the hole.

***Solution*** The cross-sectional area of the tank and the hole are found. 1/8 in. = 0.01042 ft:

$$A_s = \frac{\pi D^2}{4} = \frac{\pi (8\ ft)^2}{4} = 50.3\ ft^2$$

$$A = \frac{\pi (0.01042\ ft)^2}{4} = 8.52 \times 10^{-5}\ ft^2$$

Knowing that $h_0$ = 8 ft and $h_f$ = 1/16 in. = 5.21 × $10^{-3}$ ft, from Eq. 6.16c:

$$t = \frac{2(50.3\ ft^2)\left( (8\ ft)^{\frac{1}{2}} - (5.21x10^{-3}\ ft)^{\frac{1}{2}} \right)}{0.6(8.52x10^{-5}\ ft^2)\left( 2\left(32.2\frac{ft}{s^2}\right)\right)^{\frac{1}{2}}} = 6.76x10^5\ s = 7.82\ days$$

### 6.5.3 Weirs

A weir is an opening that allows fluid to pass, but unlike an orifice, the opening is not completely submerged. Weirs may be used to control overflow rates in SCMs and are commonly used as flow measurement devices in SCM studies. When the water level falls below the top of an orifice, it becomes a weir.

Different weir shapes have different flow equations. The weir equation for a triangular (V-notch) weir, one of the most common weir types, is given by:

$$Q = 4.28C_d tan\left(\frac{\theta}{2}\right)(h+k)^{\frac{5}{2}} \tag{6.17}$$

where $C_d$ is a discharge coefficient, θ is the angle of the notch, $h$ is the head of the water above the weir notch, and $k$ is a head correction factor. For $Q$ in $ft^3/s$, $C_d$ is approximately 0.58 and $k$ is about 0.003 ft for θ above 80°; $k$ increases to 0.009 as θ decreases from 80° to 20°. Figure 6.10 shows a diagram of a V-notch weir.

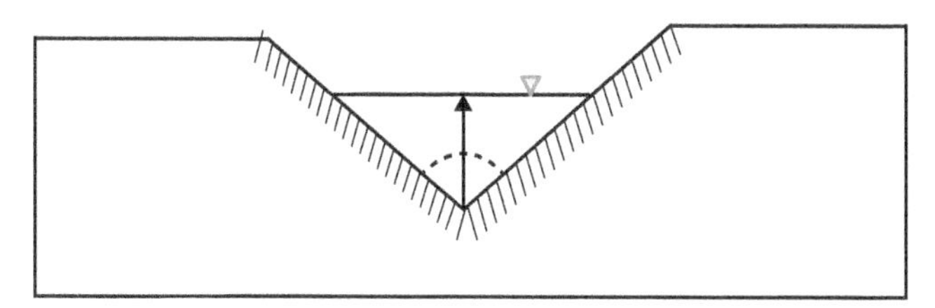

Figure 6.10 Diagram of V-notch Weir.

A set of rectangular weirs that makes up the discharge structure from a dry stormwater storage pond is shown in Figure 6.11. The weirs are designed to allow specific flow rate discharges during moderate (for the thin weir) and high flows (for the wide weir). The flow through a rectangular weir is given by:

$$Q = \frac{2}{3}C_d w\sqrt{2g}\,h^{\frac{3}{2}} \tag{6.18}$$

Where $w$ is the weir width. The discharge coefficient depends on the weir configuration, with a default value of 0.60.

Figure 6.11 Dual-Weir System Used to Control Runoff Discharge from a Dry Pond Stormwater Control Measure (Photo by Authors).

**Example 6.11** A rectangular weir, similar to that shown in Figure 6.11 is 1 ft wide. Find the weir flow at a flow depth of 6 in.

*Solution* Using Eq. 6.18:

$$Q = \frac{2}{3}(0.6)(1\ ft)\sqrt{\left(2\left(32.2\frac{ft}{s^2}\right)\right)}(0.5\ ft)^{\frac{3}{2}} = 1.135\frac{ft^3}{s}$$

## References


Albright, C.M. and Schramm, H. (2018). Improvements and applications in climate data analysis for determining reference rainfall years. *Journal of Applied Meteorology and Climatology* 57 (2): 413–420.

CNTG (2017) Curve Number Task Group, Watershed Management Technicial Committee, Environmental and Water Resources Insitute, American Society of Civil Engineers. Draft – ASCE – ASABE Proposed CN Update, September 20, 2017, https://directives.sc.egov.usda.gov/OpenNonWebContent.aspx?content=41606.wba (accessed December 2021).

Ecology (2020) *Western Washington Hydrology Model*, https://ecology.wa.gov/Regulations-Permits/Guidance-technical-assistance/Stormwater-permittee-guidance-resources/Stormwater-manuals/Western-Washington-Hydrology-Model. Accessed June 2020.

Link, T.E., Unsworth, M., and Marks, D. (2004). The dynamics of rainfall interception by a seasonal temperate rainforest. *Agricultural and Forest Meteorology* 124: 171–191.

Linsley, R.K., Kohler, M.A., and Paulhus, J.L.H. (1982). *Hydrology for Engineers*, 3rd e. McGraw Hill.

Machusic, M., Welker, A.L., and Traver, R.G. (2011). Groundwater Mounding at a Storm-Water Infiltration BMP. *Journal of Irrigation and Drainage Engineering* 137 (3): 154–160.

McCuen, R.H. (2017). *Hydrologic Analysis and Design*, 4th e. Pearson.

Merkel, W.H., Moody, H.F., and Quan, Q.D. (2015). Design rainfall distributions based on NOAA Atlas 14 rainfall depths and durations, Fifth Federal Interagency Hydrologic Modeling Conference (FIHMC), Reno, Nevada.

NEH (2004) United States Department of Agriculture, natural resources conservation service, hydrology. National Engineering Handbook, Chapter 9, 2004.

NRCS (2010) *National Engineering Handbook, Part 630 Hydrology*. Chapter 15: Time of Concentration, Natural Resources Conservation Service, United States Department of Agriculture.

NRCS (2019) *National Engineering Handbook, Part 630 Hydrology*. Chapter 4: Storm Rainfall Depth and Distribution, Natural Resources Conservation Service, United States Department of Agriculture.

Philadelphia Water Department (2009) *Philadelphia Combined Sewer Overflow Long Term Control Plan Update, Volume 5: Precipitation Analysis*. http://water.phila.gov/pool/files/Vol05_Precip.pdf (Accessed January 2021).

Schueler, T. (1987). *Controlling Urban Runoff: A Practical Manual for Planning and Designing Urban BMPs*. Washington, DC: Metropolitan Washington Council of Governments.

Tholin, A.L. and Keifer, G.J. (1960). Hydrology of Urban Runoff. *Transactions of the American Society of Civil Engineers*, 125 (1), 1317–1319.

Urban Drainage and Flood Control District (2008). *Drainage Criteria Manual*, Vol. 1. Chapter 5: Runoff. Denver, CO. June 2001, Revised April 2008.

USDA-United States Department of Agriculture (1986). *Urban Hydrology for Small Watershed, TR-55*. Natural Resources Conservation Service.
Viessman, W., Jr. and Lewis, G.L. (1996). *Introduction to Hydrology*. Chapter 3: Interception and Depression Storage. New York, 40–51.
Viessman, W., Jr and Lewis, G.L. (2003). *Introduction to Hydrology*, 5th e. Pearson.
Xiao, Q., McPherson, E.G., Ustin, S.L., and Grismer, M.E. (2000). A new approach to modeling tree rainfall interception. *Journal of Geophysical Research* 105: 29,173–29,188.

## Problems

6.1 Create a 1-h storm 5-year design storm for Baltimore, MD.
6.2 Create a 2-h storm 2-year design storm for Philadelphia, PA.
6.3 Create a 1-h storm 5-year design storm for Philadelphia, PA.
6.4 Create a 2-h storm 2-year design storm for Chicago, IL.
6.5 Create a 1-h storm 5-year design storm for Chicago, IL
6.6 Create a 2-h storm 2-year design storm for Jackson, MS.
6.7 Create a 1-h storm 5-year design storm for Jackson, MS
6.8 Create a 2-h storm 2-year design storm for Las Cruces, NM.
6.9 Create a 1-h storm 5-year design storm for Las Cruces, NM
6.10 What do you see as the strengths and weaknesses for the design storm as compared to a continuous storm record?
6.11 An urbanized watershed is shown in the figure below. Three types of flow conditions exit from the furthermost point of the watershed to the outlet. Compute the travel time ($T_t$) and the time of concentration ($T_c$) based on the following data:

| Reach | Description of Flow | Length (ft) | Slope |
|---|---|---|---|
| A to B | Overland (forest) | 500 | 7.0% |
| B to C | Overland (shallow gutter) | 900 | 2.0% |
| C to D | Storm drain with manhole cover, inlets, etc. ($n = 0.015$; diameter 3 ft) | 2000 | 1.5% |
| D to E | Open channel, gunite, trapezoidal ($b$=5 ft; $d$=3 ft; $z$=1:1; $n$=0.019) | 3000 | 0.5% |

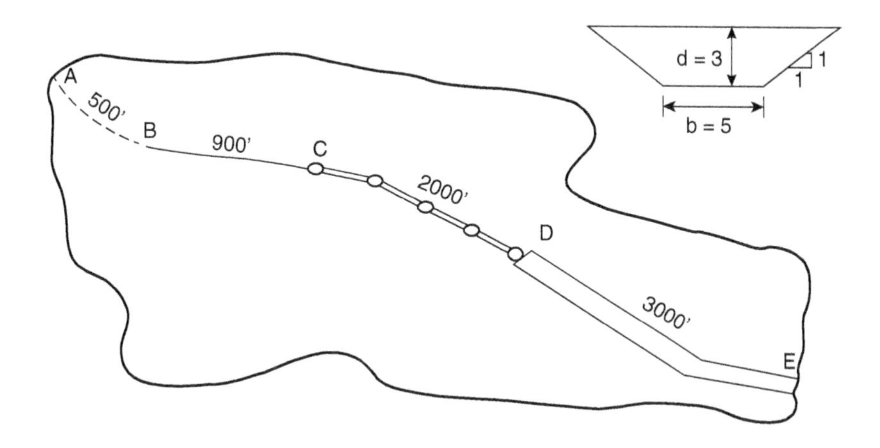

Use the table below and the Simple Method for Problems 6.12 to 6.26

| Land Use | Average Impervious | Area of Land Use (acres) | | | | |
|---|---|---|---|---|---|---|
| | | A | B | C | D | E |
| Unimproved | 0% | 5 | 3 | 0 | 0 | 0 |
| Urban green areas | 5% | 6 | 5 | 1 | 2 | 2 |
| Single family residential | 35% | 6 | 6 | 5 | 0 | 0 |
| Multifamily detached residential | 45% | 6 | 6 | 6 | 2 | 0 |
| Multifamily attached residential | 68% | 0 | 0 | 4 | 6 | 0 |
| Apartment | 70% | 0 | 6 | 6 | 5 | 6 |
| Neighborhood Commercial | 60% | 1 | 1 | 6 | 6 | 4 |
| Heavy commercial | 80% | 0 | 4 | 4 | 1 | 6 |

**6.12** For land development A, find the runoff volume for a 0.5 in. storm.
**6.13** For land development A, find the runoff volume for a 0.75 in. storm.
**6.14** For land development A, find the runoff volume for a 1.0 in. storm.
**6.15** For land development B, find the runoff volume for a 0.5 in. storm.
**6.16** For land development B, find the runoff volume for a 0.75 in. storm.
**6.17** For land development B, find the runoff volume for a 1.0 in. storm.
**6.18** For land development C, find the runoff volume for a 0.5 in. storm.
**6.19** For land development C, find the runoff volume for a 0.75 in. storm.
**6.20** For land development C, find the runoff volume for a 1.0 in. storm.
**6.21** For land development D, find the runoff volume for a 0.5 in. storm.
**6.22** For land development D, find the runoff volume for a 0.75 in. storm.
**6.23** For land development D, find the runoff volume for a 1.0 in. storm.
**6.24** For land development E, find the runoff volume for a 0.5 in. storm.
**6.25** For land development E, find the runoff volume for a 0.75 in. storm.
**6.26** For land development E, find the runoff volume for a 1.0 in. storm.

Use the table below and the Rational Method for Problems 6.27 to 6.46.

| Land Use | Area of Land Use (acres) | | | | |
|---|---|---|---|---|---|
| | A | B | C | D | E |
| Unimproved | 5 | 3 | 0 | 0 | 0 |
| Urban green areas | 6 | 5 | 1 | 2 | 2 |
| Single family residential | 6 | 6 | 5 | 0 | 0 |
| Multifamily detached residential | 6 | 6 | 6 | 2 | 0 |
| Apartment | 0 | 6 | 6 | 5 | 0 |
| Neighborhood commercial | 1 | 0 | 6 | 6 | 0 |
| Light industrial | 0 | 1 | 0 | 6 | 5 |
| Multifamily attached residential | 0 | 0 | 4 | 6 | 6 |
| Heavy industrial | 0 | 4 | 4 | 0 | 6 |
| Downtown | 0 | 0 | 0 | 1 | 6 |

6.27 For land development A, find the peak runoff flow for a 10-year storm in Baltimore, MD at a time of concentration of 30 min.

6.28 For land development A, find the peak runoff flow for a 25-year storm in Baltimore, MD at a time of concentration of 30 min.

6.29 For land development A, find the peak runoff flow for a 10-year storm in Philadelphia, PA at a time of concentration of 30 min.

6.30 For land development A, find the peak runoff flow for a 10-year storm in Baltimore, MD at a time of concentration of 60 min.

6.31 For land development B, find the peak runoff flow for a 10-year storm in Baltimore, MD at a time of concentration of 30 min.

6.32 For land development B, find the peak runoff flow for a 25-year storm in Baltimore, MD at a time of concentration of 30 min.

6.33 For land development B, find the peak runoff flow for a 10-year storm in Philadelphia, PA at a time of concentration of 30 min.

6.34 For land development B, find the peak runoff flow for a 10-year storm in Baltimore, MD at a time of concentration of 60 min.

6.35 For land development C, find the peak runoff flow for a 10-year storm in Baltimore, MD at a time of concentration of 30 min.

6.36 For land development C, find the peak runoff flow for a 25-year storm in Baltimore, MD at a time of concentration of 30 min.

6.37 For land development C, find the peak runoff flow for a 10-year storm in Philadelphia, PA at a time of concentration of 30 min.

6.38 For land development C, find the peak runoff flow for a 10-year storm in Baltimore, MD at a time of concentration of 60 min.

6.39 For land development D, find the peak runoff flow for a 10-year storm in Baltimore, MD at a time of concentration of 30 min.

6.40 For land development D, find the peak runoff flow for a 25-year storm in Baltimore, MD at a time of concentration of 30 min.

6.41 For land development D, find the peak runoff flow for a 10-year storm in Philadelphia, PA at a time of concentration of 30 min.

6.42 For land development D, find the peak runoff flow for a 10-year storm in Baltimore, MD at a time of concentration of 60 min.

6.43 For land development E, find the peak runoff flow for a 10-year storm in Baltimore, MD at a time of concentration of 30 min.

6.44 For land development E, find the peak runoff flow for a 25-year storm in Baltimore, MD at a time of concentration of 30 min.

6.45 For land development E, find the peak runoff flow for a 10-year storm in Philadelphia, PA at a time of concentration of 30 min.

6.46 For land development E, find the peak runoff flow for a 10-year storm in Baltimore, MD at a time of concentration of 60 min.

6.47 Determine and plot the NRCS Unit Hydrograph for a time period unit hydrograph precipitation excess of 15 min, a time of concentration of 2.0 h, and an area of 0.3 square miles.

6.48 Determine and plot the NRCS Unit Hydrograph for a time period unit hydrograph precipitation excess of 18 min, a time of concentration of 1.8 h, and an area of 0.6 square miles.

**6.49** Determine and plot the NRCS Unit Hydrograph for a time period unit hydrograph precipitation excess of 24 min, a time of concentration of 2.8 h, and an area of 0.7 square miles.

Use the incremental rainfall data given below to solve Problems 6.50 through 6.55.

| Time (min) | 0 | 15 | 30 | 45 | 60 | 75 | 90 |
|---|---|---|---|---|---|---|---|
| Incremental Rainfall (in) | 0 | 0.30 | 0.75 | 1.1 | 0.95 | 0.65 | 0.30 |

**6.50** Find the incremental runoff volume in 15-min increments. The curve number of the watershed is 76 and the drainage area is 5 acres.

**6.51** Find the incremental runoff volume in 15-min increments. The curve number of the watershed is 86 and the drainage area is 10 acres.

**6.52** Find the incremental runoff volume in 15-min increments. The curve number of the watershed is 96 and the drainage area is 5 acres.

**6.53** Determine the watershed hydrograph for a time of concentration of 2.0 h, and an area of 0.3 square miles.

**6.54** Determine the watershed hydrograph for a time of concentration of 1.8 h, and an area of 0.6 square miles.

**6.55** Determine the watershed hydrograph for a time of concentration of 2.8 h, and an area of 0.7 square miles.

**6.56** The flow through a concrete trapezoidal channel is 6 in. deep. The trapezoid width is 3 ft, the side slopes are 1:3 and the longitudinal slope is 0.3%. What is the flow rate?

**6.57** The flow through a concrete vee-shaped, 120° angle channel is 6 in. deep. The channel longitudinal slope is 0.3%. What is the flow rate?

**6.58** A design flow of 12 $ft^3/s$ must be conveyed through a firm gravel trapezoidal channel. The trapezoidal side slopes are 1:3 and the channel longitudinal slope is 0.2%. What is the width of the flat swale bottom required to keep the water depth at 8 in.?

**6.59** A design flow of 9.5 $ft^3/s$ must be conveyed through a firm gravel trapezoidal channel. The trapezoidal side slopes are 1:3 and the channel longitudinal slope is 0.8%. What is the width of the flat swale bottom required to keep the water depth at 6 in.?

**6.60** Find the flow through a round orifice with a 3 in. diameter, at a constant head of 4 ft.

**6.61** Find the diameter of a round orifice that will drain 100 $m^3$ over 2 days. The height of the water above the orifice centerline is constant at 1 m.

**6.62** Find the diameter of a round orifice that will drain 100 $m^3$ over 2 days. The height of the water above the orifice centerline is constant at 0.6 m.

**6.63** Find the diameter of a round orifice that will drain 50 $m^3$ over 4 days. The height of the water above the orifice centerline is constant at 1 m.

**6.64** A pond is drained by a 2 cm diameter round orifice. Find the time for the pond to drain from 0.8 to 0.4 m above an orifice centerline. The water storage area is 200 $m^2$.

**6.65** A pond is drained by a 2 cm diameter round orifice. Find the time for the pond to drain from 0.8 to 0.2 m above an orifice centerline. The water storage area is 200 $m^2$.

6.66 Find the diameter of a round orifice that will drain 100 $m^3$ over 2 days. The water storage area is 20 $m^2$ and the height of the water above the orifice centerline fall from 1 m to 0.6 m.

6.67 The water height above a 120° triangular weir is 6 in. Find the flow through the weir.

6.68 The water height above a 120° triangular weir is 8 in. Find the flow through the weir.

6.69 The water height above a 100° triangular weir is 6 in. Find the flow through the weir.

6.70 A flow of 0.5 $ft^3$/s is expected through a 100° triangular weir. What is the expected water height?

6.71 A flow of 0.8 $ft^3$/s is expected through a 125° triangular weir. What is the expected water height?

6.72 A rectangular weir must discharge 3.5 $ft^3$/s at a height of 8 in. What should be the weir width?

# 7

# SCM Hydrologic Unit Processes

## 7.1 Introduction

Building upon our understanding of watershed hydrology from Chapter 6, the focus changes to considering runoff as input to stormwater control measures (SCMs) and how an individual or group of SCMs manage the runoff. The hydrologic unit processes commonly used for SCM analysis and design include soil physics (which determines infiltration and storage), evapotranspiration, and routing of the water through a system. Consider a simple bioinfiltration SCM. Water enters and infiltrates through the surface medium. Routing is needed to determine if and when any overflow occurs. Infiltration and evapotranspiration processes are needed to understand the recovery of the soil/medium water holding capacity. If instead a pervious pavement site is considered, again soil physics considerations and routing are used, but evapotranspiration can be neglected.

Hydrologic unit processes are discussed in this chapter primarily based on their role on the management of flows and water volume in SCMs. However, most of the same processes will impact stormwater production and conveyance over the watershed.

The hydrologic fate of stormwater in an SCM or SCM system is built around a water balance. Incoming stormwater and impinging rainfall enter the SCM. The water can remain at this point and become stored or it can infiltrate into the soil/medium (Figure 7.1). If storage capacity is exceeded, overflow/bypass of the SCM will occur. Infiltration represents a critical pathway for GI SCMs, as exfiltration and removal of stored stormwater into the surrounding soils will allow additional stormwater to enter, and be managed, by the SCM (Figure 7.2). This pathway can often be more impactful than the soil storage. While evapotranspiration is another pathway, it is much more important to GSI following the rainfall event.

While these water balance pathways seem rather simple and straightforward, they, along with the stormwater characteristics, can change significantly with time, place, and of course change in climate. For example, increased temperature will increase infiltration and evapotranspiration. As a result, stormwater spatial and temporal characteristics can be difficult to predict.

*Green Stormwater Infrastructure Fundamentals and Design*, First Edition. Allen P. Davis, William F. Hunt, and Robert G. Traver.

Companion Website: www.wiley.com/go/davis/greenstormwater

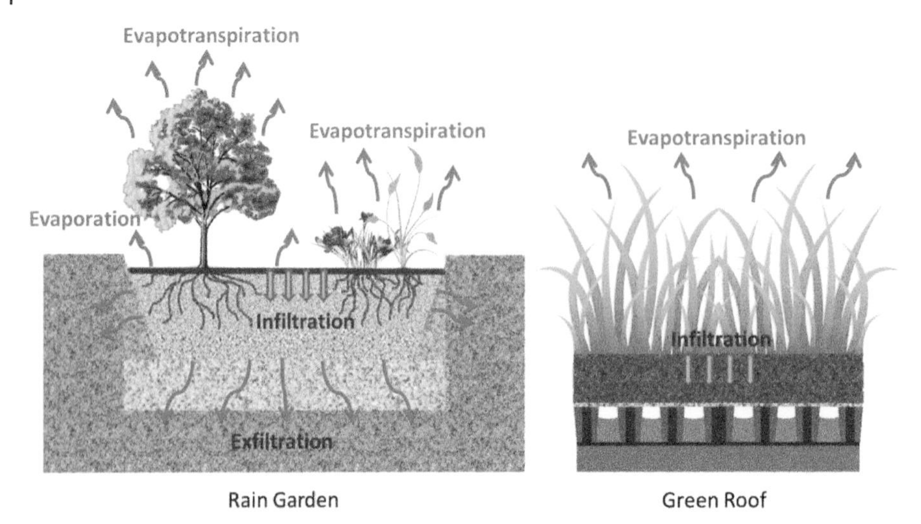


Figure 7.1 Water Balance Pathways in a Rain Garden and on a Green Roof (Ebrahimian et al. 2019).

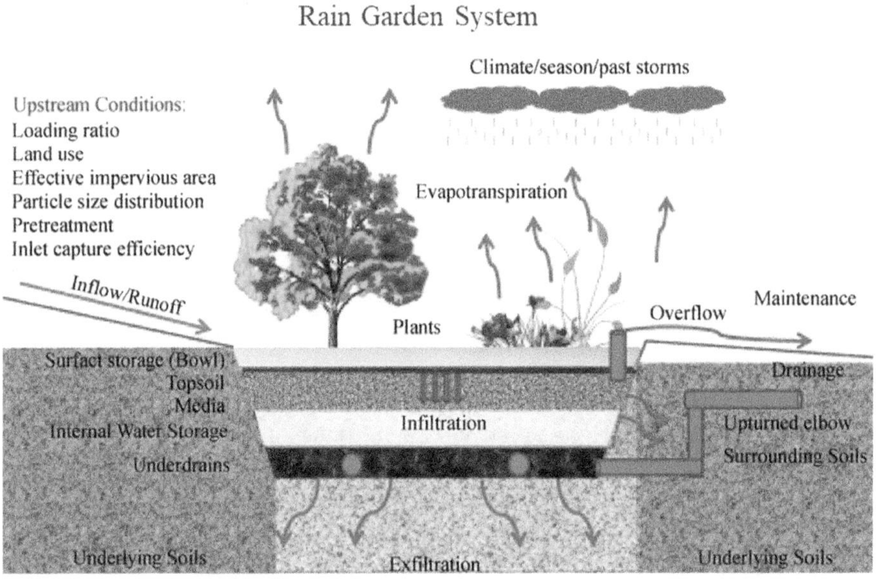

Figure 7.2 Water Balance in a Rain Garden System Showing Exfiltration to Surrounding Soils (Modified from Traver and Ebrahimian 2017).

## 7.2 SCM Soil Physics and Infiltration

Infiltration will occur as the water moves into the soil or into the SCM treatment medium. The rate of infiltration will depend on the characteristics of the soil/medium, the degree of water saturation, depth of surface storage, the occurrence of macropores, and temperature. Water that infiltrates can be retained in the soil for use by vegetation and ultimately returned to the atmosphere by evapotranspiration. Infiltration into local surrounding soils represents a loss in the water balance pathways, reducing the net stormwater volume, transferring it into subsurface water pathways which can lead to deep ground water and/or baseflow of nearby streams. These processes take time, which becomes an important design element.

The *Soil Science Society of America* defines soil as (SSSA 2008):

> (i) The unconsolidated mineral or organic material on the immediate surface of the Earth that serves as a natural medium for the growth of land plants. (ii) The unconsolidated mineral or organic matter on the surface of the Earth that has been subjected to and shows effects of genetic and environmental factors of: climate (including water and temperature effects), and macro- and micro-organisms, conditioned by relief, acting on parent material over a period of time. A product-soil differs from the material from which it is derived in many physical, chemical, biological, and morphological properties and characteristics.

Soils are generally considered to be natural, in situ materials. "Engineered" soil can be constructed and/or modified to perform a desired SCM function and is typically termed "media" or "soil media."

Soils and soil processes are key elements in many aspects of urban hydrology. Soils and soil-based media are also critical components in the design and operation of most GI SCMs. Soils provide storage, infiltration, a structural foundation, and a nutrient reservoir for vegetation, and support for biogeochemical processes, among other things. Physical, chemical, and biological characteristics of soils are also important with respect to water quality management and are discussed in Chapter 8.

### 7.2.1 Soil Texture

The texture of a soil refers to the size distribution of its various particles. The US Department of Agriculture (USDA) classifies soil texture based on the dispersed size fractions of particle diameters. Larger particles, between 50 and 2000 μm diameter are classified as *sand*. Particles between 2 and 50 μm are classified as *silt*, and *clay* particles have diameter < 2 μm. Particles larger than coarse sand (>2000 μm; 2 mm) are denoted as *gravel* and are not considered as part of the soil.

Twelve soil texture categories are created based on the mix (by mass) of the fractions of sand, silt, and clay. These categories are represented by the Soil Textural Triangle, as shown in Figure 7.3. The names given to the soil textures represent their dominant particle size(s), with a *loam* being a relatively equal (and usually agronomically desired) distribution of particle sizes.

The soil triangle is read by knowing the percentage of sand, silt, and clay. (If two are known, the third is known by difference.). In moving through the triangle, the percentage lines are parallel to the leg of the triangle representing the 0 value for that particle size.

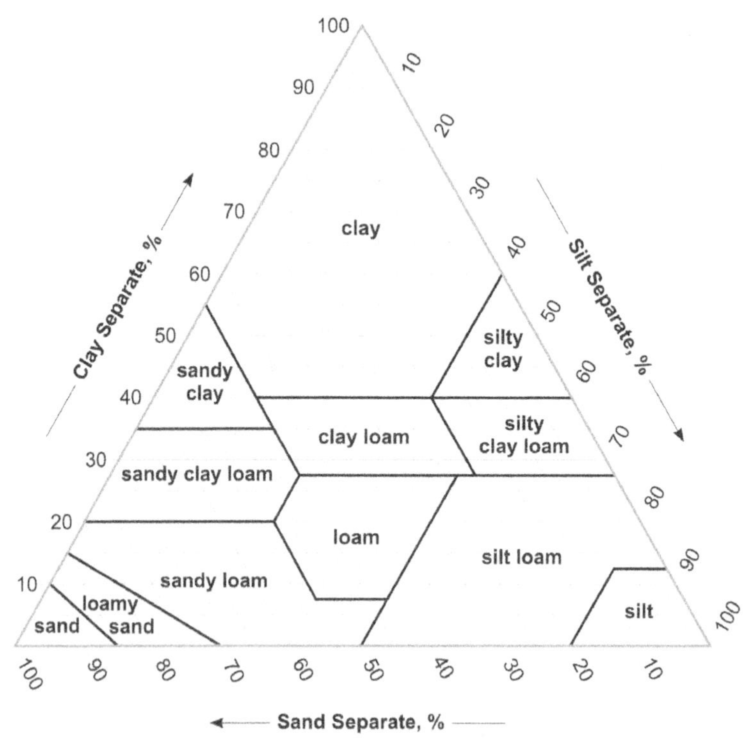


Figure 7.3 Soil Textural Triangle Showing Delineation of the 12 Different Soil Texture Categories.

### 7.2.2 Soil–Water Interactions

Critical to understanding hydrologic impacts throughout the watershed and in GI SMCs are the interactions between soil and water. Soil–water interactions are important in understanding water fate in natural and engineered media emplaced in an SCM. Similar analyses can be applied to the water balance in the landscape. Two soil–water relationships drive the performance and thus the design of GI: water storage characteristics, and the rate of water movement through the soil.

A soil system consists of three phases, the solid soil particles, water, and air. The amount of water and air present in the void space represents the *porosity* of the soil, volume that does not consist of the soil particles. The porosity, ε, is defined as:

$$\varepsilon = V_V / V_T = V_V / (V_V + V_S) \tag{7.1}$$

where V is the volume and the subscripts T, V, and S represent the total, voids, and solid, respectively.

The soil volumetric moisture content indicates the amount of water present in the soil. The soil moisture content can be expressed in percentage of volume or water depth per soil depth. For example, 0.20 $m^3$ of water found in 1 $m^3$ of soil corresponds to a moisture content of 20%. This same situation would correspond to 200 mm of water present in a depth of one meter of soil, resulting in a soil moisture content of 200 mm/m (Figure 7.4). The amount of water stored in a soil will vary depending on water and energy inputs and outputs.

**Example 7.1** Find the USDA texture of a soil that is made up of 65% sand and 25% silt.

*Solution* Knowing only these 2 fractions is necessary to define the soil texture. It is a trivial calculation to see that the clay content is 10%.

The large blue line on the triangle below represents the 65% sand line running parallel to the triangle side at 0% sand; the dashed red line is the 25% silt line (parallel to 0% silt). The intersection classifies this soil as a sandy loam. Note also that the intersection occurs at 10% clay (light green line).

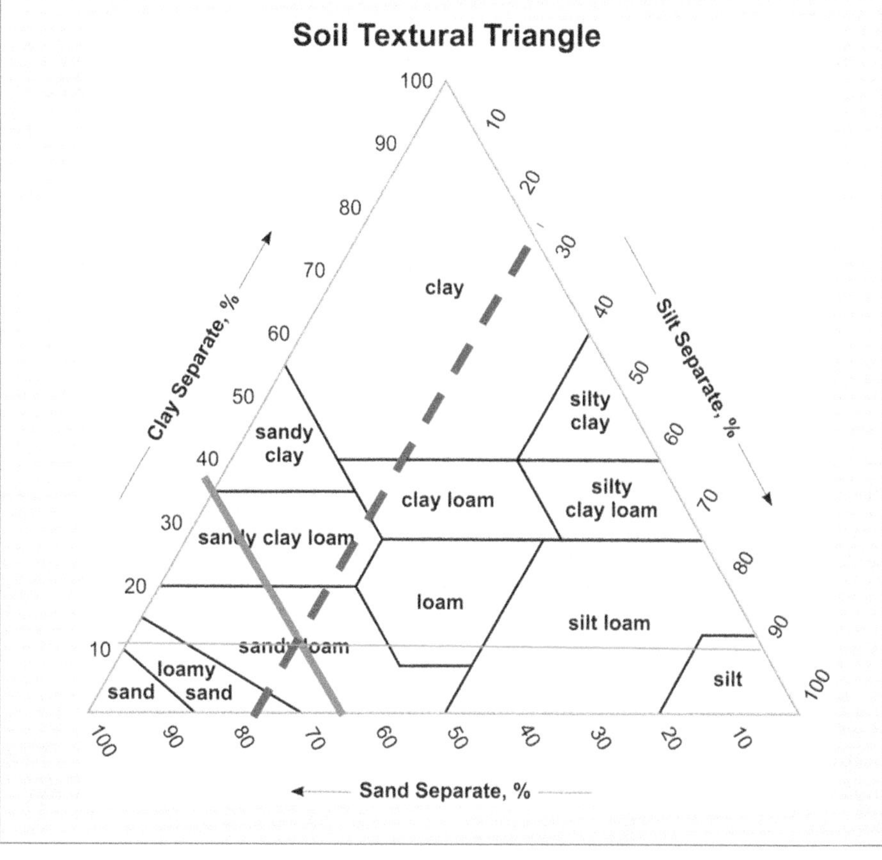


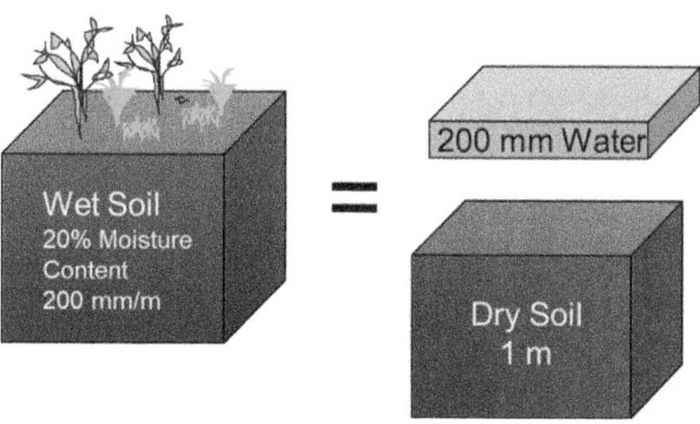


Figure 7.4 A Volumetric Soil Moisture Content of 200 mm/m, Corresponding to 20% Moisture Content.

During a water input (rain, runoff, irrigation), the soil pores will fill with water, displacing air in the voids. If all soil pores are filled with water so there is no air remaining, the soil is said to have reached *saturation* or to be saturated. (Figure 7.5a). Therefore, theoretically, the maximum volume of water that can be held by a soil is the volume of the void space, equal to the total volume times the porosity (Eq. 7.1). When the soil system reaches saturation, infiltration will occur at a rate controlled by the underlying soil.

In natural systems, the force of gravity will pull water from the soil system. This process is called drainage or percolation. The water drained from the pores is replaced by air, or filled with water infiltrating from the surface. In coarse-textured (sandy) soils, drainage is completed within a period of a few hours after the cessation or rainfall. In fine-textured (clayey) soils, drainage may take some (2–3) days.

During gravity drainage, not all of the water is removed from the soil. Some of the water is held by capillary and other forces to the soil matrix. The large soil pores will remain filled with both air and water while the smaller pores remain full of water. At this stage, the soil is said to be at *Field Capacity* (FC) (Figure 7.5b). The field capacity is the lower limit of water content (% by volume) of the soil matrix that has been allowed to freely drain by gravity. Scientifically, field capacity is defined as the amount of water remaining after the soil has been exposed to a hydraulic tension of 33 kPa (0.33 Bar).

Water that will not gravity drain is still available for uptake by plants, or it can evaporate. The dryer the soil becomes, the more tightly the remaining water is retained and the more difficult it is for the plant roots to extract it. Some of the water will be strongly held to the soil as it is occluded within soil pores and may be chemically bound to soil and organic matter. At a certain stage, the uptake of water is not sufficient to meet the needs of the plant. The soil water content at the stage is called the permanent *Wilting Point* (WP) (Figure 7.5c). The soil still contains some water, but it is too difficult for the roots to pull it from the soil. Different plant species have different abilities to extract water, so a physical measurement is used to define this value. Scientifically, the wilting point is defined as the water remaining in the soil after exposure to a hydraulic tension of 1500 kPa (15 Bar).

The values of field capacity and wilting point are related to soil particle size distributions and soil texture. Clayey soils will bind water tightly and the wilting point will occur even at relatively high volumetric water contents. Sandy soils will readily drain most of the soil water and both field capacity and wilting point will occur at low water contents (Table 7.1).

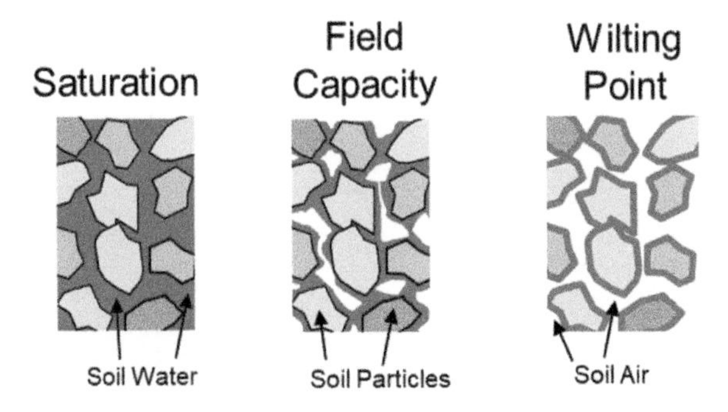


Figure 7.5 Diagrams of Water Storage in Soil Systems.

Table 7.1 Wilting Point (WP), Field Capacity (FC), and Saturation (Sat) Percentages for Various Soil Textures

| Texture Class | WP | FC | Sat |
|---|---|---|---|
| Sand | 5.0 | 9.4 | 46.3 |
| Loamy sand | 5.7 | 12.1 | 45.7 |
| Silt | 6.3 | 31.6 | 48.2 |
| Sandy loam | 8.1 | 17.9 | 45.0 |
| Loam | 12.6 | 26.7 | 45.8 |
| Silty loam | 13.7 | 32.1 | 48.2 |
| Sandy clay loam | 18.3 | 28.3 | 43.2 |
| Silty clay loam | 21.0 | 37.9 | 51.0 |
| Clay loam | 21.3 | 35.0 | 47.2 |
| Sandy clay | 26.0 | 37.1 | 44.0 |
| Silty clay | 27.8 | 41.6 | 53.2 |
| Clay | 29.9 | 42.0 | 48.8 |

At field capacity, the water and air contents of the soil are considered to be ideal for crop growth. The *Plant Available Water* (PAW) is defined as the difference between field capacity (gravity drainage) and wilting point, as the two limits in water holding capacity.

$$\text{PAW} = (\text{FC} - \text{WP})(\text{Medium Volume}) \tag{7.2}$$

As will be discussed in subsequent chapters, the PAW represents the practical volumetric storage available in an underdrained media-based SCM (such as a green roof or bioretention system). The PAW represents the maximum media storage volume during a precipitation event.

Plants employed in SCMs should be carefully selected to match the soil/media and moisture conditions. The media can be considered as the water reservoir for the plants. When the media is saturated, the reservoir is full. Most plants need both air and water in the soil; at saturation, no air is present, and the plants will suffer. Conversely, the lack of water in the media between events can also stress the vegetation. A sand-based media has the benefit of a higher infiltration rate and volume but will adversely affect the health of most vegetation, limiting the development of roots and macropores needed to maintain SCM performance over time. Plant health is important for maintenance of infiltration as the root systems develop, promote, and maintain infiltration pathways.

It should be recognized that the period of media saturation usually does not last long. Most precipitation falls in storms that are smaller than the design rainfall and immediately after the stormwater inflow has stopped, water present in the larger pores will move downward due to gravity drainage. This will soon be followed by evapotranspiration during dry weather. Taken together, these water balance pathways create powerful hydrologic restorative forces in the SCM after a storm event.

**Example 7.2** Find the Plant Available Water volume for a green roof. The roof medium has a texture of a loamy sand soil and is 10 cm deep. The roof dimensions are 20 m by 10 m.

*Solution* From Table 7.1, the field capacity and wilting points for a loamy sand soil are 12.1% and 5.7%, respectively. These are percentages that must be converted to fractions. Therefore, using Eq. 7.2:

$$\text{PAW} = (0.121 - 0.057)(20\text{ m})(10\text{ m})(0.1$$

$$\text{PAW} = 1.28\text{ m}^3 = 1280\text{ l}$$

### 7.2.3 Soil Hydraulic Properties

The rate of movement of water through a soil or soil-based medium is a critical parameter in understanding both runoff from the watershed and water management performance in a media-based SCM. Water moves through soils under the influence of energy gradients and forces that act on the water in the soils. When the soils are saturated, gravity is the dominant force. In unsaturated systems, capillary and water/adsorption forces dominate.

The impact of the external forces on the water velocity depends on the characteristics of the soil. The *hydraulic conductivity*, K, describes the capability of water to move within the soil matrix driven by matrix and gravitational potentials, dependent on soil texture (particle and pore size distribution) and moisture content. In general, water will easily move through coarse soils, which have high values of K. The units of hydraulic conductivity are length per time, therefore, cm/s; mm/hr; in/hr; etc. Hydraulic conductivity relates the speed of transmission of water from the surface environment into the soil/media, and then through the media.

Under saturated flow conditions, the movement of water through the soil occurs when all void spaces are full, with gravity providing the energy gradient via a water head. Saturated flow is given by Darcy's Law

$$v = K_{sat}\frac{dh}{dl} \tag{7.3}$$

where $v$ is the infiltrating velocity (not the velocity through the soil pores) and $dh/dl$ is the hydraulic gradient. In the hydraulic gradient, $h$ represents the water head between the points of interest and $l$ is the length (height) of the soil through which the flow is occurring. In this case, $K_{sat}$ is the *saturated hydraulic conductivity* of the soil. Because unsaturated flow is more rapid than saturated flow, saturated flow represents the minimum infiltration/percolation rate. Saturated hydraulic conductivity varies over several orders of magnitude for different soil textures. $K_{sat}$ for fine soils are orders of magnitude lower than coarse soils. Table 7.2 shows approximate saturated hydraulic conductivities for various soil textures.

Table 7.2 Saturated Hydraulic Conductivities of Various Soil Textures (Clapp and Hornberger 1978).

| Soil Texture | $K_s$ (cm/min) | Soil Texture | $K_s$ (cm/min) |
|---|---|---|---|
| Sand | 1.6 | Silty clay loam | 0.010 |
| Loamy sand | 0.94 | Clay loam | 0.015 |
| Sandy loam | 0.21 | Sandy clay | 0.013 |
| Silt loam | 0.043 | Silty clay | 0.0062 |
| Loam | 0.042 | Clay | 0.0077 |
| Sandy clay loam | 0.038 | | |

**Example 7.3** Estimate the infiltration rate of a media-based SCM under saturated conditions. The medium has characteristics of a loamy sand soil. The media depth is 2 ft, with an underdrain. The water is pooled 4 in. above the medium surface.

***Solution*** From Table 7.2, the saturated hydraulic conductivity of a loamy sand soil is approximately 0.94 cm/min. The hydraulic head is the sum of the water pool and the media depth since it is saturated. Equation 7.3 can be used assuming that the underdrain does not create a preferential flow path:

$$v = \left(0.94\frac{cm}{min}\right)\frac{\left(2+\frac{4}{12}ft\right)}{2\ ft} = 1.1\frac{cm}{min} = 25.9in/hr$$

When the soil is unsaturated, the *matric suction* draws the water in. Matric suction is the suction exerted by the soil material (matrix) that induces water to flow in unsaturated soil. It is a force that results from the combined effects of adsorption and capillarity due to the soil matrix. Water flows from a soil with low matric suction (a wet soil) to soil with a high matric suction (a dry soil).

As the water content rises, gravitational forces become dominant and the flow begins to reach saturated flow. The *Soil Water Characteristic Curve* (SWCC) shows the relationship between soil moisture content and the matric suction (Figure 7.6). When the volumetric water content is high, the matrix suction is low. Field capacity is reached as the curve sharply decreases downward. The vertical transition of the curve at low water content corresponds to the wilting point, at which very high forces are necessary to further reduce the water content. The SWCC is a function of the soil pore size distribution, and several curve fitting models such as that of van Genuchten are available for development of this curve.

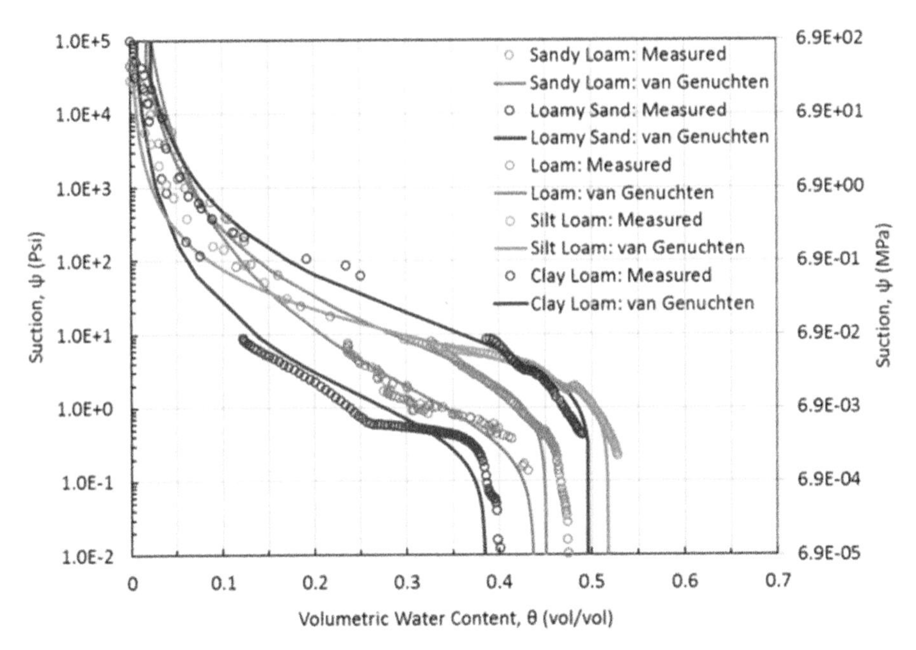


Figure 7.6 Typical Soil Water Characteristic Curves for Different Soils (Hess 2017).

The Richards Equation represents the movement of water under unsaturated conditions (Richards 1931).

$$\frac{\partial \theta}{\partial t} = \frac{\partial}{\partial z}\left[K(\theta)\left(\frac{\partial \psi}{\partial z}+1\right)\right] \tag{7.4}$$

where $K$ is the soil hydraulic conductivity, $\psi$ is the matrix suction, and $z$ is the elevation of the water above a vertical datum; $\theta$ is the water content and $t$ is time. Note that both the soil matrix suction and hydraulic conductivity are functions of soil moisture. This equation is nonlinear and does not have a closed-form solution. Other methods are routinely used to approximate solutions to Eq. 7.4, such as the Green and Ampt, which is considered later in the chapter.

Hydraulic conductivity is a function of temperature, due to its dependence on the viscosity of water:

$$K = k\frac{\rho g}{\mu} \tag{7.5}$$

where $K$ is the hydraulic conductivity, $k$ is the intrinsic permeability of the soil, $\rho$ is the density of water, $g$ is gravitational acceleration, and $\mu$ is the water dynamic viscosity. While $k$ and the density of water are relatively independent of temperature, dynamic viscosity is not. As a result, water infiltration rates in various SCMs will be highly seasonal dependent, increasing during warm months and decreasing during cold (Emerson and Traver 2008). In addition to temperature, seasonal effects on infiltration rates in the presence of vegetation have been noted. These changes have been attributed to root growth, which can increase the hydraulic conductivity through creation of macropores in the soils.

Contrary to common belief the soil infiltration rate never goes to zero, which is one of the hydrologic benefits of GSI and not considered within the NRCS curve number

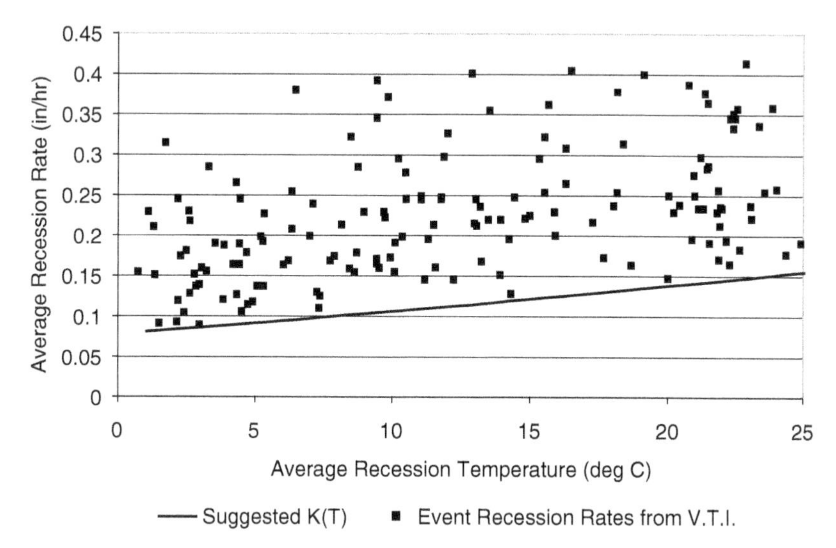


Figure 7.7 Plot of Average Event Recession Rates from the Villanova Traffic Island from 2003 to 2007.

approach. For example, if a soil has a saturated hydraulic conductivity of 0.25 in./h (0.64 cm/h), it would continue to infiltrate this rate (depending on the water head) no matter how long the storm lasted. This is why many infiltration SCMs routinely exceed their design standards, which frequently do not consider infiltration losses. Figure 7.7 shows measured surface recession rate data from the Villanova Bioinfiltration Traffic Island. The solid line represents the lower ranges of the hydraulic conductivity as it varies with temperature, while the spread in the measured infiltration data reflects the variabilities in surface depth of water, initial moisture content, suction matric potential, and temperature.

It is clear that soil/water properties are highly dependent on the soil texture. The USDA — Agricultural Research Service and Washington State University have developed a generalized tool that provides a wealth of soil/water information. The *Soil Water Characteristics Hydraulic Properties Calculator* was developed by Saxton and Rawls and "estimates soil water tension, conductivity and water holding capability based on the soil texture, organic matter, gravel content, salinity, and compaction" (https://hrsl.ba.ars.usda.gov/soilwater/Index.htm).

This tool is used often to estimate the soil water properties needed for GSI Design (Saxton and Rawls 2006). A screen shot of the Calculator is shown in Figure 7.8. Note that at the bottom it produces an SWCC for the general soil type.

### 7.2.4 Green and Ampt Model

The Green and Ampt model is a one-dimensional approximation to the Richards Equation (Eq. 7.4). It can be used for estimates of infiltration in the calculation of both watershed runoff as well as performance of a GSI SCM (e.g., Heasom et al. 2006; Lee et al. 2013). The model assumes that the soil is homogeneous, and that runoff enters through the surface of the unsaturated (Vadose Zone) soil. A linear wetting front is created normal to the flow direction. The soil area above the wetting zone becomes

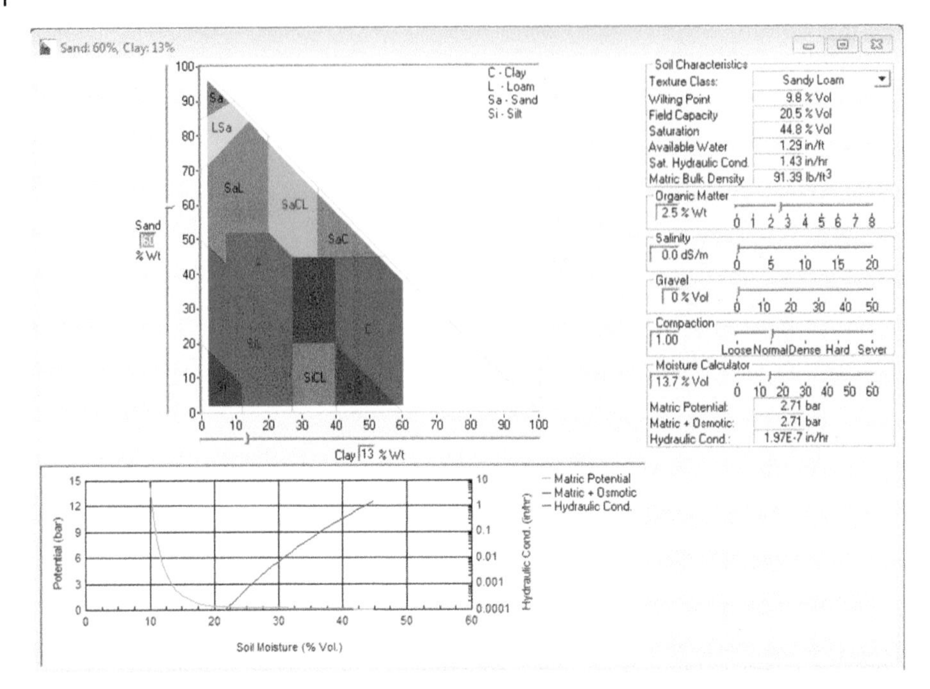


Figure 7.8 Soil Water Characteristics Hydraulic Properties Calculator Screen Shot. (https://hrsl.ba.ars.usda.gov/soilwater/Index.htm).

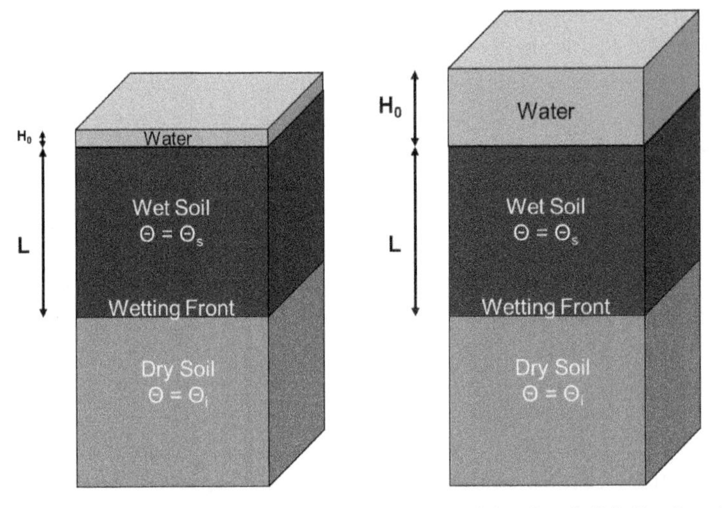


Figure 7.9 Diagram of Wetting Front Proceeding Downward in a Dry Soil Following the Green and Ampt Model. Pooled Water Is Considered Negligible for a Watershed but May Be Important for an SCM.

saturated, and that below is at some initial level, commonly assumed as the wilting point. Thus, the distance L to the wetting point divided by the available void space can be considered the infiltrated water volume (Figure 7.9)

The Green and Ampt model is applied using a backward difference finite difference approach, stepping through time as infiltration is limited to the rainfall or runoff experienced. The length of the wetting front (L) and possible ponding depth ($H_0$) represent the driving force for infiltration, with the inclusion of the soil suction head ($\varphi$).

$$f_p = K_{sat}\left(L + H_0 + \varphi\right) / L \tag{7.6a}$$

$$L = F / \varnothing \tag{7.6b}$$

Equations 7.6a and 7.6b are applied in step progression over the duration of the storm. For these equations, $f_p$ is the rate of potential infiltration and F is the cumulative loss up to the point of analysis; $K_{sat}$ (saturated hydraulic conductivity), the suction head, and $\varnothing$ (available void space) are soil properties. The ponding depth, $H_0$, is usually considered to be negligible for watershed infiltration but is important for many SCM applications where significant pooling is encouraged. The L term is tracked using the filling of the void space, as represented in Eq. 7.6b. For the first step, it usually assumed that the initial rainfall is infiltrated, starting the process. Note that no land cover terms are included, so that the model requires a depression storage term.

**Example 7.4** A bioretention cell is planned to treat stormwater from an asphalt parking area (assume all rainfall becomes runoff). The bioretention surface area is 1/20$^{th}$ that of the drainage area. The bioretention media void fraction is 0.25.the soil suction head is 8 in.,and the saturated hydraulic conductivity is 0.3 in./h. The maximum bowl storage is 15 in. For the rainfall below (given in 15-min increments), evaluate the infiltration, bowl storage depth, and overflow of the bioretention cell.

| Time (h) | Rainfall Intensity (in./h) |
|---|---|
| 0 | |
| 0.25 | 0.65 |
| 0.50 | 0.40 |
| 0.75 | 0.70 |
| 1.00 | 0.40 |
| 1.25 | 0.40 |
| 1.50 | 0.20 |
| 1.75 | 0.10 |
| 2.00 | 1.20 |
| 2.25 | 0.60 |
| 2.50 | 0.10 |
| 2.75 | 0.20 |
| 3.00 | 0.25 |
| 3.25 | 0.10 |

*Solution* The problem is solved incrementally over each 0.25-h time step ($\Delta t$, column 2). Column 4 gives the stormwater depth input to the bioretention cell as the product of the rainfall intensity and the area ratio of 20. In the first iteration at 0.25 h, all inflow (13 in./h) is assumed for potential infiltration ($f_p$ = inflow, column 6). The wetting front depth is calculated as the depth L (from Eq. 7.6b). No water is pooled during the first increment.

In the second timestep, the SCM inflow is 8.0 in./h. In this case $f_b$ is calculated as 0.485 in. from Eq. 7.6a. It is checked that $f_p$ is greater than the SCM inflow (column 5 > column 4) to give the actual $f_p$ in column 6. (In this example, this is always the case.) The change in L is calculated using Eq. 7.6b and added to the L in the previous increment (13.0 + 0.485 = 13.485 in.). Non-infiltrated water will begin to pool, with Δh calculated as the input depth (intensity *x* time) minus the infiltrated depth. The total pool depth is the prior depth + Δh. In a more complete model, when the infiltration is larger than the inflow, negative changes in H would result.

The maximum pool depth is 15 in. Any water above this value is incrementally overflowed.

| Time (h) | Δt (h) | Rainfall Intensity (in./h) | SCM Inflow (in./h) | fp – calculated (in./h) | fp – actual (in./h) | L (in.) | Δh (in.) | $H_0$ (in.) | Overflow Increment (in.) |
|---|---|---|---|---|---|---|---|---|---|
| 0 | | | | | | 0 | | 0 | 0 |
| 0.25 | 0.25 | 0.65 | 13.0 | | 13.00 | 13.00 | | 0.000 | 0.000 |
| 0.50 | 0.25 | 0.40 | 8.0 | 0.485 | 0.485 | 13.49 | 1.879 | 1.879 | 0.000 |
| 0.75 | 0.25 | 0.70 | 14.0 | 0.520 | 0.520 | 14.00 | 3.370 | 5.249 | 0.000 |
| 1.00 | 0.25 | 0.40 | 8.0 | 0.584 | 0.584 | 14.59 | 1.854 | 7.103 | 0.000 |
| 1.25 | 0.25 | 0.40 | 8.0 | 0.611 | 0.611 | 15.20 | 1.847 | 8.950 | 0.000 |
| 1.50 | 0.25 | 0.20 | 4.0 | 0.635 | 0.635 | 15.83 | 0.841 | 9.792 | 0.000 |
| 1.75 | 0.25 | 0.10 | 2.0 | 0.637 | 0.637 | 16.47 | 0.341 | 10.13 | 0.000 |
| 2.00 | 0.25 | 1.20 | 24.0 | 0.630 | 0.630 | 17.10 | 5.842 | 15.00 | 0.975 |
| 2.25 | 0.25 | 0.60 | 12.0 | 0.703 | 0.703 | 17.80 | 2.824 | 15.00 | 2.824 |
| 2.50 | 0.25 | 0.10 | 2.0 | 0.688 | 0.688 | 18.49 | 0.328 | 15.00 | 0.328 |
| 2.75 | 0.25 | 0.20 | 4.0 | 0.673 | 0.673 | 19.17 | 0.832 | 15.00 | 0.832 |
| 3.00 | 0.25 | 0.25 | 5.0 | 0.660 | 0.660 | 19.83 | 1.085 | 15.00 | 1.085 |
| 3.25 | 0.25 | 0.10 | 2.0 | 0.648 | 0.648 | 20.47 | 0.338 | 15.00 | 0.338 |

At the end of the rainfall, the wetting front extends 20.47 in. into the media, the storage pool is at its maximum depth of 15 in., and 1.595 in. of water overflowed (last column incrementally multiplied by the 0.25-h Δt).

### 7.2.5 Karst Areas

In some areas the geology is predominantly *karst*. Karst areas are dominated by rocks that are somewhat water soluble. These minerals include limestone, dolomite, and gypsum. As water infiltrates these rocks, the rocks slowly dissolve. Because of the easy dissolution of the subsurface material, sinkholes are common in these areas. Such areas also tend to have large cracks and crevices that can lead deep into the ground. Confounding this issue is that sealing areas to avoid infiltration, can also cause sinkholes to occur.

As a rule, implementation of infiltration technologies in karst areas should be reviewed by a licensed geologist. Clearly, greatly increasing the water load to karst zones will increase the rock dissolution, leading to greater probability of sinkholes. Karst areas can also have direct conduits to groundwater. As such, poor-quality untreated water from stormwater can lead to groundwater contamination

## 7.3 Evapotranspiration

Evapotranspiration (ET) is the volatilization loss of water to the atmosphere from a combined plant/soil system. Evaporation occurs from the soil surface. Transpiration occurs as water is lost through the stomata of vegetation leaves during the exchange of $CO_2$ and oxygen. Evaporation can occur from ponded/standing water or from moisture in the soil. In most systems it is difficult and not necessary to separate evaporation from transpiration and they are thus combined overall as ET. ET is generally described in terms of depth of water evaporated per unit time, e.g., mm/day or in/day.

ET is a critical element in stormwater management and regulating the performance of SCMs. ET represents one of the few available pathways in which water can be removed from an SCM, regenerating volumetric storage capacity. It also is a key component in continuous simulation models when used to restore available void space. In some SCMs, such as green/blue roofs, it represents the primary or even only pathway of water removal.

Transpiration is generally a more efficient water removal pathway than evaporation. This is one of the major advantages of incorporating vegetation into SCMs. However, it also adds a degree of complexity as water removal via ET can be complex and dependent on a number of vegetation factors that can make it hard to predict.

ET is operative during the dry conditions between rainfall events. In a media-based SCM the amount of ET that occurs after a rainfall event defines the initial soil/media water content, and subsequent stormwater storage, for the next event. The water lost to ET will depend on many variables, including temperature and the days between events.

Many ET models have been developed over the years. Some are relatively simple, requiring only a few predictor parameters. Others are more fundamentally based and complex, requiring many input parameters. An important factor with all ET models is that they predict *ET potential.* That is, the maximum ET that can be expected under the given environmental conditions. This assumes that adequate moisture is available to meet the predicted ET rate. If little or no water is present, obviously no ET can occur.

The Penman-Monteith model (Eq. 7.7) is recommended by the American Society of Civil Engineers (ASCE) as the most complete model to represent agricultural ET, but it is also the most complex. Penman-Monteith requires multiple weather and vegetative parameters, and adjustments for water stress and crop factors (ASCE 2005).

$$ET = \frac{0.0408\Delta(Rn - G) + \left(\gamma \frac{C_n}{T + 273}\right) u_2 (e_s - e_a)}{\Delta + \gamma(1 + Cdu_2)} \quad (7.7)$$

where $ET$ is the potential reference $ET$ [mm/d], $\Delta$ is slope of the vapor pressure curve [kPa/°C], $Rn$ is net solar radiation on the crop surface [MJ/m$^2$/d], $G$ is soil heat flux density [MJ/m$^2$/d], $\gamma$ is the psychometric constant [kPa/°C], $T$ is average daily temperature at 2 m from ground level [°C], $e_s$ is saturation vapor pressure [kPa], and $e_a$ is the actual vapor pressure [kPa]. Crop and water availability adjustments are required. This method is presented here for the reader to understand the complexity interactions of the climate and vegetation required to estimate ET.

Compared to the Penman-Monteith equation, the 1985 Hargreaves equation is a much simpler model (Hargreaves and Allen 2003):

$$ET = 0.0023R_a\left(T + 17.8\right)\left(Tmax - Tmin\right)^{0.5}) \tag{7.8}$$

where $R_a$ is extraterrestrial radiation measured in terms of evaporation rate [mm/day], $T$ is the mean daily temperature at 2 m elevation [°C], and $Tmax–Tmin$ is the difference between the mean daily maximum and minimum temperature [°C]. The $R_a$ factor can be estimated from meteorological data, or from Table 7.3.

**Example 7.5** Using the Hargreaves equation, find the average potential ET for the month of August at 20°N. The average maximum and minimum temperatures are 30.6 and 21.1°C, respectively.

***Solution*** The value for the extraterrestrial radiation at 20°N is found from Table 7.3 as 15.9 mm/day. Using the mean of the maximum and minimum temperatures, the average daily temperature is found as 25.9°C. Using the Hargreaves equation (Eq. 7.8) and substituting the appropriate values:

$$ET = 0.0023\left(15.9\frac{mm}{day}\right)\left(25.9 + 17.8\right)\left(30.6 - 21.1\right)^{0.5}$$

$$= 4.93\ mm/day$$

Since August has 31 days, the total potential ET is:

$$depth = ET\left(time\right) = \left(4.9\frac{mm}{day}\right)\left(31\ days\right) = 153\ mm$$

The simplest ET model is the Blaney-Criddle equation, given as:

$$\text{ET} = \rho\left(0.46\ T_m + 8\right) \tag{7.9}$$

where ET is the potential evapotranspiration in mm/day and $\rho$ is the percentage of a 24-h day that is daylight. $T_m$ is the mean daily temperature, calculated as the average between the daily high and daily low temperature. The Blaney Criddle model is derived as an estimate for ET averaged over a month. Table 7.4 gives values of $\rho$ for various latitudes for the months of the year. Air temperature is the most important factor in controlling ET and is the basis for the Blaney-Criddle model.

Because it is so simple, the Blaney-Criddle model can only be considered an estimate for ET.

Table 7.3 Extraterrestrial Radiation ($R_a$) Expressed in Equivalent Evaporation (in mm/day) (Hargreaves 1994, with Permission from ASCE)

| January (1) | February (2) | March (3) | April (4) | May (5) | June (6) | July (7) | August (8) | September (9) | October (10) | November (11) | December (12) | Latitude (degrees) (13) |
|---|---|---|---|---|---|---|---|---|---|---|---|---|
| (*a*) Northern Hemisphere | | | | | | | | | | | | |
| 3.8 | 6.1 | 9.4 | 12.7 | 15.8 | 17.1 | 16.4 | 14.1 | 10.9 | 7.4 | 4.5 | 3.2 | 50 |
| 4.3 | 6.6 | 9.8 | 13.0 | 15.9 | 17.2 | 16.5 | 14.3 | 11.2 | 7.8 | 5.0 | 3.7 | 48 |
| 4.9 | 7.1 | 10.2 | 13.3 | 16.0 | 17.2 | 16.6 | 14.5 | 11.5 | 8.3 | 5.5 | 4.3 | 46 |
| 5.3 | 7.6 | 10.6 | 13.7 | 16.1 | 17.2 | 16.6 | 14.7 | 11.9 | 8.7 | 6.0 | 4.7 | 44 |
| 5.9 | 8.1 | 11.0 | 14.0 | 16.2 | 17.3 | 16.7 | 15.0 | 12.2 | 9.1 | 6.5 | 5.2 | 42 |
| 6.4 | 8.6 | 11.4 | 14.3 | 16.4 | 17.3 | 16.7 | 15.2 | 12.5 | 9.6 | 7.0 | 5.7 | 40 |
| 6.9 | 9.0 | 11.8 | 14.5 | 16.4 | 17.2 | 16.7 | 15.3 | 12.8 | 10.0 | 7.5 | 6.1 | 38 |
| 7.4 | 9.4 | 12.1 | 14.7 | 16.4 | 17.2 | 16.7 | 15.4 | 13.1 | 10.6 | 8.0 | 6.6 | 36 |
| 7.9 | 9.8 | 12.4 | 14.8 | 16.5 | 17.1 | 16.8 | 15.5 | 13.4 | 10.8 | 8.5 | 7.2 | 34 |
| 8.3 | 10.2 | 12.8 | 15.0 | 16.5 | 17.0 | 16.8 | 15.6 | 13.6 | 11.2 | 9.0 | 7.8 | 32 |
| 8.8 | 10.7 | 13.1 | 15.2 | 16.5 | 17.0 | 16.8 | 15.7 | 13.9 | 11.6 | 9.5 | 8.3 | 30 |
| 9.3 | 11.1 | 13.4 | 15.3 | 16.S | 16.8 | 16.7 | 15.7 | 14.1 | 12.0 | 9.9 | 8.8 | 28 |
| 9.8 | 113 | 13.7 | 15.3 | 16.4 | 16.7 | 16.6 | 15.7 | 14.3 | 12.3 | 10.3 | 9.3 | 26 |
| 10.2 | 11.9 | 13.9 | 15.4 | 16.4 | 16.6 | 16.5 | 15.8 | 14.5 | 12.6 | 10.7 | 9.7 | 24 |
| 10.7 | 12.3 | 14.2 | 15.5 | 16.3 | 16.4 | 16.4 | 15.8 | 14.6 | 13.0 | 11.1 | 10.2 | 22 |
| 11.2 | 12.7 | 14.4 | 15.6 | 16.3 | 16.4 | 16.3 | 15.9 | 14.8 | 13.3 | 11.6 | 10.7 | 20 |
| 11.6 | 13.0 | 14.6 | 15.6 | 16.1 | 16.1 | 16.1 | 15.8 | 14.9 | 13.6 | 12.0 | 11.1 | 18 |
| 12.0 | 13.3 | 14.7 | 15.6 | 16.0 | 15.9 | 15.9 | 1S.7 | 15.0 | 13.9 | 12.4 | 11.6 | 16 |

*(Continued)*

Table 7.3 (Continued)

| January (1) | February (2) | March (3) | April (4) | May (5) | June (6) | July (7) | August (8) | September (9) | October (10) | November (11) | December (12) | Latitude (degrees) (13) |
|---|---|---|---|---|---|---|---|---|---|---|---|---|
| 12.4 | 13.6 | 14.9 | 15.7 | 15.8 | 15.7 | 15.7 | 15.7 | 15.1 | 14.1 | 12.8 | 12.0 | 14 |
| 12.8 | 13.9 | 15.1 | 15.7 | 15.7 | 15.5 | 15.5 | 15.6 | 15.2 | 14.4 | 13.3 | 12.5 | 12 |
| 13.2 | 14.2 | 15.3 | 15.7 | 15.5 | 15.3 | 15.3 | 15.5 | 15.3 | 14.7 | 13.6 | 12.9 | 10 |
| 13.6 | 14.5 | 15.3 | 15.6 | 15.3 | 15.0 | 15.1 | 15.4 | 15.3 | 14.8 | 13.9 | 13.3 | 8 |
| 13.9 | 14.8 | 15.4 | 15.4 | 15.1 | 14.7 | 14.9 | 15.2 | 15.3 | 15.0 | 14.2 | 13.7 | 6 |
| 14.3 | 15.0 | 15.5 | 15.5 | 14.9 | 14.4 | 14.6 | 15.1 | 15.3 | 15.1 | 14.5 | 14.1 | 4 |
| 14.7 | 15.3 | 15.6 | 15.3 | 14.6 | 14.2 | 14.3 | 14.9 | 15.3 | 15.3 | 14.8 | 14.4 | 2 |
| 15.0 | 15.S | 15.7 | 15.3 | 14.4 | 13.9 | 14.1 | 14.8 | 15.3 | 15.4 | 15.1 | 14.8 | 0 |
| (*b*) Southern Hemisphere | | | | | | | | | | | | |
| 17.5 | 14.7 | 10.9 | 7.0 | 4.2 | 3.1 | 3.5 | 5.5 | 8.9 | 12.9 | 16.5 | 18.2 | 50 |
| 17.6 | 14.9 | 11.2 | 7.5 | 4.7 | 3.5 | 4.0 | 60 | 9.3 | 13.2 | 16.6 | 18.2 | 48 |
| 17.7 | 15.1 | 11.5 | 7.9 | 5.2 | 4.0 | 4.4 | 6.5 | 9.7 | 13.4 | 16.7 | 18.3 | 46 |
| 17.8 | 15.3 | 11.9 | 8.4 | 5.7 | 4.4 | 4.9 | 6.9 | 10.2 | 13.7 | 16.7 | 18.3 | 44 |
| 17.8 | 15.5 | 12.2 | 8.8 | 6.1 | 4.9 | 5.4 | 7.4 | 10.6 | 14.0 | 16.8 | 18.3 | 42 |
| 17.9 | 15.7 | 12.5 | 9.2 | 6.6 | 5.3 | 5.9 | 7.9 | 11.0 | 14.2 | 16.9 | 18.3 | 40 |
| 17.9 | 15.8 | 12.8 | 9.6 | 7.1 | 5.8 | 6.3 | 8.3 | 11.4 | 14.4 | 17.0 | 18.3 | 38 |
| 17.9 | 160 | 13.2 | 10.1 | 7.5 | 6.3 | 6.8 | 8.8 | 11.7 | 14.6 | 17.0 | 18.2 | 36 |
| 17.8 | 16.1 | 13.5 | 10.5 | 8.0 | 6.8 | 7.2 | 9.2 | 12.0 | 14.9 | 17.1 | 18.2 | 34 |
| 17.8 | 162 | 13.8 | 10.9 | 8.5 | 7.3 | 7.7 | 9.6 | 12.4 | 15.1 | 17.2 | 18.1 | 32 |
| 17.8 | 16.4 | 14.0 | 11.3 | 8.9 | 7.8 | 8.1 | 10.1 | 12.7 | 15.3 | 17.3 | 18.1 | 30 |

Table 7.3 (Continued)

| January (1) | February (2) | March (3) | April (4) | May (5) | June (6) | July (7) | August (8) | September (9) | October (10) | November (11) | December (12) | Latitude (degrees) (13) |
|---|---|---|---|---|---|---|---|---|---|---|---|---|
| 17.7 | 16.4 | 14.3 | 11.6 | 9.3 | 8.2 | 8.6 | 10.4 | 13.0 | 15.4 | 17.2 | 17.9 | 28 |
| 17.6 | 16.4 | 14.4 | 12.0 | 9.7 | 8.7 | 9.1 | 10.9 | 13.2 | 15.5 | 17.2 | 17.8 | 26 |
| 17.5 | 16.5 | 14.6 | 12.3 | 10.2 | 9.1 | 9.5 | 11.2 | 13.4 | 15.6 | 17.1 | 17.7 | 24 |
| 17.4 | 16.5 | 14.8 | 12.6 | 10.6 | 9.6 | 10.0 | 11.6 | 13.7 | 15.7 | 17.0 | 17.5 | 22 |
| 17.3 | 16.5 | 15.0 | 13.0 | 11.0 | 10.0 | 10.4 | 12.0 | 13.9 | 15.8 | 17.0 | 17.4 | 20 |
| 17.1 | 16.5 | 15.1 | 13.2 | 11.4 | 10.4 | 10.8 | 12.3 | 14.1 | 15.8 | 16.8 | 17.1 | 18 |
| 16.9 | 16.4 | 15.2 | 13.5 | 11.7 | 10.8 | 11.2 | 12.6 | 14.3 | 15.8 | 16.7 | 16.8 | 16 |
| 16.7 | 16.4 | 15.3 | 13.7 | 12.1 | 11.2 | 11.6 | 12.9 | 14.5 | 15.8 | 16.5 | 16.6 | 14 |
| 16.6 | 16.3 | 15.4 | 14.0 | 12.5 | 11.6 | 12.0 | 13.2 | 14.7 | 15.8 | 16.4 | 16.5 | 12 |
| 16.4 | 16.3 | 15.5 | 14.2 | 12.8 | 12.0 | 12.4 | 13.5 | 14.8 | 15.9 | 16.2 | 16.2 | 10 |
| 16.1 | 16.1 | 15.5 | 14.4 | 13.1 | 12.4 | 12.7 | 13.7 | 14.9 | 15.8 | 16.0 | 16.0 | 8 |
| 15.8 | 16.0 | 15.6 | 14.7 | 13.4 | 12.8 | 13.1 | 14.0 | 15.0 | 15.7 | 15.8 | 15.7 | 6 |
| 15.5 | 15.8 | 15.6 | 14.9 | 13.8 | 13.2 | 13.4 | 14.3 | 15.1 | 15.6 | 15.5 | 15.4 | 4 |
| 15.3 | 15.7 | 15.7 | 15.1 | 14.1 | 13.5 | 13.7 | 14.5 | 15.2 | 15.5 | 15.3 | 15.1 | 2 |
| 15.0 | 15.5 | 15.7 | 15.3 | 14.4 | 13.9 | 14.1 | 14.8 | 15.3 | 15.4 | 15.1 | 14.8 | 0 |

Table 7.4 Mean Daily Percentage (p) of Annual Daytime Hours for Different Latitudes

| Latitude | North | Jan | Feb | Mar | Apr | May | June | July | Aug | Sept | Oct | Nov | Dec |
|---|---|---|---|---|---|---|---|---|---|---|---|---|---|
| | South | July | Aug | Sept | Oct | Nov | Dec | Jan | Feb | Mar | Apr | May | June |
| 60° | | 0.15 | 0.20 | 0.26 | 0.32 | 0.38 | 0.41 | 0.40 | 0.34 | 0.28 | 0.22 | 0.17 | 0.13 |
| 55° | | 0.17 | 0.21 | 0.26 | 0.32 | 0.36 | 0.39 | 0.38 | 0.33 | 0.28 | 0.23 | 0.18 | 0.16 |
| 50° | | 0.19 | 0.23 | 0.27 | 0.31 | 0.34 | 0.36 | 0.35 | 0.32 | 0.28 | 0.24 | 0.20 | 0.18 |
| 45° | | 0.20 | 0.23 | 0.27 | 0.30 | 0.34 | 0.35 | 0.34 | 0.32 | 0.28 | 0.24 | 0.21 | 0.20 |
| 40° | | 0.22 | 0.24 | 0.27 | 0.30 | 0.32 | 0.34 | 0.33 | 0.31 | 0.28 | 0.25 | 0.22 | 0.21 |
| 35° | | 0.23 | 0.25 | 0.27 | 0.29 | 0.31 | 0.32 | 0.32 | 0.30 | 0.28 | 0.25 | 0.23 | 0.22 |
| 30° | | 0.24 | 0.25 | 0.27 | 0.29 | 0.31 | 0.32 | 0.31 | 0.30 | 0.28 | 0.26 | 0.24 | 0.23 |
| 25° | | 0.24 | 0.26 | 0.27 | 0.29 | 0.30 | 0.31 | 0.31 | 0.29 | 0.28 | 0.26 | 0.25 | 0.24 |
| 20° | | 0.25 | 0.26 | 0.27 | 0.28 | 0.29 | 0.30 | 0.30 | 0.29 | 0.28 | 0.26 | 0.25 | 0.25 |
| 15° | | 0.26 | 0.26 | 0.27 | 0.28 | 0.29 | 0.29 | 0.29 | 0.28 | 0.28 | 0.27 | 0.26 | 0.25 |
| 10° | | 0.26 | 0.27 | 0.27 | 0.28 | 0.28 | 0.29 | 0.29 | 0.28 | 0.28 | 0.27 | 0.26 | 0.26 |
| 5° | | 0.27 | 0.27 | 0.27 | 0.28 | 0.28 | 0.28 | 0.28 | 0.28 | 0.28 | 0.27 | 0.27 | 0.27 |
| 0° | | 0.27 | 0.27 | 0.27 | 0.27 | 0.27 | 0.27 | 0.27 | 0.27 | 0.27 | 0.27 | 0.27 | 0.27 |

**Example 7.6** Estimate the available ET from a green roof located in New York City in July using the Blaney-Criddle equation. The maximum and minimum temperatures are 30°C and 22°C, respectively.

*Solution* From the temperatures given, $T_m$ is calculated as 26°C. Using a map, such as Google Maps, the latitude of New York is just over 40°N, therefore from Table 7.4 ρ for July is 0.33. Using Eq. 7.9:

$$ET = 0.33(0.46(26+8)) = 5.2 \text{ mm/day}$$

## 7.4 Soil Moisture Accounting

To interrelate the rainfall, infiltration, evapotranspiration, and soil–water unit processes requires a soil moisture accounting approach, tracking the movement of the water through the soil vadose zone over the duration of the rain event. This accounting is incorporated in many stormwater hydrologic models, including SWMM, HEC-HMS, and Recarga. The accounting is applied as a series of incremental time steps through spatial elements or cells, tracking the movement of the water through the system, changing the water storage in each cell as it infiltrates or evaporates using conservation of mass. Most models use similar accounting methods; however, some of the assumptions used in the models are different, usually to reduce computational effort. Figure 7.10 shows how the system is partitioned into several layers, using SWMM and Recarga as examples.

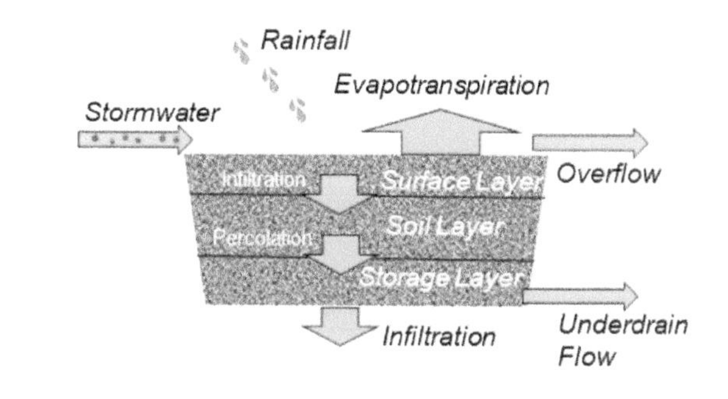


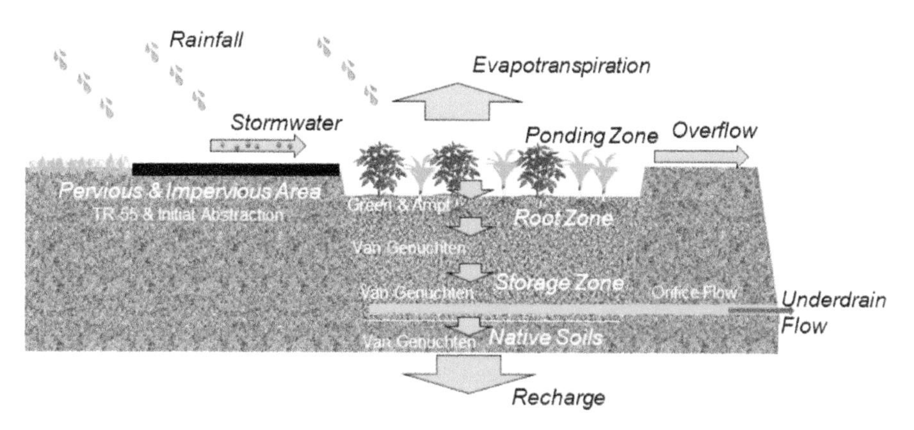


Figure 7.10 Soil Moisture Accounting Diagrams, SWMM LID (Top) and Recarga (Bottom).

The surface layer consists of the surface bowl storage, receiving runoff from the watershed and direct rainfall. Depth of storage is normally tracked, and usually the Green and Ampt equation is used to estimate the infiltration over the time step. Overflow is simply calculated when the inflow exceeds the available storage volume and infiltration capacity. Some models will apply ET losses to the surface bowl.

The soil layer is either modeled as a linear reservoir, using a simplified percolation (Darcy) approach to water movement, or by a more computationally intensive use of the Soil Water Characteristic Curve discussed earlier. Some models allow multiple soil layers. Inflow enters from the surface layer Green and Ampt application, with outflow "percolation" moving to the next soil layer governed by Darcy flow. When no ponding is observed, some models will include water loss via ET in the soil layers down to the root depth. The advantage of the SWCC approach is that soil properties such as suction head can be related to the available void space.

The storage layer refers to a gravel or sand storage area, normally where an underdrain would be placed if used. Inflow to the storage layer occurs from the upper soil layer; discharge will occur via recharge to the in situ soils, or through an underdrain characterized by an orifice representation.

## 7.5 Storage Indication Routing

Storage routing is used to determine the impact of a storage basin on a flow stream. The storage will fill and discharge. The discharge will be a function of the depth of the storage as the water head will drive the discharge rate.

The change in storage (S) as a function of time is the difference between the input flow (I) and output flow (O), rigorously presented as:

$$\frac{dS(t)}{dt} = I(t) - O(t) \tag{7.10a}$$

For use, this equation is integrated, then solved for incremental steps from time $t_i$ to $t_{i+1}$ over the time of interest.

$$\frac{2S(t_{i+1})}{\Delta t} + O(t_{i+1}) = \left[I(t_{i+1}) + I(t_i)\right] + \left[\frac{2S(t_i)}{\Delta t} - O(t_i)\right] \tag{7.10b}$$

To use this expression, the change in the storage depth as a function of the storage volume must be known. This relationship is usually complex and will depend on the contours and topography of the storage landscape. The outflow discharge will depend on the height of the storage, as flow is commonly controlled by a weir, orifice, or some combination.

Routing is used in stormwater wetlands and routing examples are provided in Chapter 20.

## 7.6 Computer-Based Stormwater Models

Several computer-based models have been developed over the past decades for use in evaluating stormwater production and the predicted impacts of SCMs. The models have become more sophisticated over the years as mechanistic understanding of stormwater hydrology and water quality has advanced and greater computing power has become available. A few of the most common models are briefly described below. No particular model is endorsed here; the reader is advised to become educated on the applications, benefits, and limitations of each model.

The NRCS curve number procedure for runoff estimation has been compiled by the USDA into a process document entitled *Urban Hydrology for Small Watersheds, TR (Technical Report) 55* (USDA 1986). This process has been coded into software computerized by USDA (*WinTR-55)* and also interfaced via several commercial firms. It is widely used throughout the United States to examine small urbanizing watersheds.

*SWMM* (Storm Water Management Model) was developed by the US EPA in 1971 and has undergone four major version upgrades, currently at version 5. It has been widely used for evaluation and planning in stormwater management. It includes components for precipitation and contaminant buildup, runoff and contaminant washoff, infiltration, conveyance and storage, and LID controls and stormwater management. SWMM can analyze design storms or continuous simulation. A review of SWMM has been provided by Niazi et al. (2017).

*WinSLAM* (Source Loading and Management Model for Windows) was developed in the 1970s to estimate stormwater pollutant loads generated in urban drainage areas. It is based on small-scale hydrology and routing. *WinSLAM* considers pollutant source loads for different urban areal components (e.g., roofs, pavements) and urban land uses. It allows the incorporation of typical structural SCMs along with some nonstructural SCMs such as street sweeping. *WinSLAM* has been updated through the years and has continuous simulation capabilities. It will not work for non-urban land uses, such as agricultural uses or to evaluate predevelopment conditions.

*DRAINMOD* has been utilized specifically for modeling bioretention system performance. This program was originally developed for evaluating water flow and quality in agricultural tile drainage systems. A number of specific examples have been presented showing the efficacy of this model with bioretention (e.g., Brown et al. 2013).

*RECARGA* is a model developed for bioretention, raingardens, and infiltration basins. It allows surface ponding and several different layers of media. Runoff prediction is based on the NRCS curve number process. SCM hydrologic performance mechanisms include infiltration, overflow, underdrain flow, and ET. *RECARGA* uses the Green and Ampt model for describing the infiltration, a surface water balance, the van Genuchten equation for drainage, and orifice flow through the underdrain. The model can evaluate single events or continuous simulation.

## References

ASCE. (2005) *The ASCE Standardized Reference Evapotranspiration Equation*, Allen, R.G, Walter, I.A., Elliott, R., Howell, T., Itenfisu, D., and Jensen,M., Eds. Alexandria, VA: American Society of Civil Engineers.

Brown, R.A., Skaggs, R.W., and Hunt, W.F. (2013). Calibration and validation of DRAINMOD to model bioretention hydrology. *Journal of Hydrology* 486: 430–442.

Clapp, R.B. and Hornberger, G.M. (1978). Empirical equations for some soil hydraulic properties. *Water Resources Research* 14 (4): 601–604.

Ebrahimian, A., Wadzuk, B., and Traver, R. (2019). Evapotranspiration in green stormwater infrastructure systems. *Science of the Total Environment* 688: 797–810.

Emerson, C.H. and Traver, R.G. (2008). Multiyear and seasonal variation of infiltration from storm-water best management practices. *Journal of Irrigation and Drainage Engineering-Asce* 134 (5): 598–605.

Hargreaves, G.H. (1994). Defining and using reference evapotranspiration. *Journal of Irrigation and Drainage Engineering* 120 (6): 1132–1139.

Hargreaves, G.H. and Allen, R.G. (2003). History and evaluation of hargreaves evapotranspiration equation. *Journal of Irrigation and Drainage Engineering* 129 (1): 53–63.

Heasom, W., Traver, R.G., and Welker, A. (2006). Hydrologic modeling of a bioinfiltration best management practice. *Journal of the American Water Resources Association* 42 (5): 1329–1347.

Hess, A.J. (2017). *Rain Garden Evapotranspiration Accounting*. Villanova University: Doctorate of Philosophy Dissertation.

Lee, R.S., Traver, R.G., and Welker, A.L. (2013). Continuous modeling of bioinfiltration storm-water control measures using green and ampt. *Journal of Irrigation and Drainage Engineering* 139 (12): 1004–1010.

Niazi, M., Nietch, C., Maghrebi, M., Jackson, N., Bennett, B.R., Tryby, M., and Massoudieh, A. (2017). Storm water management model: Performance review and gap analysis. *Journal of Sustainable Water in the Built Environment* 3 (2): 04017002.

Richards, L.A. (1931). Capillary conduction of liquids through porous mediums. *Physics* 1 (5): 318–333.

Saxton, K.E. and Rawls, W.J. (2006). Soil water characteristic estimates by texture and organic matter for hydrologic solutions. *Soil Science Society of America Journal* 70 (5): 1569–1578.

SSSA-Soil Science Society of America. (2008). *Glossary of Soil Science Terms 2008*. Madison, WI: Soil Science Society of America.

Traver, R.G. and Ebrahimian, A. (2017). Dynamic design of green stormwater infrastructure. *Frontiers of Environmental Science & Engineering* 11 (4): 15.

USDA-United States Department of Agriculture. (1986). *Urban Hydrology for Small Watershed, TR-55*. Natural Resources Conservation Service, Conservation Engineering Division. Washington DC.

## Problems

**7.1** Find the soil texture for a soil that is 75% sand and 10% clay.

**7.2** Find the soil texture for a soil that is 65% sand and 25% clay.

**7.3** Find the soil texture for a soil that is 35% sand and 35% clay.

**7.4** Find the soil texture for a soil that is 25% sand and 10% silt.

**7.5** Find the Plant Available Water volume for an SCM media with texture of a sand. The media dimensions are 24 m by 4 m by 0.3 m deep.

**7.6** Find the Plant Available Water volume for an SCM media with texture of sandy loam. The media dimensions are 14 m by 8 m by 0.2 m deep.

**7.7** Estimate the infiltration rate of a media-based SCM under saturated conditions. The medium has characteristics of a sandy loam soil. The media depth is 1 m, with an underdrain. The water is pooled 4 cm above the medium surface.

7.8 An SCM infiltrates at 10 in./h when the water pool is 6 in. above medium surface. The medium depth is 2.5 ft with an underdrain. Estimate the saturated hydraulic conductivity of the medium.

7.9 Estimate the potential ET for your location in July using the Hargreaves model.

7.10 Estimate the potential ET for your location in July using the Blaney-Criddle model.

7.11 Estimate the potential ET for your location in April using the Hargreaves model.

7.12 Estimate the potential ET for your location in April using the Blaney-Criddle model.

7.13 Estimate the average potential ET in July using the Hargreaves model for the following cities:

a. Boston, MA
b. Charlotte, NC
c. Minneapolis, MN
d. Austin, TX
e. Los Angeles, CA
f. Seattle, WA
g. Paris, France
h. Beijing, China
i. Sydney, Australia

# 8

# Unit Processes for Stormwater Quality Mitigation

## 8.1 Introduction

Improvements in water quality will occur through Stormwater Control Measures (SCMs) via one or more unit processes or operations. These unit processes act upon a specific property of the pollutant, allowing it to be removed from the water. These unit processes are implemented specifically as part of the design of the SCM or occur as part of another operational component of the SCM. In general, most SCMs use the same treatment unit processes as do water treatment and wastewater treatment. These unit processes have been studied for a many years and much is known about how they operate and how they can be optimized. Additionally, since many of the nature-based GI SCMs include soils and vegetation, the basic science behind pollutant removal by these components has been developed by agricultural researchers, and much can be inferred from this research. Taken together, a large amount of information can be extrapolated about the physical, chemical, and biological behavior of pollutants in GI SCMs by considering the soil and vegetation science, and comparing them to water/wastewater treatment facilities.

Most water and wastewater treatment plants, in general, operate under steady-state conditions, that is, a relatively constant flow rate. This clearly is not true for SCMs, as the flow will respond to the significant variability of rainfall and runoff, both during an event and from event to event. This greatly complicates the performance analysis of the unit processes inherent to the SCM. Nonetheless, use of steady-state treatment equations can be helpful in understanding and analyzing SCM performance.

Water-quality unit processes common to SCMs include sedimentation, filtration, adsorption, and several biological processes. The biological processes can act upon hydrocarbons, nitrogen, and other bio-active pollutants; additionally, the fate of microorganisms themselves in SCMs is an important consideration to water quality. Generally, these unit processes can be broadly categorized as either separation or transformation processes. With separation, the pollutant is removed from the water but becomes accumulated in the treatment process. From the perspective of the SCM, this can lead to long-term concerns due to pollutant accumulation, associated maintenance requirements, and estimations of facility lifetimes. Transformation processes will change the pollutant into a different form. The transformed species may also be considered a pollutant or be benign. This form may also accumulate or be released.

*Green Stormwater Infrastructure Fundamentals and Design*, First Edition. Allen P. Davis, William F. Hunt, and Robert G. Traver.

Companion Website: www.wiley.com/go/davis/greenstormwater

Some transformations may be slow and occur within the SMC in the time between storm events. Heat transfer can be considered as a unit process as well. The unit processes and some of the engineering and science behind the effectiveness of these processes are discussed in this chapter. Some SCMs will include many unit processes, while others may only include one or two.

Because of the importance of particulate matter as a stormwater pollutant, measured as total suspended solids (TSS) or suspended sediment concentration (SSC) (see Chapter 3), most SCMs include some unit process for particulate matter removal. For SCMs that do not target particulate matter, it is common that particles will clog or be otherwise detrimental to their performance. In general, most particulate pollutants can be easily removed from runoff, especially larger particulate matter. Very small particles and dissolved pollutants can be much more difficult to remove.

## 8.2 Reactions, Reactors, and Reactor Engineering

Mass balance tools are commonly used to evaluate the water-quality unit processes that take place in SCMs. With a mass balance analysis, the effect of a physical, chemical, or biological process occurring in a control volume (such as a SCM) can be quantified. Detailed mass balance development for environmental reactor systems can be found in most introductory environmental engineering texts (e.g., Davis and Masten 2004; Mihelcic and Zimmerman 2014; Vesilind et al. 2010). An overview is provided here to support the use of mass balance tools in stormwater quality improvement.

A mass balance considers the fates of pollutants or other substances of interest in the control volume. Pollutant mass can enter the control volume with the incoming water and leave with the effluent. Pollutants can also be removed from or transformed (or added) in the control volume. For stormwater purposes, the control volume is the SCM (or some part of the SCM).

For chemical reactions, or processes that can be modeled similar to chemical reactions, the removal of the pollutant is generally based on the some dependency on the concentration of chemical in the control volume, $C$. The reaction rate, $r$, is given as

$$r = kC^x \tag{8.1}$$

where $k$ is known as the reaction rate constant and $x$ is the reactor order. In many cases, the reactor order is an integer and is commonly equal to 1. This is known as a first-order reaction. Other common orders are the zero order ($x = 0$) and second order ($x = 2$). The reaction order depends on the characteristics of the reaction taking place. The concentration of the pollutant will change in a control volume due to the various reactions that occur and spatially due to mixing (or lack of) in the reactor. Since reaction rates commonly depend on the concentration of the reactant, the reactor mixing will affect the performance of the reactor.

Two ideal mixing regimes exist for treatment reactors. In one ideal case, the reactor is assumed to be completely mixed. In this case, the concentration of the substance of interest is assumed to be the same at every spatial point in the reactor. This is known as a completely mixed flow reactor (CMFR). At the opposite extreme, a plug flow reactor (PFR) is assumed to have no mixing and a spatial concentration gradient usually exists (Figure 8.1).

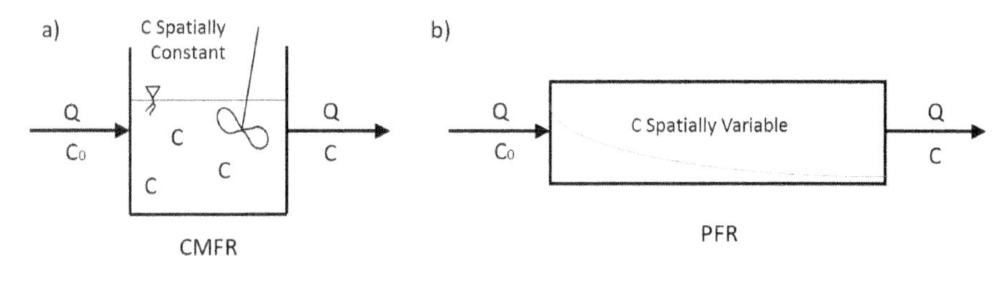


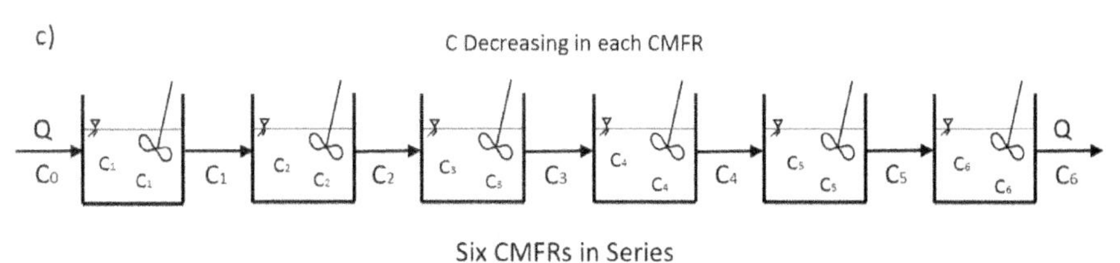


Figure 8.1 Ideal Treatment Reactors: (A) Completely Mixed Flow Reactor, (B) Plug Flow Reactor, and (C) Six CMFRs in Series.

A third consideration in reactor analysis is how the operating conditions may change with time. Variations with time add a high degree of complexity to the mass balance analysis, yet most stormwater processes operate under non-steady-state conditions. Nonetheless, for simple system analyses steady-state assumptions can provide insight to the operation of SCMs.

With consideration of all these factors, simple design/analysis equations are available for evaluating performance of various systems, including stormwater systems. For example, for a first-order reaction in a steady-state CMFR, the resulting equation is

$$C = C_0 \left( \frac{1}{1 + \frac{k_1 V}{Q}} \right) \tag{8.2}$$

where $C$ is the concentration of the pollutant leaving the CMFR. $C_0$ is the input concentration. $k_1$ is the first-order rate constant. $V$ is the reactor volume and $Q$ is the volumetric flow rate through the reactor.

For first-order, steady-state plug flow, the equation is

$$C = C_0 \exp^{\left(-k_1 \frac{V}{Q}\right)} \tag{8.3}$$

The ratio of the volume and steady-state flow rate through a reactor is defined as the *hydraulic retention time*, $t_R$. This represents the average time the fluid spends in the reactor.

$$t_R = \frac{V}{Q} \tag{8.4}$$

Table 8.1 gives the steady-state equations for CMFRs and PFR for zero-, first-, and second-order reactions.

Table 8.1 Mass Balance Reactor Equations for CMFR and PFR under Steady-state Conditions.

| Reaction order | CMFR | PFR |
|---|---|---|
| Zero | $C = C_0 - k_0 t_R$ | $C = C_0 - k_0 t_R$ |
| First | $C = C_0 \left( \frac{1}{1 + k_1 t_R} \right)$ | $C = C_0 \exp^{(-k_1 t_r)}$ |
| Second | $C = \frac{(1 + 4k_2 t_R C_0)^{1/2} - 1}{2k_2 t_R}$ | $C = \frac{C_0}{1 + k_2 t_R C_0}$ |

It can be shown that putting CMFRs in series, where the discharge of one CMFR becomes the input to the next, reduces the degree of mixing in the overall system. This approach can be used for *modeling* a system that is nonideal and has mixing characteristics between that of a CMFR and a PFR. In this modeling exercise, the total reactor volume does not change; the size of each individual reactor becomes correspondingly smaller.

For a first-order, steady-state system, the design equation is

$$C = C_0 \left( \frac{1}{1 + \frac{k_1 t_R}{n}} \right)^n \tag{8.5}$$

where $n$ is the number of CMFRs in series. For $n = 1$, the reactor is modeled as 1 CMFR, and Equation 8.5 becomes Equation 8.2. The larger the value of $n$, the less mixing in the reactor. It can be mathematically shown that as $n$ goes to $\infty$, Equation 8.5 approaches Equation 8.3, the PFR reaction, indicating no mixing.

**Example 8.1** Denitrification occurs in a small stormwater wetland under steady-state conditions. The wetland is 3 ft deep and covers an area of 0.75 ac. The inflow/outflow rate is 1.25 ft$^3$/s. The denitrification rate is first order with a rate constant of 0.1 h$^{-1}$. For an influent nitrate concentration of 4 mg N/L, find the effluent concentration:

a) Assuming the wetland is well mixed and acts as a CMFR.
b) Assuming the wetland has minimal mixing and acts as a PFR.
c) Assuming the wetland is moderately mixed and has mixing characteristics equal to 6 CMFRs in series.

*Solution* The wetland volume is found as

$$V = (3 \text{ ft})(0.75 \text{ ac})(43{,}560 \text{ ft}^2/\text{ac}) = 98{,}000 \text{ ft}^3$$

From Equation 8.4, the hydraulic retention time is found:

$$t_R = \frac{V}{Q} = \frac{98{,}000 \text{ ft}^3}{1.25 \text{ ft}^3/\text{s}} = 78{,}400 \text{ s} = 21.8 \text{ h}$$

a) Assuming a CMFR, from Equation 8.2, the discharge concentration of nitrogen is found:

$$C = 4\ \text{mg/L}\left(\frac{1}{1+\left(0.1\ \text{h}^{-1}\right)\left(21.8\ \text{h}\right)}\right) = 1.26\ \text{mg/L}$$

b) For a PFR, from Equation 8.3, the discharge concentration of nitrogen is found:

$$C = 4\ \text{mg/L}\exp^{\left(-\left(0.1\text{h}^{-1}\right)\left(21.8\text{h}\right)\right)} = 0.452\ \text{mg/L}$$

c) And for six CMFRs in series, from Equation 8.5:

$$C = 4\ \text{mg/L}\left(\frac{1}{1+\frac{\left(0.1\ \text{h}^{-1}\right)\left(21.8\ \text{h}\right)}{6}}\right)^6 = 0.622\ \text{mg/L}$$

Note that the predicted concentrations are different depending on the mixing characteristics. The 6-CMFR-in-series model gives a result between the other two extremes of complete (CMFR) and no (PFR) mixing.

**Example 8.2** Water enters and leaves a pond at 0.12 $m^3/s$. The water in the pond contains pesticide ZZZ at a concentration of 0.28 mg/L. The ZZZ concentration in the water leaving the pond is 0.18 mg/L. The average pond depth is 1.4 m and it is approximately round, with a diameter of 120 m. Find the first-order rate constant for the degradation of ZZZ in the pond. Assume the pond is well mixed.

*Solution* The pond volume is found as

$$V = \left(1.4\ \text{m}\right)\left(\pi\left(120\ \text{m}\right)^2/4\right) = 1.583\times10^4\ \text{m}^3$$

From Equation 8.4, the hydraulic retention time is found:

$$t_R = \frac{V}{Q} = \frac{1.583\times10^4\ \text{m}^3}{0.12\ \text{m}^3/\text{s}} = 1.319\times10^5\ \text{s} = 36.7\ \text{h}$$

Assuming a CMFR, from Equation 8.2, the rate constant is found:

$$k_1 = \frac{\left(\frac{0.28\ \text{mg/L}}{0.18\ \text{mg/L}}\right)-1}{\left(36.7\ \text{h}\right)} = 0.0151\ \text{h}^{-1}$$

## 8.3 Removal of Particulate Matter

Particulate matter is removed from water via physical separation processes, specifically sedimentation, driven by gravity, and filtration, via particle attachment to a porous medium.

### 8.3.1 Sedimentation

Sedimentation results from the gravity separation of particulate matter due to its density difference from that of water. Because most particulate matter has a greater density than water, the particulate matter will gradually sink in the water. If flow conditions are adequately quiescent, and adequate time is available, this simple mechanism can result in the removal of a significant amount of particulate matter from incoming runoff.

Sedimentation will occur in SCMs where water becomes stored or ponded, or the flow path is widened, so that the horizontal velocity becomes slowed (Figure 8.2). The downward settling velocity of a particle in a fluid can be described using Stokes' law (Masters and Ela 2008).

$$v_s = \frac{\left(\gamma_p - \gamma_w\right)d^2}{18\mu} = \frac{\left(\rho_p - \rho_w\right)gd^2}{18\mu} \tag{8.6}$$

where $v_s$ is the terminal settling velocity of the particle; $\gamma$ is the specific weight and $\rho$ is the density of the particle, $p$, and the water, $w$. The particle diameter is $d$, the water viscosity is $\mu$, and g is gravitational acceleration. The terminal settling velocity occurs as a constant velocity is reached, when the sedimentation force (downward) is balanced by the buoyant and drag forces (upward).

In Equation 8.6, it is seen that particles with large densities and diameters will settle faster than those with smaller values. Use of Equation 8.6 shows that coarse particulate matter will settle a few centimeters within a few seconds and removal of these types of particles is relatively easy. Settling of smaller particles will require more time, which may not be available in many SCMs. Silt- and clay-sized particles can take hours or longer to settle and cannot be expected to settle in a typical SCM. Also note that particles that have settled in a SCM may be re-mobilized at a later time if a high-velocity flow scours the surface.

Analysis of sedimentation in flow systems can be made assuming steady-state plug flow conditions (Figure 8.3). A particle can be considered to be removed via sedimentation if it settles the depth ($H$) of the ponded zone within the hydraulic retention time of the SCM, $t_R$. This defines the theoretical settling velocity for this reactor, $v_c$.

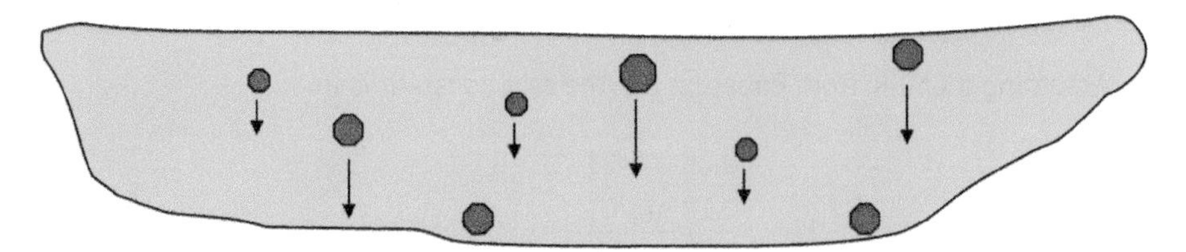

Figure 8.2 Sedimentation of Particles in a Quiescent Pond. Larger and Denser Particles Will Settle Faster.

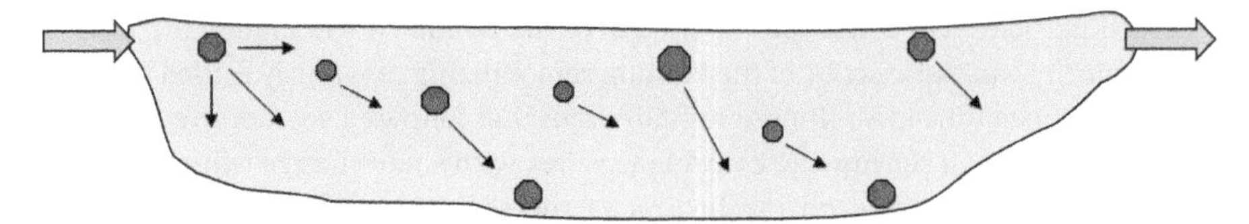

Figure 8.3 Sedimentation of Particles in a Flowing Pond. Particles Will Move Horizontally with the Flow and Settle Downward.

$$v_c = \frac{H}{t_R} \tag{8.7a}$$

Using the definition of $t_R$,

$$v_c = \frac{H}{t_R} = \frac{HQ}{V} = \frac{Q}{A} \tag{8.7b}$$

where $A$ is the bottom area of the ponded zone ($V/H$).

Equation 8.7 provides a link between sedimentation design and particulate matter removal. The design is given by the $Q/A$ term. Any particle that has a settling velocity greater than or equal to $v_c$ ($=Q/A$) will be removed via sedimentation. Particles that have settling velocities less than $v_c$ will be partially removed, at an efficiency equal to the ratio $v_s/v_c$.

Equation 8.7 indicates that for removal of small particles, that is those with small settling velocities, large areas and small flows are needed. Therefore, sedimentation will be most effective in SCMs where a large ponded area is present, such as ponds and wetlands. Sedimentation will also be most effective for removal of large particles and during low intensity rainfall events that produce low flows into the SCM. Forebays can be effective as a pretreatment practice, removing large particulate matter via sedimentation, preventing them from entering the SCM.

**Example 8.3** A small stormwater wetland covers an area of 0.75 ac. Find the sediment particle size that can be completely removed by the wetland during a storm event that produces a 25 $ft^3/s$ runoff flow. The particle has a specific gravity of 2.5; the temperature is 50°F.

*Solution* A flow rate of 2.5 $ft^3/s$ is equal to 2.5 ac-in/h into the wetland.

Knowing the flowrate, from Equation 8.7, the settling velocity of the completely captured particle is

$$v_s = \frac{Q}{A} = \frac{25 \text{ ac} - \text{in/h}}{0.75 \text{ ac}} = 33.3 \text{ in/h} = 7.72 \times 10^{-4} \text{ ft/s}$$

At 50° F, water viscosity is $2.74 \times 10^{-5}$ lbs/$ft^2$

Rearranging Stokes' law, Equation 8.6, gives particle size:

$$d = \sqrt{\frac{18 \mu v_s}{(\gamma_p - \gamma_w) g}} = \sqrt{\frac{18(2.74 \times 10^{-5} \text{ lbs/ft}^2)(7.72 \times 10^{-4} \text{ ft/s})}{(2.5(62.4) - 62.4 \text{ lb/ft}^3)}}$$

$$= 6.38 \times 10^{-5} \text{ ft} = 0.019 \text{ mm}$$

Settled particles will collect on the bottom of the ponded area. Sudden increases in water velocity can cause scour of the bottom, remobilizing previously settled particles. This can happen during an intense rainfall event that follows a smaller event and can be very important in stormwater control measures with small storage volumes. Settled sediment will accumulate on the bottom of the sedimentation area, increasing in depth. This sediment layer can reduce the storage volume of the pond, decreasing subsequent sedimentation efficiency by decreasing retention time.

For multiple particle sizes, the removal of each size is evaluated separately, then the total removal is determined. Since the settling velocity and removal of particulate matter depends on the square of the particle diameter, large particles are readily removed, while smaller ones are more difficult to settle from the flow.

**Example 8.4** As in Example 8.3, a small stormwater wetland covers an area of 0.75 ac. Find the removal of particles from the wetland during a storm event that produces a 25-ft$^3$/s runoff flow. The sediment particle size distribution is given in the first two columns of the following table. The particles have a specific gravity of 2.5; the temperature is 20°C.

*Solution* From Example 8.3, the critical settling velocity is found as $7.72 \times 10^{-4}$ ft/s = $2.35 \times 10^{-4}$ m/s. Since a distribution of particle sizes is provided, the average particle size (Column 4) is calculated for each size fraction (Column 3). For each average particle size, the particle settling velocity is calculated from Stokes' law (Equation 8.6); the results are presented in the Column 5, as $v_S$. The removal of each particle size is given as the ratio of $v_s/v_c$, Column 6.

The last column multiplies the fractional removal of the particle (Column 6) by the particle size fraction (Column 3) to get the removal of each particle size. The total removal is the sum of all fractions, 98.6%. Not surprisingly, based on the answer from Example 8.3, most of the particles are removed; all of the particles larger than 19 μm are removed.

| Particle size distribution ($d$) | Particle size (μm) | Particle size fraction | Average particle size (μm) | Settling velocity (m/s) | Removal fraction ($v_s/v_c$) | Removal of particle fraction |
|---|---|---|---|---|---|---|
| 100 | 970 | 0.2 | 895 | 0.500 | 1.00 | 0.2 |
| 80 | 820 | 0.2 | 705 | 0.310 | 1.00 | 0.2 |
| 60 | 590 | 0.1 | 360 | 0.0809 | 1.00 | 0.1 |
| 50 | 130 | 0.2 | 93 | 0.00540 | 1.00 | 0.2 |
| 30 | 56 | 0.2 | 40 | $9.99 \times 10^{-3}$ | 1.00 | 0.2 |
| 10 | 24 | 0.1 | 18 | $2.02 \times 10^{-4}$ | 0.859 | 0.859 |
| 0 | 12 | | | | | |
| | | | Cumulative removal efficiency | | | 0.986 = 98.6% |

### 8.3.2 Filtration

Filtration is a particle-removal unit process that occurs as the water flows through a porous medium, such as sand. The medium will trap particulate matter at the surface via straining mechanisms, where the particles cannot penetrate the pores of the media (Figure 8.4). Also, depth filtration can occur as small particles travel with the flow into the media and become attached to the media surfaces via colloid transport and attachment processes. SCMs such as sand filters and bioretention cells allow flow through porous media and will act as particulate matter filters. Some filtration may also occur during horizontal flow through thick vegetation as with grass swales.

Rapid and slow sand filters have been used for many years to remove colloids (small particles) and microorganisms from water supplies to produce very low turbidity, potable water. The performance efficiency depends on the characteristics of the filtering media and that of the particulate matter to be captured. Particulate matter in stormwater runoff must be transported to the surface of the filter media, whereby it can become attached to the media.

A simple model for estimating filtration efficiency under steady-state plug flow conditions is given as

$$\frac{N}{N_0} = \exp\left(-1.5\frac{(1-\epsilon)\alpha\eta L}{d_c}\right) \tag{8.8}$$

where $N$ and $N_0$ are the number-based concentration (e.g., number) of particles exiting and entering the filter media, respectively. (For a single particle size, $N$ is proportional to $c$, the mass-based concentration.) $\varepsilon$ is the porosity of the porous treatment media. $\alpha$ is the particle attachment parameter, theoretically ranging in value from 0 to 1, and $\eta$ is the particle transport parameter; both $\alpha$ and $\eta$ are dimensionless. $L$ is the travel path length of the media (filter depth) and $d_c$ is the diameter of the media (collector). For a mixed media, the value of $d_{10}$ is commonly used to represent the media diameter.

The parameter $\alpha$ represents the fraction of particle attachments to the media surface relative to the number of times particles strike/contact the surface, representing the transport mechanisms of the particles to the media collector surface. Three transport mechanisms are possible, as shown in Figure 8.5 and listed in Table 8.2. Smaller particles will be dominated by diffusional transport. Larger particles are transported by the sedimentation and interception mechanisms. The total transport, $\eta$, used in Equation 8.8, is the sum of the three $\eta$ values calculated using the equations in Table 8.2.

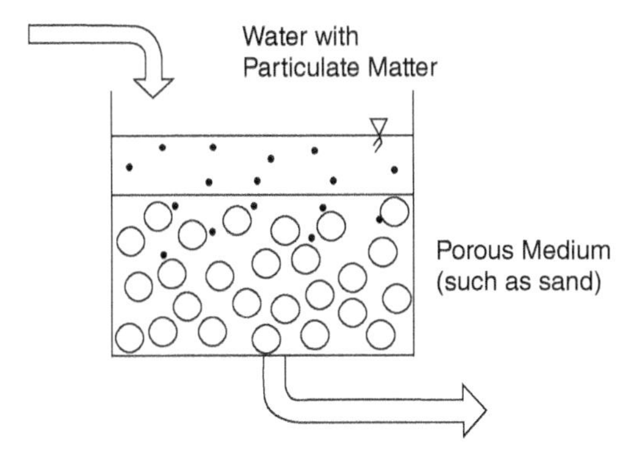


Figure 8.4 Removal of Particulate Matter Via Filtration through a Sand Filter.

*Transport Mechanisms of Particles to Collector*

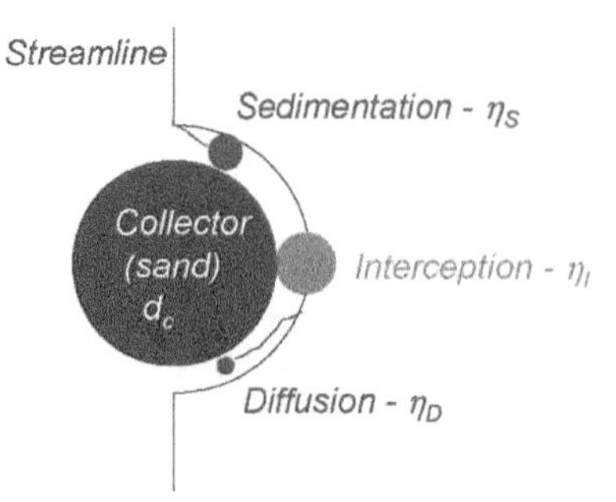


**Figure 8.5** Transport Mechanisms for Particles to a Collector During Filtration.

**Table 8.2** Transport Mechanisms for Filtration Model, Equation 8.8.

| Filtration transport mechanism | Equation |
|---|---|
| Sedimentation | $\eta_s = \frac{(\rho_p - \rho_w) g d_p^{\,2}}{18 \mu v}$ |
| Interception | $\eta_I = \frac{3}{2}\left(\frac{d_p}{d_c}\right)^2$ |
| Diffusion | $\eta_D = 0.9\left(\frac{kT}{\mu d_p d_c v}\right)^{\frac{2}{3}}$ |

**Example 8.5** Estimate the filtration removal efficiency for a 10–μm particle (specific gravity of 2.0) via a 0.5-m depth of sand during 16 in/h infiltration. The sand $d_{10}$ is 0.8 mm and the porosity is 0.4. The particle attachment efficiency is 0.05 and the temperature is 15°C.

***Solution*** Each transport term from Table 8.2 must be calculated. The infiltration rate of 16 in/h corresponds to $1.13 \times 10^{-4}$ m/s:

$$\eta_s = \frac{(\rho_p - \rho_w) g d_p^{\,2}}{18 \mu v} = \frac{\left(2.0(1000) - 1000\ \text{kg/m}^3\right) 9.8\ \text{m/s}^2 \left(10 \times 10^{-6}\ \text{m}\right)^2}{18\left(1.14 \times 10^{-3}\ \text{kg/ms}\right)\left(1.13 \times 10^{-4}\ \text{m/s}\right)} = 0.422$$

$$\eta_I = \frac{3}{2}\left(\frac{d_p}{d_c}\right)^2 = \frac{3}{2}\left(\frac{0.01\ \text{mm}}{0.8\ \text{mm}}\right)^2 = 1.34 \times 10^{-4}$$

$$\eta_D = 0.9\left(\frac{kT}{\mu d_p d_c v}\right)^{\frac{2}{3}} =$$

$$= 0.9\left(\frac{\left(1.38 \times 10^{-23}\ \text{J/K}\right)(298\ \text{K})}{\left(1.14 \times 10^{-3}\ \text{kg/ms}\right)\left(10 \times 10^{-6}\ \text{m}\right)\left(0.8 \times 10^{-3}\ \text{m}\right)\left(1.13 \times 10^{-4}\ \text{m/s}\right)}\right)^{\frac{2}{3}}$$

$$= 2.26 \times 10^{-4}$$

This sum of the three transport values is 0.422, dominated by the sedimentation transport.

From Equation 8.8, the particle-removal fraction is

$$\frac{C}{C_0} = \exp\left(-1.5\frac{(1-0.4)(0.1)(0.422)(0.5\,\text{m})}{(0.8\times10^{-3}\,\text{m})}\right) = 4.8\times10^{-11}$$

This states that the concentration leaving the media is expected to be very small and 100% particle removal via filtration can be assumed.

This filtration model is only valid for "clean" media under steady-state flow conditions. As captured particulate matter accumulates in the filtration media, Equation 8.8 will no longer be valid as porosity and other factors are changed. Nonetheless, this equation can provide estimates of filtration performance and impacts of design variables under various operational conditions.

Because of the relatively fine media used in many SCMs (such as bioretention) and the relatively low flow rates compared to water treatment rapid sand filtration, the filtration mechanism can be very effective for removal of particulate matter from runoff. Captured particles will accumulate at the media surface where straining occurs and in the upper zones of the media. Incoming runoff first contacts the media upper zone and the majority of the treatment takes place here. As collected solids build up on the media surface and the upper layers, clogging may become a concern. Incorporating vegetation in the media may assist in minimizing clogging. Otherwise, some type of maintenance may be necessary, as will be discussed in later chapters.

The capture of bacteria and other microorganisms via filtration mechanisms should also follow filtration theory discussed above (Parker et al. 2017). Microbial transport to the media surface will follow the same mechanisms as inorganic particles. However, the attachment processes will be different, as manifest in the $\alpha$ parameter in Equation 8.8. Microorganisms have a number of different mechanisms available for attachment. Additionally, once attached, microorganisms, as living beings, can multiply, die-off, or become dormant, depending on a host of environmental conditions.

## 8.4 Removal of Dissolved Pollutants: Adsorption

Adsorption is a physicochemical process where a dissolved chemical (adsorbate) becomes accumulated at the surface of a solid (adsorbent). Adsorption is employed in water treatment processes for the removal of dissolved pollutants such as phosphorus species, metals, and various toxic organic compounds. Adsorption also commonly occurs in nature; soils and other geochemical media are composed of organic and inorganic constituents that have high affinities for many dissolved substances, including many common pollutants.

Adsorption can occur in SCMs where the water will flow through or very near some adsorptive media. Dissolved compounds common to stormwater that can be adsorbed include heavy metals, various toxic organic compounds, dissolved phosphorus, and ammonium. The engineering analysis of the transport, removal, and accumulation of these different pollutants is similar. However, the chemistry of how these compounds interact with the media can be very different. As a result, different media properties can result in very different performance for removal of different dissolved pollutants.

### 8.4.1 Adsorption Equilibrium Models

Adsorption equilibrium can be quantitatively described using a number of simple models. These models quantify the relationship between the concentration of the target chemical/pollutant in the water phase, $C$, and that adsorbed to the surface, $q$, as diagrammed in Figure 8.6. The units of $C$ may typically be mg/L and those of $q$, mg of adsorbate/g adsorbent. These models describe the equilibrium relationship between $C$ and $q$, with all other conditions (e.g., temperature, pH, etc.) held constant and are known as *isotherm* models. These models usually (but not always) have a functional relationship that allows $q$ to approach a maximum value at high $C$ due to the saturation of adsorption sites on the adsorbent surface.

The *Langmuir* isotherm has a strong fundamental foundation. It is given as

$$q = \frac{QK_LC}{1 + K_LC} \tag{8.9}$$

where $Q$ is the maximum adsorption density, having the same units as $q$. $K_L$ is the Langmuir adsorption constant, describing the strength of the adsorption. It has units reciprocal to C. Equation 8.9 exhibits a linear increase in $q$ with increase in $C$ at low concentrations. However, as $C$ becomes large, $q$ will asymptote to $Q$, the maximum amount of adsorbate achievable on the adsorbent.

Equation 8.9 can be linearized in a number of ways. However, most of them result in some kind of reciprocal or self-correlation, which can lead to inaccuracies in parameter determination. For example, the most common linearization is the reciprocal equation:

$$\frac{1}{q} = \frac{1}{QK_L}\frac{1}{C} + \frac{1}{Q} \tag{8.10}$$

In this expression $1/q$ is a linear function of $1/C$. However, taking the reciprocal of both the variables in an expression can lead to errors being greatly magnified. As a result, linearizations for the Langmuir isotherm must be used very carefully. It is

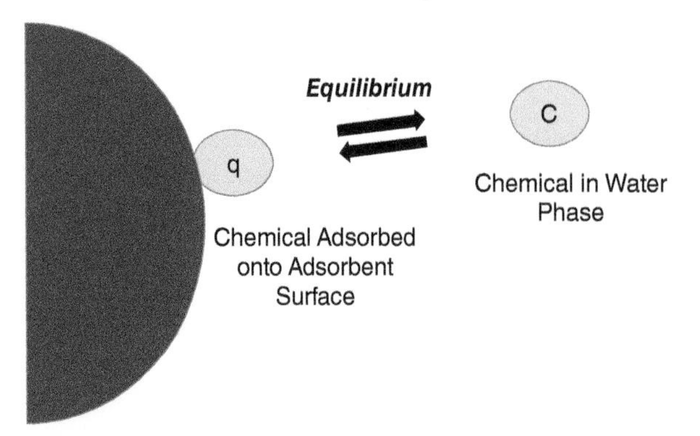


Figure 8.6 Adsorption Equilibrium between a Substance Dissolved in Water Phase and a Substance Adsorbed onto Surface of Adsorbent.

always good practice to check values obtained from linearizations using the non-linearized data, that is, using Equation 8.9.

A second common adsorption equilibrium model is the *Freundlich* isotherm. In this case, the isotherm relationship is a power model:

$$q = K_F C^{\frac{1}{n}} \tag{8.11}$$

where $K_F$ is the Freundlich adsorption constant and $1/n$ is the power function exponent, related to the energy of adsorption. The Freundlich isotherm does not asymptote to a fixed value at high adsorbate concentrations, but the slope decreases with increasing adsorption coverage of the surface. The Freundlich isotherm has been found to do a very good job in describing the adsorption of many pollutants to different natural and synthetic adsorbents. As a result, it is commonly used in many environmental engineering applications.

Taking the log of both sides of Equation 8.11 results in a linearized equation.

$$\log q = \frac{1}{n}\log C + \log K_F \tag{8.12}$$

Data that can be described by the Freundlich isotherm will plot as a straight line on the log $q$ - log $C$ plot. Values of $K_F$ and $1/n$ can be determined by the intercept and slope, respectively, from this plot.

### 8.4.2 Batch Adsorption

Adsorption isotherm information and values of isotherm constants are usually determined in laboratory or pilot studies using batch systems. In this case, the adsorbent is added to multiple water batches with different ratios of adsorbate and adsorbent concentrations. The amount adsorbed is determined by difference via mass balance analysis.

The overall adsorbate mass is conserved in the batch system; it just transfers from the water to the surface of the adsorbent (or the reverse). The input to the system is the mass added in the water, $C_0V$, and that added on the adsorbent, $q_0M$ (Figure 8.7). $C_0$ is the concentration of the adsorbate in the water, $V$ is the volume of water, $q_0$ is the

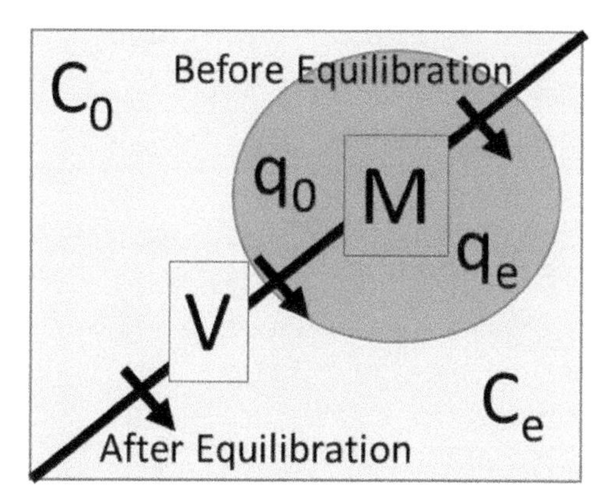


Figure 8.7 Mass Balance for Batch System Adsorption Equilibrium.

amount of adsorbate initially adsorbed on the adsorbent (zero for a "clean" adsorbent, may not be for used adsorbents or natural materials), and $M$ is the mass of adsorbent.

After contacting, the system will equilibrate based on the system characteristics; the adsorbate will equilibrate between the water and adsorbent phases. At this point, the mass in the water is $C_eV$ and that on the adsorbent is $q_eM$, where the $e$ subscripts indicate the equilibrium concentrations. The mass balance states that

$$C_0V + q_0M = C_eV + q_eM \tag{8.13}$$

**Example 8.6** Laboratory studies have been completed to evaluate the adsorption of the solvent trichloroethylene (TCE) from water onto a (clean) soil mixture. Eight batches of soil are made, each with 0.05 g of soil in 0.5 L of water. Various concentrations of TCE are synthesized in the eight batches, ranging from 0.1 to 30 mg/L. The TCE and soil are mixed overnight so that adsorption equilibrium is reached. The concentrations of TCE before and after equilibrium are given in the following table. (a) Evaluate the adsorption of TCE onto the soil mixture as described by a Freundlich isotherm, and (b) determine the Freundlich isotherm constants.

| Initial TCE (mg/L) | Final TCE (mg/L) | Initial TCE (mg/L) | Final TCE (mg/L) |
|---|---|---|---|
| 0.1 | 0.0044 | 3 | 0.72 |
| 0.3 | 0.03 | 5 | 1.5 |
| 0.5 | 0.06 | 10 | 3.7 |
| 1 | 0.15 | 30 | 14 |

***Solution*** For each TCE concentration, the amount adsorbed, $q_e$ (mg/g), is found using the batch adsorption mass balance, Equation 8.13, with $q_0 = 0$ for a clean soil.

$$q_e = \frac{(C_0 - C_e)V}{M}$$

These values are shown in Column 3 of the following table:

| Initial TCE (mg/L) | Final TCE (mg/L), $C_e$ | Adsorbed TCE (mg/g), $q_e$ | Log $C_e$ | Log $q_e$ |
|---|---|---|---|---|
| 0.1 | 0.0044 | 0.956 | -2.36 | -0.0195 |
| 0.3 | 0.03 | 2.7 | -1.52 | 0.431 |
| 0.5 | 0.06 | 4.4 | -1.22 | 0.643 |
| 1 | 0.15 | 8.5 | -0.82 | 0.929 |
| 3 | 0.72 | 22.8 | -0.143 | 1.36 |
| 5 | 1.5 | 35 | 0.176 | 1.54 |
| 10 | 3.7 | 63 | 0.568 | 1.80 |
| 30 | 14 | 160 | 1.15 | 2.20 |

A plot of TCE adsorbed, $q_e$, as a function of TCE in solution at equilibrium, $C_e$, is given below, using a log scale.

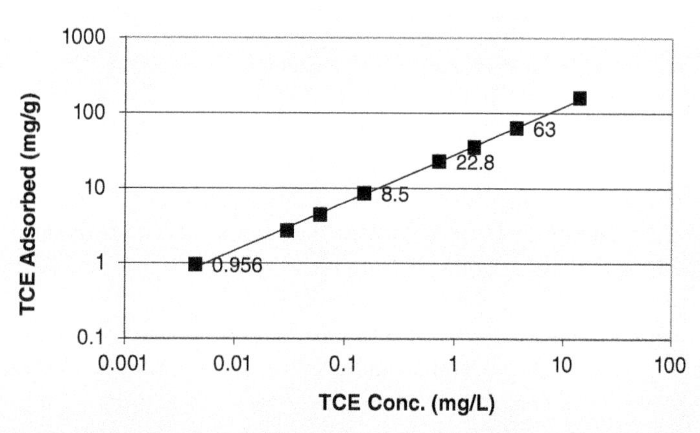


An excellent linear relationship is found between log $q_e$ and log $C_e$, indicating that the Freundlich model provides an excellent description to the data.

The same data after taking the logs are shown below. From this plot, or a linear regression, it can be seen that the slope is equal to 0.64, which is $1/n$ in Equations 8.11 and 8.12. The intercept is 1.445 which is log $K_F$, and $K_F$ is equal to 27.9.

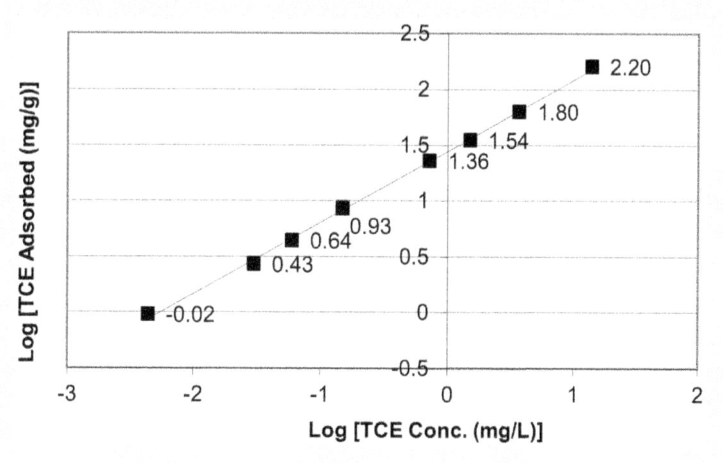


Finally, a fit of the Freundlich model to the experimental data, without the log transformation, is shown below, again indicating excellent fit of the model.

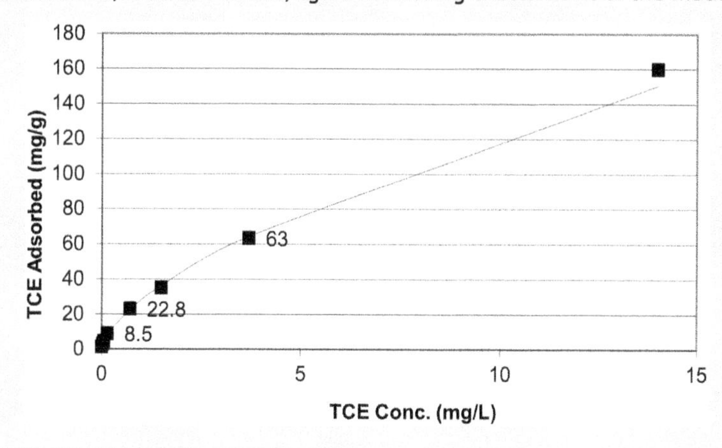

### 8.4.3 Adsorption Column Dynamics

As water flows through the media in a SCM, the behavior can be considered as similar to an adsorption column. As shown in Figure 8.8, the contact with the media will be at the top of the "column." The dissolved chemicals will be adsorbed in this zone, based on their affinity with the media. Therefore, the accumulation of pollutants will occur in the upper zones of the media, with this burden working downward as adsorption sites on the media become occupied with various chemicals. Once the available sites are filled, no further pollutant removal from the water will occur. Some degree of competition for sites may be expected in some situations.

The dynamics of pollutant removal through adsorption columns can be very complex, especially with the highly non-steady conditions and multiple adsorbing chemicals inherent to stormwater. The simplest approach to quantifying adsorption in columns is to assume steady-state and local equilibrium. Under these conditions, when the effluent concentration equals the influent at saturation, the entire media must be in equilibrium with the influent (and effluent) concentration. This will provide an estimate of the adsorption capacity of the media under the conditions of interest.

Similar to batch adsorption, a mass-balance approach is used, equating the mass removed from the water to that accumulated onto the adsorbent. The mass of adsorbate (pollutant) adsorbed onto the media at startup is given by the product of $Mq_0$, where $M$ is the mass of adsorbent in the column and $q_0$ is the amount of adsorbate present on the media *before* any adsorption takes place. Again, $q_0$ may be effectively 0 for a clean medium; however, it may be much larger for a ubiquitous compound like phosphorus. Similarly, the mass of adsorbate after adsorption equilibrium, corresponding to the media capacity, is given by $Mq_e$, where $q_e$ is the amount of adsorbate adsorbed on the media *in equilibrium with the influent* adsorbate concentration.

The mass of pollutant removed from the incoming water is $C_0V$ or $C_0Qt$, where $C_0$ is the incoming adsorbate concentration. $V$ is the volume of water treated by the media before the total adsorption capacity is used, which is equal to the product of the incoming flow rate, $Q$, and the time before the capacity is used. $C_C$ is the concentration of adsorbate discharged from the column, which is frequently assumed as 0. At the point

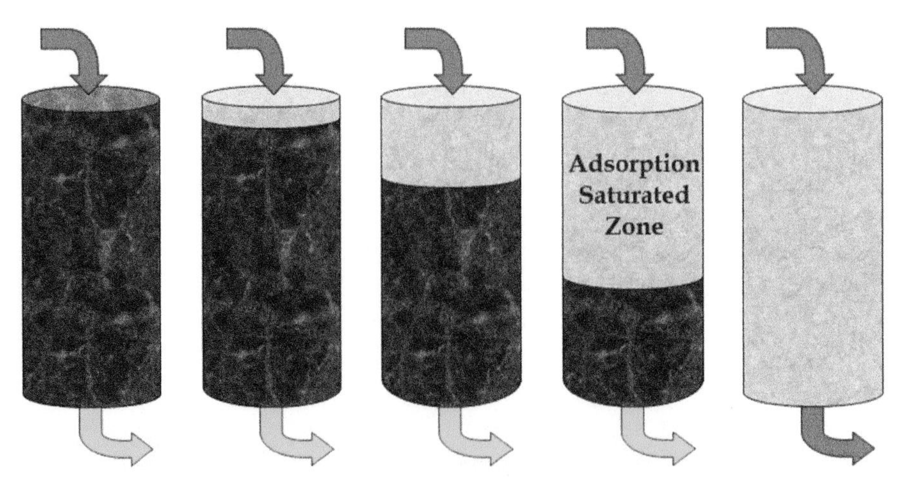


Figure 8.8 Adsorption Zone of Pollutant in an Adsorbent Media Column.

at which the capacity of the media is used up, no additional mass is removed from the water and the pollutant mass balance gives

$$(C_0 - C_C)Qt \approx C_0 Qt = Mq_e - Mq_0 \tag{8.14}$$

or

$$\frac{M}{Qt} = \frac{C_0}{(q_e - q_0)} \tag{8.15}$$

The value of $q_e$ is found as the amount adsorbed in equilibrium with $C_0$ using an isotherm relationship, such as that given in Equation 8.9 or Equation 8.11.

Note that if an adsorbent is used that has a large amount of the pollutant already adsorbed onto it, the treatment lifetime can be small. Also, desorption from the adsorbent can occur if the influent concentration is less than the equilibrium concentration. This can be estimated using the same mass balance (Equation 8.15), but with $q_0$ greater than $q_e$, which would give $C_C > C_0$.

**Example 8.7** Estimate the lifetime of a soil adsorption column treating stormwater containing 0.5 mg/L TCE. The stormwater flow is 5 L/min and the column contains 1000 g of (clean) soil. Use the Freundlich isotherm derived in Example 8.6.

***Solution*** Equation 8.15 will be used to solve this problem. It is rearranged to solve for the breakthrough time:

$$t = \frac{M(q_e - q_0)}{QC_0}$$

From Equation 8.11 and Example 8.5 results:

$$q = K_F C^{\frac{1}{n}} = 27.9(0.5\ \text{mg/L})^{0.64} = 17.9\ \text{mg/g with } q_0 = 0$$

$$\text{therefore: } t = \frac{1000\ \text{g}(17.9\ \text{mg/g})}{(5\ \text{L/min})(0.5\ \text{mg/L})} = 7160\ \text{min} = 4.97\ \text{days}$$

### 8.4.4 Adsorption of Hydrophobic Organic Compounds

Hydrocarbons typically found in urban runoff (fuel components, PAH) will generally adsorb strongly to natural geomedia such as soil. This binding mostly occurs to the organic matter in the media via hydrophobic partitioning interactions. These hydrocarbons are only sparingly soluble, so they will partition into the organic matter in and on the soil, where the energetics are more favorable. As a result, adsorbent media with high organic matter content will tend to have high capacity for hydrocarbons. In the same manner, the more hydrophobic the organic pollutant, the greater affinity for the organic matter in the media.

The adsorption of hydrophobic organic chemicals is commonly described using a linear isotherm, also called a partitioning model. In this case,

$$q = K_d C \tag{8.16}$$

This is a simplification of both the Langmuir and Freundlich isotherm models, where $K_d$ is the linear partitioning coefficient. As noted above, $K_d$ is generally a function of the hydrophobicity of the adsorbate and the organic matter content of the adsorbing medium (Schwarzenbach et al. 2016).

### 8.4.5 Adsorption of Heavy Metals

The adsorption of metals, such as lead, copper, cadmium, and zinc, generally involve a specific binding reaction between the metal and adsorbent surface. Adsorbent media containing organic matter, clays, and metal oxides have reactive sites that can bind the metals. Therefore, media containing these materials will be most active in removing metals. The binding reaction can be represented as

$$\underline{S} - OH + M^{2+} \leftrightarrow \underline{S} - OM^{+} + H^{+} \tag{8.17}$$

where $\underline{S}$–OH represents a specific active site at the surface of the adsorbent medium; $M^{2+}$ represents a free divalent metal. Correspondingly, $\underline{S}$–$OM^{+}$ is the metal adsorbed to the media surface (Stumm and Morgan 1996).

Note from Equation 8.17 that the proton, $H^{+}$, is involved in the metal adsorption reaction. Therefore, metal adsorption is strongly dependent on pH. Metals will adsorb strongly at higher pH, but much less so at lower pH. Under neutral pH conditions, most metals of interest in urban stormwater management will be adsorbed onto soil or a soil-based media.

Arsenic (which is technically a metalloid), chromium, and a few other pollutants form oxyanionic species in water. That is, the metal(loid) species contains oxygen and carries a negative charge. For example, hexavalent chromium commonly exists as $HCrO_4^{-}$ and arsenic as $AsO_4^{3-}$ and $H_2AsO_3^{-}$. Oxyanions will adsorb onto sorbing media similar to metals, but via a different mechanism:

$$\underline{S} - OH + M'O_4^{2-} \leftrightarrow \underline{S} - M'O_4^{-} + OH^{-} \tag{8.18}$$

where $M'O_4^{2-}$ and $\underline{S}$–$M'O_4^{-}$, respectively, represent the oxyanion in water and adsorbed on the media surface. Note from Equation 8.18 that the adsorption releases $OH^{-}$. Oxyanion adsorption is favored at lower pH, but not at higher pH, the opposite of that for cationic metals.

A medium that has a high content of organic matter and media fines will generally have a high capacity for the adsorption of cationic and oxyanionic metals. Media with excessive sand content will likely have lower metal adsorption capacity.

### 8.4.6 Adsorption of Phosphorus

Dissolved phosphorus species will also adsorb onto SCM media. For orthophosphorus, $H_2PO_4^{-}$ and $HPO_4^{2-}$ are the dominant species, and the adsorption will proceed via a reaction similar to that presented for oxyanions in Equation 8.18:

$$\underline{S} - OH + HPO_4^{2-} \leftrightarrow \underline{S} - PO_4^{2-} + H_2O \tag{8.19}$$

where $\underline{S}-PO_4^{2-}$ is the adsorbed phosphate species. Phosphate adsorption occurs at pH near neutral and will decline at higher and lower pH.

Research has shown that phosphate is strongly adsorbed by oxides of iron and aluminum. Iron and aluminum oxides are a common component of soils and related media. These oxides can exist in many forms. Freshly precipitated Al and Fe oxides generally have high surface areas and very high affinities for adsorption of phosphate. So in Equation 8.19, if the surface represented by $\underline{S}$–OH is aluminum or iron oxide, the degree of phosphate adsorption is expected to the strong. Specifying media with high natural Fe and/or Al content, or adding supplemental Fe and/or Al to the media, can lead to media with significant phosphorus removal capacity.

### 8.4.7 Adsorption of Ammonium

Ammonium, $NH_4^+$, is a cation. In general, it is strongly adsorbed (or ion-exchanged) onto negatively charged soil/media. Clays commonly carry a negative charge and ammonium adsorption/ion exchange is expected on high-clay media. However, other cations, including $Ca^{2+}$, $K^+$, and high concentrations of $Na^+$, can compete with ammonium and prevent adsorption (or cause it to desorb, Khorsha and Davis 2017).

## 8.5 Leaching Processes

Some materials used in the construction of SCMs can leach pollutants. The most serious concerns with leaching have been found for nitrogen and phosphorus leaching from the use of natural nutrient-rich organic materials, such as compost. Compost is created via the microbial degradation/stabilization of many different natural organic materials, including leaves and grass, food waste, manure, and wastewater biosolids. The result is the release of readily available nutrients for plant growth. In most cases, use of compost in SCMs should be avoided. Other forms of organic matter such as mulch, peat, and wood chips have lower nitrogen and phosphorus contents and are more resistant to microbial decomposition. They may release nitrogen and phosphorus, but usually not to the great extent as compost. Nutrient leaching may continue for a long time in SCMs, and the media will act as a source, until the leaching compounds are completely leached out or otherwise reach a steady state based on the runoff conditions.

Other media additives and amendments may be considered as inexpensive materials (commonly waste materials) for use in SCM media to enhance pollutant removals. Leaching of undesirable substances must be considered before the use of these materials.

## 8.6 Microbiological Processes

Microorganisms have the ability to transform pollutants and other substances to different forms, frequently as they utilize them for some growth function. Microorganisms generally are classified to include bacteria, viruses, and protozoa. Bacteria are small organisms, approximately 1–2 μm in size, and are ubiquitous in the environment. They exist naturally in soils and other media. Ecologically many of these microorganisms are considered as decomposers, important components of natural ecosystems that break down complex materials, including waste materials, into simple, usually inorganic products.

Bacteria have a number of growth requirements for survival. These include compounds that act as electron donors and acceptors so that the bacteria can gain energy through metabolic processes. Every electron donor must be coupled with an electron acceptor. For survival and successful population growth, the proper environmental conditions must be maintained, which vary, sometimes significantly, for different bacteria. Moisture and nutrients are necessary. Other factors include proper temperature and pH.

### 8.6.1 Microbial/Pathogen Survival

The physicochemical capture of microbial pathogens via media filtration was discussed in the *Filtration* section of this chapter. However, the fate of captured microorganisms will depend greatly on the ambient environmental conditions to which they are exposed. Proper moisture, oxygen, pH, temperature, and nutrients may allow them to thrive and proliferate. The absence of the appropriate conditions will cause the microbe to die or transition to some type of dormant state.

Conditions that can prove detrimental to the survival of microbial pathogens include lack of moisture and exposure to UV light, including sunlight. Additionally, the presence of predator organisms, such as protozoa for bacteria, can reduce microbial populations.

### 8.6.2 Organic Matter Degradation

Many natural bacteria and other microorganisms use organic compounds as electron donors, resulting in organic matter decomposition. Microorganisms break down complex organic matter into simpler compounds, eventually mineralizing it to $CO_2$. Oxygen is the electron acceptor; thus the overall process is a respiration.

$$\text{Organic matter} + O_2 \xrightarrow{\text{microbes}} CO_2 + H_2O + \text{energy} \tag{8.20}$$

The use of oxygen as an electron acceptor classifies this as an aerobic process.

From an urban stormwater runoff perspective, organic compounds of interest may include hydrocarbons, possibly pesticides, and maybe other compounds. It also includes natural organic matter. Aerobic microbial degradation of various hydrocarbons has been documented in many natural circumstances. The process may be slow (on the order of days) and include several intermediate organic compounds but eventually should lead to the production of $CO_2$. Hydrocarbon degrading organisms exist naturally; they just need to acclimate to specific hydrocarbon compounds once exposure of the microbes to the hydrocarbons occurs.

In stormwater treatment, microbial transformation processes are not limited to the timeframe of the runoff event. Generally, the treatment will be a coupled process where first the target hydrocarbon or other organic compound is adsorbed onto the media, or it enters the SCM attached to particulate matter. Once it is held within the treatment system, biological processes can occur slowly, but effectively, over several days or longer as long as the environmental conditions are conducive to the process of interest (LeFevre et al. 2012). Biodegradation of organic pesticides or other more recalcitrant compounds is more complex. Some will biodegrade, but others are resistant. Intermediate degradation compounds are also more likely to accumulate.

Natural organic matter from vegetation will biodegrade as part of the ecological decomposition process. As this material is degraded, nitrogen, phosphorus, and other compounds that were part of the original biomolecules will be released. Intermediates in the process will likely include dissolved organic compounds containing N and P. Eventually, the nitrogen and phosphorus will be released, typically as ammonium–N and orthophosphate, as noted in Section 8.5 on leaching.

### 8.6.3 Nitrification

Nitrification is the microbial conversion of ammonium to nitrate. It is an aerobic process with ammonium as the electron donor. The overall reaction is

$$NH_4^+ + 2O_2 \xrightarrow{\text{nitrifying microbes}} NO_3^- + H_2O + 2H^+ \tag{8.21}$$

The nitrification process is sensitive to pH, with a maximum rate at a pH of about 8.4. As noted in Equation 8.21, the nitrification process will decrease the pH, although not significantly at low nitrogen concentrations and in environments with significant pH buffering.

The nitrification process occurs via two discrete steps. The first is the oxidation of ammonium to nitrite.

$$NH_4^+ + 1.5O_2 \xrightarrow{\textit{Nitrosomonas}} NO_2^- + H_2O + 2H^+ \tag{8.22}$$

This reaction can be very slow and frequently limits the nitrification process. *Nitrosomonas* species have difficulty competing in an environment high in organic matter, where heterotrophic organisms are degrading organic matter. Three oxidation steps are required in this reaction. The first is carried out by the enzyme *ammonia monooxygenase* (amo) and the second by *hydroxylamine oxidoreductase* (hao).

The second step in the nitrification process is the conversion of nitrite to nitrate, with the key enzyme *nitrite oxidoreductase* (nxr):

$$NO_2^- + 0.5O_2 \xrightarrow{\textit{Nitrobactor}} NO_3^- \tag{8.23}$$

This process is usually faster than the first step and little accumulation of nitrite is usually found in active nitrifying environments.

Evidence of nitrification has been found in several stormwater SCMs, suggesting that ammonium can be held by SCM media and that the conditions are appropriate for *Nitrosomonas* and *Nitrobactor* to nitrify the ammonium. However, the product, nitrate, is not readily adsorbed or otherwise affiliated with media or components of an SCM. Therefore, nitrate washout can frequently be found in initial flows exiting many SCMs.

### 8.6.4 Denitrification

In the denitrification process, nitrate is the electron acceptor. Therefore, the system must be void of oxygen for the process to proceed. The electron donor is some form of organic matter (or less frequently, an inorganic compound such as sulfur):

$$2NO_3^- + \text{organic matter} \xrightarrow{\text{denitrifying microbes}} N_2(g) + CO_2 + OH^- \quad (8.24)$$

This process takes four steps, catalyzed by the enzymes *nitrate reductase* (Nar), *nitrite reductase* (Nir), *nitric oxide reductase* (Nor), and *nitrous oxide reductase* (Nos).

In stormwater treatment, the organic matter is likely some form of natural organic matter, such as that from leaves, grasses, roots, or wood. In most natural systems, abundant organic matter should be present for the denitrification process.

For effective denitrification, a zone or water storage pool must be created such that it will remain anoxic (void of oxygen) during the majority of the stormwater treatment. Typically, this will require that the pool zone remains saturated for long periods and be fairly isolated from the atmosphere. Forced submerged water storage areas will meet these conditions. Any oxygen in the water is used quickly by aerobic organisms metabolizing organic matter and the system will rapidly become anoxic. Once the system becomes anoxic, denitrifying microorganisms, which exist naturally, should be able to initiate the denitrification process.

The denitrification reaction is typically slow, on the order of hours to days. Therefore, special hydraulic conditions must be created to allow adequate treatment time. Time may be the limiting factor in stormwater denitrification (Halaburka et al. 2017). Temperature will be important and higher temperatures will increase the reaction rate. The product of the denitrification reaction is nitrogen gas, the primary component of the atmosphere and a benign nitrogen compound.

**Example 8.8** A water contains 5 mg/L (as N) nitrate. What concentration of organic matter, represented empirically as $C_6H_6O_4$, is necessary for the stoichiometric denitrification of this nitrate?

***Solution*** Equation 8.24 must be balanced to solve this problem, with $H_2O$ added to balance the H and O. Using constants as the stoichiometric coefficients:

$$A NO_3^- + B\, C_6H_6O_5 \xrightarrow{\text{denitrifying microbes}} D N_2(g) + E\, CO_2 + F\, OH^- + G H_2O$$

First, setting $B = 1$, gives $E = 6$ to balance the carbon. Balancing $H$ gives $F + 2\,G = 6$. A charge balance gives $A = F$. The $N$ balance gives $A = D/2$. Finally, the oxygen balance is

$$3A + 5 = (6 \times 2) + F + G$$

Substituting for $A$ and $G$ give:

$$3F + 5 = 12 + F + ½(6 - F)$$

giving

$$2.5F = 10 \text{ and } F = 4.$$

Therefore, $A = 4$, $D = 2$, and $G = 1$.

The balanced equation is

$$4NO_3^- + C_6H_6O_5 \xrightarrow{\text{denitrifying microbes}} 2N_2(g) + 6CO_2 + 4OH^- + H_2O$$

Therefore, 4 × 14 = 56 mass units of N (using only the mass of N, since our nitrate concentration is as N) reacts with 6 × 12 + 6 × 1 + 5 × 16 = 158 mass units of organic matter. Thus:

$$5\ \text{mgN/L}\left(\frac{158\ \text{mg Organic matter}}{56\ \text{mg N}}\right) = 14.4\ \text{mg/L organic matter required}$$

**Example 8.9** Calculate the amount of organic matter needed to denitrify stormwater runoff from a 0.4 ha drainage area for 20 years. Assume 1 m/year rainfall, 80% rainfall to runoff, and a total N concentration of 1.2 mg/L. Use the stoichiometry of Example 8.8 for the denitrification reaction.

***Solution*** The stormwater volume over 20 years is calculated as

$$V = (0.8)(1\ \text{m/year})(20\ \text{years})(0.4\ \text{ha})\left(10^4\ \text{m}^2/\text{ha}\right) = 6.4\times10^4\ \text{m}^3 = 6.4\times10^7\ \text{L}$$

Using the stoichiometry of Example 8.8:

$$\text{Organic matter mass} = \left(6.4\times10^7\ \text{L}\right)(1.2\ \text{mgN/L})\left(\frac{158\ \text{mg Organic matter}}{56\ \text{mgN}}\right)$$

$$= 2.09\times10^8\ \text{mg OM} = 209\ \text{kg OM}.$$

Two hundred and nine kilograms of organic matter is required for the denitrification.

## 8.7 Phytobiological Processes

Many stormwater control measures rely on vegetated processes and vegetation can provide a number of pathways for improving stormwater characteristics. Green vegetation contains carbon, nitrogen, phosphorus, potassium, and other elements. Vegetation is approximately 2–5% nitrogen (by dry weight, Shuman 2000), suggesting that plant uptake of this nutrient could be an important stormwater treatment mechanism. Phosphorus is about 0.3–0.5% dry weight.

Foremost, vegetation can play a major role controlling the water balance. ET is a critical component of green roofs, especially, but can be important in other vegetated SCMs as well. Growth and death of plant roots can open pores and pathways for water flow in the growth media. Through this mechanism, vegetation can prevent SCM media clogging when it becomes loaded with sediment.

During plant growth and survival, pollutants, including metals, hydrocarbons, nitrogen, and phosphorus, may be taken up into the plant matter, including the roots and shoots. This mechanism removes these pollutants from the SCM media, but not from the SCM system. Removal of the vegetation may be necessary for removal of the phytoaccumulated pollutants.

Additionally, the plant root zone area, known as the rhizosphere, is an environment that typically hosts enhanced microbial activity. This microbial activity can lead to heightened rates of various microbial processes, resulting in hydrocarbon degradation, and nitrogen transformations, as discussed in the previous section.

## 8.8 Heat Transfer

High temperatures in stormwater can create stream habitats that are unacceptable to many desirable fish species, especially trout and salmon. In some cases, heat can be considered as a water pollutant and reduction in stormwater temperature can be a desirable SCM attribute.

Heat will be transferred from warmer to cooler materials. The cooler material will warm and the warmer will cool until both materials reach the same temperature, a process known as thermal equilibrium. The amount of heat transferred will depend on the mass of each material, the initial temperatures, and the specific heat capacity of each material. The specific heat capacity describes the amount of heat energy it takes to raise the temperature of a material by one degree, normalized by the mass of the material. Water has a high specific heat capacity of 4.19 kJ/(kg K) at 10°C; it takes 4.19 kJ of heat energy to increase the temperature of 1 kg of water by 1°C. The specific heat capacity will vary somewhat with temperature, but usually not significantly over temperatures common to stormwater systems.

Thermal equilibrium can be described by a heat (energy) balance, where the total energy of the system remains constant. Energy moves from one material to the other and it is assumed that no heat escapes to the surroundings.

Initial heat, $E$, is given by the product of the mass ($m$), temperature ($T$), and specific heat capacity ($c_p$):

$$E = mTc_p \tag{8.25}$$

For two materials, 1 and 2, at two temperatures, initial ($I$, which will be different for each) and final ($F$, which will be the same for each):

$$m_1 T_{1,I} c_{p,1} + m_2 T_{2,I} c_{p,2} = m_1 T_F c_{p,1} + m_2 T_F c_{p,2} \tag{8.26}$$

Stormwater can be heated by the transfer of heat to the water from a hot asphalt; the asphalt will cool. Runoff can be cooled in SCMs as it comes into contact with cooler sand or soil media. Ponded water, such as in a stormwater wetland, may be heated by the sun during warm months.

**Example 8.10** A sand filter is 12 m long, 3 m wide, and 1 m deep. The sand bulk density if 1400 kg/m$^3$ and the specific heat capacity of the sand is 0.80 kJ/(kg K). The initial temperature of the sand is 18°C (64°F). During a summer day, runoff at 32°C (90°F) flows through the filter. Assuming that the flow is slow enough to reach thermal equilibrium, find temperature of the water and sand (a) after the passage of 10 m$^3$ of water through the filter and (b) after the passage of 100 m$^3$ of water through the filter.

*Solution* Equation 8.26 will be used.
The mass of sand is found as

$$m_1 = (12\ \text{m})(3\ \text{m})(1\ \text{m})\left(1400\ \text{kg/m}^3\right) = 50{,}400\ \text{kg}$$

a) 10 $m^3$ of water has a mass of 104 kg. Solving Equation 8.26 for $T_F$:

$$T_F = \frac{(m_1)(T_{1,I})(c_{p,1}) + (m_2)(T_{2,I})(c_{p,2})}{(m_1)(c_{p,1}) + (m_2)(c_{p,2})}$$

Because the temperature shows up on both sides of the equation and that K and °C both have the same incremental increase, °C can be used in these equations. (If $K$ is used, the result is the same.)

$$\frac{(50{,}400\ \text{kg})(18°\text{C})(0.80\ \text{kJ/(kgK)}) + \left(10^4\ \text{kg}\right)(32°\text{C})\left(4.19\ \text{kJ/(kgK)}\,c_{p,2}\right)}{(50{,}400\ \text{kg})(0.80\ \text{kJ/(kgK)}) + \left(10^4\ \text{kg}\right)\left(4.19\ \text{kJ/(kgK)}\,c_{p,2}\right)}$$

$$T_F = 25.1°\text{C}$$

b) 100 $m^3$ of water has a mass of 105 kg.

$$\frac{(50{,}400\ \text{kg})(18°\text{C})(0.80\ \text{kJ/(kgK)}) + \left(10^4\ \text{kg}\right)(32°\text{C})\left(4.19\ \text{kJ/(kgK)}\,c_{p,2}\right)}{(50{,}400\ \text{kg})(0.80\ \text{kJ/(kgK)}) + \left(10^4\ \text{kg}\right)\left(4.19\ \text{kJ/(kgK)}\,c_{p,2}\right)}$$

$$T_F = 30.8°\text{C}$$

The higher the volume of water treated, the less impact the sand has on the water temperature.

Heat will be transferred via thermal conduction, convection, and radiation. For stormwater treatment using a porous medium, conduction is expected to be the dominant mechanism. The greater the temperature difference between the two materials, the faster the transfer. Also, more rapid heat transfer will occur when the contact surface area is larger.

## References

Davis, M.L. and Masten, S.J. (2004). *Principles of Environmental Engineering and Science.* Boston: McGraw Hill.

Halaburka, B.J., LeFevre, G.H., and Luthy, R.G. (2017). Evaluation of mechanistic models for nitrate removal in woodchip bioreactors. *Environmental Science & Technology* 51 (9): 5156–5164.

Khorsha, G. and Davis, A.P. (2017). Characterizing clinoptilolite zeolite and hydroaluminosilicate aggregates for ammonium removal from stormwater runoff. *Journal of Environmental Engineering* 143 (2): 04016082.

LeFevre, G.H., Hozalski, R.M., and Novak, P.J. (2012). The role of biodegradation in limiting the accumulation of petroleum hydrocarbons in raingarden soils. *Water Research, Special Issue on Stormwater in Urban Areas* 46 (20): 6753–6762.

Masters, G.M. and Ela, W.P. (2008). *Introduction to Environmental Engineering and Science*, 3rd Ed. Upper Saddle River, NJ: Prentice Hall.

Mihelcic, J.R. and Zimmerman, J.B. (2014). *Environmental Engineering; Fundamentals, Sustainability, Design*, 2nd Ed. Hoboken, NJ: Wiley.

Parker, E.A., Rippy, M.A., Mehring, A.S., Winfrey, B.K., Ambrose, R.F., Levin, L.A., and Grant, S.B. (2017). Predictive power of clean bed filtration theory for fecal indicator bacteria removal in stormwater biofilters. *Environmental Science & Technology* 51 (10): 5703–5712.

Schwarzenbach, R.P., Gschwend, P.M., and Imboden, D.M. (2016). *Environmental Organic Chemistry*, 3rd Ed. Hoboken, NJ: Wiley.

Shuman, L.M. (2000). Mineral nutrition. In: *Plant-Environment Interactions*, 2nd Ed. (ed. R.E. Wilkinson), 65–109. New York: Marcel Dekker, Inc.

Stumm, W. and Morgan, J.J. (1996). *Aquatic Chemistry: Chemical Equilibria and Rates in Natural Waters*, 3rd Ed. Hoboken, NJ: Wiley.

Vesilind, P.A., Morgan, S.H., and Heine, L.G. (2010). *Introduction to Environmental Engineering*, 3ed Ed. Boston, MA: Cengage Learning.

## Problems

**8.1** A stormwater control measure has a volume of 4000 $ft^3$. Pollutant $A$ enters at 12 mg/L and undergoes a first-order reaction in the SCM, with $k_1 = 0.02\ min^{-1}$. Find the concentration of $A$ leaving the SCM at a steady-state flow rate of 50 $ft^3$/min, assuming completely mixed conditions.

**8.2** A stormwater control measure has a volume of 4000 $ft^3$. Pollutant $A$ enters at 12 mg/L and undergoes a first-order reaction in the SCM, with $k_1 = 0.02\ min^{-1}$. Find the concentration of $A$ leaving the SCM at a steady-state flow rate of 50 $ft^3$/min, assuming plug flow conditions.

**8.3** A stormwater control measure has a volume of 4000 $ft^3$. Pollutant $A$ enters at 12 mg/L and undergoes a first-order reaction in the SCM, with $k_1 = 0.02\ min^{-1}$. Find the concentration of $A$ leaving the SCM at a steady-state flow rate of 50 $ft^3$/min, assuming mixing conditions described by four CMFRs in series.

**8.4** A stormwater control measure has a volume of 4000 $ft^3$. Pollutant $A$ enters at 12 mg/L and undergoes a first-order reaction in the SCM, with $k_1 = 0.02\ min^{-1}$. Find the concentration of $A$ leaving the SCM at a steady-state flow rate of 25 $ft^3$/min, assuming completely mixed conditions.

**8.5** A stormwater control measure has a volume of 10,000 $ft^3$. Pollutant $A$ enters at 12 mg/L and undergoes a first-order reaction in the SCM, with $k_1 = 0.02\ min^{-1}$. Find the concentration of $A$ leaving the SCM at a steady-state flow rate of 50 $ft^3$/min, assuming completely mixed conditions.

8.6 A stormwater control measure has a volume of 4000 $ft^3$. Pollutant $A$ enters at 12 mg/L and undergoes a second order reaction in the SCM, with $k_2 = 0.02$ L/(mg-min). Find the concentration of $A$ leaving the SCM at a steady-state flow rate of 50 $ft^3$/min, assuming completely mixed conditions.

8.7 A stormwater control measure has a volume of 4000 $ft^3$. Pollutant $A$ enters at 12 mg/L and undergoes a zero order reaction in the SCM, with $k_0 = 0.02$ mg/(L min). Find the concentration of $A$ leaving the SCM at a steady-state flow rate of 50 $ft^3$/min, assuming completely mixed conditions.

8.8 A stormwater control measure has a volume of 4000 $ft^3$. Pollutant $A$ enters at 12 mg/L and undergoes a first-order reaction in the SCM. The concentration of $A$ leaving the SCM at a steady-state flow rate of 50 $ft^3$/min is 2.4 mg/L. Find the value of $k_1$ assuming plug flow conditions.

8.9 Pollutant $A$ enters a SCM at 12 mg/L and undergoes a first-order reaction with $k_1 = 0.08\ min^{-1}$. The desired concentration of $A$ leaving the SCM at a steady-state flow rate of 50 $ft^3$/min is 0.12 mg/L Find the SCM volume required, assuming plug flow conditions.

8.10 Stormwater control measure 1 has a volume of 2000 $ft^3$ and is completely mixed. SCM 2 has a volume of 800 $ft^3$ and has plug flow conditions. Pollutant $A$ enters SCM 1 at 12 mg/L and the discharge from SCM 1 enters SCM 2. Pollutant $A$ undergoes a first-order reaction in both of the SCMs, with $k_1 = 0.02\ min^{-1}$. Find the concentration of $A$ leaving SCM 2 at a steady-state flow rate of 50 $ft^3$/min.

8.11 What is the Stokes settling velocity of a particle with a diameter of 0.05 mm and specific gravity of 1.8, at 32°F?

8.12 How long would it take a bacterium, diameter = 2 μm, specific gravity = 1.1, to settle 1 m at 15°C?

8.13 A sediment forebay is being designed to remove a target particle of 0.5 mm at a flow of 15 $ft^3$/s. The specific gravity is 2.4 and the design temperature is 50°F. What should be the bottom area of the forebay?

8.14 A sedimentation pond is approximately round, with a diameter of 25 m. What size particle (specific gravity = 2.5) should be completely removed by this pond at a flow rate of 25,000 L/min and 10°C?

8.15 A retention pond is used to manage flows and to improve water quality in the stormwater runoff from a large roadway interchange. The flows and concentrations into and out of these ponds are typically highly dynamic but can be buffered by flow control devices. Steady-state conditions must be assumed to allow simple evaluation of performance, which we will do here. The bottom of the pond has a width of 30 ft and a length of 50 ft. The sidewalls are sloped at 1:3 (vertical:horizontal). Consider three different steady-state storm conditions for small, medium, and large storm events. Assume a design temperature of 50°F, and that the particles have a density of 2.1 $g/cm^3$. Steady-state characteristics of the flow, pond, and input runoff water are given below.

| | Small | Medium | Large |
|---|---|---|---|
| Flow rate | 5 cfs | 15 cfs | 50 cfs |
| Water depth | 2 ft | 3 ft | 4 ft |
| Sediment conc. | 75 mg/L | 180 mg/L | 440 mg/L |

**Sediment size distribution (by mass)**

| dia. (μm) | % of total | % of total | % of total |
|---|---|---|---|
| <2 | 10 | 1 | 6 |
| 2–6 | 12 | 2 | 10 |
| 6–10 | 12 | 5 | 7 |
| 10–20 | 14 | 10 | 2 |
| 20–40 | 9 | 20 | 1 |
| **dia. (μm)** | **% of total** | **% of total** | **% of total** |
| 40–50 | 14 | 15 | 2 |
| 50–80 | 15 | 10 | 5 |
| 80–100 | 10 | 12 | 10 |
| 100–200 | 8 | 6 | 20 |
| 200–300 | 1 | 10 | 15 |
| 300–400 | 1 | 7 | 10 |
| 400–500 | 5 | 2 | 12 |

For each storm type, answer the following questions.

a. What is the hydraulic retention time for the pond (min)?
b. What is the hydraulic loading on the pond (gpd/ft$^2$)?
c. What are the expected sediment removal (%) and the discharge sediment concentration (mg/L)?
d. Discuss your results.

**8.16** Consider that the bioretention media has a $d_{10}$ (collector diameter) of 0.65 mm, a depth of 0.6 m, and a porosity of 51%. For a TSS particle size of 0.01 mm (specific gravity of 2.5) and a sticking coefficient of 0.02, what is the predicted TSS removal fraction $(1 - N/N_0)$ exiting the bioretention facility during the stormwater treatment at a superficial velocity of 50 cm/h at 10°C? Assume that the dominant filtration transport mechanism is sedimentation.

**8.17** A filter manufacturer is designing a carbon filter for home drinking water use. The Freundlich adsorption isotherm constants for lead are $K_F = 3.4$ and $1/n = 0.96$ (for $C$ in mg/L and $q$ in mg/g). The filter should last 6 months (180 days) before lead breakthrough, which is the recommended replacement lifetime. Assumptions include a family of four, each person drinking 1:l of water per day from the filter.

a. The lead concentration in the untreated water is 0.008 mg/L. How much carbon should be added to the filter assuming it operates in a column configuration (g)?
b. Assume that the filter contains 200 g of carbon. If a family of eight drinks 1.5:l of water each per day from the filter, how many days would the filter last until breakthrough?
c. A modified carbon with higher lead adsorption capacity is under study by the manufacturer. During a batch laboratory study, 0.04 mg/L Pb is reduced to 0.004 mg/L by the addition of 2 g/L of the new carbon. The value of the Freundlich $1/n$ remains unchanged at 0.96. What is the value of the new $K_F$?

8.18 A large pond near a residential area receives runoff from a surrounding agricultural field. Because of the excessive phosphorus levels in the pond, frequent algal blooms occur, leading to aesthetically unacceptable conditions and odors. Measured concentrations of P are about 1.8 mg/L. A consulting firm is hired and one of the recommendations is to add an aluminum oxide-based adsorbent to the pond to adsorb excess phosphorus. The adsorbent would sink and carry the phosphorus to the sediments. The following table provides lab-scale batch data on phosphorus adsorption onto the adsorbent at 20°C. The adsorbent was added at 0.25 g in 250 mL of pond water.

| Initial (P) (mg/L) | Final (P) (mg/L) | Initial (P) (mg/L) | Final (P) (mg/L) |
|---|---|---|---|
| 4 | 0.95 | 0.7 | 0.085 |
| 3 | 0.66 | 0.3 | 0.032 |
| 2 | 0.33 | 0.1 | 0.005 |
| 1 | 0.11 | | |

a. Find the Freundlich isotherm parameters, $K_F$ and $1/n$, based on these laboratory data.
b. Find the amount of adsorbent required (kg) to be added to the pond to reduce the P concentration to less than 0.1 mg/L. The pond is approximately elliptical, with one length equal to 350 ft, the width equal to 110 ft, and an average 4 ft depth. **Clearly** note any assumptions that you make.
c. What do you expect the concentration of P to be after the addition of 5000 kg of adsorbent to the pond (mg/L)?
d. In the winter, the pond contains 2.1 mg/L P. If 1250 kg of adsorbent is added to the pond in winter, the equilibration pond P concentration decreases to 0.8 mg/L; if the adsorbent addition is increased to 2500 kg (total), the P concentration decreases to 0.4 mg/L. What is the expected P concentration after 7500 kg adsorbent addition?
e. Explain the differences in adsorbent performance between the results of Part b and Part d. Why is the performance different? Be specific.

8.19 Several streams contain a highly toxic compound known as CaPsToNe. The following needs for a strong adsorbent of this compound are known:

a. Wastewater is occasionally produced that contains 62 mg/L CaPsToNe. This water is stored until the volume reaches 2000 gallons. It is desired to add 1.0 kg of adsorbent in this batch system to reduce the CaPsToNe concentration to 3 mg/L.

b. Also a continuous wastewater flow is produced at 2 gpm, containing CaPsToNe at 24 mg/L. This flow is to be treated via an adsorption column. This column is a cylinder, length of 34 in, diameter of 10 in. The adsorbent in the column should last 180 days before replacement.

c. Another continuous wastewater flow is produced at 6 gpm, containing CaPsToNe at 18 mg/L. This flow is to be treated in a slightly larger cylindrical column, length of 34 in, diameter of 12 in. The adsorbent in the column should last 100 days before replacement. Find the technical specifications for this adsorbent. The adsorption process should follow a Freundlich isotherm, so specify $K_F$ and $1/n$. Also specify the adsorbent bulk density, either in lb/ft$^3$ or kg/m$^3$.

**8.20** A stormwater control measure contains 100 kg of organic matter. Find the volume of water that can be denitrified by this organic matter. The water contains 4.4 mg/L total N. Use the stoichiometry of Example 8.8 for the denitrification reaction.

**8.21** 80,000 kg of sand is present in a sand filter at 15°C. Find the volume of water, at 22°C, that can be treated (at thermal equilibrium), before the system temperature increases above 18°C.

**8.22** 40,000 kg of sand is present in a sand filter at 15°C. Find the volume of water, at 22°C, that can be treated (at thermal equilibrium), before the system temperature increases above 18°C.

**8.23** 80 m$^3$ of water, at 28°C, must be treated (at thermal equilibrium) to 22°C. Find the mass of sand (at 15°C) required.

**8.24** 80,000 kg of sand is present in a sand filter at 15°C. About 80 m$^3$ of water must be treated (at thermal equilibrium) to below 18°C. Find the maximum water temperature allowed.

# 9

# Stormwater Performance Measures and Metrics

## 9.1 Introduction

A major challenge in the management of urban stormwater is defining appropriate metrics for success. For example, what is the appropriate degree of water quality or runoff management that is necessary for implementation into an urban development area to protect the surrounding environment and ecosystems? This is not a simple question to answer because of the complexity of the landscapes, both urban and that of the pervious natural, undeveloped lands. Added to this are the stochastic nature of rainfall/runoff, and natural geologic and climatic differences, and different water body designated uses, which contribute additional layers of complexity.

The ideals of measures and metrics are continuously evolving. Several tools and techniques are available that have grounding in historical hydrologic analyses, wastewater treatment, or other disciplines. Many of these tools are imperfect for GI use. However, metrics are needed if we are to address the environmental detriments caused by urban development. These topics are explored in this chapter with the caveat that more refined measures and metric are expected in the future.

In establishing stormwater management requirements, many different criteria are considered by various jurisdictions depending on the needs of the local water environments. Regulatory requirements and retrofit goals are linked to changes in the runoff characteristics from the watershed or sewer-shed (in the case of highly channelized and piped systems) and their effects on regional water bodies. Water-quality improvement, groundwater recharge, and stream protection are common criteria. Criteria are related to the climate, soils, land form, vegetation, and surroundings and vary from state to state and within states. Thus, requirements in Washington State established for rock-bed salmon rivers are different from those in Maryland and Pennsylvania with longer soil–river systems, and from nitrogen issues in North Carolina. Generally, the effects from the hydrologic perspective can be related to runoff volume and erosive flow.

Stormwater control measures provide the link between the existing conditions of the watershed and the desired hydrologic and water-quality characteristics leaving the watershed. The selection, design, construction, and maintenance of the stormwater control measures (SCMs) will determine how well they meet specific desired performance metrics.

*Green Stormwater Infrastructure Fundamentals and Design*, First Edition. Allen P. Davis, William F. Hunt, and Robert G. Traver.

Companion Website: www.wiley.com/go/davis/greenstormwater

## 9.2 Reference Conditions and Defining Thresholds

In many cases, a desired goal of urban stormwater management is to replicate some hydrogeological parameter representative of undeveloped lands. This may be as simple as reduction in a peak flow during a design storm or as complex as "replicating pre-development hydrology and water quality."

Nevertheless, in order to meet even the simplest comparison to predevelopment conditions necessitates the selection of the type of undeveloped land to be replicated, as well as the design storm(s) to be considered. What is the predevelopment land use that should be replicated? Is this the same for every case? Land uses may have changed several times over the years, so what baseline is used? Maybe a forest is considered as the most pristine land use. Should all developments aim to replicate forest hydrology? What if the land has been in agricultural use for the past 100 years? Even agricultural lands have many uses, including row crops, pasture, hay, and orchard. The reference condition specified is critical to addressing SCM treatment goals. Similarly, should the hydrologic performance be replicated over a specific storm size? Or many sizes?

Additionally, when selecting reference conditions, what parameters should be considered in the stormwater management process? Hydrologically, parameters include matching runoff volume and surface discharge, flow rates, infiltration rates, groundwater recharge, and ET. A whole suite of water-quality parameters could be selected. Spatial and temporal differences in these parameters throughout the drainage catchment may require evaluation as well.

As part of the reference condition analysis, thresholds are being developed and specified in many jurisdictions. These thresholds are selected from water-quality or hydrologic parameters typical to what may be found in a selected undeveloped land. These thresholds may be established to limit erosive flows, aquatic toxicity, pollutant concentrations, or the mass loadings of pollutants to a water body.

## 9.3 Volume Control

### 9.3.1 Runoff Depth

Runoff volume criteria are usually expressed as a depth of rainfall or runoff over the drainage area. The volume required for capture or management in the SCM is then obtained by multiplying the depth by the established drainage area. However, the diversity of defining a volume to capture or recreate can vary widely among different jurisdictions. Some regions, such as Pennsylvania, require capture of larger volumes by specifying an extreme storm event (the 2-year, 24-h storm). By specifying this large volume, recharge, water quality, and stream geomorphology concerns are being addressed. Recent US federal regulations use the 90% annual daily storm event as a design requirement. This is the annual 90th percentile of average daily precipitation. Many east coast regions use 1 in (2.5 cm) as the volume requirement to mitigate for recharge and water quality.

Figure 9.1 shows a diagram of the cumulative rainfall depth from 70 years of data at Thurgood Marshall Baltimore-Washington Airport, similar to that of Figure 2.2. The lower black curve shows the fraction of the total volume contributed by storm events

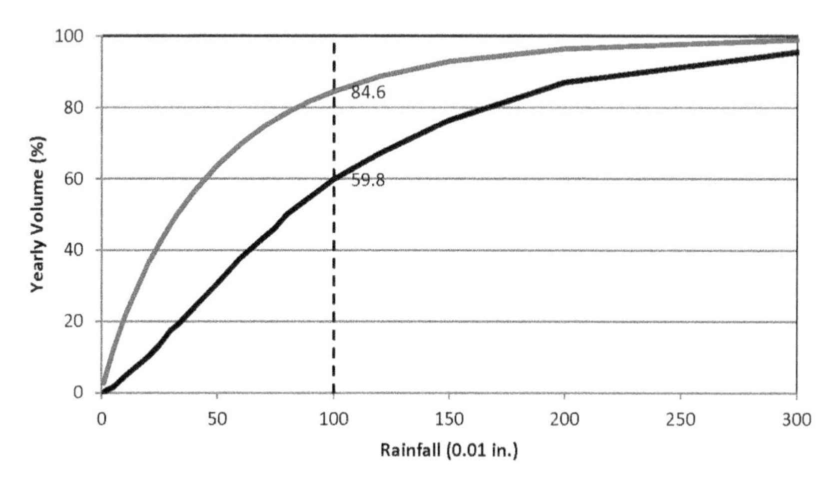


Figure 9.1 Cumulative Rainfall Depth from 70 years of Data at Thurgood Marshall Baltimore-Washington Airport.

of the posted size and lower. Based on this analysis, 59.8% of the rainfall depth was from events 1 in (2.5 cm) and smaller. The upper red curve shows the fraction of cumulative volume from events of the posted size and lower, *plus the posted depth subtracted from the larger events*. So, for this analysis, 84.6% of the total rainfall volume occurs in storms of 1 in or less, plus the first inch of rainfall from all larger storms. This type of information provides the basis for the use of a fixed rainfall/runoff volume as the basis for a treatment metric.

Frequently, when a water-quality requirement is added, it is based upon the "first" flush concept and the capture and treatment of an initial runoff volume. A *water-quality volume* is defined in this manner. As was discussed in Chapter 5, with a first flush the majority of the pollutant mass load is found in the initial fractions of the total runoff volume. If a first flush is considered, a pollutant mass curve would be even higher than the upper curve in Figure 9.1.

### 9.3.2 Curve Number Reduction

The NRCS Curve Number method for runoff depth prediction is described in Chapter 6. The curve number is based on the imperviousness of the land surface, with a value of 100 representing complete imperviousness and values becoming lower with increasing perviousness. Curve number reduction has been used as a metric, albeit imperfect, to describe effective stormwater hydrologic management. Through the addition of a single, or some combination of SCMs, the curve number of a developed impervious area can be reduced to a value that may be representative of a more pervious drainage area. From the perspective of a metric, matching a post-development curve number to that estimated for the pre-developed area has been suggested.

Curve number reduction has been employed to reference the hydrologic impact of several SCMs, including green roofs (Carter and Rasmussen 2006; Fassman-Beck et al. 2016), pervious pavements (Bean et al. 2007; Martin and Kaye 2014; Schwartz 2010), and bioretention (Olszewski and Davis 2013). These studies will be discussed in detail in the respective SCM chapters.

The use of curve numbers and curve number reductions may be an inaccurate metric to describe SCM performance. Usually the curve number analysis does not accurately capture the hydrologic action of the SCM. The SCM usually has an initial storage component, where no runoff discharge will occur, followed by a discharge once the storage capacity is reached. The discharge will depend on the water-holding capacity and infiltration characteristics of the SCM and the surrounding soils. This volumetric storage performance of SCMs is discussed in Chapter 11.

## 9.4 Peak Flow, Flow, and Geomorphology

As discussed in Chapter 1, the speed of transport of the runoff is a key challenge in urban stormwater management. The direct connection of roofs to driveways, streets, curb and gutter, to the drainage network, directly to the receiving waters create a water superhighway, causing the water to enter streams and rivers at a high volume and rate of flow. This is demonstrated in Figure 9.2, where the peak flow of the urban hydrograph is much higher and earlier than that of a nearby undeveloped watershed for the same storm. This high flow rate and velocity can severely erode the banks of the streams. This erosion is damaging to the streams directly, eroding stream banks and removing habitat and nutrient-processing ability. Also the eroded sediment is transported downstream, causing other problems.

The amount of sediment carried by a stream can be related to the power of the stream (flow and slope). Adding more flow from a storm event increases the stream power and allows it to carry more sediment and larger sediment particles (Lane 1954).

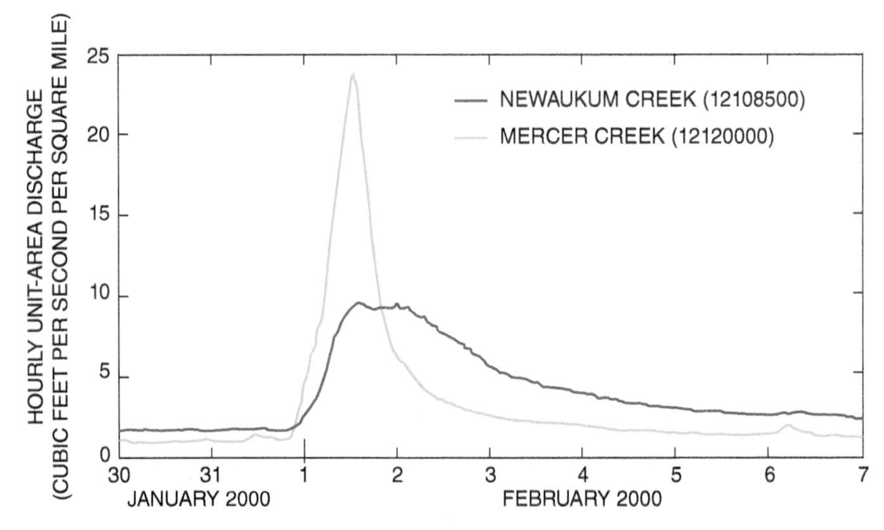


**Figure 9.2** Streamflow in Mercer Creek, an Urban Stream in Western Washington, Increases More Quickly, Reaches a Higher Peak Discharge, and Has a Larger Volume During a One-Day Storm on February 1, 2000, Than Streamflow in Newaukum Creek, a Nearby Rural Stream, That Drains a Basin of Similar Size. (Credit: US Geological Survey, http://water.usgs.gov/edu/watercyclerunoff.html).

In response, peak flow became a primary hydrologic metric for stormwater control. A common metric is that the flow peak runoff rate for the post-development condition has to be reduced to that of the predevelopment for a specific storm event(s), such as the 2-year and/or 10-year average return period. This could be done by building stormwater ponds with orifices and weirs that would release the flow at the predevelopment peak rate.

Nonetheless, a focus only on flow peaks and peak reductions can still lead to channel erosion. This is depicted in Figure 9.3. The predevelopment hydrograph is depicted as the lower blue curve; the upper red curve is that for the post-development. Detention is used to manage the runoff, storing the water volume produced during the peak flow. This stored water volume is discharged at (maximum) the predevelopment peak as given by the dashed black line. However, the predevelopment peak flow can still be erosive, but it only occurs for a short time predevelopment. In the modified condition, the lower predevelopment peak is held for a much longer duration. This can result in long periods of erosive flows (McCuen 1979).

The combination of volume and flow duration, when as a continuous flow measure, adds perspective to comparisons between developed and undeveloped lands. Figure 9.4 shows the flow of a forested stream in Maryland over about 6 years, normalized by the drainage area. The runoff is a log scale, so some reasonable variation in the flow was experienced. During one drought period, flow decreased significantly. Overall, the flow averaged about 0.8 mm/day. Maximum flows ranged about 3–8 mm/day.

Figure 9.4 can be compared to Figure 9.5, which shows flows from an urban stream in Maryland over the same time period. Although the overall average is about the same at approximately 0.8 mm/day, the flow variation was significant on both the high and low sides. Flows were frequently above 10 mm/day and in a few cases approached 100 mm/day. Frequent lower excursions were documented also. The urban landscape prevents shallow infiltration, which provides stream baseflow. During dryer periods, the baseflow was reduced in the stream.

The curves shown in Figures 9.4 and 9.5 can provide targets for addressing runoff flows. Flow above a threshold value, such as 10 mm/day, should be minimized or

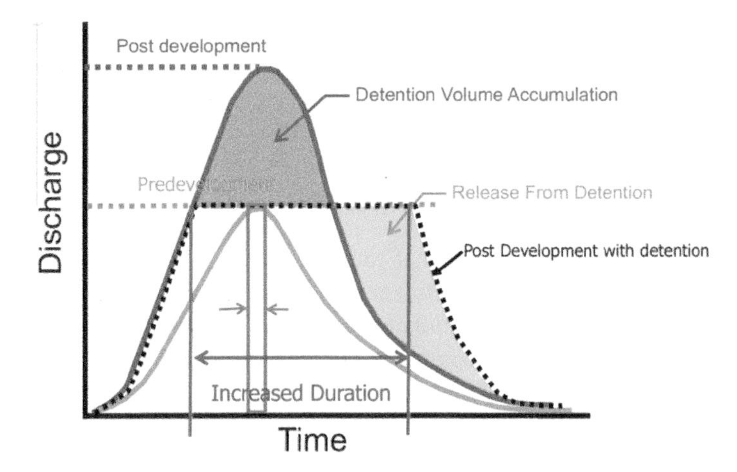


Figure 9.3 Water Quantity Measures Based on Pre- and Post-development Hydrographs, and a Constant Release Detention Facility.

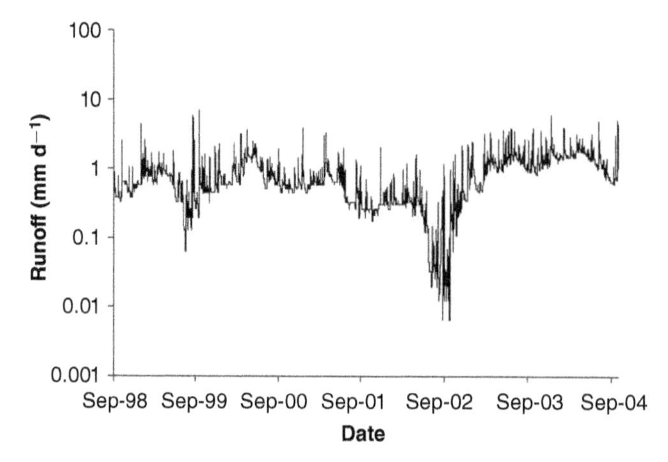


Figure 9.4 Continuous Flow Measured from a Forested Stream in Maryland (Normalized by Drainage Area) (Shields et al. 2008).

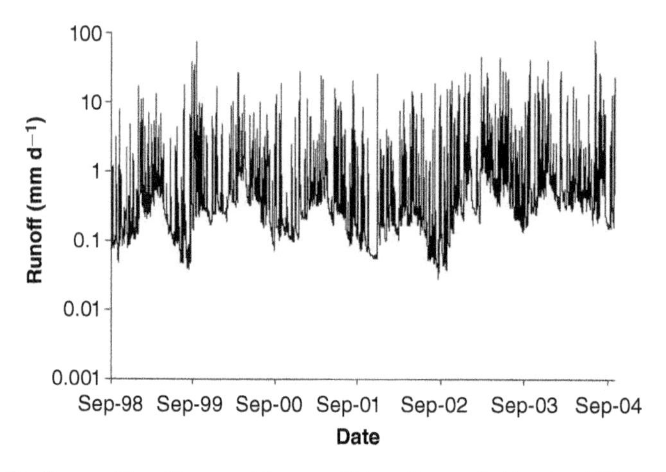


Figure 9.5 Continuous Flow Measured from an Urban Stream in Maryland (Normalized by Drainage Area) (Shields et al. 2008).

eliminated, as these can lead to erosion. Obviously, very high flows will lead to flooding. Also, lower flows should be avoided so as to maintain adequate flows to support diverse aquatic life. No aquatic life can survive conditions where the stream baseflow is eliminated.

Building upon the consideration that the duration in which the stream exceeds some threshold value can lead to detrimental impacts on streams, flow duration information can provide a metric for runoff evaluation and treatment success. Flow duration curves can be prepared by taking a hydrograph, or series of hydrographs, and ranking the data from highest flow to lowest. This gives the duration of each flow. Figure 9.6 provides examples of two stream flow duration curves from a subbasin discharge point. The blue curve shows data before urbanization of the watershed and the red curve represents more recent flows after urbanization. The duration of high flows

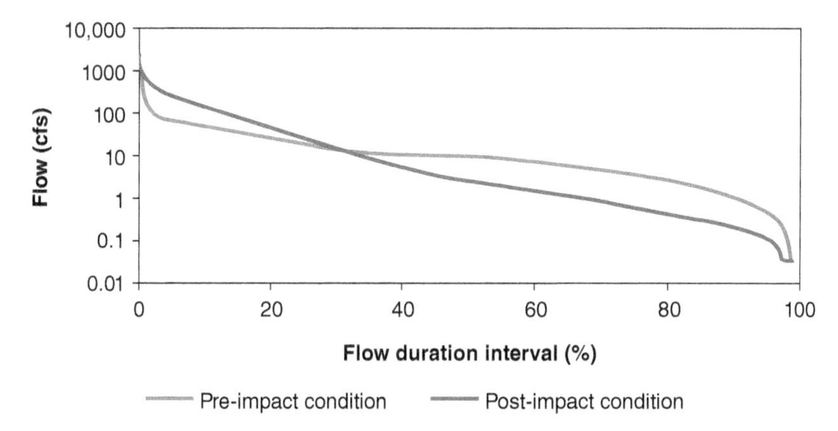


Figure 9.6 Stream Flow Duration Curves Showing Impact of Urbanization. Predevelopment Conditions Include Years 1962–1979 and Post-development Was 1985–2008 (Kannan et al. 2018).

is much greater for the fully developed site conditions. In addition, the urbanized watershed has smaller duration of low flows.

In combined sewer systems, maximum stormwater release rates are being enforced. This will minimize overflows in the combined systems. For example, stormwater inflow to the combined sewer may be limited to maximum release rate of 10% of the system capacity. Stormwater infrastructure would need to be designed to meet this requirement.

## 9.5 Pollutant Percent Removal

A number of water-quality metrics are now being used for defining SCM performance. These include metrics based on removal, concentration, and annual mass loads. Because of the variability inherent to rainfall and runoff, duration and exposure of the water body to high pollutant concentrations may be important. Also, rather than having a hard limit, probability and exceedance parameters can be included as metrics.

In search of simple metrics, many regulations have been promulgated with a focus on a simple parameter, such as percent reduction of some water-quality parameter employing a SCM. This follows conventional usage in wastewater treatment where steady-state conditions are common. However, due to the complexity and variability of runoff systems, percent removal generally does not provide a meaningful metric for environmental protection. Fraction or percent removals, $R^0$, are calculated as

$$R^0 = \left(1 - \frac{C}{C_0}\right) \tag{9.1}$$

where $C$ is the SCM output concentration (typically the EMC) and $C_0$ is the input concentration. (Multiply by 100 to get percent.) Therefore, this metric relies on the input concentration to calculate the performance efficiency. The input pollutant

Table 9.1 Example of percent removal as a SCM performance metric

| Storm event | Input concentration (mg/L) | Output concentration (mg/L) | Removal (%) |
|---|---|---|---|
| A | 500 | 50 | 90 |
| B | 50 | 10 | 80 |
| C | 5 | 3 | 40 |

concentration varies greatly in stormwater, as discussed in Chapter 5. Therefore, percent removal from an SCM will vary within storms and for different storms. Storms with high input pollutant concentrations will commonly produce higher percent removals since SCM output concentrations tend not to vary as much as input.

Table 9.1 shows a simple data set that demonstrates this perspective. The 50-mg/L output from a 500-mg/L input is an impressive 90% removal. However, the best water-quality result is the 3 mg/L output, yet the removal is only 40%.

## 9.6 Chesapeake Bay Retrofit Curves

The Chesapeake Bay Program provides credit for pollutant removal for the implementation of SCMs in retrofit situations based on performance curves estimated from SCM data (CSN 2012). The amount of runoff depth captured (per impervious acre), RD, is given by

$$RD = \frac{12RS}{IA} \tag{9.2}$$

where RS = runoff storage volume is the volume treated by the SCM (ac ft), and IA = impervious area (ac).

SCMs are divided into those that provide treatment only and those that provide treatment and runoff reduction (via infiltration, etc.). The pollutant removal credit, as a percent removal, is given based on the RD, with a different curve for treatment SCMs and for runoff reduction SCMs. Curves have been developed for phosphorus, nitrogen, and sediment because these are the three pollutants regulated under the Bay TMDL. As an example, for a treatment SCM with RD equal to 0.5 in, credits for phosphorus, nitrogen, and sediment are 41%, 26%, and 52%; these removals increase to 52%, 44%, and 55%, respectively, for runoff reduction SCMs.

## 9.7 Target Effluent Concentrations

One of the simplest and most desired targets is a threshold pollutant concentration for in-stream water quality. A limit can be used based on aquatic toxicity, eutrophication potential, or other concerns. Water-body threshold concentrations have been collected from a number of sources for use in stormwater management; some of these thresholds are summarized in Table 9.2. These values are based on typical concentrations regulated in wastewater discharges, ambient water-quality

Table 9.2 Fresh water concentration thresholds that can be used as SCM treatment metrics.

| Water-quality parameter | Units | Value | Type of threshold |
|---|---|---|---|
| TSS | mg/L | 25 | Common Wastewater Discharge Limit |
| $NO_2 + NO_3$ | mg/L as N | 0.2 | Potomac River Ambient Water Quality Goal |
| TN | mg/L as N | 0.32–0.87 | Ambient Water Quality Criteria Recommendations (USEPA 2000, USEPA 2001) |
| TP | mg/L as P | 0.03125 | Ambient Water Quality Criteria Recommendations (USEPA 2000, USEPA 2001) |
| *E. coli* | MPN/100 mL | 126 | Recreational Water Limit |
| Fecal coliform | MPN/100 mL | 200 | Recreational Water Limit |
| Cd–T | μg/L | 2.0 acute, 0.25 chronic | MD Fresh Water Toxic Substances Criteria for Ambient Surface Waters |
| Cr(VI) | μg/L | 16 acute, 11 chronic | MD Fresh Water Toxic Substances Criteria for Ambient Surface Waters |
| Cu–T | μg/L | 13 acute, 9 chronic | MD Fresh Water Toxic Substances Criteria for Ambient Surface Waters |
| Pb–T | μg/L | 65 acute, 2.5 chronic | MD Fresh Water Toxic Substances Criteria for Ambient Surface Waters |
| Ni–T | μg/L | 470 acute, 52 chronic | MD Fresh Water Toxic Substances Criteria for Ambient Surface Waters |
| Zn–T | μg/L | 120 acute, 120 chronic | MD Fresh Water Toxic Substances Criteria for Ambient Surface Waters |
| Chloride | mg/L | 250 | Secondary Drinking Water Limit |

standards for nutrients, and toxicity standards for heavy metals. While these values are not regulatory standards, they do provide reasonable goals for SCMs that discharge to natural waters and allow for comparisons between SCM input and output in meeting these goals.

The value for TSS is a typical NPDES permit discharge for a municipal wastewater treatment plant. The value for $NO_x$ was published as water-quality goals for the Potomac River to protect against eutrophication and to provide "excellent" water quality. More recently, a number of states are setting numeric criteria for surface water quality, specifically for total N and total P (NRC 2012). The TN and TP values are a range for rivers, lakes, and reservoirs in EPA nutrient region XIV, which considers the US eastern coastal plain; other regions have similar values. Values for N and P can be quite stringent.

The *coliform* concentrations are based on federal recreational water limits that are used to direct beach closures. The heavy metals concentrations are taken from acute and chronic limits for aquatic life in Maryland fresh waters. The chloride limit is the federal secondary drinking water standard.

## 9.8 Annual Mass Load

In a watershed TMDL application, the mass load discharge is the metric of interest. The annual pollutant mass load must be reduced below a specified value to meet the TMDL limit. Various SCMs can be implemented throughout the watershed to reduce the total annual pollutant mass below that listed in the TMDL for that specific land-use sector. SCMs that reduce runoff pollutant concentrations will reduce the overall pollutant mass loads, helping to address TMDL challenges.

The annual load, $L$, is commonly calculated using the simple method.

$$L = cRP(\text{EMC}) \tag{9.3}$$

where $c$ is the fraction of the drainage area that contributes runoff, $R$ is a factor to account for the fact that very small rainfall events do not produce runoff (abstraction), $P$ is the annual precipitation depth, and EMC is the event mean concentration of the pollutant based on the best available data. The default value for $R$ is 0.9 in urban areas. Equation 9.3 must be modified to obtain the desired set of units.

**Example 9.1** The TSS EMC from a small commercial catchment with 75% impervious area is 65 mg/L. The annual rainfall is 38 in/year, with an estimated 90% causing runoff. Calculate the annual TSS load in as lb/ac/year and kg/ha/year.

*Solution* Using Equation 9.3 and the given values:

$$L = cRP(\text{EMC}) = (0.75)(0.9)(38\ \text{in/year})(65\ \text{mg/L}) = 1667\ \text{in mg/L year}$$

Converting to SI units:

$$L = 1667\ \text{in mg/L/year}\frac{(2.54\ \text{cm/in})(1000\ \text{L/m}^3)(10^4\ \text{m}^2/\text{ha})}{(100\ \text{cm/m})(10^6\ \text{mg/kg})} = 423\ \text{kg/ha/year}$$

Converting this value to English units:

$$L = 423\ \text{kg/ha/year}\frac{(2.2\ \text{lb/kg})}{(2.47\ \text{ac/ha})} = 377\ \text{lb/ac/year}$$

The annual pollutant load is important in evaluating the contributions of different land uses to the overall pollutant input to a water body and TMDL evaluations. The watershed can be divided into different land uses and the area of each land use is multiplied by the annual load for each respective land use. The net result is the total pollutant mass load to the water body. Reductions in urban runoff loads to meet TMDLs can be obtained through the implementation of SCMs targeted to the treatment of specific pollutants. GI SCMs can be very successful in the reduction of mass loads, since these SCMs commonly act on reducing both the stormwater volume and pollutant concentrations.

**Example 9.2** A TMDL for a river requires the reduction of 7000 lb of P/year from the urban sector, from 42,000 lb P/year to 35,000 lb P/year. The urban area is 15,000 ac. An SCM is targeted that can produce treated runoff at 0.4 lb P/ac/year. On average, how many urban acres must be treated by this SCM?

*Solution* The current average annual P load by the urban area is

$$L = \frac{42{,}000 \text{ lb P/year}}{15{,}000 \text{ ac}} = 2.8 \text{ lb/ac/year}$$

Letting $x$ be the number of acres retrofit with SCMs:

$$35{,}000 \text{ lb P/year} = x(0.4 \text{ lb/ac/year}) + (15{,}000 - x)(2.8 \text{ lb/ac/year})$$

$x$ =2917 ac must be retrofit with SCMs.

## 9.9 Probability and Exceedance

Because rainfall, which is stochastic, is the driver of stormwater and stormwater management, pollutant concentrations in stormwater from various storm events can also be analyzed stochastically. One procedure for this is using a probability distribution of pollutant concentrations, in a manner similar to that done for rainfall and runoff events. In this method, a Weibel or other similar distribution is used for the plotting position, using concentration data ranked from smallest to largest, such as

$$p = \frac{i - \alpha}{(n + 1 - 2\alpha)} \tag{9.4}$$

where $i$ is the ranking number and $n$ is the total number of observations. A value for $\alpha = 3/8$ has been used in stormwater studies. Pollutant concentrations in urban stormwater typically follow a log-normal distribution on a probability plot (Van Buren et al. 1997). The range of concentrations can span over two or more orders of magnitude.

A log-normal probability plot for TSS concentrations from two parking lot/roadway areas is presented in Figure 9.7. Note from Figure 9.7 that the EMCs from each site range over almost two orders of magnitude, from 7 to 400 mg/L at the College Park site and 5 to 500 mg/L at Silver Spring. The smaller values are relatively clean, whereas the higher represent highly polluted waters. The 50% probability value corresponds to the median, 78 and 27 mg/L for College Park and Silver Spring, respectively.

The distribution of pollutant concentrations, as noted in Figure 9.7, can be evaluated in reference to a target concentration metric, as discussed in Section 9.7. Some of the runoff concentrations will be less than the target, while some will be greater. The TSS target of 25 mg/L, as noted in Table 9.2, is highlighted in Figure 9.7. Therefore, these runoff values can be discussed in terms of probability and exceedance of the target concentration metric. For the data of Figure 9.7, the 25-mg/L metric exceedance probability is about 80%, indicating that 80% of the runoff events are expected to exhibit a TSS EMC that exceeds 25 mg/L. At the Silver Spring site, the 25-mg/L exceedance is only about 50%.

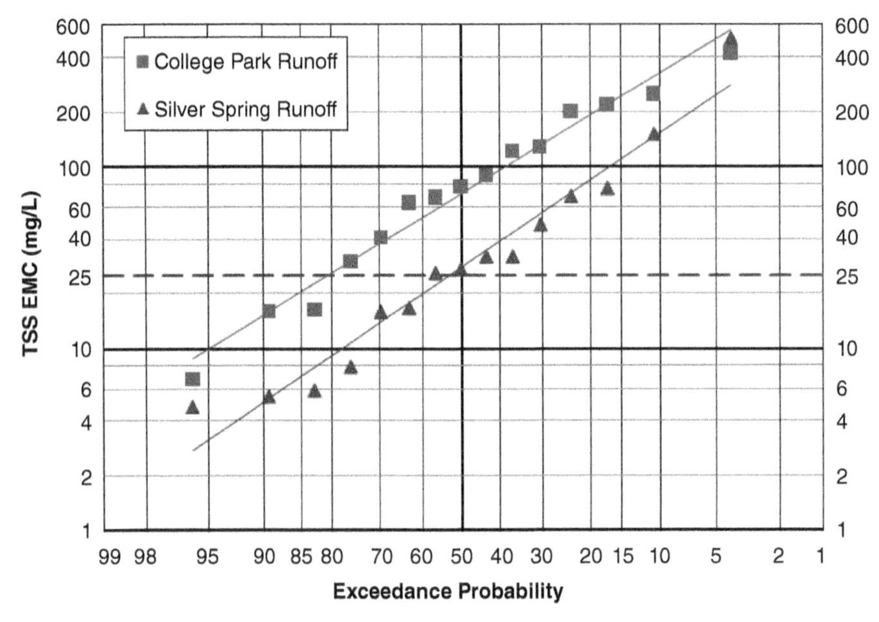


**Figure 9.7** Probability Distribution of TSS EMCs from Runoff at Two Maryland Urban Sites. Data Taken from Li and Davis (2009).

This value of exceedance probability can be extended to SCM treatment performance. The SCM will lower (or possibly raise) the pollutant concentration in the SCM discharge. The SCM discharge concentrations will also have a distributed range of values. The discharge concentration range can also be compared to a target concentration and exceedance probability determined. The effectiveness of the SCM can be represented in terms of the change in exceedance probability imparted by the SCM.

Figure 9.8 shows TSS probability plots for the two parking roadway areas presented in Figure 9.7, along with the corresponding effluent from a bioretention cell at each site. The TSS concentrations in the water exiting the bioretention underdrains are clearly lower than the input. While the treatment efficiency may vary from storm event to storm event, it can be represented as a change in the exceedance probability from the target value. A design target can be established that would allow some probability of exceedance of the target. This may be a better performance metric than a fixed standard that could not be exceeded and would allow no flexibility during extreme events. With a specified exceedance probability, for the majority of the events, the SCM would provide adequate treatment and the discharge EMC would be below the target threshold.

**Example 9.3** A water-quality standard for the College Park bioretention facility designated in Figure 9.8 states that the discharge TSS EMC can exceed 40 mg/L for only 10% of the runoff events. Does this bioretention cell meet this requirement? What is the 40-mg/L discharge exceedance probability? What was the 40-mg/L TSS exceedance probability before the bioretention treatment?

***Solution*** In Figure 9.8, the CP bioretention cell-treated TSS in the runoff is given by the circles, with the corresponding line representing the log-trend. The trend line meets 40 mg/L at about 4% probability. Therefore, the effluent is expected to exceed 40 mg/L for only 4% of the storm events, which meets and exceeds the regulatory requirement. The 40-mg/L requirement was exceeded about 70% of the time for the runoff from the roadway/parking lot, coming into the bioretention cell.

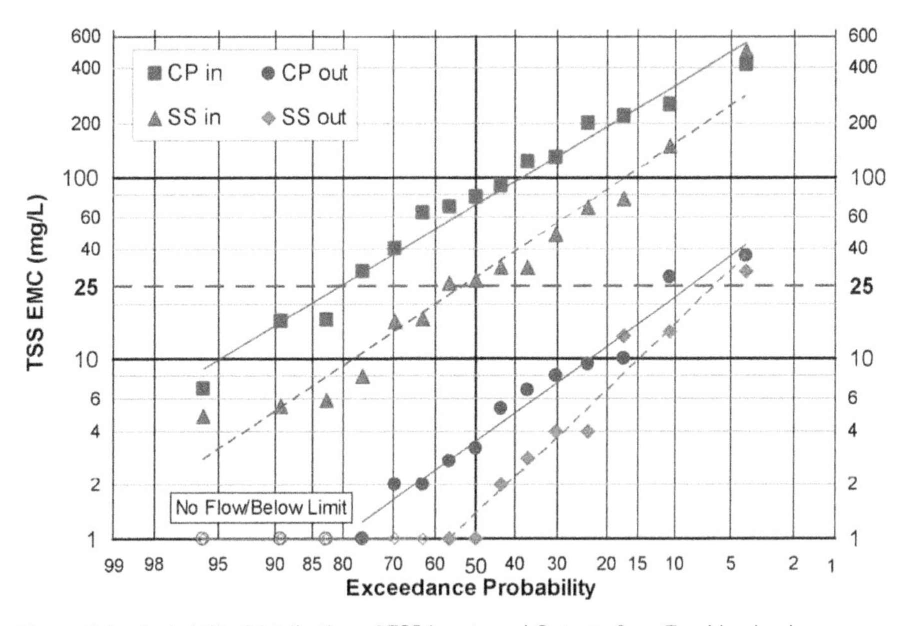


Figure 9.8 Probability Distribution of TSS Inputs and Outputs from Two Maryland Bioretention Cells (Li and Davis 2009, with Permission from ASCE).

## 9.10 Pollutant Durations

From a toxicity perspective, the duration in which an organism is exposed to concentrations exceeding some threshold value can be a critical parameter influencing organism viability. Similar to flow duration, discussed in Section 9.4, minimum duration for toxicity exposure is desired. While pollutant duration information can be difficult and costly to obtain, it can provide a water-quality metric for runoff quality evaluation and treatment success.

As the flow duration curve is related to the hydrograph, the pollutant duration curve is related to the pollutograph. To find the pollutant duration curve, discrete pollutant measurements are needed as a function of time during a runoff event. These concentrations are assumed to be representative of runoff for a fixed sample time corresponding to each specific sample.

To synthesize the pollutant duration curve, the sample concentrations, with their respective durations, are ranked highest to lowest. The pollutant duration curve is plotted as concentration as a function of cumulative duration. Pollutant duration curves can be created for a single runoff event, or as combinations of many events, monthly, seasonally, yearly, or more.

As a metric, a threshold water-quality concentration is selected, such as a value from Table 9.2. The cumulative duration in which the runoff concentration exceeds this value is selected as the evaluation metric. Similarly, from a treatment perspective, the discharge from an SCM can be arranged into a pollutant duration curve. The exceedance duration for the SCM discharge can be determined and compared to that from the catchment to determine a metric of water-quality performance for the SCM.

TSS pollutant duration curves for a highway runoff and output from two grass swales are shown in Figure 9.9. The TSS data are plotted on a log scale because of the large range of concentrations experienced. This range is larger than an EMC range, since individual concentration measurements have a larger range than the flow weighted composite concentrations. For the highway runoff, individual TSS concentrations greater than 1000 mg/L were noted. A threshold value of 30 mg/L was selected in this example and the highway runoff exceeded 30 mg/L TSS for 40 total hours during this study of about 97 h of measured runoff.

The maximum swale discharge TSS concentration just exceeded 100 mg/L for both swales. One swale had a pretreatment filter strip adjacent to the swale length (FS), the other did not (No-FS). The swale discharges exceeded the 30-mg/L threshold for only about 2–3 h. The overall TSS duration curve is shorter for the swales because a significant fraction of the flow was infiltrated by the swales and did not result in discharge (hydrologic modification).

The discharge from a SCM would be expected to exhibit a pollutant duration curve with lower concentrations. It may have a longer or shorter duration, depending on the effect of the SCM on flow.

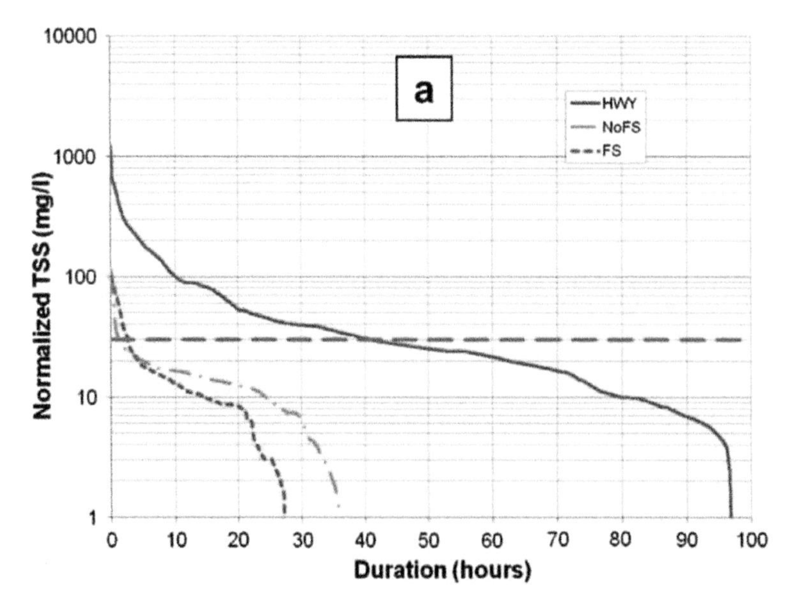


Figure 9.9 TSS Pollutant Duration Curves for the Input (HWY) and Outputs from Two Maryland Grass Swales (Stagge et al. 2012).

**Example 9.4** The following table, reproduced from Example 5.1, gives concentrations of TSS in runoff from a campus parking lot. Calculate and plot the TSS duration curve for this storm event:

| Time of day | Flow rate (L/s) | TSS (mg/L) |
|---|---|---|
| 10:00 | 0 | – |
| 11:00 | 2.8 | 175 |
| 12:00 | 4.9 | 45 |
| 13:00 | 5.7 | 75 |
| 14:00 | 20 | 280 |
| 15:00 | 15 | 145 |
| 16:00 | 5.4 | 115 |
| 17:00 | 1.2 | 21 |
| 18:00 | 0 | |

*Solution* For lack of finer data, the duration of each TSS concentration is assumed to be 1 h, extrapolating ½ h before and after each measurement. The concentrations are ranked from highest to lowest and assigned a 1-h duration. The tabulated curve and plot are as follows:

| Duration (h) | Cumulative duration (h) | TSS (mg/L) |
|---|---|---|
| 1 | 1 | 280 |
| 1 | 2 | 175 |
| 1 | 3 | 145 |
| 1 | 4 | 115 |
| 1 | 5 | 75 |
| 1 | 6 | 45 |
| 1 | 7 | 21 |

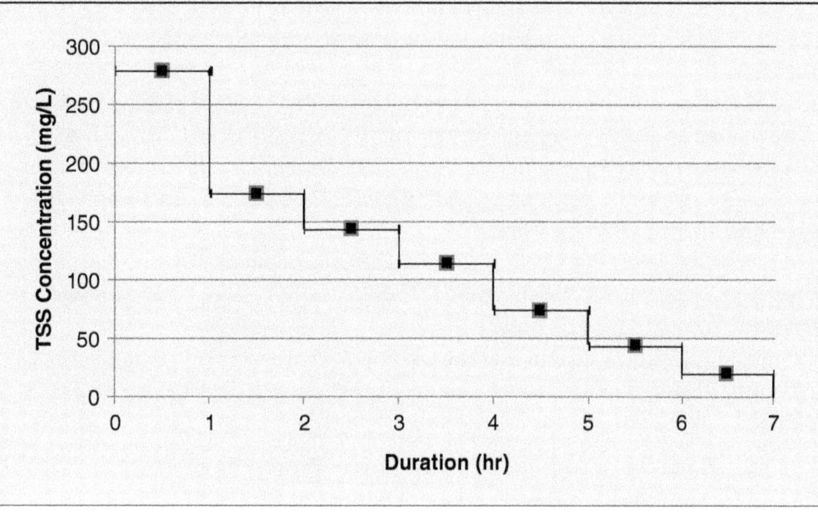

## References

Bean, E.Z., Hunt, W.F., and Bidelspach, D.A. (2007). Evaluation of four permeable pavement sites in eastern North Carolina for runoff reduction and water quality impacts. *Journal of Irrigation and Drainage Engineering-ASCE* 133 (6): 583–592.

Carter, T.L. and Rasmussen, T.C. (2006). Hydrologic behavior of vegetated roofs. *Journal of the American Water Resources Association* 42 (5): 1261–1274.

CSN, Chesapeake Stormwater Network (2012). *Recommendations of the Expert Panel to Define Removal Rates for Urban Stormwater Retrofit.* Ellicott City, MD: Chesapeake Stormwater Network.

Fassman-Beck, E., Hunt, W., Berghage, R., Carpenter, D., Kurtz, T., Stovin, V., and Wadzuk, B. (2016). Curve number and runoff coefficients for extensive living roofs. *Journal of Hydrologic Engineering* 21 (3): 04015073.

Kannan, N., Anandhi, A., and Jeong, J. (2018). Estimation of stream health using flow-based indices. *Hydrology* 5 (1), 20.

Lane, E.W. (1954). *The Importance of Fluvial Morphology in Hydraulic Engineering.* Denver, Colorado: US. Department of the Interior Bureau of Reclamation December.

Li, H. and Davis, A.P. (2009). Water quality improvement through reductions of pollutant loads using bioretention. *Journal of Environmental Engineering ASCE* 135 (8): 567–576.

Martin, W.D. and Kaye, N.B. (2014). Hydrologic characterization of undrained porous pavements. *Journal of Hydrologic Engineering* 19 (6): 1069–1079.

McCuen, R. (1979). Downstream effects of stormwater management basins. *Journal of the Hydraulics Division-ASCE* 105 (11): 1343–1356.

National Research Council (2012). *Review of EPA's Economic Analysis of Final Water Quality Standards for Lakes and Flowing Waters in Florida.* Washington DC: The National Academies Press.

Olszewski, J.M. and Davis, A.P. (2013). Comparing the hydrologic performance of a bioretention cell with predevelopment values. *Journal of Irrigation and Drainage Engineering-ASCE* 139 (2): 124–130.

Schwartz, S.S. (2010). Effective curve number and hydrologic design of pervious concrete storm-water systems. *Journal of Hydrologic Engineering* 15 (6): 465–474.

Shields, C.A., Band, L.E., Law, N., Groffman, P.M., Kaushal, S.S., Savvas, K., and Fisher, G.T. (2008). Streamflow distribution of non-point source nitrogen export from urban-rural catchments in the Chesapeake Bay watershed. *Water Res. Research* 44 (9): W09416.

Stagge, J.H., Davis, A.P., Jamil, E., and Kim, H. (2012). Performance of grass swales for improving water quality from highway runoff. *Water Research* 46 (20): 6731–6742.

USEPA (2000). *Ambient Water Quality Criteria Recommendations. Rivers and Streams in Nutrient Ecoregion XIV.* Washington, D.C: EPA 822-B-00-022 U.S. Environmental Protection Agency, Office of Water.

USEPA (2001). *Ambient Water Quality Criteria Recommendations. Lakes and Reservoirs in Nutrient Ecoregion XIV.* Washington, D.C: EPA 822-B-01-011 U.S. Environmental Protection Agency, Office of Water.

Van Buren, M.A., Watt, W.E., and Marsalek, J. (1997). Application of the log-normal and normal distributions to stormwater quality parameters. *Water Research* 31 (1): 95–104.

## Problems

9.1 Find rainfall data from a nearby NOAA weather station. Make plots of annual rainfall depth as a function of event depth, as in Figure 9.1. Find the fraction of annual rainfall captured by a SCM that captures the first inch of runoff.

9.2 Monitoring data in an urban area indicate an EMC for TSS of 142 mg/L. The annual rainfall averages 44 in/year. Assuming 95% rainfall to runoff, find the estimated annual TSS load, in lb/ac/year.

9.3 Stormwater data from a highway indicate an EMC for total P of 0.98 mg/L. The annual rainfall averages 104 cm/year. Assuming 95% rainfall to runoff, find the estimated annual TP load, in kg/ha/year.

9.4 The TSS EMC for stormwater entering a SCM is 126 mg/L and that leaving the SCM is 22 mg/L. The annual rainfall averages 48 in/year. Assuming 95% rainfall to runoff and 40% SCM volume reduction, find (a) the annual TSS load into the SCM, (b) the annual TSS load discharged from the SCM, and (c) the annual TSS load reduction, all in lb/ac/year.

9.5 The total nitrogen EMC for stormwater entering a SCM is 4.2 mg/L and that leaving the SCM is 2.8 mg/L. The annual rainfall averages 120 cm/year. Assuming 95% rainfall to runoff and 45% SCM volume reduction, find (a) the annual TN load into the SCM, (b) the annual TN load discharged from the SCM, and (c) the annual TN load reduction, all in kg/ha/year. Comment on the load reduction from TN treatment and that from stormwater volume reduction.

Use the table below for Problems 9.6–9.9.

A stormwater control measure treats runoff from a paved parking lot with a drainage area of 0.75 ac. Runoff input and output (from underdrain) data measured during a rain event are presented in the following table. The SCM media dimensions are 50 × 30 ft × 2.5 ft ($L \times W \times H$).

| Input | | | Output | | |
|---|---|---|---|---|---|
| Time of day | Flow rate (L/s) | TSS (mg/L) | Time of day | Flow rate (L/s) | TSS (mg/L) |
| 05:10 | 1.5 | 348 | 08:20 | 0.06 | 22 |
| 05:30 | 0.52 | 98 | 08:50 | 0.11 | 21 |
| 05:50 | 2.1 | 118 | 09:20 | 0.15 | 9 |
| 06:10 | 0.54 | 52 | 09:50 | 0.16 | 7 |
| 06:40 | 0.40 | 63 | 10:20 | 0.14 | 6 |
| 07:10 | 0.38 | 87 | 11:20 | 0.14 | 5 |
| 07:40 | 0.58 | 161 | 12:20 | 0.13 | 4 |
| 08:10 | 1.9 | 282 | 13:20 | 0.11 | 5 |
| 09:10 | 1.4 | 188 | 14:20 | 0.09 | 6 |
| 10:40 | 0.34 | 61 | 16:20 | 0.03 | 4 |

Note: When needed, integrate using trapezoids; extrapolate 1 $\Delta t$ at the beginning and end of the storm to 0 for each parameter.

**9.6** Calculate and plot the hydrographs and pollutographs, input and output on same plot. Discuss any trends that you see.

**9.7** Calculate the total TSS mass and the EMC for the stormwater input and for the treated stormwater discharge. Calculate the bioretention treatment efficiency for stormwater volume and for TSS, based both on EMC and on total mass.

**9.8** Plot the stormwater input and discharge flow duration curves and comment on the differences.

**9.9** Plot the stormwater input and discharge pollutant duration curves and comment on the differences.

Use the table below for Problems 9.10–9.15.

The following table presents runoff flows and pollutant concentrations measured from a highway during a storm event.

| Time of day | Flow rate (L/s) | TSS (mg/L) | TP (mg/L) | $NO_3$-N (mg/L) | TKN (mg/L) | T Cu (μg/L) | T Pb (μg/L) |
|---|---|---|---|---|---|---|---|
| 15:00 | 6.4 | 241 | 0.52 | 0.95 | 3.9 | 69 | 23 |
| 15:20 | 12.7 | 370 | 0.71 | 0.47 | 7.4 | 54 | 17 |
| 15:40 | 2.1 | 148 | 0.23 | 0.70 | 1.4 | 25 | 15 |
| 16:00 | 22.0 | 215 | 0.48 | 0.19 | 4.8 | 53 | 33 |
| 16:20 | 6.1 | 108 | 1.02 | 0.35 | 2.9 | 138 | 33 |
| 16:40 | 1.8 | 59 | 0.22 | 0.52 | 0.98 | 24 | 11 |
| 17:00 | 1.3 | 41 | 0.28 | 0.52 | 0.77 | 17 | 3 |
| 17:20 | 0.82 | 26 | 0.19 | 0.57 | 0.56 | 15 | 3 |

Note: When needed, integrate using trapezoids; extrapolate 1 $\Delta t$ at the beginning and end of the storm to 0 for each parameter.

**9.10** Plot the pollutant duration curve for TSS. Comment on the total and fraction of time that the value exceeds the target concentration listed in Table 9.2.

**9.11** Plot the pollutant duration curve for TP. Comment on the total and fraction of time that the value exceeds the target concentration listed in Table 9.2.

**9.12** Plot the pollutant duration curve for NO3–N. Comment on the total and fraction of time that the value exceeds the target concentration listed in Table 9.2.

**9.13** Plot the pollutant duration curve for TKN. Comment on the total and fraction of time that the value exceeds the target concentration listed in Table 9.2.

**9.14** Plot the pollutant duration curve for T–Cu. Comment on the total and fraction of time that the value exceeds the target concentration listed in Table 9.2.

**9.15** Plot the pollutant duration curve for T–Pb. Comment on the total and fraction of time that the value exceeds the target concentration listed in Table 9.2.

# 10

# Preventing Runoff and Stormwater Pollution

## 10.1 Introduction

From a fundamental perspective, the most effective approach to address the many problems of stormwater is to not create excess runoff in the first place. Through careful planning and implementation, this goal can be approached within the existing landscape and new development. In either case, the goal in reducing excess runoff is to promote infiltration and evapotranspiration, and reduced runoff velocity, usually at the expense of impervious surface runoff. This avoidance must occur without compromising other important aspects and functionality of the landscape. Also important to this discussion is managing water quality. Preventing pollutants from being placed, deposited, or mobilized in the urban landscape will result in higher runoff quality and reduced need for treatment. Fortunately, the focus on quantity and quality frequently can be complimentary.

## 10.2 Site Design and Low Impact Development

Traditionally, the preferred method of land development is to clear-cut all the trees and level the ground surface, as shown in Figure 10.1. Unfortunately, this operation can lead to significant changes in the water balance on the site and the water quality leaving the site. Top soil is removed and the remaining soil is compacted by the heavy construction machinery. Leveling destroys surface storage. Removal of mature trees results in the loss of their ET and other beneficial capabilities.

The concept of low impact development (LID) is to develop land with minimal impact to the local environment, with a focus on stormwater. This phrase was coined in the 1990s in Prince George's County, Maryland, to manage urban stormwater more smartly than had been previously done. Similar philosophies have different names in different areas throughout the world, including environmental site design (ESD) and sustainable urban drainage systems (SUDS).

The underlying thesis of LID is to create a developed site that will have hydrologic characteristics that mimic those of the site before development (although exactly what this standard is remains open for discussion and interpretation, as discussed in Chapter 9). To approach such predevelopment conditions, the site layout and design must be

*Green Stormwater Infrastructure Fundamentals and Design*, First Edition. Allen P. Davis, William F. Hunt, and Robert G. Traver.

Companion Website: www.wiley.com/go/davis/greenstormwater

Figure 10.1 Land Clear-Cut for Construction of a Housing Development. (Photo by Authors).

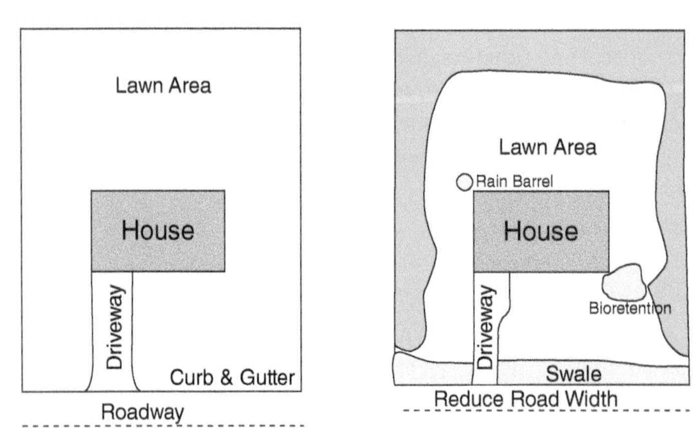


Figure 10.2 Schematic of Residential Lot with Traditional (Left) and Low Impact (Right) Development.

addressed early in the planning process. The positioning and placement of buildings, roadways, and the drainage infrastructure design will impact the degree of success in which LID goals are met. Minimizing land disturbance will maintain the advantages offered by the undeveloped land, including keeping mature trees and leaving contours on the site to allow runoff storage and to promote infiltration and ET. Runoff flow pathways should avoid concentration and be long, tortuous, and pass over infiltrating surfaces. The impervious footprint should be as small as possible, for example, focusing on smaller driveways and roadside swales instead of curb and gutter. Localized stormwater control measures should be placed adjacent to impervious surface where its use cannot be avoided (Figure 10.2). Replacement of gray infrastructure collection systems with green approaches can reduce both cost and environmental impact.

Figure 10.3 Stormwater Control Treatment Train, Consisting of Bioswales to Rain Garden to Infiltration Trench, at Villanova University. (Photo by Authors).

Incorporating LID into existing areas, or after many critical design decisions have been made, via retrofit is much more difficult and complex than addressing it early-on. Large infrastructure that is in place cannot be moved and the freedom to work within the entire landscape is lost. Options for avoiding runoff are fewer and may be less effective.

Replacement of inlets and piping (gray infrastructure) within the right-of-way with bioswales and rain gardens (green infrastructure) can reduce stormwater volume and velocity, and improve quality, all at a reduced construction costs. An example of this is shown in Figure 10.3, where the runoff from a parking garage is directed into a linear system of bioswales, rain gardens, and an infiltration trench.

## 10.3 Compacted Urban Surfaces

One of the most problematic issues with urban land development is the compaction of the soils. This occurs primarily during the construction process but can also be exacerbated by day-to-day activities on the landscape.

Undeveloped land will generally have high biological activity, which leads to a high soil permeability. Tree, shrubs, and grasses have roots that go to various depths into the soil, opening up flow pathways as they grow, die, and decompose. Worms and insects burrow throughout the soil, mixing soil layers and creating paths for water movement. Natural contours exist in the land, creating small areas for pools and water storage during rainfall.

The land development process drastically alters the characteristics of the natural soil. Top soils are removed as land is clear-cut, flattened, and shaped. Heavy equipment is used to clear the land and shape the soil. Soils that are left at the surface are compacted throughout the construction processes. Even land that may be left as green space, such as lawn and open areas, typically experience significant compaction from the large heavy wheels of construction equipment. A thin layer of topsoil may be replaced for lawn areas, overlying heavily compacted soils below. In lawns created with only new grasses, the roots will not penetrate deeply and the infiltration may only be shallow. It may take many years for the soils to again become rich in biological activity.

Compacted urban soils are problematic, creating excess runoff. Because the underlying soils may be heavily compacted, the degree of infiltration may not be as high as expected

**Example 10.1** The lawn space in commercial/institutional area is estimated at 10 ac. The soil is hydrologic soil group B, but it is compacted to the characteristics of HSG D. The grass condition is considered as fair. For a 1-in storm, find the increase in runoff depth and volume due to the compaction.

*Solution* From Table 6.6, the curve number for fair condition open space B soil is 69 and for D soil is 84. $P = 1.00$ in. The initial abstraction, potential maximum soil moisture retention, and runoff depths are found:

| Uncompacted (B soil) | Compacted (D soil) |
|---|---|
| $S = 1000/69 - 10 = 4.49$ in | $S = 1000/84 - 10 = 1.90$ in |
| $I_a = 0.2\ (4.49) = 0.90$ in | $I_a = 0.2\ (1.90) = 0.38$ in |
| $Q = (1.00 - 0.90)^2/(1.00 - 0.90 + 4.49) = 0.041$ in | $Q = (1.00 - 0.38)^2/(1.00 - 0.38 + 1.90) = 0.34$ in |
| $V = 0.041$ in (10 ac) = 0.41 ac in = 1500 $ft^3$ | $V = 0.34$ in (10 ac) = 3.4 ac in = 12,340 $ft^3$ |

The differences in runoff depth and volume are significant.

in normal grassed areas (Gregory et al. 2006; Pitt et al. 2008). Runoff models that assume significant infiltration in green spaces may be overestimating infiltration due the extreme compaction of the soils. These soils may continue to be compacted as large mowing equipment, pedestrian traffic, and occasional vehicular traffic cross the land.

### 10.3.1 Avoiding Compaction and Promoting Infiltration

Conscientious efforts to avoid compaction during construction and operational practices can minimize soil compaction. This will require careful planning and management during the construction process. Minimizing disturbances to soils will maximize infiltration, decreasing runoff.

In the developed watershed, monoculture grass will not provide the water management capabilities that more heterogeneous types of landscaping will. Incorporating trees, shrubs, and various flowers and grasses will offer a more diverse landscape which can lead to better water management, as well as aesthetic and habitat benefits. In the Chesapeake Bay watershed, the use of *BayScapes* provides landscaping alternatives to manicured grass (https://www.chesapeakebay.net/news/blog/10_chesapeake_bay_native_plants_to_plant_in_your_yard_this_spring).

### 10.3.2 Soil Restoration

Urban soil compaction can be addressed through physical and chemical soil restoration methods. In a technique borrowed from agriculture, urban soils can be *subsoiled*. In subsoiling, the soil is plowed very deeply, up to about 18 in (46 cm) below the surface. This will rip through the upper layers, but also the lower subsurface soils, either or both of which may be highly compacted. Subsoiling requires a large tractor and a heavy plow which will destroy the existing surface. Ideally, this soil will be better managed after the subsoiling and after final preparation and planting will possess much improved infiltration capacities. Nonetheless, this process is expensive, disruptive, and is not applicable to small areas where the large equipment cannot maneuver.

In one study, shallow (4-in, 10-cm) tillage of a compacted soil has shown to increase infiltration rates (Bean and Dukes 2015). NRCS curve numbers were reduced from 75 to 49 and from 87 to 71 for two soils tested. The addition of fly ash was found to be detrimental to the treatment.

Adding an amendment such as compost to a compacted soil can improve soil infiltration capabilities. The compost must be incorporated into the soil layers. The very high rich organic content of the compost gives structure to the soil, better allowing it to agglomerate and hold and infiltrate water. The organic matter and nutrients in the compost will provide a growth medium for various forms of vegetation and for fauna, both with capabilities of increasing the infiltration rates of the compacted soils (Chen et al. 2013; Olson et al. 2013).

### 10.3.3 De-paving

De-paving is the removal of pavement and converting it to vegetative space or a stormwater control measure (SCM). De-paving is generally done in small areas where the pavement is no longer needed. Before de-paving it is important to have an understanding of the site conditions after the pavement is removed. The amount of water that will flow to it must be estimated, and a determination must be made if this water flow will be a problem. Also, before removing pavement, it is important to have soils under the pavement checked for contamination. Older urban soils may be contaminated with heavy metals, such as lead and cadmium. If the underlying soils are contaminated, de-paving is probably not a good idea. Runoff entering these soils could mobilize the pollutants and opening the pavement may expose wildlife to the contamination. More problems may be created than solved in such a situation.

During de-paving, the pavement is broken up with jackhammers or manual hammers and pry bars. Pavement slabs and underlying gravel are removed (and ideally are beneficially repurposed). The underlying soils are likely compacted and should be loosened. Finally, the area can be filled with topsoil and planted with grasses or flowers, as a green space, or converted to a SCM, like a rain garden.

### 10.3.4 Removing Abandoned Housing

In some older cities, the cities are looking to buy properties of abandoned housing and subsequently demolishing the buildings. The property is greened and converted into some type of SCM or a series of SCMs. If a basement remains, it may be left, at least partially, to allow some ponding of water. While it may be costly for the city, this program removes blighted buildings, creates land for stormwater management where land is usually at a premium, and creates greenspace (and associated benefits) in areas where it is scarce. Similar to de-paving, issues related to final site water flow, contaminated soils, and re-purposing demolition materials must be addressed during the project.

## 10.4 Street Trees

Street trees can be a valuable part of the green stormwater infrastructure. Trees can provide multiple benefits for stormwater management, similar to other, more formalized SCMs. The tree leaf canopy and branches can intercept rainfall before it hits the ground. The amount intercepted is related to the leaf area index (LAI). The LAI is defined as the total (one-sided) area of all leaves divided by the area under the leaf canopy. The LAI can vary from 0 (no leaves) to approaching 10 for thick leaf cover. Information compiled by the state of Minnesota suggests that deciduous trees will intercept 0.14 in (3.6 mm) of rainfall and interception of coniferous trees is 0.40 in (10.2 mm) (stormwater.pca.state.mn.us). This intercepted water will later be evaporated so that it does not become surface runoff.

Healthy trees can also take up and transpire significant amounts of water. Therefore, conveying runoff into a tree box area can allow this water to be removed via tree transpiration. Trees can also take up a number of pollutants, including N and P for growth. As well, support of a healthy soil rhizosphere can provide other water-quality benefits.

However, healthy trees require healthy space, both above and below ground. Above the ground, trees must not interfere with utilities, roadway, and building areas. Trees also require large amount of soil for roots. A large tree (16 in (40 cm) diameter at breast height) can require 1000 $ft^3$ (28 $m^3$) of soil. This soil must be permeable and pervious to hold and transport water and to allow penetration of the tree roots. Urban soils that are severely compacted will not allow water and root penetration and will not support healthy tree growth.

Engineering modifications that can assist with healthy tree growth include pervious pavements, structural soils, and structural pillars. Permeable pavements, as discussed in Chapter 15, allow rainfall to penetrate through the pavement surface to the ground/media below. In the case of street trees, this infiltrating runoff can be available to the tree roots as they spread in all directions. Structural soils and structural pillars allow trees to be used in areas where heavy structural loads are required on the pavements and sidewalks. Pillars are used to transfer the pavement load deeper into the subsurface. This can allow a non-compacted soil to be used near the trees.

Structural soil is a mix of 70–80% angular gravel and 20–30% clay loam soil. A small amount of a hydrogel is added to keep these two materials from separating. The gravel provides structural strength to support sidewalks and even parking. However, the soil remains uncompacted and allows tree roots and water to penetrate (Bartens et al. 2009).

Exact stormwater benefits resulting from trees added to or left in the watershed can be difficult to calculate. Impact will depend on tree species, size, and climate. The

US Forest Service has developed *i-Tree*, a set of programs to estimate benefits of trees in the urban environment (*itreetools.org*). These public domain tools can provide estimates of stormwater management at the individual-tree level. Some jurisdictions are providing stormwater management credit for including trees in the urban design. A typical credit may include stormwater management of 100 $ft^2$ (9.3 $m^2$) of impervious area for each tree located within 20 ft (6.1 m) of impervious surface.

In addition to stormwater management, urban trees can provide numerous other benefits, including temperature reduction due to shading (Razzaghmanesh et al. 2021) reduction of air pollutants, increases in property values, and social benefits. Nonetheless, tree leaf-fall may lead to water-quality concerns from nutrient leaching (Bratt et al. 2017).

## 10.5 Disconnecting Impervious Surfaces

A *disconnected impervious surface (DIS)* is a built-upon area (usually a roof or a paved surface) that discharges runoff to a vegetated area that is sized and graded to reduce runoff and pollutants. Until recently, most development has been designed as *connected impervious surface*, that is, draining to gutters, pipes, and ditches that rapidly convey stormwater without runoff reduction or treatment. Using DIS instead of connected impervious surfaces can reduce stormwater flows and pollutant loadings.

Two types of DIS are common: downspout disconnection and disconnected paved areas (Figure 10.4). Design elements that apply to both downspout disconnection and disconnected paved areas are as follows:

- The vegetated receiving area would not include any impervious surface. In addition, a minimum distance between building foundation and vegetated area receiving runoff must be established (e.g., 6 ft, 2 m).
- The vegetated receiving area should have a maximum slope (e.g., 10%) with land graded to promote diffuse flow in all directions.
- Vegetative cover needs to avoid clumping species of vegetation. Typically, many turf grasses allow diffuse flow.
- Some soil preparation is advantageous for optimal performance, encouraging infiltration. For example, tilling to a depth of 8 in (20 cm) can promote better vegetation establishment.

Figure 10.4 Disconnected Downspouts and Disconnected Pavement. (Photos by Authors).

### 10.5.1 Defining Disconnected Impervious Surface

Several factors determine how well an impervious surface has been disconnected: (1) run-on to vegetated area ratio, (2) length of flow path disconnect, (3) vegetated area underlying soil type and condition, and (4) slope and grade of receiving vegetation area. Jurisdictions that employ—and give stormwater mitigation credit for—downspout disconnection typically require certain threshold conditions to be met or exceeded that account for the above factors.

Myriad options exist for downspout disconnection and for quantifying the effects of these options. Table 10.1 provides an example set of options in a summary of downspout disconnection requirements established by the state of North Carolina.

Based on these guidelines, the downspout of a roof is directed to a vegetated area that is either 6 by 12 ft (1.8 by 3.7 m) or 12 by 24 ft (3.7 by 7.3 m) depending on the space available and stormwater management credit sought. Figure 10.5 illustrates how a designer could incorporate downspout disconnection onto a residence. The maximum roof surface area directed to any downspout disconnection is 500 $ft^2$ (47 $m^2$). The red-hatched areas around the perimeter of the built-upon areas on the lot show the required 5-ft (1.5 m) setback from building foundations.

For *disconnected paved areas*, a gravel verge or other transition is provided between the edge of the paved surface and the vegetated area, as pictured in Figure 10.4b.

### 10.5.2 Calculating the Benefit of Disconnecting Imperviousness

Disconnection can be credited through model simulation of the runoff process following the impervious surfaces onto the pervious surfaces. While time and resource intensive, this process may show added benefits for larger/complex projects. A simpler means of numerically accounting for impervious surface disconnections is done by calculating runoff generated off a property. For large storm events, assigning a curve number that is a composite of the permeable and impermeable surface is one method that can be used. However, this method does not truly credit the volume mitigation of the receiving vegetation area.

**Table 10.1** Minimum Criteria for Impervious Surface Disconnection in North Carolina (NCDEQ 2020).

1. For disconnected roofs, a maximum of 500 $ft^2$ (47 $m^2$) of roof shall drain to each disconnected downspout and corresponding vegetated receiving area. The receiving vegetated area shall be a rectangle using minimum dimensions* of either 6 by 12 ft (1.8 by 3.7 m) or 12 by 24 ft (3.7 by 7.3 m; width of vegetated area by length of run in direction of flow). The entire rectangle shall not include any impervious surface to ensure that water released from the roof does not run onto another impervious surface
2. For disconnected pavement, the receiving vegetated area shall be either 10 or 15* ft (3.1 or 4.6 m) wide. The maximum length of pavement run that may discharge to the vegetated area is 100 ft (31 m) and the maximum slope of the pavement shall be 7%
3. The vegetated area shall have a maximum slope of 7% with land graded to promote sheet flow
4. The vegetated cover shall be established dense lawn with no clumping species
5. All sites built within the past 50 years shall be tilled to 8 in (20 cm) prior to vegetation establishment

* Two minimum dimensions are provided by NCDEQ (2020); each is associated with a different amount of runoff reduction credit.

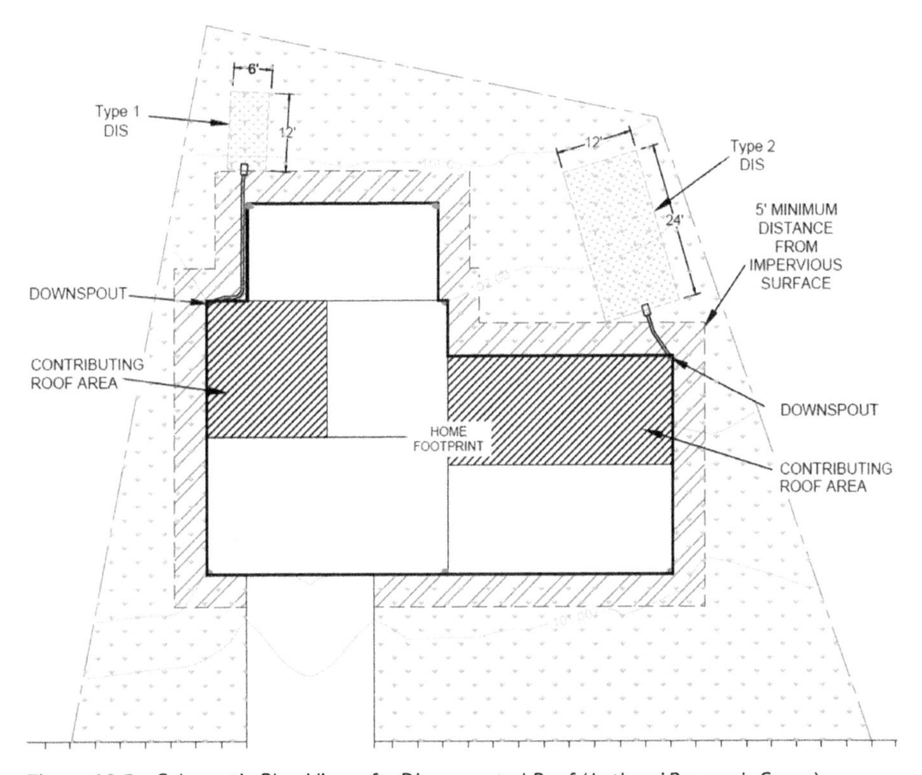


Figure 10.5 Schematic Plan View of a Disconnected Roof (Authors' Research Group).

In locations where annual hydrology has been well defined, long-term runoff reduction percentages can be assigned to different configurations of impermeable surface disconnection. Table 10.2 shows an example of this. Three categories of impervious surface disconnection (labeled Types 1, 2, and 3 in Table 10.2) have been identified to account for moderate, high, and very high performance of a DIS. As noted in Table 10.2, underlying hydrologic soil groups have an important impact on the amount of runoff that is infiltrated.

Table 10.2 Example Sizing and Credit for Disconnected Impervious Surface for North Carolina (NCDEQ 2020).

| | Type 1 DIS | | Type 2 DIS | | Type 3 DIS |
|---|---|---|---|---|---|
| Disconnected roof: vegetated area size | 6 ft × 12 ft (2 m × 4 m) | | 12 ft × 24 ft (4 m × 8 m) | | 12′ × 24′ (4 m × 8 m) and site BUA* <24% |
| Disconnected paved area: vegetated area size | 10 ft (3 m) width | | 15 (4.6 m) width | | 15′ (4.6 m) width and BUA <24% |
| Hydrologic soil group | A/B | C/D | A/B | C/D | A/B only |
| Runoff reduction credit | 45% | 30% | 65% | 50% | 100% |
| TSS reduction credit | 45% | 30% | 65% | 50% | 85% |
| TN reduction credit | 45% | 30% | 65% | 50% | 70% |
| TP reduction credit | 45% | 30% | 65% | 50% | 70% |

* BUA = Built-upon area.

### 10.5.3 Design

The design standards for DIS presented herein are based on providing a minimum loading ratio of 7:1 for (rooftop area:vegetated area) or 10:1 for (pavement area:vegetated area). Lower loading ratios may provide greater treatment efficiency. These loading ratios ensure a significant level of infiltration for stormwater and correspondingly significant pollutant reductions.

#### 10.5.3.1 Design Step 1: Consider Siting and Feasibility

Table 10.3 lists some of the many considerations that go into DIS design at a specific site.

**Table 10.3** Example Siting and Feasibility Considerations for DIS in North Carolina (NCDEQ 2020).

| | |
|---|---|
| **Installation size** | The size of disconnected roof areas is limited to a maximum of 500 $ft^2$ (47 $m^2$) per downspout. Paved areas are limited to a 100-ft (31 m) run of pavement; however, there is no limit to the length of pavement that may be disconnected. This will allow most standard roadway cross-sections to be disconnected provided that there is an adequate width of vegetated area in the right-of-way |
| **Proximity to building foundations and utilities** | As a precaution, at least 5 ft (1.5 m) of setback from building foundations should be allowed for downspout disconnection. The limit of 500 $ft^2$ (47 $m^2$) for each downspout makes it unlikely that foundations or underground utilities will be adversely affected by DIS |
| **Proximity to water supply wells** | No setback from water supply wells should be required for DIS |
| **Status of the site as high or low density** | DIS can be used on either high or low density sites. On high density sites, DIS can reduce required SCM volumes. However, DIS may be impractical on highly built-out sites that do not have vegetated areas available to receive stormwater runoff. On low density sites, as much impervious surface as possible should be handled as DIS |
| **Soil type** | DIS may be used on any soil type, although the runoff reduction (and stormwater management credit) will vary |
| **Site slopes** | The vegetated areas associated with DIS should have a slope of 7% or less. It may not be cost-effective to meet the requirement for gently sloping vegetated surface on a steep site |
| **Seasonal high water table** | No seasonal high water table requirements are associated with DIS |
| **Stormwater hotspots** | DIS should not be used to treat stormwater hotspots—areas where concentrations of pollutants such as oils and grease, heavy metals, and toxic chemicals are likely to be significantly higher than in typical stormwater runoff |
| **Redevelopment sites** | Care should be taken when implementing DIS at redevelopment sites. Stormwater should not be infiltrated into contaminated soils because this can cause mobilization and dispersion of toxic substances |
| **Maintenance access** | Because its performance depends on maintenance, DIS should be accessible by mowing equipment |

#### 10.5.3.2 Design Step 2: Design the Drainage and Outlet System

The vegetated receiving area should have a uniform slope that does not exceed 7%. The vegetated area may be graded to achieve this slope but should have additional positive grade at the end of the receiving area for possible runoff to be directed offsite and not cause ponding.

Corrugated plastic pipe (Figure 10.6) allows the designer and owner to adjust the location of the discharge point. Corrugated plastic pipe should be fitted with an outlet to better spread the flow.

Figure 10.7 shows an incorrectly designed downspout disconnection. The vegetated area is too sparsely vegetated and it is undersized. It is preferable to place the vegetated area where it will not be subject to heavy foot traffic in order to maintain a healthy vegetated cover.

#### 10.5.3.3 Design Step 3: Select Vegetation

If appropriate vegetation is not already established on site, then a seed blend application is recommended. Seed blends should be selected based on shade/sun exposure of the vegetated area and regional climate. Sod grown in a clay base should not be used, or otherwise, it should be washed.

Forested areas are not recommended as vegetated receiving areas because uneven micro-topography often causes channelization, which reduces the disconnected surface area exposed to stormwater infiltration.

Figure 10.6 Converter Joint for Directing Downspout to Vegetated Area. (Photo by Authors).

Figure 10.7 Disconnected Downspout Connected to Inadequately Sized Vegetated Area. (Photo by Authors).

### 10.5.4 Water-Quality Benefits

Water-quality benefits can be derived through DIS. In the case of DIS through hydrologic soil groups A and B, the water-quality benefit can be estimated as flow through filter strips (Chapter 19). For C and D soils, the benefit is estimated from the treatment of 1 in (2.5 cm) of runoff storage over the receiving area surface.

### 10.5.5 Performance Results

Seven roof DIS sites were studied in coastal North Carolina (Taguchi et al. 2019). The soils were sandy with saturated hydraulic conductivities ranging from 13 to 440 mm/h (0.5–17 in/h). Annual volume reduction ranged from 42% to 87% at these sites, with a mean of 64%. Concentrations of total suspended solids (TSS) and nitrogen were low in both input and output flows and pollutant mass reductions were essentially equal to volume reductions. The saturated hydraulic conductivity of the soil was the most important parameter in controlling performance, even though the roof area to infiltrating lawn area ranged from 4:1 to 14:1.

**Example 10.2** The downspout from a rooftop area of 20 ft × 20 ft is directed onto the lawn as part of a DIS project. The soils are sandy with a saturated hydraulic conductivity of 4 in/h. Find the infiltration area needed and the roof:lawn ratio necessary to completely infiltration a 1-in/h rainfall. Since the drainage area is so small, ignore any time of concentration affects.

*Solution* The flow rate is the rainfall intensity multiplied by the roof area.

$$Q = iA = \left(\frac{1}{12}\ \text{ft/h}\right)(20\ \text{ft})(20\ \text{ft}) = 33.3\ \text{ft}^3/\text{h}$$

The lawn area is found by dividing the flow rate by the infiltration rate (saturated hydraulic conductivity).

$$A = \frac{Q}{K_{sat}} = \frac{33.3\ \text{ft}^3/\text{h}}{\frac{4}{12}\ \text{ft/h}} = 100\ \text{ft}^2$$

And the roof:lawn ratio is

$$\text{roof:lawn} = \frac{(20\ \text{ft})(20\ \text{ft})}{100\ \text{ft}^2} = 4 = 4:1$$

## 10.6 Pollution Prevention

Compromised water quality in urban runoff results from the wash-off of pollutants built into or deposited onto the drainage areas. Methods to reduce the buildup of these pollutants, such as removing them before rainfall occurs, should lead to improved runoff quality. In an increasing number of instances, the use of materials in the watershed that contribute toxic pollutants is being strongly discouraged or even prohibited.

### 10.6.1 Street Sweeping

Street sweeping can be used to remove particulate matter from pavement surfaces during dry weather so that they will not be mobilized by rainfall and runoff during wet weather events. Street sweeping is accomplished via a customized truck (Figure 10.8) that sweeps and vacuums the pavement surface as the truck drives along. It will pick up larger sediment, as well as some trash, leaves, and other detritus. Street sweeping has the potential to be an effective SCM. However, the benefits contributed by street sweeping are difficult to quantify and results from performance studies are frequently inconclusive and somewhat controversial (Kang et al. 2009). Nonetheless, street sweeping may be one of the few options available to address water quality in highly urbanized areas.

To be effective for water-quality improvement, street sweeping must be done very frequently, as often as between every rainfall event. This can be prohibitively expensive. Many jurisdictions plan to sweep streets every few months for aesthetic reasons, such as in the spring to remove salt and friction materials left from winter, or in the fall to remove leaf detritus. However, particulate buildup on impervious surfaces occurs daily and sweeping will have the greatest impact during the rainfall events directly after the sweeping. Additionally, the sweeping must cover most of the streets in the

Figure 10.8 Advanced Street Sweeper Truck. (Photo by Authors).

catchment to have a water-quality impact. For climates in which rainfall is frequent, the high frequency of sweeping required cannot be economically justified. In areas that have a dry season preceding a rainy season, an aggressive street-sweeping program just before the rains may provide tangible benefits.

One of the primary controversies over street sweeping as a SCM is the efficiency for small particle size collection. Sweepers are designed to pick up sand-sized material and larger trash and detritus. They tend not to be very effective for the smaller particulate matter that may be important in affecting water quality. Research trying to link specific removals of particulate material from roadways via street sweeping does not match well with field measurements of TSS reduction in runoff. Particulate matter measured as TSS should be dominated by particles smaller than those targeted by street sweeping.

While street sweeping targets sediments and particulate matter, the impact on other pollutants in the watershed is complex and also not well understood. Many other pollutants will be affiliated with particulate matter and some particulate matter may contain pollutant constituents (e.g., zinc in tire wear particles, nutrients in leaf matter). Therefore, in addition to particulate matter removal, sweeping may also be removing phosphorus, nitrogen, and various toxic compounds. However, reductions of these pollutants via sweeping are difficult to quantify. Older mechanical street sweepers are not effective in picking up the fine particulate matter to which many pollutants of interest may be affiliated. Vacuum-assisted or regenerative-air sweepers are much more effective in picking up fine street particulate matter.

Table 10.4 Example Efficiencies of Pollutant Removals by Street Sweeping Using Vacuum-assisted or Regenerative-air Sweepers, Used in the Chesapeake Bay Watershed (Expert Panel 2015). Removals are Subtracted from Baseline Loads, as Given in the Text.

| Sweeping frequency | TSS removal (%) | TN removal (%) | TP removal (%) |
|---|---|---|---|
| 2 per week | 21 | 4 | 10 |
| 1 per week | 16 | 3 | 8 |
| Every 2 weeks | 11 | 2 | 5 |
| Every 4 weeks | 6 | 1 | 3 |
| Every 2 months | 4 | 0.7 | 2 |
| Every 3 months | 2 | 0 | 1 |

Table 10.4 gives removal credits for nutrients and sediment in the Chesapeake Bay watershed at various sweeping frequencies (Expert Panel 2015). These nutrient reductions are subtracted from standard pollutant wash-off loads from the urban areas (e.g., sediment 1300 lb/ac year [1460 kg/ha year]; nitrogen 15.5 lb/ac year [17.4 kg/ha year]; phosphorus 1.93 lb/ac year [2.16 kg/ha year]). Only roads that have curb and gutter (where the pollutants accumulate before they are swept) are credited for sweeping.

A mass loading approach has also been used to estimate pollutant load reductions from street sweeping.

$$L = 0.7(P)M \tag{10.1}$$

where $L$ is the mass of each pollutant captured in the sweeper; $M$ is the mass of collected material in the sweeper; 0.7 converts wet mass into dry mass; $P$ is a factor to account for the amount of a specific pollutant in the dry mass. For sediment, $P$ is equal to 0.3, as about 30% of the dry solids are smaller than 0.25 mm, those important for water quality. For nitrogen and phosphorus, $P = 0.0025$ and 0.001, respectively, the fractions of these nutrients typical in street sediments. $P$-values for Cu, Pb, and Zn are $8 \times 10^{-5}$, $1 \times 10^{-4}$, and $1.8 \times 10^{-4}$, respectively (Expert Panel 2015).

After the collected pollutant mass is calculated, this is divided by the number of curbed lanes miles (km) that was swept. The collected matter yield for a typical sweeper yield is about 600 lb/curb mile (170 kg/km).

**Example 10.3** A street-sweeping program sweeps 80 mi of roadway once per month. Using the default values for sediment capture and characteristics, find the yearly removal of sediment, nitrogen, and phosphorus.

*Solution* The total roadway miles is

$$80 \text{ mi}(1 \text{ sweep/month})(12 \text{ months/year}) = 960 \text{ mi}$$

The default total sediment capture is 600 lb/mile. Therefore, the total sediment is

$$960 \text{ mi}(600 \text{ lb/mi}) = 5.76 \times 10^5 \text{ lb}.$$

Equation 10.1 is used to convert the total sediment into water-quality parameters. For sediment:

$$0.7(0.3)(5.76\times10^5\ \text{lbs})=1.21\times10^5\ \text{lbs of sediment}$$

For nitrogen and phosphorus:

$$0.7(0.0025)(5.76\times10^5\ \text{lbs})=1000\ \text{lbs of N}$$

$$0.7(0.001)(5.76\times10^5\ \text{lbs})=400\ \text{lbs of P}$$

### 10.6.2 Product Prohibition

Anthropogenic products that are used in the urban landscape may dissolve, corrode, or leach during wet weather events and cause downstream environmental problems. Fertilizers are employed in many of our urban green areas and excess fertilizer can run off these sites. A number of toxic compounds are used for a variety of reasons throughout our urban infrastructure and during long-term weathering, wear, corrosion, and attrition may be washed into waterways. When water quality reaches a point of significant concern, local governments have acted to ban or limit use or application of some materials in the urban environment. If enough small jurisdictions act to ban a product component, it is likely that the manufacturer will have to modify the product to maintain its market share and to prevent banned material from being shipped to the wrong jurisdiction.

#### 10.6.2.1 Fertilizers

A number of states have begun limiting the amount of nitrogen and/or phosphorus that can be sold in off-the-shelf home use lawn fertilizers. These bans are established to prevent homeowners from over-fertilizing lawns, leading to nitrogen and phosphorus wash-off.

#### 10.6.2.2 Asphalt Coal Tar Sealants

Sealants are used on asphalts to prolong the life and color of the asphalt. Coal tar sealants were very common for many years. However, research in the early 2000s noted high concentrations of polycyclic aromatic hydrocarbons, PAH, in sediments near sealed parking lots; the research was able to attribute the PAH to coal tar sealants (Mahler et al. 2005; Yang et al. 2010). Some PAH are carcinogens and toxicity in the sediments was noted. The city of Austin TX banned coal tar sealants in 2006 and bans have extended to the state level.

#### 10.6.2.3 Copper in Brake Pads

Excessive levels of copper in sediments on the US west coast have been attributed to vehicle brake dust (and other sources). During the braking process, the brake pad wears, releasing small particles of the pad material. During wash-off events, this dust is transported to local waterways. Brake pads have typically contained several percent copper and other metals. Research has suggested that brake dust is a major source of the accumulated copper (Davis et al. 2001; Hur et al. 2003). As a result, several west coast states are limiting the copper content that can be used in automobile brakes.

Figure 10.9 Sign to Encourage Pet Owners to Clean up after Their Pet. Such Efforts Can Prevent Pathogens from Pets from Being Mobilized by Stormwater and Washed into Local Streams. (Photo by Authors).

## 10.7 Education

Education is and will continue to be a critical component of stormwater management. Most watersheds are dominated by privately owned lands and educated landowners can make smart decisions in managing the water that falls on and leaves their property. Runoff reduction techniques, such as the use of permeable surfaces, disconnected downspouts, rain barrels, and rain gardens, all can be readily implemented at the individual homeowner level. Conscientious use of fertilizers and hazardous materials will limit chemical wash-off from the property. If practiced by many, informed decisions at the local parcel level can have impact at larger scales.

An important public education program is to encourage the removal of pet wastes from open areas. This is an issue with private lands but can be more of a problem on public green spaces, especially in high density urban areas. Pet wastes can contain very high concentrations of pathogenic microorganisms and can be readily washed into receiving streams during subsequent rains. Education of the public and diligence is the best way to deal with this pollutant source (Figure 10.9).

## References

Bartens, J., Day, S.D., Harris, J.R., Wynn, T.M., and Dove, J.E. (2009). Transpiration and root development of urban trees in structural soil stormwater reservoirs. *Environmental Management* 44 (4): 646–657.

Bean, E.Z. and Dukes, M.D. (2015). Effect of amendment type and incorporation depth on runoff from compacted sandy soils. *Journal of Irrigation and Drainage Engineering* 141 (6).

Bratt, A.R., Finlay, J.C., Hobbie, S.E., Janke, B.D., Worm, A.C., and Kemmitt, K.L. (2017). Contribution of leaf litter to nutrient export during winter months in an urban residential watershed. *Environmental Science & Technology* 51 (6): 3138–3147.

Chen, Y., Day, S.D., Wick, A.F., Strahm, B.D., Wiseman, P.E., and Daniels, W.L. (2013). Changes in soil carbon pools and microbial biomass from urban land development and subsequent post-development soil rehabilitation. *Soil Biology & Biochemistry* 66: 38–44.

Davis, A.P., Shokouhian, M., and Ni, S.B. (2001). Loading estimates of lead, copper, cadmium, and zinc in urban runoff from specific sources. *Chemosphere* 44 (5): 997–1009.

Expert Panel (2015). *Recommendations of the Expert Panel to Define Removal Rates for Street and Storm Drain Cleaning Practices Chesapeake Bay Program* (Sebastian Donner, Bill Frost, Norm Goulet, Marty Hurd, Neely Law, Tom MaGuire, Bill Selbig, Justin Shafer, Steve Stewart, Jenny Tribo), September 18, 2015.

Gregory, J.H., Dukes, M.D., Jones, P.H., and Miller, G.L. (2006). Effect of urban soil compaction on infiltration rate. *Journal of Soil and Water Conservation* 61 (3): 117–124.

Hur, J., Yim, S., and Schlautman, M.A. (2003). Copper leaching from brake wear debris in standard extraction solutions. *Journal of Environmental Monitoring* 5 (5): 837–843.

Kang, J.-H., Debats, S.R., and Stenstrom, M.K. (2009). Storm-water management using street sweeping. *Journal of Environmental Engineering-Asce* 135 (7): 479–489.

Mahler, B.J., Van Metre, P.C., Bashara, T.J., Wilson, J.T., and Johns, D.A. (2005). Parking lot sealcoat: An unrecognized source of urban polycyclic aromatic hydrocarbons. *Environmental Science & Technology* 39 (15): 5560–5566.

NCDEQ (2020) North Carolina Department of Environmental Quality Stormwater Design Manual, C-10 Disconnected Impervious Surface. https://deq.nc.gov/media/17544/download (accessed December 2021).

Olson, N.C., Gulliver, J.S., Nieber, J.L., and Kayhanian, M. (2013). Remediation to improve infiltration into compact soils. *Journal of Environmental Management* 117: 85–95.

Pitt, R., Chen, S.-E., Clark, S.E., Swenson, J., and Ong, C.K. (2008). Compaction's impacts on urban storm-water infiltration. *Journal of Irrigation and Drainage Engineering* 134 (5): 652–658.

Razzaghmanesh, M., Borst, M., Liu, J., Ahmed, F., O'Connor, T., and Selvakumar, A. (2021). Air temperature reductions at the base of tree canopies. *Journal of Sustainable Water in the Built Environment* 7 (3): 04021010.

Taguchi, V.J., Carey, E.S., and Hunt, W.F. (2019). Field monitoring of downspout disconnections to reduce runoff volume and improve water quality along the North Carolina Coast. *Journal of Sustainable Water in the Built Environment* 5 (1): 04018018.

Yang, Y., Van Metre, P.C., Mahler, B.J., Wilson, J.T., Ligouis, B., Razzaque, M.M., Schaeffer, D.J., and Werth, C.J. (2010). Influence of Coal-Tar Sealcoat and other carbonaceous materials on polycyclic aromatic hydrocarbon loading in an urban watershed. *Environmental Science & Technology* 44 (4): 1217–1223.

## Problems

**10.1** Eight deciduous trees, each with a projected canopy area of 600 $ft^2$, are planted in a 1-ac parking lot. Estimate the reduction in runoff provided by the tree canopy in a 0.4-in rainfall, assuming 95% runoff from rainfall on the parking surface.

**10.2** Find the number of deciduous trees, each with a projected canopy area of 50 $m^2$, needed to reduce the runoff by 5% in 1-cm rainfall over a 0.4-ha impervious plaza. Assume 95% runoff from rainfall on the plaza surface.

10.3 The area of lawn space in a commercial/institutional area is estimated at 5 ac. The soil is hydrologic soil group B, but it is compacted to the characteristics of HSG C. The grass condition is considered as fair. For a 0.75-in storm, find the increase in runoff depth and volume due to the compaction.

10.4 The area of lawn space in a commercial/institutional area is estimated at 3 ha. The soil is hydrologic soil group B, but it is compacted to the characteristics of HSG C. The grass condition is considered as good. For a 2-cm storm, find the increase in runoff depth and volume due to the compaction.

10.5 The downspout from a rooftop area of 30 ft × 20 ft is directed onto the lawn as part of a DIS project. The lawn soils have a saturated hydraulic conductivity of 4 in/h. Find the infiltration area needed, and the roof:lawn ratio necessary to completely infiltrate a 1-in/h rainfall.

10.6 A DIS project has a roof:lawn area of 10:1. Find the rainfall intensity that can be managed if the lawn soils have a saturated hydraulic conductivity of 2 in/h.

10.7 A DIS project has a roof:lawn area of 8:1. Find the rainfall intensity that can be managed if the lawn soils have a saturated hydraulic conductivity of 20 mm/h.

10.8 A rooftop area is 25 ft × 25 ft and an available DIS lawn area is 5 ft wide by 15 ft long. Find the soil infiltration rate required to fully manage a 1.5-in/h rainfall.

10.9 Using Chesapeake Bay guidelines, find the sediment wash-off load from an urban area that is vacuum swept once per week.

10.10 Using Chesapeake Bay guidelines, find the nitrogen wash-off load from an urban area that is vacuum swept twice per week.

10.11 Using Chesapeake Bay guidelines, find the phosphorus wash-off load from an urban area that is vacuum swept every 4 weeks.

10.12 Using Chesapeake Bay guidelines, what is the yearly sediment wash-off load from a 60-ac urban area that is vacuum swept every 2 weeks.

10.13 Using Chesapeake Bay guidelines, find the yearly nitrogen wash-off load from a 150-ac urban area that is vacuum swept every 4 weeks.

10.14 A street-sweeping program sweeps 40 mi of roadway twice per month. Find the yearly removal of sediment.

10.15 A street-sweeping program sweeps 80 mi of roadway twice per month. Find the yearly removal of nitrogen.

10.16 A street-sweeping program sweeps 120 mi of roadway once per month. Find the yearly removal of phosphorus. If the lane is 12-ft wide, find the P reduction in lb/ac year.

10.17 A street-sweeping program sweeps 100 mi of roadway every 2 months. Find the yearly removal of copper.

# 11

# Green Infrastructure Stormwater Control Measures

## 11.1 Introduction

To mitigate the many problems created by excess stormwater runoff from impervious surfaces, various control measures have been developed, designed, and implemented. A primary function of most of these control measures is to contain or slow some of the excess water. Initially, these systems were primarily stormwater ponds that were installed to store water during rainfall events and to allow this water to drain at a slow, controlled rate. More recently, a range of engineered and hybrid natural facilities have become mainstream.

Stormwater management facilities have historically been designated as stormwater best management practices (simplified as BMPs), following the terminology originally developed for wastewater treatment. This terminology is still very common for describing any type of stormwater management device or practice. More recently, a report by the National Academies has suggested the more descriptive name of stormwater control measures (SCMs) (NRC 2009). This terminology was changed to reflect the advancement of knowledge to an engineered solution approach and is slowly gaining momentum.

SCMs can be physical structural systems or devices, such as a green roof or bioretention cell. They can also be practices designed for improving stormwater conditions, such as street sweeping or a public education programs. These latter practices are designated as nonstructural SCMs.

## 11.2 Fundamentals of Stormwater Control Measures

Structural SCMs incorporate various unit processes for water quantity and quality management discussed in previous chapters. For hydrology control SCMs affect the water balance of the stormwater to be managed. Processes that affect the water balance include storage during the storm event (Figure 11.1). Natural fates of incoming and stored water include infiltration into the surrounding soils and evapotranspiration of stored water into the atmosphere. This stored water can also be used beneficially if a water demand is found that can come close to matching the supply (based on both quantity and quality). The more pathways that are available in an SCM, the more effective the SCM in reducing the hydrologic impacts of the stormwater. The

*Green Stormwater Infrastructure Fundamentals and Design*, First Edition. Allen P. Davis, William F. Hunt, and Robert G. Traver.

Companion Website: www.wiley.com/go/davis/greenstormwater

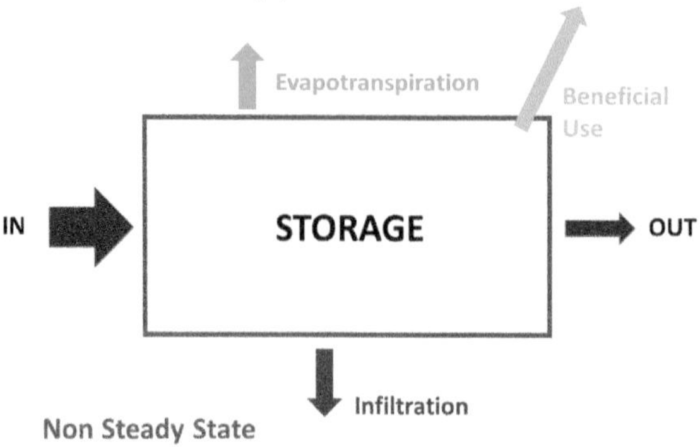


Figure 11.1 Water Balance Approach to Understanding SCM Performance.

magnitude of each pathway will depend on the size of the SCM, either volume or surface area normal to the pathway. Therefore, larger SCMs will have a greater impact relative to smaller ones.

Water-quality improvement requires the incorporation of one or more water treatment unit processes into the SCM (Chapter 8, Figure 11.2). In many of the first-generation SCMs, water quality was not a concern and limited water-quality improvement occurred. Later, as more jurisdictions regulated for water quality, more unit processes were incorporated and the toolbox of different types of SCMs became larger.

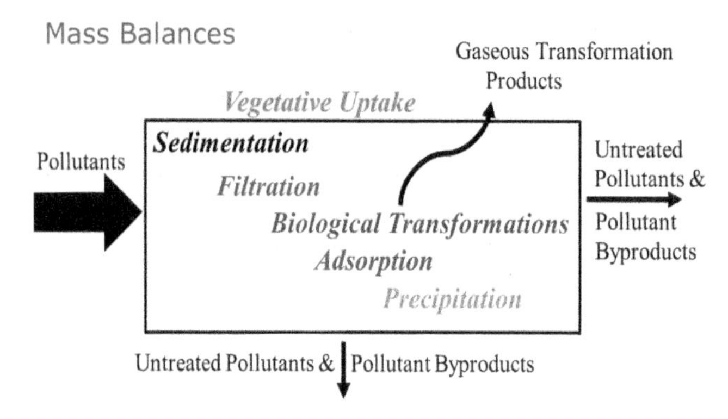


Figure 11.2 Water Treatment Unit Processes Available in SCMs for Stormwater Quality Improvement.

## 11.3 Designing to Climate and the Watershed

Different climates lead to different rainfall patterns, temperatures, and other factors important to SCM design and performance. As noted in Chapter 6, rainfall depth, intensity, and duration are important in considering hydrologic performance. Being aware of the expected rainfall patterns in an area will lead to more effective SCM designs. SCMs at locations with long duration and low intensity events can be designed differently from locations with short duration and high intensity events, even if the total rainfall or runoff depth may be identical.

In areas that have dry/rainy seasons, the first rainfall of the season may be extremely polluted (a seasonal first flush) due to the buildup of pollutants on the watershed during the dry season. The stormwater quality in this event will be very different from that of subsequent events. Vegetated SCMs will need special care during dry seasons to ensure that the vegetation survives.

It must be recognized that not all drainage areas will exert the same pollutant loads onto SCMs. Industrial and highly urbanized areas may have higher pollutant loads than less intensively developed areas. Intersections may have higher loads than straight highways due to vehicle braking and turning. Stormwater from rooftops may have very low pollutant loads since they are primarily impacted by atmospheric deposition only and will not generally experience loads contributed by vehicles, fertilizers, road deicers, and many other urban pollutant sources.

## 11.4 Types of Stormwater Control Measures

A wide range of structural SCMs now exist, known under a variety of names. All of these SCMs have many common elements. Table 11.1 lists some of the many common SCMs. The naming of various SCMs can be regional dependent and has not been standardized. Slight variations in a particular SCM also may lead to a different name; as a result, the same SCM may have different names in different areas. Adding to this complexity, various combinations and hybridizations of SCMs are becoming more common, again resulting in naming confusion. For this text, the authors have focused naming on primary definitions, with particular emphasis on the unit process(es) inherent to the SCM.

The selection and design of SCMs will depend primarily on the characteristics of the land area developed, rainfall patterns, the space available for the SCM, and regulatory requirements, goals, and crediting for stormwater management and treatment. Other factors under consideration include costs, safety, and aesthetics. Local geologic factors may control the SCM employed, such as locally high water tables and karst underlying soils.

Table 11.1 Common Stormwater Control Measures, Associated Practices, and the Chapter in Which These SCMs Are Discussed.

| SCM | Associated practices | Chapter |
|---|---|---|
| Manufactured devices | Separation/Filter, hydroseparator | 12 |
| Green roofs | Living roof, blue roof | 13 |
| Rainwater harvesting | Cisterns, rain barrels | 14 |
| Permeable pavements | Pervious pavers, pervious pavements | 15 |
| Infiltration basins | Infiltration trench | 16 |
| Sand filters | Specific designs for specific areas, e.g., Austin sand filter, Delaware sand filter | 17 |
| Bioretention | Bioinfiltration, bioswale, biofilter, rain garden | 18 |
| Swales and filter strips | Vegetated swale, grass swale, dry swale, vegetated filter strips | 19 |
| Stormwater wetlands | Pocket wetland, submerged gravel wetland | 20 |

Increasingly, treatment trains, combinations of SCMs in series, are being employed for comprehensive management of hydrologic and water-quality concerns. Diversion of excessive and erosive flows that exceed the design of smaller volume and water-quality SCMs can help protect infiltrating surfaces and reduce maintenance. Additionally, even more modifications, including design enhancements and amendments to media, are being developed to provide improved removal of specific pollutants. While SCMs are grouped under specific names in Table 11.1 and throughout this book, most the SCMs have a number of design options and even though they may be the same type of SCM, performances can vary greatly among systems with different designs (and placed in different areas).

## 11.5 Nonvegetated Stormwater Control Measures

Although many are not "green," a number of nonvegetated SCMs are commonly included as part of the Green Infrastructure toolbox. These SCMs are considered under the spectrum of GI since they possess the characteristics important to GI stormwater management, that is, they utilize natural processes and reduce volumetric and quality impacts. Infiltration basins and trenches and rock beds, permeable pavements, and water harvesting technologies are included here. All have aspects of advanced hydrology management. These SCMs focus primary on stormwater volume reduction. Nonetheless, water-quality benefits that are related to volume reduction, namely, infiltration and filtration, are achieved, though they are usually a secondary issue with these SCMs.

A short overview is provided for these SCMs with more details given in subsequent chapters dedicated to them.

### 11.5.1 Infiltration Basins and Rock Beds

Infiltration basins and rock beds are designed with runoff storage and infiltration as the major water management operations. Runoff is directed into the basin/bed. Some amount of storage will occur as the bed fills, water becoming stored between the rocks and/or sand of the bed. With this prolonged storage, infiltration into the surrounding soils is expected. These SCMs are expected to be most successful when the incoming runoff is relatively free of suspended solids that can clog the bed (or a pretreatment is used to remove suspended solids) and when surrounding soils have high permeability.

### 11.5.2 Permeable Pavements

Permeable pavements can be considered more of a runoff reduction SCM, as opposed to a facility that receives and manages runoff from other areas. By making the pavement permeable, the normal generation of runoff from the paved surface will not happen. The rainfall will pass through the pavement/pavers to the support media below. The media below the pavement will have some degree of storage and, when placed in pervious surrounding soils, will allow infiltration. Pervious pavements are susceptible to clogging from suspended solids and must be periodically cleaned, usually via vacuuming. Since dissolved pollutants may be captured by previously collected sediment in the pavement, permeable pavements also address some water-quality issues. Aspects of water-quality improvements are also possible during the storage/infiltration, depending on the design of the media and underdrain presence and/or configuration below the pavements.

Permeable pavements can be employed in a number of different areas but are not recommended for use in areas of high vehicle load, high speeds, high slopes, and high sediment loads. Parking areas, bike lanes, and especially pedestrian walkways appear to be excellent areas for use of permeable pavements.

### 11.5.3 Cisterns and Rain Barrels

Cisterns and rain barrels allow for direct storage of runoff draining from building roofs. Rain barrels are smaller and tend to be used on residential houses and smaller structures. Cisterns are large storage tanks that can hold thousands of gallons (liters) of water. Cisterns can be placed on the ground surface or buried below ground. Water is directed into the cistern from the roof downspout(s).

To be effective, the cistern must have available storage at the beginning of the storm. Any stored water must be drained before each next rain event, so that the maximum volume is available for storage. Several strategies can be employed to drain the storage tank. The water can simply be pumped or allowed to flow out of the cistern into the surrounding land area without beneficial use. This would likely occur during dry weather and the water would infiltrate into the soil or run onto impervious landscape. In the latter situation, since the flow is occurring during a dry period, the environmental impact should be minor.

Of course, a more sustainable mode of action for the cistern/rain barrel is to use the captured and stored water in a beneficial manner. This could simply be for local irrigation or combined with another SCM such as a green roof. Higher level uses (e.g., toilets, laundry, vehicle washing) may be explored if adequate volumes are available and appropriate water quality can be ensured.

### 11.5.4 Sand Filters

Sand filters, as the name implies, use a layer of sand to remove particulate matter from incoming stormwater. Sand filters have been used for decades in the treatment of water for drinking and, mechanistically, the treatment is similar. Particulate matter is captured on the sand media, primarily at the surface of the sand layer. As the capture continues, the headloss through the filter increases and clogging begins. Maintenance of sand filters is critical to sustain their performance; once a filter becomes clogged, the flow through the sand is minimal and no further treatment will occur. In most cases, sand filters target water-quality improvement, and will slow flows, but usually are not designed for stormwater volume control.

## 11.6 Vegetated Stormwater Control Measures

Integrating nature-based components and corresponding natural treatment into the urban landscape can enhance the overall management of urban stormwater. Such enhancements will also provide other benefits, including habitat and attractive aesthetics. Natural treatment systems have a diverse makeup and correspondingly can offer many different unit processes that have the ability to beneficially affect the water balance and water quality. Natural systems tend to be resilient and robust. However, emplacing them in a harsh urban environment will typically add stresses to components of the system that may not be found in more natural settings. System designs must consider these stresses.

Vegetated SCMs include vegetated swales and filter strips, bioretention and related technologies, green roofs, and stormwater wetlands. As noted in Table 11.1, other vegetated SCMs can be named that could be added to this list, but most are some form or composite of the ones listed above.

Vegetative SCMs can include many of the same unit processes as nonvegetated. However, the vegetation (and frequently the support media) lends a number of unique aspects to these SCMs. Clearly, the vegetation can affect the water balance by promoting ET and enhance infiltration through maintenance of infiltration passageways. However, the vegetation can provide a number of other beneficial services to the SCM. In green roofs, ET is the only pathway in which stored water can be removed from the roof, since infiltration is not an option. Additionally, dense and changing root systems can maintain porosity in media that receives sediment-laden runoff. The rhizosphere in the plant root zone can create and nurture an environment in which various microorganisms can thrive, leading to opportunities for biological transformations of captured runoff pollutants. The vegetation itself can take up nutrients and some toxic compounds. Thus, the design of the media must incorporate the needs of the vegetation.

The aesthetic benefits of the vegetation should not be underestimated. Public acceptance and embracement of vegetated SCMs is important to the successful implementation of green infrastructure technologies as part of the developed infrastructure. Planting plans that consider aesthetics, and include well-placed flowers, shrubs, and ornamental grasses, will go a long way in promoting acceptance, pride of ownership, and litter-free systems (Figure 11.3).

Vegetated SCMs can also create terrestrial habitat. Studies have noted significantly higher measures of biodiversity in bioretention systems with complex

**Figure 11.3** SCM Treatment Train at Villanova University Showing Diversity of Attractive Vegetation. (Photo by Authors).

vegetation coverage as compared to lawn-type green spaces (Kazemi et al. 2009a, 2009b, 2011). A discussion of ecosystem services provided by SCMs is the topic of Chapter 4.

Frequently, consensus does not exist on which vegetation species and designs may prove best for vegetated SCMs; different situations and climatic regions may demand different designs. Research and field experience continue to refine plant palette recommendations.

A great advantage of using vegetated SCMs is that their performance is mechanistically related to rainfall, infiltration, and evapotranspiration. More rainfall results in a higher rate of infiltration and ET, as infiltration and ET require available rainfall. For example, the highest ET rates recorded at Villanova University's green roof site occurred following high rainfall events. Similarly, the greatest volume infiltrated at Villanova's infiltration trench occurred during Superstorm Sandy in 2012 (Lewellyn et al. 2016).

### 11.6.1 Vegetation Challenges

Nonetheless, the use of vegetation generates a number of design and maintenance challenges. Vegetation must be able to thrive in urban stormwater management environments. The local environmental conditions must be carefully considered for design, construction, operation, and maintenance to ensure vegetation survivability. In some cases, the environmental conditions can be harsh. Coarse-growing media are commonly selected to promote rapid stormwater infiltration. However, coarse media may not support many forms of vegetation due to a lack of available moisture during dry days following rain events, even a short time after it rains. Rooftops or highly urbanized areas can be exposed to direct sunlight and high temperatures for long periods during summer days. On the other hand, during rainfall events, vegetation in some SCMs can be inundated with water and, in some cases, submerged for hours or even days. Runoff water is polluted, even with high concentrations of deicing salts in winter in colder climates. The costs associated with purchase and planting, and maintenance of vegetation, can be significant and, in some cases, are among the largest costs for some SCMs. Traditional methods to keep vegetation alive during times of stress, such as supplemental watering and fertilizing, can both prove detrimental to the overall goals of stormwater management. As a result, ensuring plant survival is crucial to ensure long-lasting cost-effective stormwater management.

The use of vegetation as part of a green infrastructure adds many new complexities to the design, construction, operation, and maintenance of the traditional urban infrastructure. Plants (as well as many other aspects of vegetated SCMs) are living entities. They have environmental needs for survival and will evolve over time. Species will grow and die. Many traditional engineers, designers, and contractors are not experienced in working with plants in green infrastructure and horticultural experts need to be consulted throughout the project lifetime. The initial establishment of plants in a new vegetated SCM will require care that may include watering and possibly some form of supplemental nutrients, such as mulch, fertilizer, and/or compost. Again, these materials must be used judiciously (more is not always better) so as to not create a stormwater problem. Plant function will also vary seasonally, leading to different SCM performances at different times of the year. Figures 11.4–11.6 show parking lot bioretention cells in summer and winter, showing the stark differences in the

Figure 11.4 Bioretention Cell in Summer in North Carolina. (Photo by Authors).

Figure 11.5 Bioretention Cell in Winter in Maryland. (Photo by Authors).

vegetation characteristics. It should be noted that in many areas of the country, higher rainfall occurs during the growing season.

Many aspects of maintenance of green infrastructure will differ from that of traditional "gray" stormwater infrastructure, with the added concern of keeping vegetation alive. The maintenance, however, may be similar in nature to traditional

Figure 11.6 Bioinfiltration Cell at Villanova University Filled with Plowed Snow. (Photo by Authors).

landscaping that is commonly done seasonally by landscaping contractors. This can include:

- trash cleanup
- removal of volunteer vegetation
- replacement of dead vegetation
- cutback of grasses and pruning of shrubs and trees
- replacement and/or addition of mulch or a similar ground surface cover

Thus, a separate GI maintenance industry need not be created.

### 11.6.2 Green Roofs

Green roofs have a shallow layer of storage/growing media on the roof, supporting a collection of plants. As with permeable pavements, green roofs are not receiving runoff but are preventing the formation of runoff. A fixed amount of rainfall is stored on the roof during the rain event. This stored water is then evapotranspired from the media and by the plants during dry periods. Heavy weight loads on the roof challenge the media and plants to be lightweight and many different types of green roof media have been examined. Also, keeping green roof vegetation alive and thriving can be a challenge when water and nutrients become scarce and when roof temperatures become very high. Green roof vegetation should be shallow-rooted, drought tolerant, and have high ET potential.

In addition to stormwater management, green roofs may offer other benefits, such as lower summer cooling costs due to the cooling provided by the evaporation of the stored water.

### 11.6.3 Bioretention

Bioretention uses a soil-based medium and vegetation. Bioretention is becoming increasingly popular due to its versatility is addressing both hydrologic and water-quality issues. The primary component is a sandy/soil medium that can hold and infiltrate water. Several unit processes, including sedimentation, filtration, adsorption, and biological transformations, can contribute to water-quality improvement. Bioretention vegetation can be varied in type and density and can add to the management of runoff volume, through ET and water-quality improvement.

A number of hybrid bioretention designs are being promoted in many areas. Different media types and design modifications target specific pollutants or hydrologic goals. The presence, placement, or absence, of an underdrain, may depend on local soil and groundwater conditions, and stormwater quality and treatment requirements.

### 11.6.4 Vegetated Swales and Filter Strips

Swales and filter strips are used for the conveyance and infiltration of runoff, along with some basic treatment. Swales are primarily conveyance SCMs, while filter strips are employed predominantly for infiltration. Swales are vegetated channels, usually trapezoidal in shape, because they are easy to construct. Filter strips are mildly sloped land areas that receive sheet flow. Both are planted predominantly with grasses. This helps with maintenance and allows unobstructed flow paths. Full coverage of vegetation is important so that bare soil is not exposed to erosive runoff.

### 11.6.5 Stormwater Wetlands

Stormwater wetlands are designed to mimic natural wetlands. They have both shallow and deep water pools and a mix of vegetation types. The pools permit runoff storage and the different depths allow the presence of water even under drought conditions and can support the variable biogeochemical reactions that occur under different oxygen concentration. Stormwater wetland vegetation must be able to survive under continuous wet conditions. Stormwater wetlands can host a variety of unit processes for water-quality improvement. However, because they are so biologically active, they may not be as effective as other SCMs in nutrient control.

## 11.7 Selecting the SCM Site

Commonly, the area of a site relegated to stormwater management has been the land of least value to the development, with little consideration given to selection of the site for stormwater performance. Careful consideration of site characteristics could be overall beneficial to the development of the site because selection of an area with good stormwater performance characteristics could lead to a smaller SCM footprint and greater area available for other aspects of the development.

The SCM should be installed away from current and anticipated utilities, both aboveground and subsurface. All SCMs will require maintenance at some point and some may require frequent inspection and maintenance, along with occasional repair. The SCM should be place in the area that allows these tasks to be done efficiently and effectively.

Of course, topography is critical to the stormwater infrastructure. The SCM must be placed so that the flow from the drainage area easily is conveyed into the SCM. That does not mean that all SCMs must be placed at the lowest points in the drainage area. In fact, good stormwater management requires systems distributed throughout the watershed. The SCM should be placed at the lowest point of the drainage area border.

As noted in Figure 11.1, infiltration is a desired pathway for stormwater removal for many SCMs. The native soils must have characteristics to allow effective infiltration to manage stormwater volume and soil infiltration rates will determine the area of infiltration SCMs.

Evaluation of the infiltration rate or hydraulic conductivity of soils, while sounding simple, can be a complicated process. Measurement of hydraulic conductivity can be affected by heterogeneities in the soils and by flow that is assumed one-dimensional (downward) but actually occurs in all three directions.

Site measurements of infiltration generally involve driving a pipe of some size into the soil, filling the pipe with water, and timing the rate of infiltration into the soil. In a simple test, a hole is dug and a casing is placed in the soil to the depth of interest for the infiltration. The casing is filled with water and allowed to drain for 24 h. After 24 h, the casing is again filled with water and the drop of water level is timed to determine the infiltration rate. This test may be repeated several times and an average taken.

A standardized device for measuring infiltration at soils surfaces is a double-ring infiltrometer. The double-ring infiltrometer consists of two concentric cylindrical tubes that are driven into the ground. Both tubes are filled with water. The infiltration rate of the inner tube is measured. The idea of the outer ring is to saturate the soils near the inner ring so that the flow will be more closely one dimensional. New instrumentation is now available that can provide estimates of infiltration rates with the use of less water. Infiltration rate/hydraulic conductivities can vary widely within a short special distance. Variation of an order of magnitude within the site of an SCM would not be unusual.

The infiltration rates will determine the volume of stormwater that can be infiltrated at the site. Where infiltration is restricted or undesirable, underdrain systems can be installed to provide a flow path for slow release of the treated runoff.

## 11.8 Stormwater Treatment Media

Many of the GI SCMs include some provision for runoff to flow/infiltrate through an engineered media. This is true for permeable pavements, green roofs, sand filters, and bioretention. Additionally, infiltration into the surrounding native soils involves interactions of water with a media. These interactions can have important ramifications on water quality.

### 11.8.1 Rock, Gravel, and Coarse Sand

Rock, gravel, and coarse sand may be used in rock beds or infiltration basins. In these SCMs, the purpose of the media is for structural support only. The presence of these coarse media should have no impact on water quality. Coarse sand may be employed in a sand filter or infiltration trench. Some gravel and coarse sand may be fine enough to exert some degree of filtration to the water, resulting in water-quality improvement. These media are not expected to be chemically active and no adsorption or other chemical treatment is expected. Sand is typically mixed into bioretention media to yield higher infiltration rates.

### 11.8.2 Silts and Clays

Silt is defined by US Department of Agriculture (USDA) as soil particle size between 2 and 75 μm; clays have particle size <2 μm. These finer particles, especially the clays, have surface chemical reactivities that can promote a number of reactions important to water-quality improvement. Clays can carry a negative surface charge resulting from a slight imbalance of charges in the mineral matrix. Additionally, iron, aluminum, and manganese oxides can be part of the fine soil components, directly or as coatings on other media. These oxides have reactive chemical groups on their surfaces that can adsorb cations and anions, dependent on ambient environmental conditions, including pH. These reactions are described in Chapter 6.

Soils with some limited amount of fines will be used in bioretention media, allowing this SCM to have high treatment capabilities. However, a tradeoff will occur as higher levels of fines will compromise the infiltration rate, and thus the hydraulic treatment capacity of the bioretention cell. A careful balance must be struck between allowing fines for high treatment reactivity and restricting fines to allow high infiltration rates. Inclusion of fines will increase water-holding capacity and support plants that lead to greater water loss via ET.

Soil characteristics and texture are also important in defining the reactivity of the ambient soils in a watershed. Soils with high infiltration rates, such as sandy soils in the coastal plains, will produce less runoff. Also, sandy soils will rapidly infiltrate water that is stored in a SCM. The hydrologic effectiveness of infiltration basins, permeable pavements, bioretention, swales, and filter strips will all depend on the permeability of the surrounding soils. These soils will allow collected runoff to be drained and transported away from the SCM. These soils can also contribute additional water-quality improvement. Pollutants that were not removed via treatment in the SCM may be removed as the water migrates through the surrounding soils. Soils tend to have a high affinity for the removal of metals, phosphorus, and many organics (Chapter 6). However, nitrate and chloride are not expected to be attenuated by the native soils and may cause further subsurface water-quality concerns.

### 11.8.3 Organic Media

Organic matter can play a major role in the performance of an SCM, especially one that is vegetated. The organic matter can bind pollutants, including heavy metals and hydrophobic organics. It can supply nutrients to plants and assist in moisture retention for the vegetation. Surface organic layers can add aesthetic benefit. However, readily degraded organic matter can contribute excess nutrients to stormwater. More discussion on nutrient leaching from organic matter is provided in the individual SCM chapters.

## 11.9 Volumetric Storage

Important to the hydrologic performance of all SCMs is the volumetric storage capacity of the SCM and the volumetric performance once this capacity is reached. Understanding the volumetric capacity is critical to the selection and design of specific SCMs. From a static performance perspective, the volumetric relationship can be described via the curves shown in Figure 11.7. The plot presents the volume exiting the SCM as a function of the stormwater volume entering the SCM. The horizontal line at zero SCM discharge represents the SCM storage volume. For a storage SCM like a cistern, this volume represents the volume of the cistern. For a bioretention system or a green roof, this volume will include the volumetric storage in the porosity of the media, as described in Chapter 7, and the respective chapters on these SCMs. Some ponding storage may be included also. Some SCMs, like swales or sand filters, will have limited storage.

The overall volumetric performance of SCMs will include the continuous infiltration of stored water during the storm event. Unless the SCM is completely blocked and isolated from the surrounding soils, exfiltration into the surrounding soils will always occur, as described by the soil physics and infiltration analysis presented in Chapter 7. This continuous infiltration will recover some available storage beyond that calculated using a static analysis, a process known as *dynamic storage* (Traver and Ebrahimian 2017). Consideration of dynamic storage will allow the use of smaller SCMs compared

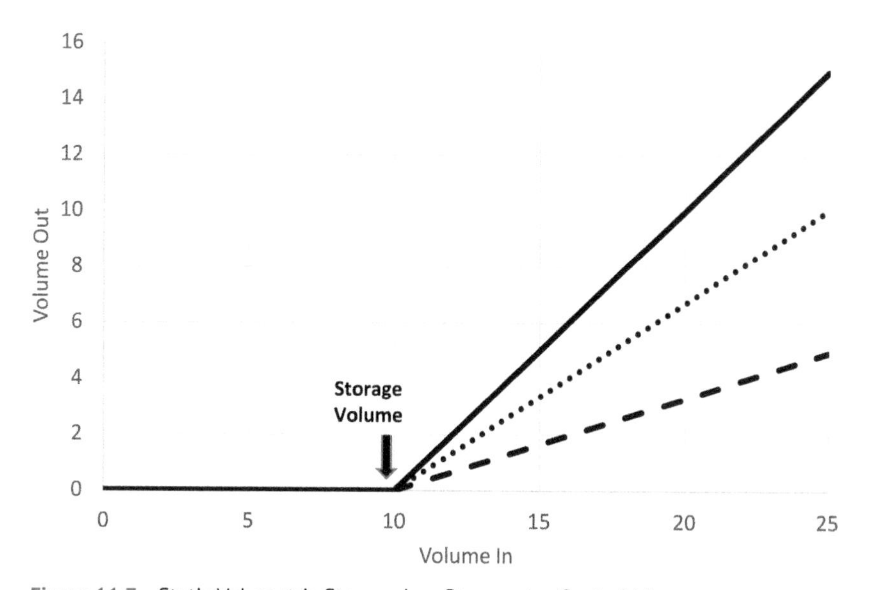


Figure 11.7 Static Volumetric Storage in a Stormwater Control Measure.

to designs based on static storage alone. Dynamic storage considerations will also allow SCM design to be tailored to the local climate and soils.

The size of the SCM will determine the storage volume. The storage volume that is needed to meet the needs of the watershed can be used to size the appropriate SCM. Multiple SCMs may be placed in a treatment train arrangement to obtain the proper volumetric storage (and treatment). This simple analysis, of course, assumes that the entire storage volume is available at the beginning of a rainfall event; this may not be true if rain occurred a short time prior. For effective volume management, some of the SCM storage volume must be recovered soon after a rain event, although dynamic storage will increase the volume managed by the SCM. This means maximizing ET and infiltration between storm events and emptying any storage. Reliance on natural processes like ET and infiltration means that available storage volume will be seasonal, as both can be slowed greatly by cooler weather. This also means that SCM volumetric performance will vary with climate.

The sloped lines after the storage volume represent the SCM volumetric reduction after the storage capacity is met. These relationships will differ based on the type of SCM. A cistern or green roof will provide no additional volumetric impact after reaching capacity. However, any SCM that has an infiltration component will continue to reduce volume. In these cases, the relationship will depend on the infiltration characteristics of the surrounding soils. SCMs in areas with permeable sandy soils can still provide significant volume reduction, even after reaching their storage capacity, as suggested by the dashed line in Figure 11.7.

## 11.10 Drains and Underdrains

A critical component to all SCMs is the drain or underdrain (or absence of one). As noted above and to be emphasized in specific SCM chapters, rainfall/runoff storage is a critical operation in runoff management. For a SCM to perform as designed, most available storage should be available for stormwater during an event. This means that after (and possibly during) an event, stored water must be removed. Ideally, this will occur via infiltration, ET, and/or beneficial use. However, if these processes are inadequate to recover the storage, a drain system may be placed in the SCM.

Ponded and harvested water will be drained through orifices and weirs specifically designed to release a fixed volume of water at a controlled rate. In media-based SCMs, a perforated pipe, known as an underdrain, is installed that extends underneath the SCM, collecting water after it has passed through the media. The drains and underdrain will remove the water stored above them and convey the water to the stormwater infrastructure or to a natural water body.

In areas where infiltration and/or ET is moderate to high, underdrains may not be necessary as these water pathways may have adequate flows to recover storage. Placement of the underdrain at different levels in the media can also create opportunities for increased infiltration and ET. The use of an underdrain system with a raised discharge point can impart several desirable characteristics to an SCM. In this case, a subsurface volumetric storage is created in the SCM media, commonly known as an internal water storage (IWS) zone (Figure 11.8). The IWS can hold water between events to allow for greater infiltration or enhanced ET. It may also create anoxic

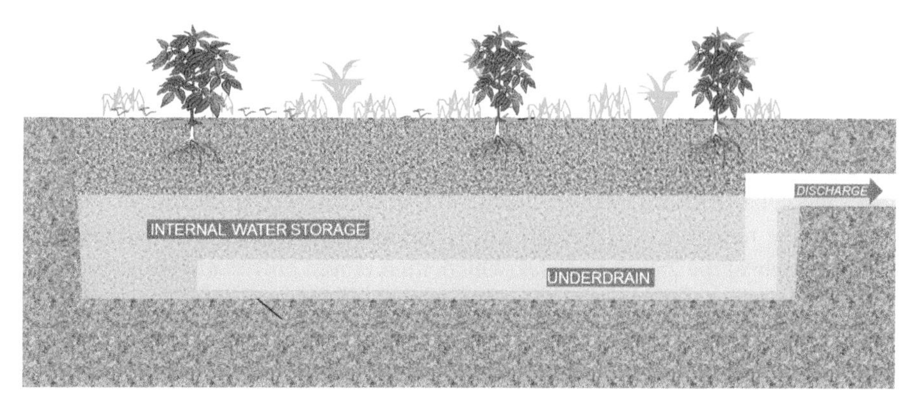


Figure 11.8 SCM with Underdrain and Internal Water Storage.

conditions conducive to denitrification and other beneficial biological pollutant transformations.

## 11.11 "Irreducible Concentrations"

Frequently in SCM studies, the concentration of a pollutant is found to be reduced to a low concentration by the SCM, but not to zero. In fact, many of the same types of SCMs may reduce a pollutant concentration to approximately the same value, such as for TSS in sand filters (Barrett 2005). This phenomenon has been termed as the "irreducible concentration," the concentration below which the SCM cannot be more effective. However, this terminology is somewhat misleading.

By carefully examining the characteristics of stormwater flows and water quality, in conjunction with the applicable unit processes, greater fundamental understanding of the "irreducible" concentrations, can be offered. Specific reasons for the appearance of an irreducible concentration include broad particle size distributions in stormwater, the dependencies of particulate matter removal on particle size (squared), adsorption equilibrium established by SCM treatment media, and vegetative nutrient needs.

An example can be presented focused on particulate matter removal. As noted in Chapter 8, the efficiencies of both sedimentation and filtration processes depend on particle size, usually the square of the particle diameter. Therefore, large particles are much easier to remove than smaller ones. During the SCM treatment, the large particles will be readily removed, but the smaller ones may be generally unaffected. Thus, effectively, a constant effluent TSS concentration results, essentially equal to the concentration of the smaller particles. This situation can be illustrated in the following example:

Consider a runoff flow with three particle sizes: 10 mg/L of 25 μm particles, 50 mg/L of 100 μm particles, and 200 mg/L of 250 μm particles, for a total of 260 mg/L TSS. This runoff is treated in a sand filter SCM. From Chapter 8, the filtration equation is of the form:

$$C = C_0 \exp\left(-kd^2\right) \tag{11.1}$$

where $d$ is the particle diameter and $C$ and $C_0$ represent the SCM effluent and influent particulate matter concentrations. $k$ represents the combination of all of the treatment terms and includes parameters such as flow velocity, sand media diameter, and sand filter length.

Figure 11.9 shows the results of this analysis as $k$ increases from 0 to 0.005 $\mu m^{-2}$. The large 250-μm particles are removed at low values of $k$. As $k$ increases, the 100-μm particles become removed. At $k > 0.003\ \mu m^{-2}$, the total TSS concentration is made up almost entirely of the 25-μm particles, which are removed only to a small extent. The TSS concentration, at just under 10 mg/L, appears "irreducible" and independent of design and operational conditions (as represented by variable $k$). A similar analysis can be done using a SCM and sedimentation theory (Chapter 8).

Irreducible concentrations for dissolved pollutants leaving media-based SCMs have been discussed by Li and Davis (2016). When discharge pollutant concentrations are controlled by adsorption processes, the equilibrium adsorbed concentration on the media will control the effluent runoff concentration. When the concentration coming into the SCM is higher than the equilibrium adsorbed concentration, pollutant removal will occur. When the influent concentration is lower than the equilibrium concentration, desorption may occur. The result is that most effluent concentrations will be near the same value, the "irreducible" concentrations. This concentration is controlled by the characteristics of the media. Such data are indicated by a relatively flat probability plot curve for SCM effluent concentrations, as compared to influent concentrations (Figure 11.10). An intersection point, representative of the equilibrium concentration, may also be present, as shown in figure 11.10 for dissolved phosphate removal in a modified bioretention facility.

Changing or modifying the media will change the equilibrium concentration for a water-quality parameter in a media. This has been explored by several researchers, especially for enhancing phosphorus removal by adding or increasing the amorphous iron and aluminum content of the media.

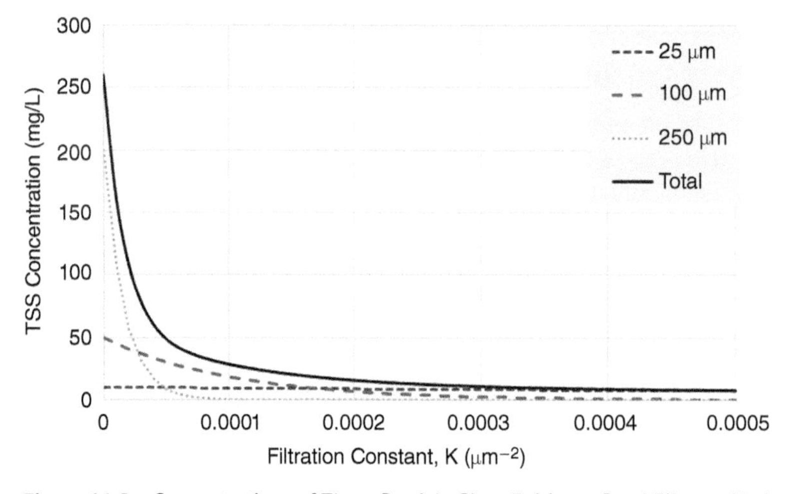


Figure 11.9 Concentrations of Three Particle Sizes Exiting a Sand Filter at Various Values of the Treatment Parameter $k$ (Equation 11.1). At Large $k$, Only the Smallest Particles Remain and the TSS Concentration Appears "Irreducible.".

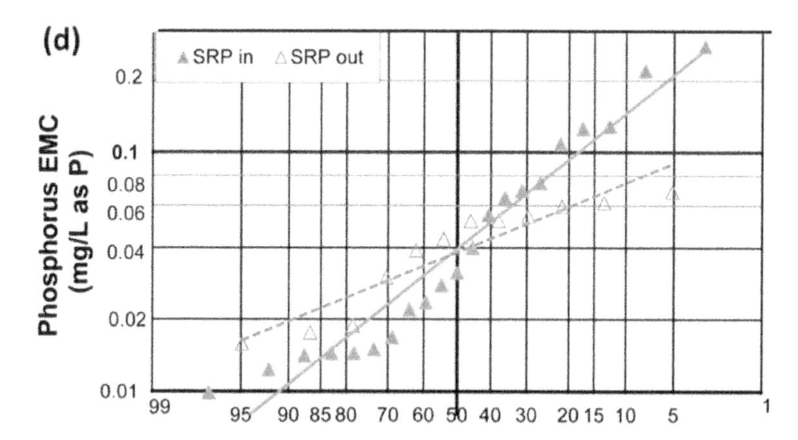


Figure 11.10 Input and Outpoint Phosphate Concentrations (SRP) for a Modified Bioretention Facility, with an Intersection Point and with Output Concentrations Varying Much Less than Input Concentrations, Indicative of an "Irreducible Concentration."

Reprinted with Permission from Liu, J., and Davis, A. P. (2014). "Phosphorus Speciation and Treatment Using Enhanced Phosphorus Removal Bioretention." *Environmental Science & Technology*, 48(1), 607–614. Copyright 2014 American Chemical Society.

Densely vegetated SCMs, such as stormwater wetlands, will demonstrate irreducible concentrations of nutrients in the water. The cycling of nutrients through growing, dormancy, dying, and leaf drop of vegetation will maintain a finite level of macro- and micronutrients in the wetland.

## References

Barrett, M.E. (2005). Performance comparison of structural stormwater best management practices. *Water Environment Research* 77 (1): 78–86.

Kazemi, F., Beecham, S., and Gibbs, J. (2009a). Streetscale bioretention basins in Melbourne and their effect on local biodiversity. *Ecological Engineering* 35 (2009): 1454–1465.

Kazemi, F., Beecham, S., and Gibbs, J. (2011). Streetscape biodiversity and the role of bioretention swales in an Australian urban environment. *Landscape and Urban Planning* 101 (2011): 139–148.

Kazemi, F., Beecham, S., Gibbs, J., and Clay, R. (2009b). Factors affecting terrestrial invertebrate diversity in bioretention basins in an Australian urban environment. *Landscape and Urban Planning* 92 (2009): 304–313.

Lewellyn, C., Lyons, C.E., Traver, R.G., and Wadzuk, B.M. (2016). Evaluation of seasonal and large storm runoff volume capture of an infiltration green infrastructure system. *Journal of Hydrologic Engineering* 21 (1): 04015047.

Li, J. and Davis, A.P. (2016). A unified look at phosphorus treatment using bioretention. *Water Research* 90: 141–155.

Liu, J. and Davis, A.P. (2014). Phosphorus speciation and treatment using enhanced phosphorus removal bioretention. *Environmental Science & Technology* 48 (1): 607–614.

NCR, National Research Council of the National Academies (2009). *Urban Stormwater Management in the United States*. Washington, D.C: The National Academies Press.

Traver, R.G. and Ebrahimian, A. (2017). Dynamic design of green stormwater infrastructure. *Frontiers of Environmental Science & Engineering* 11 (4): 15.

## Problems

**11.1** Find a SCM type listed in your state stormwater manual. Compare and contrast this SCM to the most similar SCM listed in Table 11.1.

**11.2** Take a walk through your neighborhood, around your university, or in a commercial area. Note the stormwater control measures. Are they green infrastructure SCMs? Photograph and describe them, noting the season. Describe the hydrologic and water-quality benefits they provide.

**11.3** Visit a bioretention cell or similar nearby SCM soon after a rainfall. Note the pooled water level. Come back and check again a few hours later. If possible and can be done safely, measure the pooled water levels with a ruler. Calculate the infiltration rate.

**11.4** Use sedimentation theory (Chapter 8) to demonstrate how an irreducible concentration can occur in a sedimentation basin. Like the filtration example discussed earlier in this chapter, assume 10 mg/L of 25 μm particles, 50 mg/L of 100 μm particles, and 200 mg/L of 250 μm particles, for a total of 260 mg/L TSS. Try different values of hydraulic loading ($Q/A$) and evaluate total TSS removal. Discuss your results.

**11.5** Using sedimentation theory (Chapter 8), compare the removal of 300 μm particles to that of 30 μm at the flow rate of 10,000 L/min into a basin with bottom area of 350 $m^2$. Discuss this result in terms of an irreducible concentration.

# 12

# Inlets, Bypasses, Pretreatment, and Proprietary Devices

## 12.1 Introduction

This chapter discusses the connections between the green infrastructure (GI) stormwater control measure (SCM) and the drainage area through which the runoff is generated. Inlets provide hydraulic connectivity. They are designed to handle the hydraulic load and properly route it to the SCM. Pretreatment may be provided by the inlet or a separately installed treatment device. Usually pretreatment is based on the removal of trash and of heavy sediments and sediment loads from the runoff so that the trash and sediments do not clog the SCM. The SCM will be sized to treat a stormwater event of a specific size. The inlet system or SCM must include some provision to allow high flows and excess stormwater volume to bypass the SCM so that flooding will not occur in the watershed. Proprietary devices are briefly discussed as part of the urban SCM toolbox.

## 12.2 Inlets

Inlets provide the flow pathway from the runoff source to the SCM. Inlets can come in many forms; a number of different types of inlets are shown in Figure 12.1. In some cases, there is no specific inlet area. In Figure 12.1a, the surrounding walkway area leads directly into the bioretention cell via sheet flow and a concrete step.

One of the most common types of inlets is a curb cut, as shown in Figure 12.1b. The curb cut is placed into the curb at specific points along the curb line. Curb cuts can also be installed into existing curbs in a GI retrofit situation. Figure 12.1c and 12.1d both show the use of more traditional storm drain inlets, which are connected to the GI SCM. Figure 12.1e and 12.1f are photos of inlets that are placed along roadways in retrofit designs.

Inlets collect flow from the roadway or other runoff source. The inlet must be designed with adequate capacity so that the drainage area will not become flooded during events smaller than the design event. Detailed hydraulic design of inlet structures can be found in other texts (e.g., Linsley et al. 1992).

Several of the inlets shown in Figure 12.1 have rip-rap behind the inlet to dissipate some of the runoff energy and prevent erosion of the media in the SCM. Depending on the inlet design, some collection of trash and large sediment may occur in/near the inlet.

*Green Stormwater Infrastructure Fundamentals and Design*, First Edition. Allen P. Davis, William F. Hunt, and Robert G. Traver.

Companion Website: www.wiley.com/go/davis/greenstormwater

Figure 12.1 (a–f) Various Types of Inlets Employed in Green Infrastructure Stormwater Control Measures (Photos by Authors).

## 12.3 Stormwater Bypass

Details of SCM hydraulic design were presented in Chapter 7. SCMs are designed for a specific volumetric capacity and may have some degree of dynamic storage that must be considered. Most GI SCMs are not designed for extreme rainfall events and such events will overload the storage capacity of the SCM, potentially causing erosion and increasing maintenance costs. Bypass provisions are included with SCM design so that once the volumetric capacity of the SCM is reached, excess volume will bypass the

Figure 12.2 Various Types of Bypass Systems Employed in Green Infrastructure Bioretention Facilities. (a) Overflow Pipe and Cage. (b) Concrete Weir Strip Overflow. (c) Bypass in Action; Excess Flow Will Not Enter Bioretention and Bypasses to Storm Drain Inlet (Photos by Authors).

SCM and be discharged directly to the outlet system. This will prevent stored water from backing up into the drainage area and causing flooding.

The bypass can be constructed in a number of ways (Figure 12.2). For many systems, the bypass consists of some type of overflow outlet. This may occur in rainwater harvesting, bioretention, green roofs, and sand filters. An outlet is constructed at a bypass height in the SCM. Once the storage capacity is reached in these SCMs, any additional volume will increase the level of the pool over the design value and this excess stormwater will flow directly into the outlet system. This stormwater will be untreated, with the exception of some minor particulate matter removal that may have occurred in the inlet and with the initial flow of water into the SCM storage. In some systems, the pooling may prevent stormwater from entering the SCM (Figure 12.2c). When the storm event exceeds the capacity of a permeable pavement, the excess water will run off and not enter the pavement—not much different than with a standard pavement.

## 12.4 Catch Basin and Inlet Filters

A number of proprietary filters are available for installation in inlets and catch basins. These filters are constructed with some type of filter fabric or geotextile. They are designed to fit into an inlet so many different configurations are available. They will filter particulate matter from incoming stormwater. The particulate matter will accumulate in the filter and periodically the filter must be removed and washed or replaced. The filters have some design function built into them to allow flow to bypass during high flows resulting from extreme events, or when the filter fills with sediment.

## 12.5 Pretreatment

Nearly every stormwater practice will benefit from some type or degree of pretreatment. In some instances, runoff is specifically treated prior to arriving to the practice and other times pretreatment is accomplished internal to the SCM. Pretreatment, because it often requires a second device or design, is part of a special, and frequently used, treatment train system of operation.

The goal of pretreatment is typically to collect gross solids and the coarser fraction of sediment before these pollutants reach the principal SCM. Pretreatment devices are specifically designed to be easily maintained, so that the maintenance burden in the principal SCM is (often greatly) reduced.

Pretreatment systems are focused on only one main treatment mechanism, but it manifests itself in multiple ways: sedimentation. Thus, the governing equation for pretreatment devices is Stokes' law (Chapter 8). Types of pretreatment SCMs vary per the landscape and type of principal SCM they are pretreating. Most common among pretreatment SCMs are forebays, short filter strips, and swales. Many manufactured devices are specifically designed to capture gross solids and sediment.

## 12.6 Forebays

As the name implies, forebays are platforms or pools placed at the entrance to SCMs. They vary in size depending upon the catchment treated. Forebays were initially used as a part of wet ponds. Now, they can be incorporated in much smaller scale practices such as bioretention and level spreader-vegetated filter strip SCMs (Figure 12.3) to

Figure 12.3 Forebays for (a) Wet Ponds, (b) Bioretention, (c) Sand Filter, and (d) Level Spreader-vegetated Filter Strips (Photos by Authors).

reduce maintenance needs. Forebays can remain wet (Figure 12.3a and 12.3b) or dewater between events (Figure 12.3c). From a pollutant-removal perspective, wet forebays are expected to perform better, because they will limit resuspension of captured particulate matter. However, forebays that allow infiltration may be preferred due to the reduced mosquito risk. Forebays can be lined with rip-rap, which is selected to limit internal scour. Designers may choose to hold the rip-rap in place with concrete. Rip-rap, however, can complicate clean out/maintenance of the forebay.

The inlet shown in Figure 12.4 has a small removable weir at the discharge point. This weir is designed to collect trash and sediment, incorporating some forebay performance to the simple inlet.

The surface area, depth, and liner of the forebay are determined by two factors: (1) reduction of flow velocities to rates that are nonerosive and (2) the expected gross solids and coarse sediment load, which vary with the watershed size. The portion of the forebay closest to the inlet is dedicated to energy dissipation. The balance of the forebay is sized to collect sediment and gross solids (Figure 12.5).

Figure 12.4 Street Inlet Structure with Removable Weir at Back to Collect Trash and Sediment before Conveyance into the Bioretention Cell (Photos by Authors).

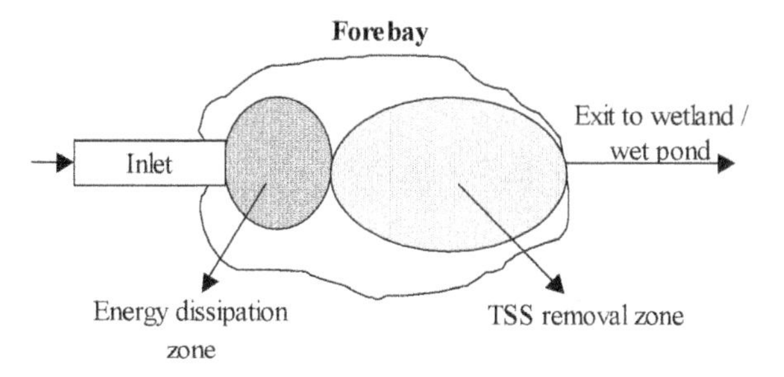


Figure 12.5 Plan View Schematic of Forebay Highlighting the Energy Dissipation Zone and the Sedimentation Zone (Credit: Authors' Research Group).

### 12.6.1 Forebay Design

Many design manuals only give cursory guidance to forebay surface area requirements, suggesting simply that the forebay surface area should be 10–20% of the footprint of the SCM. A more rigorous design approach is suggested.

Step 1—Calculate inflow rate to be dampened to nonerosive velocities. This velocity is a reflection of how frequently the property owner is expected to make "structural" repairs to the forebay due to internal scouring. It is suggested that a 10–25-year average return interval be used.

Step 2—Calculate the depth and surface area needed to reduce flow velocities. The use of a rip-rap or concrete liner is determined here. This surface area, however, is only a fraction of the total surface area.

Step 3—Estimate annual quantity of gross solids in runoff and load of coarse sediment (sand) to be captured by the forebay.

Step 4—Using Stokes' law (Chapter 8), determine the required length and depth of flow for particles to settle.

Step 5—Choose a maintenance interval for dredging (removal of collected sediment and gross solids). A more frequent schedule allows the forebay's surface area to be smaller. A reasonable assumption is that solids removal will be done every 5–10 years.

The shape of a forebay does not have "hard and fast" guidance; forebays in general must allow access to all accumulated sediment and gross solids. Their width is determined by the furthest reach of an excavator arm. Because forebays are intended to serve two purposes: (1) still water and (2) store sediment and gross solids, their profile should initially be relatively deep, and as the water progresses, it shallows out (Figure 12.6).

Following Stokes' law, a falling distance of 0.3 m (1 ft) allows sediment to reach the forebay bottom and be captured three times more quickly than a 0.9-m (3 ft) depth. Forebays do not need to be hydraulically connected to the balance of the practice. They can be an initial pool that is "stepped down" into the main body.

Many forebays are unlined. As long as a sufficient depth is provided at the forebay entrance, a liner is not necessary. One option is to line the forebay with rip-rap. An advantage of this approach is that the bottom of the forebay is clearly established, which is useful during restorative maintenance. In some areas, rip-rap-lined forebays are also cemented in place, allowing the removal of sediment and gross solids from forebays using a vacuum truck. Cementing rip-rap in place prevents rocks from getting stuck and/or clogging the vacuum hose.

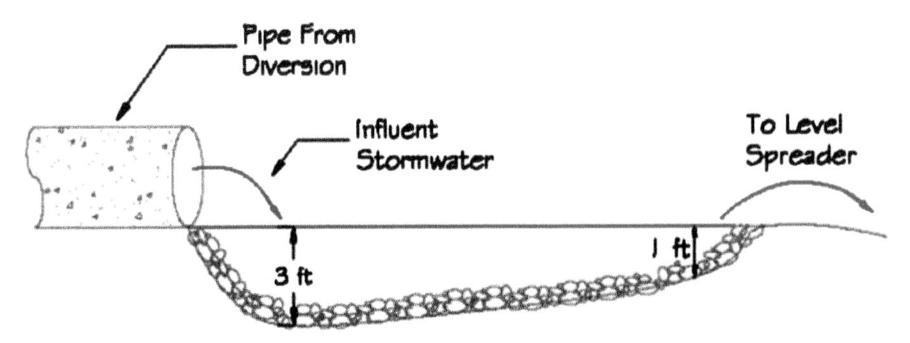


Figure 12.6 A Typical Cross-section for the Forebay. Deeper Area near Inlet Allows for Energy Dissipation, While the Shallower Portion of the Forebay Allows for Easier Accumulation of Fine Particles due to Sedimentation (Credit: Authors' Research Group).

**Example 12.1** Find the forebay area needed to settle a 0.15-mm (and greater) sediment particles from flow of 1 $ft^3/s$.

**Solution** Using English units, 0.15 mm = $4.92 \times 10^{-4}$ ft. Assuming 50°F and a particle specific gravity of 2.6, Stokes' law gives

$$v_c = \frac{(\gamma_p - \gamma_w)d^2}{18\mu} = \frac{(2.6(62.4) - 62.4\ \text{lb/ft}^3)(4.92 \times 10^{-4}\ \text{ft})^2}{18(2.74 \times 10^{-5}\ \text{lb s/ft}^2)} = 0.049\ \text{ft/s}$$

The forebay footprint area is given by the flow rate divided by the particle settling velocity:

$$A_b = Q/v_c = (1\ \text{ft}^3/\text{s})/(0.049\ \text{ft/s}) = 20.4\ \text{ft}^2$$

While there are understandable concerns with adding "excess cement" to an ecosystem-based stormwater practice, because the design of the forebay requires it to be regularly cleaned out, having a definitive and stable bottom (provided by the cemented rip-rap) is a recommended design feature. This is particularly true when the forebay is located so that a vacuum truck can easily access the forebay.

### 12.6.2 Forebay Maintenance

Gross solids and sediment will be continuously accumulated in a forebay. As such, removing collected solids is the primary maintenance task associated with forebays. The frequency of this maintenance is watershed dependent and varies from extremely infrequent (once per 30 years) in stable and litter-free watersheds to very frequent (once per year) in unstable watersheds with large sediment loads. Trash removal in urban settings can be much more frequent. For planning purposes, a study conducted by Johnson (2007) found a typical forebay cleanout frequency to be once per 5–10 years.

The trigger for sediment removal is when the storage capacity of the forebay is reached. The point at which the forebay's storage volume is 50% occupied by sediment and gross solids is a good target. In forebays where no design plans are available, a surrogate is needed to determine the point when maintenance is needed. In the forebays of most small-to-medium size wetlands, when the average (not maximum) accumulated soil depth is within 1 ft (0.3 m) of the surface, the forebay needs to be cleaned out.

Larger forebay capacity can be measured using a boat with simple sonar. For shallow forebays, a person in a hip or chest waders walks a few transects with either a rod or a yardstick. In the most obvious cases, the need to dredge a forebay can be visually determined (such as sediment presence at the surface at normal pool).

## 12.7 Proprietary Devices

Because of the sizable demand for stormwater control measures, a large number of proprietary devises are available on the market from various vendors. These devices are commonly used in areas where space is limited and many are installed completely subsurface. A few of these devices focus on a specific unit process for hydrology management; many include provisions for stormwater quality treatment. Examples of hydrology management include underground structural storage devices. For treatment, these devices can promote sedimentation (or another similar particle separation process), filtration, or adsorption.

Underground structural storage is generally some type of structural plastic framework that can be used to store runoff during a storm event. The structure is commonly placed beneath a parking lot, with the framework strong enough to support the parking lot structural load, or whatever surface use is desired (Figure 12.7). Frequently, the bottom of the storage is left open as native soil or covered with stone to allow the stored water to infiltrate between storms, replenishing the storage over time. Part of the bottom near the storage inlet may be kept impervious so that sediment that enters the storage and settles quickly can be easily cleaned from the storage zone.

Many proprietary devices are available for stormwater quality improvement (Figure 12.8). The most common type of treatment is particle removal by gravity separation, or by vortex separation. In gravity separation, the device consists of some type of storage, baffles, weirs, and walls, creating quiescent conditions for particle settling. In vortex separation, the water is directed into a circular device, where at high flows, the rotation of the flow creates centrifugal forces that drive the particles to the edge walls, allowing them to be removed.

Figure 12.7 Heavy Duty Plastic Structures to be used for Subsurface Stormwater Storage (Photo by Authors).

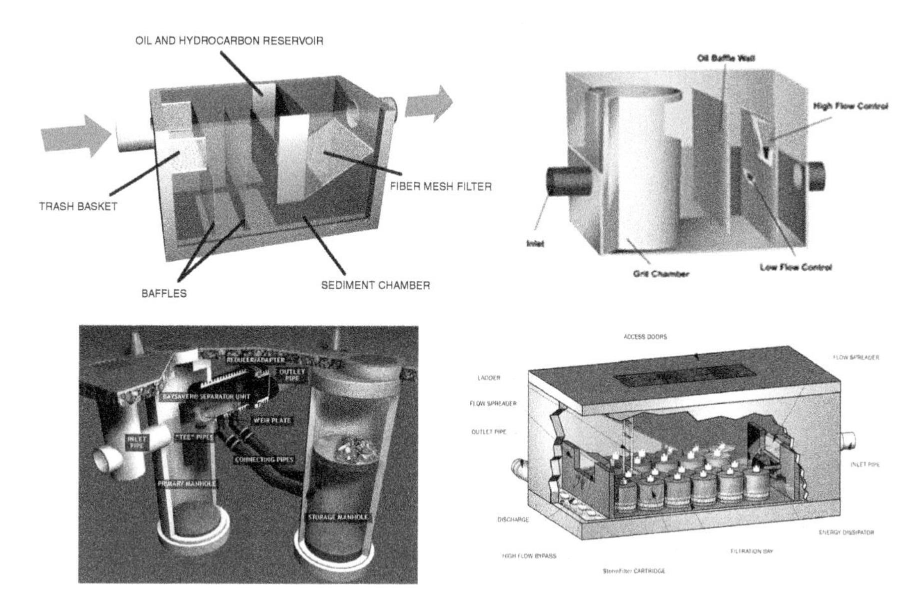


Figure 12.8 Proprietary Stormwater Treatment Devices (USEPA 2005a, 2005b, 2005c, 2005d).

The removal effectiveness of particulate matter by hydrodynamic devices will depend on the flow rate of the stormwater and the size of the particulate matter (Cates et al. 2009; Wilson et al. 2009). This relationship is based on having adequate retention time for particle separation. The largest particles will be readily forced out of the water, while the devices are minimally effective on the removal of fine particulate matter. Some separators have two sumps, a settling chamber and a floatables trap, which can result in higher particle removal efficiency.

Wilson et al. (2009) used a dimensionless Péclet number to describe the efficiency of hydrodynamic separators. The Péclet number, **P**, for the separator is defined as

$$P = \frac{v_s HD}{Q} \tag{12.1}$$

where $v_s$ is the particle settling velocity, found using Stokes' law (or a modification to account for turbulent flows), and $Q$ is the flow rate. $H$ is height, or settling distance, and $D$ is the diameter of the separator. The removal efficiency of particulate matter through the separator was found to be approximated by

$$\eta = \left( \frac{1}{R^b} + \frac{1}{(aP)^b} \right)^{-\left(\frac{1}{b}\right)} \tag{12.1}$$

where $a$, $b$, and $R$ are parameters specific to each separator. Values of $a$ were found to range from 0.41 to 1.77 for the settling chamber only and 1.42 to 2.20 for total system performance. Values for $b$ were 0.62–4.15 and 1.22–2.32, respectively; those for $R$ were 0.74–0.99 and 0.97–1.0 (Wilson et al. 2009). Under the conditions evaluated, sand and fine sand-sized particles were effectively removed, while fine particulate matter was not.

**Example 12.2** Find the removal of 0.15 mm sediment particles from flow of 2.2 ft³/s through a hydrodynamic separator system. The separator has a 10-ft depth and 12-ft diameter; values of $a$, $b$, and $R$ are 2.02, 1.22, and 1.0, respectively.

***Solution*** From Example 12.1, the settling velocity is 0.049 ft/s. The Péclet number is

$$P = \frac{(0.049\ \text{ft/s})(10\ \text{ft})(12\ \text{ft})}{2.2\ \text{ft}^3/\text{s}} = 2.67$$

Using Equation 12.2, the efficiency is calculated:

$$\eta = \left(\frac{1}{1.0^{1.22}} + \frac{1}{\left(2.02(2.67)\right)^{1.22}}\right)^{-\left(\frac{1}{1.22}\right)} = 0.906 = 90.6\%$$

Proprietary stormwater filtration devices come in a variety of forms. In these devices, the flow is directed through a filter cartridge, plate, cloth, or some other device that will filter and separate suspended solids (Figure 12.8d). Similarly, adsorption systems use a proprietary media and configuration, usually such that the stormwater is directed through the media for the adsorption of a targeted dissolved pollutant, such as phosphorus or a heavy metal (or metals). The filters are designed to be removed and replaced after a designated service life, which is commonly based on a fixed amount of runoff treated (and usually translated into a maintenance time interval).

Proprietary device manufacturers will provide design information and system requirements for the use of their devices. Regular maintenance of the device will be necessary; this may be done by the manufacturer and may be included with the purchase.

In many applications, proprietary devices may not be used for complete management of a stormwater runoff stream but may be part of a treatment train. Proprietary devices are commonly used for pretreatment of runoff before entering a GI SCM, similar to the use of forebays, as discussed in the previous section.

## 12.8 Accumulated Trash and Sediment

As discussed, the primary objective of pretreatment and many proprietary devices is the removal of gross solids and large sediment. These materials will accumulate in the SCM over time. Regular maintenance is required to remove these solids, where they are vacuumed out or somehow otherwise removed. If this maintenance is not done, accumulated sediment can be scoured and washed out later during a high-flow storm event, pushing these sediments out of the device and into the subsequent SCM or receiving waters. Usually, the device efficiency is reduced when the storage is filled with solids, since the residence time in the device is reduced.

A study conducted by McNett and Hunt (2011) examined the toxicity of sediments and solids collected in 30 forebays. The heavy metals Cd, Cr, Cu, Ni, and Zn were specifically studied. No forebay metal sediment concentration reached the point where it was too toxic to spread on the landscape and seed. In fact, the worst case example (for Ni from a wet pond at a commercial site) still had 2.4 times excess capacity available to accumulate this metal.

## References


Cates, E.L., Westphal, M.J., Cox, J.H., Calabria J., and Patch, S.C. (2009). Field evaluation of a proprietary storm-water treatment system: Removal efficiency and relationships to peak flow, season, and dry time. *Journal of Environmental Engineering* 135 (7): 511–517.

Johnson, J.L. (2007). *Evaluation of Stormwater Wetland and Wet Pond Forebay Design and Stormwater Wetland Pollutant Removal Efficiency*. MS Thesis. Raleigh, NC: North Carolina State University.

Linsley, R.A., Franzini, J.B., Freyberg, D.L., and Tchobanoglous, G. (1992). *Water-Resources Engineering*, 4th Ed. New York: McGraw-Hill.

McNett, J.K. and Hunt, W.F. (2011). An evaluation of the toxicity of accumulated sediments in forebays of stormwater wetlands and wetponds. *Water, Air, & Soil Pollution* 218 (1–4): 529–538.

U.S. EPA (2005a) Environmental Technology Verification Report: Stormwater Source Area Treatment Device — The Stormwater Management StormFilter® using Perlite Filter Media. August 2005. EPA/600/R-05/137. Washington DC.

U.S. EPA (2005b) Environmental Technology Verification Report: Stormwater Source Area Treatment Device — BaySaver Technologies, Inc., BaySaver Separation System, Model 10K. September 2005. EPA/600/R-05/113. Washington DC.

U.S. EPA (2005c) Environmental Technology Verification Report: Stormwater Source Area Treatment Device — Vortechnics, Inc., Vortechs® System, Model 1000. September 2005. EPA/600/R-05/140. Washington DC.

U.S. EPA (2005d) Environmental Technology Verification Report: Stormwater Source Area Treatment Device — Practical Best Management of Georgia, Inc. CrystalStream ™ Water Quality Vault, Model 1056. June 2005. EPA/600/R-05/085. Washington DC.

Wilson, M.A., Mohseni, O., Gulliver, J.S., Hozalski, R.M., and Stefan, H.G. (2009). Assessment of hydrodynamic separators for storm-water treatment. *Journal of Hydraulic Engineering* 135 (5): 383–392.


## Problems

**12.1** Find the volume of water that can be treated by an inlet filter before it reaches its sediment collection capacity, assuming 100% TSS capture. The filter has dimensions of 18 in square on the bottom and can fill to 8 in with sediment. The sediment (dry) bulk density is 82 lb/ft$^3$. The average TSS concentration entering the filter is 60 mg/L.

**12.2** An inlet filter treats 1100 m$^3$ of stormwater at 75 mg/L TSS. The filter captures 80% of the TSS. The filter has dimensions of 46 cm by 30 cm on the bottom. Assuming a sediment (dry) bulk density is of 1200 kg/m$^3$, find the depth of sediment accumulated in the filter.

**12.3** Find the particle size that is 90% removed in a hydrodynamic separator system at 2.0 $ft^3/s$ (particle density = 2.6 $g/cm^3$; $T = 50°F$). The separator has a 10-ft depth and 12-ft diameter; values of $a$, $b$, and $R$ are 2.02, 1.22, and 1.0, respectively.

**12.4** Find the flow rate that will remove 90% of a 0.20-mm particle in a hydrodynamic separator system (particle density = 2.6 $g/cm^3$; $T = 50°F$). The separator has a 10-ft depth and 12-ft diameter; values of $a$, $b$, and $R$ are 1.42, 1.75, and 0.97, respectively.

**12.5** Find the diameter of the hydrodynamic separator system that will remove 90% of a 0.12-mm particle at 2.3 $ft^3/s$ (particle density = 2.6 $g/cm^3$; $T = 50°F$). The separator has a 12-ft depth; values of $a$, $b$, and $R$ are 2.20, 1.85, and 0.99, respectively.

**12.6** The particle-size distribution for a runoff is given below. Find the overall particle removal by treating this runoff through a hydrodynamic separator system at 2.0 $ft^3/s$. The separator has a 10-ft depth and 12-ft diameter; values of $a$, $b$, and $R$ are 2.02, 1.22, and 1.0, respectively.

| Average particle size (μm) | Size fraction (%) |
|---|---|
| 16 | 9 |
| 17.5 | 15 |
| 94 | 11 |
| 188 | 22 |
| 375 | 18 |
| 750 | 14 |
| 1500 | 11 |

Particle density = 2.6 $g/cm^3$; $T = 50°F$

**12.7** The particle size distribution for a runoff is given in Problem 12.6. Find the overall particle removal by treating this runoff through a hydrodynamic separator system at 2.0 $ft^3/s$. The separator has an 8-ft depth and 10-ft diameter; values of $a$, $b$, and $R$ are 2.02, 1.22, and 1.0, respectively.

**12.8** The particle size distribution for a runoff is given in Problem 12.6. Find the overall particle removal by treating this runoff through a hydrodynamic separator system at 2.2 $ft^3/s$. The separator has a 10-ft depth and 12-ft diameter; values of $a$, $b$, and $R$ are 1.86, 2.32, and 0.99, respectively.

# 13

# Green Roofs

## 13.1 Introduction

Rooftops are one of the major sources of stormwater runoff in developed areas and thus are also a major opportunity for incorporation of green stormwater infrastructure. Simply stated, the green roof is designed to capture and hold some volume of rainfall for later removal by evaporation and transpiration, along with the slow possible release of overflows. In highly urbanized areas, space is at a premium and land is costly. Managing stormwater on rooftops can be one of the few options available to reduce the amount of stormwater produced. Like permeable pavements, green roofs can be classified more as a stormwater control measure (SCM) that reduces the production of runoff from rainfall, rather than a runoff treatment system.

Green roofs, also called living roofs, eco-roofs, and landscaped roofs, are vegetated. Blue roofs are not vegetated and act as SCMs through rainfall storage and subsequent evaporation only. There are many ancillary advantages to green roof adoption that include reducing heat island effects, improving air quality, increased roof life, and simply to beautify our urban living space.

Often in the urban environment, the rooftop may be the SCM location of choice, due to the high density of impervious surfaces and underground utilities. An example of a large-scale green roof project built by the PECO Energy Company in Philadelphia is shown in Figure 13.1. The dominance of the rooftop in the urban area is clearly seen in the photograph.

## 13.2 Climate and Green Roofs

Some climatic areas may not be suited for green roofs. A building rooftop is a harsh environment and roof restrictions typically limit the amount and type of media that can be placed on the roof. As a result, hot and dry climates may not be suitable for green roof use. In these areas, lack of consistent precipitation may prevent natural growth of green roof vegetation. Survival of such vegetation may require irrigation of the roof, which will greatly increase costs and can be the antithesis of sustainable urban water use.

*Green Stormwater Infrastructure Fundamentals and Design*, First Edition. Allen P. Davis, William F. Hunt, and Robert G. Traver.

Companion Website: www.wiley.com/go/davis/greenstormwater

Figure 13.1 PECO Green Roof in Philadelphia. (Credit: Google Maps/Google Earth, 2015).

Related arguments can be made against green roofs in other areas. In areas where large storms are common, the green roof may not have adequate capacity for management of the rainfall volume. In extremely cold climates, annual evapotranspiration (ET) may be so small as to limit the benefits of green roofs.

## 13.3 Types of Roofs

### 13.3.1 Green Roofs

The types or categories of green roofs are growing. Green roofs are traditionally separated into extensive and intensive roofs. Intensive roofs are similar to gardens, where the soil is deep enough for trees, bushes, and pedestrian plazas. Intensive green roofs are very thick, consisting of over 30 cm (1 ft) of media and other components. While being versatile, intensive green roofs are expensive, heavy and are therefore uncommonly used when stormwater management is the motivating force for selection.

Most green roofs are extensive. Extensive green roofs have shallower depths and are commonly built for shallow-rooted, drought resistant plants. Extensive green roofs are the norm for green roof installation, especially when a green roof's primary purpose is stormwater management and will be the focus of this chapter.

Figures 13.2–13.4 show examples of how green roofs fit into the framework of urban space. Figure 13.2 shows a green roof on the second floor of an apartment complex. Note the doors opening onto patios surrounding the green roof, and the views of the upper residences. Figure 13.3 shows a green roof integrated around a patio of a row

Figure 13.2 Paseo Verde Apartment Green Roof in Philadelphia. (Photo by Authors).

home in Philadelphia, PA, combining form and function. The green roof at the North Carolina State University library is shown in Figure 13.4. It is visible from many vantage points within the library and, in addition to being functional in stormwater management, is an integral part of the building's architectural character.

As urban land area is always at a premium, other locations are increasingly becoming attractive for application of green roof technology, to reduce heat island effects in the urban environment, and to take advantage of unoccupied space in the urban landscape. Figure 13.5 shows three less-conventional green roof technology applications.

### 13.3.2 Blue Roofs

Blue or gravel roofs are also becoming more common in urban areas for stormwater management. For these roofs, the outflow is restricted to store water on the roof, allowing for evaporation between rain events (Figure 13.6). Blue roofs can be less costly then green roofs but are less efficient as they do not possess the runoff volume reduction advantage gained from transpiration by the green roof vegetation.

Figure 13.3 Row Home Green Roof in Philadelphia. (Photo by Authors).

Figure 13.4 North Carolina State University Library Green Roof. (Photo by Authors).

Figure 13.5 Uncommon Green Roof Applications: (A) Gasoline Station with Green Roof (Photo by Authors); (B) Green Roof Picnic Shelter at Villanova University; and (C) Birdhouse with Green Roof (Photos by Authors).

Figure 13.6 Blue Roof Media and Storage Trays. (Photo by Authors).

(A)

(B)

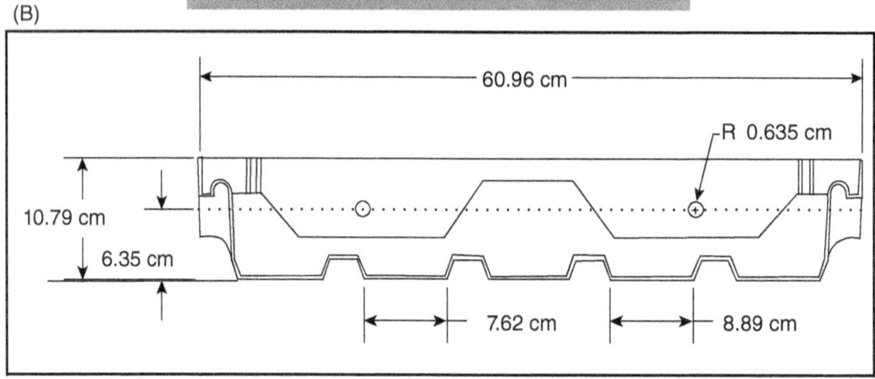


Figure 13.7 Blue Roof Overflow Device to Control Stored Water Level on the Roof: (A) Photograph (Photo by Authors) and (B) Side View of a Blue Roof Containerized System with Drawdown Orifice Depth and Diameter Illustrated.

Designing a blue roof has similarities to designing a very shallow pond. The outlet structure protruding from the roof (Figure 13.7) restricts release of a water quality/ design event for up to 2 days. This allows for evaporation to occur during subsequent hours after the precipitation event. Frequently, the roof is hot during this time, increasing the evaporation rate. Often the discharge rate is controlled by an orifice or small weir. Use of orifices and weirs for flow control is described in Chapter 7. Alternatively, if containerized blue roofs (such as Figure 13.6) are implemented, the tray itself will have very small orifices drilled along the side. The height of the holes is dictated by design storm capture and expected evaporation rates.

## 13.4 Extensive Green Roof Components

Green roofs reduce runoff by detaining some volume of rainfall (principally in the media) for subsequent evapotranspiration. During intense and/or large rainfall events, the media serves to mitigate flow rates, once the media's plant available water (PAW) has been filled. Unless roof drainage has been clogged, green roofs rarely (if ever) saturate.

A typical cross-section of an extensive green roof is presented in Figure 13.8. Working from the bottom up, the bottom green roof component is some type of waterproof membrane. This layer is necessary to prevent any leakage of water from the roof into the building. Above this layer is generally the storage layer or drainage layer. This can consist of an "egg carton" type of material or other proprietary blanket-like layer. Excess precipitation can be stored in this layer.

Above the storage/drainage layer is the green roof media. Media selection is arguably the key design component. Many different types of media continue to be used and investigated for use in green roofs. Green roof media have several important criteria for consideration. The media should be able to support vegetation growth and survival, be sufficiently coarse so that water pooling does not occur at the media surface, and importantly, be lightweight and durable. Yet, the media will ideally have properties so that it supplies ample water storage capacity and supports plant health during dry weather. It must structurally support plant growth and must not become airborne during major wind storms. Nonproprietary mixes include expanded aggregate (such as clay, shale, slate) and pumice, which are lightweight and porous.

Some amount of organic matter is important for the media in order to support the plant survival. As described by Fassman and Simcock (2012), the amount and type of organic matter to be added to green roof media is highly variable. Approximately 20% by volume has been common, but recent design standards have limited this to 10%. The organic matter adds some water-holding capacity and acts as a supply of nutrients for the plants. However, too much organic matter or unstable forms of organic matter present in the media has been repeatedly shown to lead to (often excessive) nutrient discharges from green roofs. A wide range of organic materials are used, including peat and various types of compost. Organic matter can be problematic as it decomposes, reducing media depth (a critical design component, as discussed later in this chapter, for volume retention).

Green roof vegetation promotes ET to remove the detained water. The vegetation must be tolerant of both very wet conditions and prolonged dry periods. Because the

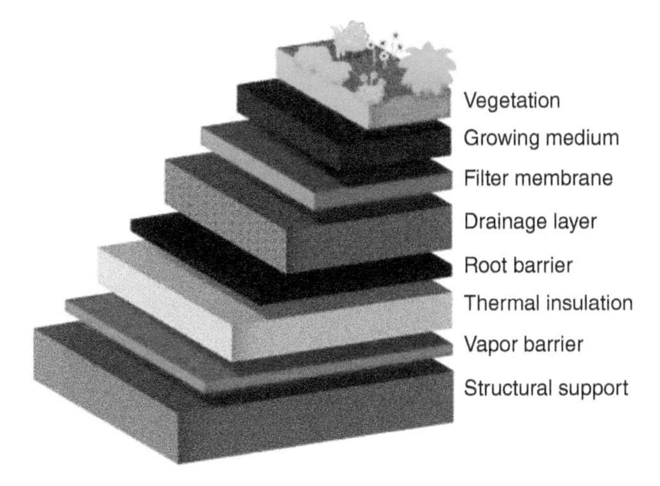


Figure 13.8 Cross-section of Typical Green Roof. (Credit: US EPA).

Figure 13.9 Sedums on the Green Roof at Villanova University. (Photos by Author's Research Team).

green roof environment is harsh, vegetation selection is critical and tends to be dominated by succulents. Succulent plants have leaves, stems, and/or roots that are thick and fleshy to store water under arid conditions. Properly chosen plants should require minimal maintenance. Nutrient needs should be supplied mostly by the falling rain, though some suppliers recommend some minimal additives. The most common maintenance task on a green roof is weeding.

The most common plant used in green roofs is some species of the sedum genus (Figure 13.9). Sedums store water in their leaves, can promote high rates of ET, and have been the vegetation of choice for most green roof studies. Other environments can support grasses, but they usually have a deeper soil storage area and may require more maintenance and watering. Figure 13.10 shows a meadow green roof at the Merion Golf Course (Haverford PA) built to fit in with the golf course and a turf green roof situated on the visitor center of the Newgrange, Ireland, neolithic historic site. Green roof designers are strongly encouraged to consult green roof plant experts as part of the plant selection process.

Many manufacturers are now producing green roof tray systems. Trays often arrive ready for deployment (media, plants incorporated). The tray systems are able to store water on roofs with higher pitches (Figure 13.11). The waterproof membrane is needed, but typical roof drainage infrastructure does not change when tray systems are utilized. When structurally sufficient, tray systems can be used on existing buildings as a green roof retrofit. An example of one product is shown in Figure 13.12.

Figure 13.10 Green Roof Photographs: (A) Green Meadow Roof–Merion Golf Course (Roof Is to Right of Gravel Channel) and (B) Turf Green Roof at Newgrange, Ireland, Neolithic Site (Photos by Authors).

## 13.5 Hydrologic Design Strategies

### 13.5.1 Rainfall Capture

Capture and storage of rainfall will directly reduce the amount of runoff yielded by a roof. This storage occurs primarily within the pore spaces of the selected media. The presence and type of vegetation plays a very minor role in water storage and control during the rain event. Some small amount of storage may also occur in the drainage layer.

The green roof media should have a high permeability so that ponding of water on top of the media does not occur. It has been recommended that the saturated hydraulic conductivity, $K_S$, of the media be 10 times the peak rainfall intensity of the design storms for the location. Pumice-based media provide $K_S$ values around 0.1 cm/s (Liu and Fassman-Beck 2016).

The discussion in Chapter 11 describes estimation of water storage in green roof media. The available water can be estimated as the difference in volumes between the field capacity of the media (gravity drainage), and the wilting point, the amount of

Figure 13.11 A Newly Planted Green Roof Utilizing Container Trays on a High Pitched Roof in Asheville, NC. (Photo by Authors).

Figure 13.12 A Tray Green Roof System. (Photo by Authors).

water that cannot be utilized by the plants. For green roofs, this water storage value been termed as the PAW (Fassman and Simcock 2012).

The PAW can be estimated from media characteristics, such as particle size distribution. For any media, the PAW can be found from laboratory testing of moisture content at field capacity (10 kPa) and wilting point (1500 kPa). Table 13.1 shows values of PAW from several green roof mixes in Auckland New Zealand. An expanded slate-based aggregate media used across the Southeast and Mid-Atlantic USA has a PAW ranging between 20% and 25%, which is in line with the values reported in New Zealand.

Table 13.1 Moisture Storage Potential for Green Roof Substrates Examined at the University of Auckland (Modified from Fassman and Simcock 2012; Fassman-Beck and Simcock 2013).

| Media composition (% by volume) | PAW (%) |
|---|---|
| 20% 1–7-mm pumice, 60% 4–10-mm pumice, 20% composted pine bark fines | 24.2 |
| 50% 4–10 mm pumice, 30% 1–8 mm zeolite, 20% composted pine bark fines | 23.1 |
| 70% 4–10-mm pumice, 10% zeolite < 3 mm, 15% pine bark fines + mushroom compost, 5% sphagnum peak | 28.9 |
| 20% 1–4 mm pumice, 40% 4–8 mm pumice 20% expanded clay 20% commercial garden mix | 20.2 |

With knowledge of the PAW, the media depth, $D_m$, is determined based on the design rainfall depth, $D_s$, that must be captured (Fassman-Beck and Simcock 2013):

$$D_m = \frac{D_s}{\text{PAW}} \tag{13.1}$$

Generally, the minimum value of for green roof media depth is 100 mm (4 in). Looking at Table 13.1, a typical value for green roof media PAW may be 25%. Correspondingly, green roof media of 100-mm media thickness should be able to store (if previously dry), about 25 mm (1 in) of rainfall.

This simple storage calculation is only valid with dry media. Once field capacity is reached, rainfall will no longer be detained and will therefore flow through the media. If insufficient time has lapsed between rainfall events, the entirety of PAW will not be available. While seasonally dependent, 4–7 dry days may be needed for the green roof media to "recharge" and be "ready" to capture a 25-mm design event. Furthermore, if the drainage infrastructure clogs, the media will become saturated and runoff and/or horizontal flow will occur, especially on sloped roofs.

Over time, the porosity of the media and the organic matter content can change. This will occur as some media may undergo attrition and biological decay. Growth and death of vegetation can affect porosity. Limiting organic content and ensuring regular maintenance of the roof help control these factors.

A recent work evaluated runoff data from multiple green roofs from various countries, with an attempt to provide a measure of the overall hydrologic performance. The use of the rational formula to describe green roof runoff volumetric performance was generally unsuccessful. The value of the rational method $C_v$ varied with the rainfall depth and from roof to roof (Fassman-Beck et al. 2016), with a resulting expression:

$$C_v = a\exp\left(b / P\right) \tag{13.2}$$

where $P$ is the precipitation depth and $a$ and $b$ are both empirical parameters. The values of $a$ and $b$ were both found to be highly site dependent.

National Resources Conservation Service (NRCS) curve numbers are commonly employed to determine runoff from sites based on rainfall (Chapter 6). Because of this, there is interest in using curve numbers to evaluate the effectiveness of green roofs in mitigating hydrologic impacts. A curve number (CN) value equal to 86 has been observed for a green roof (Carter and Rasmussen 2006). This value is equal to that of poor condition open space on a hydrologic soil group C (Chapter 6). This can be compared to a CN of 98 for impervious rooftops.

While attempts by Fassman-Beck et al. (2016) to determine a universal curve number for green roof hydrologic performance were unsuccessful, there was a central tendency toward a typical CN in the range of 84–86. Data fitted to curve number relationships did not fit well and the CN varied from 75 to 96. Different green roof designs resulted in different hydrologic performances. Roof slope, media type, vegetation type and density, and climate played a large role in the determining hydrologic performance.

More specifically, based on the storage of water in the media, Fassman-Beck et al. (2016) recommend that Equation 13.1 be used to determine the volumetric storage on a green roof. For rainfall depth less than the storage, no runoff will occur. For events that exceed the storage threshold defined by Equation 13.1, a curve number of 84 is recommended. This combination of storage with no runoff, followed by some degree of hydrologic management, is common to most noninfiltration SCMs and is described in Chapter 11. Computer simulation models such as the Storm Water Management Model (SWMM) have the ability to route water movement and storage mechanistically (Peng and Stovin 2017).

### 13.5.2 Evapotranspiration

After the rainfall event, ET will regenerate the roof water storage capacity. The greater the ET, the faster this will occur, so regeneration will be greatest during summer, up to a factor of about four. The presence of plants is important, accounting for 20–48% of the lost water (Voyde et al. 2010).

Green roofs in Philadelphia, PA, show about 3–4 mm of ET for the first 2 days following a significant rainfall event in April through November (Wadzuk et al. 2013); the first 2 days after rain represents the time when the most water stored in the green roof is available for ET. The range was 1–10 mm. Cumulatively, the Philadelphia green roof had 783–980 mm of ET loss, out of 817–1351 mm of monitored rainfall during a 3-year period.

Because actual ET is limited by the availability of water on a green roof, modifications to the roof drainage design can increase the overall ET. Altering a green roof design so that the drainage occurred at the level of the media surface, as opposed to drainage at the bottom of the media, greatly increased the ET (Zaremba et al. 2016). In the case of the bottom drainage, the media could only reach the point of field capacity and ET accounted for 62% and 56% of (April–November) precipitation in the 2-year study (Figure 13.13A). In this figure, the dark shading represents the cumulative rainfall over the study and the light shading represents the overflow drainage from the roof. The undrained green roof allowed the media to become saturated and 68–88% of the precipitation was evapotranspired (Figure 13.13B—hatched shading). Keep in mind that saturated media weighs more, and therefore the additional weight must be accounted for during the structural design.

While several formulas have been developed to estimate ET (as discussed in Chapter 7), it is important to understand that these equations predict *potential* ET. This is the ET that occurs when adequate water is available for the plants. However, once ET begins after a rain stop, the process of ET decreases available water. The potential ET may still be high, but as the stored water in the media decreases from field

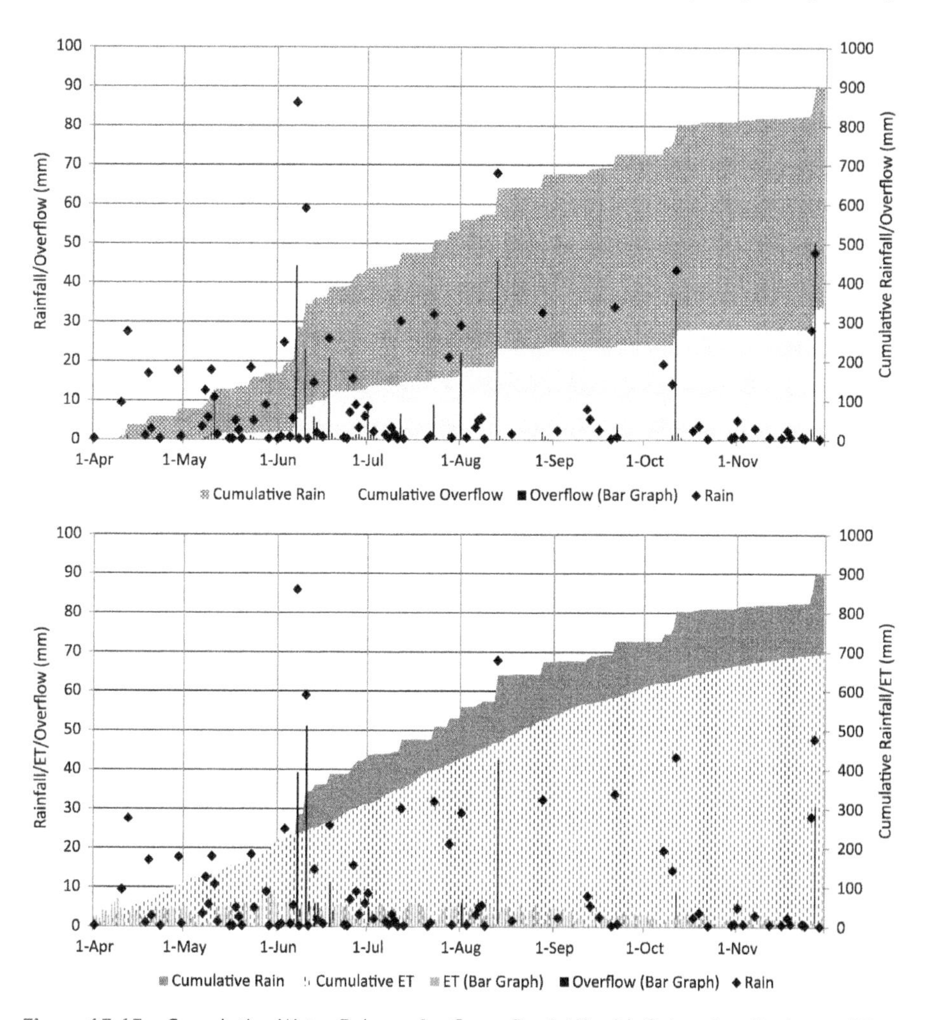


Figure 13.13 Cumulative Water Balance for Green Roof: (A) with Subsurface Drainage; (B) with Overflow Drainage Only (Zaremba et al. 2016, with Permission from ASCE). Daily Values of ET and Outflow Are Represented by Bar Graphs in Gray and Black, Respectively, and Daily Rainfall Totals Are Represented by Black Diamonds with Values on the Left Axis; Cumulative Curves for Rainfall Are Shaded in Black Hatch Marks and Cumulative Measured ET Is Represented by Black Dashes.

capacity back to the wilting point and the PAW decreases to zero, the *actual* ET will decrease. This process follows an approximate exponential decay with time (Fassman and Simcock 2012; Poe et al. 2015), with rates as high as 0.29 mm/h in New Zealand (Voyde et al. 2010) and 0.41 mm/h near Philadelphia (Wadzuk et al. 2013) when adequate water is available.

Even though it is one of the simplest models, the Blaney–Criddle model (Chapter 7) was found to provide acceptable estimates of potential ET from a green roof in Philadelphia, PA (Wadzuk et al. 2013). The Hargreaves equation (Chapter 7) can also provide acceptable ET estimates. The ASCE Penman–Monteith provided similar predictions, but this model is very parameter intensive.

**Example 13.1** The media of a green roof (located at 35°N latitude) has a thickness of 12 cm, with a PAW of 26%. Assuming that a large rainfall fills the media, find the number of days required for this stored water to be evapotranspired in July. Use the Hargreaves equation; the average maximum temperature is 31°C and average minimum temperature is 21°C.

*Solution* Equation 13.1 is rearranged to solve for the rainfall depth stored in the media:

$$D_s = D_m(\text{PAW}) = (12 \text{ cm})(0.26) = 3.12 \text{ cm} = 31.2 \text{ mm}$$

The average temperature is 26°C. From Table 7.3, the extraterrestrial radiation, $R_a$, is 16.75 mm/day (equivalent evaporation), as the average value for 34 and 36°N latitude. Find ET from the Hargreaves equation.

$$\begin{aligned} ET_0 &= 0.0023 R_a (\text{TC} + 17.8)(T_{\max} - T_{\min})^{0.5} \\ &= 0.0023(16.75 \text{ mm/day})(26 + 17.8)(31 - 21)^{0.5} \\ &= 5.34 \text{ mm/day} \end{aligned}$$

Therefore, the time to evapotranspire is:

$$\text{time} = \frac{31.2 \text{ mm}}{5.34 \text{ mm/day}} = 5.85 \text{ days}$$

Insight into plant metabolism under green roof conditions is important to understanding and optimizing green roof vegetation for water management. Once plants become water stressed, they begin to conserve water and the transpiration rate decreases (Voyde et al. 2010). Recent studies into the behavior of sedum species, specifically *S. album*, suggests that these plants may not be the best overall choice for green roof vegetation. *S. album* is drought tolerant, which is important. Yet, this drought tolerance may be related to crassulacean acid metabolism (CAM). With CAM, the plants fix carbon dioxide at night, storing it. This allows the plants to close their stomata during the day, limiting water loss. This mechanism may limit the daytime ET of *S. album* (Starry et al. 2014). Needless to say, optimizing vegetation selection remains an important research topic and insufficient information is available currently to inform design. Overall, designers still primarily select vegetation based upon appearance and survival.

## 13.6 Water Quality Design

Water quality improvement has not been an important design parameter for green roofs; however, because early green roof media formulations tended to export nutrients, the use of green roofs in certain nutrient sensitive watersheds was restricted. Green roofs capture rainfall, which may contain some low levels of nitrogen, but is generally considered of good water quality. The primary water quality issue with green roofs is to minimize the leaching of pollutants from the media used in the roof. The major challenge here is the common use of compost-amended media to support the vegetation and when the media comprises large proportions of organic matter. This results from the desire of the designer and need to establish the plants for aesthetic purposes. While high organic matter is beneficial for initial plant growth, leaching of nutrients is common, especially during the first few years of the roof. Compost can leach nitrogen and phosphorus and should be avoided, especially in nutrient-sensitive areas. However, important is the amounts of nitrogen and phosphorus relative to the runoff in the surrounding areas and the receiving streams. Many cities have combined sewers and the stormwater is conveyed to and treated by the wastewater treatment plant. In these cases, hydrologic management is critical to minimize overflows, but water quality concern is minimal.

### 13.6.1 Phosphorus

P concentrations in rainfall are very low, around 0.01 mg/L or less (Berndtsson et al. 2006). However, several studies have found substantially higher P concentrations in green roof discharge. Total P concentrations have ranged from around 0.5–1.5 mg/L (Berndtsson et al. 2006) to 2–3 mg/L (Bliss et al. 2009). About half of the P found was in the form of phosphate (Berndtsson et al. 2006). Barr et al. (2017) measured a median concentration of 1.36 mg/L total P from a green roof in Pennsylvania, with 1.28 mg/L (as P) orthophosphorus. Even with substantial hydrologic mitigation, it is rare that green roofs can decrease phosphorus loading to receiving waters (vis-à-vis a conventional roof); but this depends on the media used in the green roof and any regular fertilizer applications. Some jurisdictions have placed a strict maximum proportion of organic matter present in media in order to limit P discharge from green roofs. For example, North Carolina limits organic matter content to 10%, with 5% being recommended.

An annual green roof discharge load of 0.91 g/m$^2$ year (9.1 kg/ha year) total P was calculated by Barr et al. (2017). This is larger than the value of 2.25 kg/ha year total P load estimated for urban impervious area in the Chesapeake Bay watershed, as presented in Chapter 5.

**Example 13.2** The total P event mean concentration (EMC) measured from a green roof is 1.2 mg/L. About 1.2 m of rainfall falls on the roof, but 70% of the rainfall is held and evapotranspired. Calculate the annual total P load from this roof.

*Solution* The load can be calculated using the simple method, Equation 8.3. With the green roof, 100% of the roof area contributes to the load, and $c$ = 1.0. The green roof abstracts 70% of the rainfall, giving $R$ = 0.3. Therefore:

$$L = cRP(\text{EMC}) = (1.0)(0.3)(1.2\ \text{m/year})(1.2\ \text{mg/L}) =$$
$$= 0.432(\text{m mg})/(\text{L year})$$

Converting to the appropriate load units:

$$L = 0.432(\text{m mg})/(\text{L year})\frac{(1000\ \text{L/m}^3)(10^4\ \text{m}^2/\text{ha})}{(10^6\ \text{mg/kg})} = 4.32\ \text{kg/(ha year)}$$

### 13.6.2 Nitrogen

In contrast to P, nitrogen concentrations in rainfall can be significant, due to nitrogen oxides produced during combustion processes. Berndtsson et al. (2006) found green roofs to take up nitrate, but to release organic N, with little net change in N concentrations. Total N concentrations in green roof discharges ranges from about 0.5 up to 4 mg/L. Barr et al. (2017) measured a median total N concentration of 2.7 mg/L from a green roof. TKN was the dominant N species at 1.7 mg/L. Total N in the precipitation was 0.85 mg/L. Unlike phosphorus, it seems that nitrogen mass loads discharged from green roofs may be comparable to those of conventional roofs.

The total N load from the green roof studied by Barr et al. (2017) was 1.9 $g/m^2$ year (19 kg/ha year). The comparable total N load for urban impervious area in the Chesapeake Bay watershed is 11.8 kg/ha year total N (Chapter 5).

### 13.6.3 Metals

Concentrations of metals, specifically, Cr, Cu, Pb, and Zn, were not significantly changed from rainfall to discharge in green roof systems (Berndtsson et al. 2006). This, therefore, results in substantial decreases in metal loads discharged from green roofs relative to those of conventional roofs due to the runoff reduction.

## 13.7 Inspection and Maintenance

The two principal needs for green roof maintenance are (1) maintain vegetation health and (2) prevent roof drainage clogging. Vegetation must be kept from all drainage infrastructure to prevent roots and vegetation detritus from interfering with roof drainage. Broad coverage of healthy plants is necessary to maximize green roof ET. Plant health is more easily managed when proper plants are selected initially. Invasive vegetation can migrate onto green roofs and will need to be removed; invasive species might even be native species, but many species native to North America struggle to survive in the long term and can become dormant or die during dry periods. This vegetation can even become a fire hazard if people have access to the roof. During prolonged drought, some choose to irrigate their green roof.

Care should be taken to ensure that excess nutrients are not applied to green roofs through fertilization. While ensuring a vigorous stand of vegetation is important to proper green roof function, excess nutrients can leach from the roof media and may lead to high discharge of nutrient loads.

## 13.8 Other Green Roof Benefits

Green roofs unlike many other stormwater practices offer a surprising amount of ancillary benefits, many of which may be more important to the structure and surroundings than the water quality and hydrologic benefits discussed in this text. For example, green roofs are believed to increase roof life by a factor of perhaps three, because green roof media protect roof membranes from ultraviolet light and dampen temperature fluctuations. For many long-term owners of properties (such as government-owned facilities), the prospect of holding off significant roof repairs for 50–60 years is sufficient reason for a green roof.

A study in Toronto, ON, demonstrated a peak summer temperature difference of 65–30°C between a traditional roof and green roof, respectively. The vegetation, media, and filter membrane layers associated with green roofs reduce the amount of heat transferred to the building below, thus moderating temperatures inside the building and producing energy savings during periods of warm weather. As reviewed by Oberndorfer et al. (2007), heat fluxes through green roofs have been found to be 10–95% lower than those through conventional roofs, depending upon the local climate. The cooling benefits of green roofs have been evaluated at a citywide scale. Banting et al. (2005) estimated that a widespread network of green roofs in Toronto, Canada, could reduce ambient summertime air temperatures by up to 2°C.

Green roofs have been shown to improve air quality as well. Getter and Rowe (2006) summarized reductions in diesel engine air pollution, sulfur dioxide and nitrous oxides, and suspended particulates for these systems, with the total air quality benefit directly related to the leaf area index of the species used on the roof.

Increasingly, green roof food provisioning is an important reason for their use. For example, a 200-$m^2$ intensive green roof planted with herbs on the Fairmont Waterfront Hotel in downtown Vancouver, Canada, was utilized in the hotel's restaurant. While the roof cost $25,000 to install, the annual value of the herbs was estimated at over $20,000 (Paladino 2004). A variety of vegetables and melons have been successfully produced on green roof gardens in the United States, as has rice in China and India (Velazquez and Kiers 2007), though deeper media depths and supplemental irrigation are typically required. Other benefits include biodiversity (with green roofs providing habitat to fauna, a unique benefit in urban environments), and the establishment of "an urban oasis" when the roof scape is accessible, or at least visible, to tenants and owners.

## References


Banting, D., Doshi, H., Li, J., Missios, P., Au, A., Currie, B.A., Verrati, M. (2005) *Report on the Benefits and Costs of Green Roof Technology for the City of Toronto*. Toronto: Ryerson University. https://www.coolrooftoolkit.org/wp-content/uploads/2012/05/Green-Roofs-Toronto.pdf (accessed December 2021).

Barr, C.M., Gallagher, P.M., Wadzuk, B.M., and Welker, A.L. (2017). Water quality impacts of green roofs compared with other vegetated sites. *Journal of Sustainable Water in the Built Environment* 3 (3): 04017007.

Berndtsson, J.C., Emilsson, T., and Bengtsson, L. (2006). The influence of extensive vegetated roofs on runoff water quality. *Science of the Total Environment* 355 (1-3): 48–63.

Bliss, D.J., Neufeld, R.D., and Ries, R.J. (2009). Storm water runoff mitigation using a green roof. *Environmental Engineering Science* 26 (2): 407–417.

Carter, T.L. and Rasmussen, T.C. (2006). Hydrologic behavior of vegetated roofs. *Journal of the American Water Resources Association* 42 (5): 1261–1274.

Fassman-Beck, E., Hunt, W., Berghage, R., Carpenter, D., Kurtz, T., Stovin, V., and Wadzuk, B. (2016). "Curve Number and Runoff Coefficients for Extensive Living Roofs." *Journal of Hydrologic Engineering*, 21(3), 04015073.

Fassman, E. and Simcock, R. (2012). Moisture measurements as performance criteria for extensive living roof substrates. *Journal of Environmental Engineering* 138 (8): 841–851.

Fassman-Beck, E.A. and Simcock, R. (2013). Living roof review and design recommendations for stormwater management. Prepared by Auckland UniServices for Auckland Council. Auckland Council technical report TR2013/045.

Getter, K.L. and Rowe, D.B. (2006). The role of extensive green roofs in sustainable development. *HortScience* 41 (5): 1276–1285.

Liu, R. and Fassman-Beck, E. (2016). Effect of composition on basic properties of engineered media for living roofs and bioretention. *Journal of Hydrologic Engineering* 21 (6): 06016002.

Oberndorfer, E., Lundholm, J., Bass, B., Coffman, R., Doshi, H., Dunnett, N., Gaffin, S., Kohler, M., Liu, K., and Rowe, B. (2007). Green roofs and urban ecosystems: Ecological structures, functions, and services. *BioScience* 57 (10): 823–833.

Paladino & Company, Inc. (2004). Green roof feasibility review. Paladino, Seattle, WA.

Peng, Z. and Stovin, V. (2017). Independent validation of the SWMM green roof module. *Journal of Hydrologic Engineering* 22 (9): 04017037.

Poe, S., Stovin, V., and Berretta, C. (2015). Parameters influencing the regeneration of a green roofs retention capacity via evapotranspiration. *Journal of Hydrology* 523: 356–367.

Starry, O., Lea-Cox, J.D., Kim, J., and van Iersel, M.W. (2014). Photosynthesis and water use by two Sedum species in green roof substrate. *Environmental and Experimental Botany* 107: 105–112.

Velazquez, L. and Kiers, H. (2007). Hot trends in design: Chic sustainability, unique driving factors & boutique Greenroofs. Proc. 5th annual greening rooftops for sustainable communities Conference, Minneapolis, MN.

Voyde, E., Fassman, E., Simcock, R., and Wells, J. (2010). Quantifying evapotranspiration rates for New Zealand green roofs. *Journal of Hydrologic Engineering* 15 (6): 395–403.

Wadzuk, B.M., Schneider, D., Feller, M., and Traver, R.G. (2013). Evapotranspiration from a Green-roof storm-water control measure. *Journal of Irrigation and Drainage Engineering* 139 (12): 995–1003.

Zaremba, G.J., Traver, R.G., and Wadzuk, B.M. (2016). Impact of drainage on Green roof evapotranspiration. *Journal of Irrigation and Drainage Engineering* 142 (7): 04016022.

## Problems

**13.1** How much rainfall (depth) can be held by a 4-in green roof with a PAW equal to 28%?

**13.2** How much rainfall (depth) can be held by a 4-in green roof with a PAW equal to 32%?

**13.3** How much rainfall (depth) can be held by a 120-mm green roof with water-holding characteristics similar to a sandy loam soil?

**13.4** How much rainfall (depth) can be held by a 120-mm green roof with water-holding characteristics similar to a loamy sandy soil?

**13.5** Compare the runoff volume from green roof 40 ft by 80 ft, with a NRCS curve number of 86 to a conventional roof with a curve number of 98 for rainfall depths of 0.25, 0.5, 1, and 2 in.

**13.6** A 5-in green roof has a PAW of 26% and an effective CN of 86 beyond the roof storage capacity. Draw the green roof runoff discharge depth for rainfall ranging from 0 to 4 in.

13.7 What average ET rate (mm/day) is necessary to regenerate the water-holding capacity of a 120-mm green roof with water-holding characteristics similar to a sandy loam soil?

13.8 A blue roof tray system is intended to completely store a 20-mm event. Provide key design elements, including: media porosity, media depth, and orifice location. Calculate this for a flat roof and one on a 10% pitch.

13.9 A blue roof comprises 2 ft × 2 ft modular trays filled with media in Raleigh, NC. It is intended to store 1.0 in of rain. During a 5-year, 24-h event, a 5-year, 5-min, rainfall intensity occurs midway through the storm. If one orifice is located on each side of the tray (for 4 total), what is the minimum orifice diameter assuming water level rise is restricted to 2 in above the orifice invert?

13.10 The extraterrestrial radiation for a city is 15 mm/day; the average maximum temperature is 30°C; and average minimum temperature is 19°C. How long will it take to evapotranspire 40 mm of stored water from a green roof?

13.11 The media of a green roof has a thickness of 10 cm, with a PAW of 24%. The extraterrestrial radiation for a city in April is 14 mm/day; the average maximum temperature is 17°C and average minimum temperature is 6°C. How long will it take to evapotranspire all stored water from a green roof?

13.12 A green roof with storage to hold 1 in of rain must be drained in 3 days. The extraterrestrial radiation is 14.5 mm/day and the average temperature is 76°F. Find the maximum and minimum temperature to meet this requirement.

13.13 The stored water on a green roof takes 1.9 days to evapotranspire in July; extraterrestrial radiation, maximum, and minimum temperature are 15 mm/day, 30°C, and 20°C, respectively. Find the time to evapotranspire this water in April at 14 mm/day, 18°C, and 7°C, respectively.

13.14 The stored water on a green roof takes 1.9 days to evapotranspire in one city; extraterrestrial radiation, maximum, and minimum temperature are 16 mm/day, 30°C, and 20°C, respectively. Find the time to evapotranspire in a higher latitude city at 14 mm/day, 24°C, and 13°C, respectively.

13.15 The total P EMC measured from a green roof is 1.4 mg/L. About 44 in of rainfall falls on the roof and 65% of the rainfall is held and evapotranspired. Calculate the annual total P load from this roof in lb/ac year and kg/ha year.

13.16 The total N EMC measured from a green roof is 2.4 mg/L. About 38 in of rainfall falls on the roof and 75% of the rainfall is held and evapotranspired. Calculate the annual total N load from this roof in lb/ac year and kg/ha year.

13.17 The total P EMC measured from a green roof is 0.8 mg/L. The desired total P discharge mass load from the roof is 1.8 kg/ha year. For an annual rainfall of 38 in, find the fraction of rainfall that must be evapotranspired from the roof.

13.18 The total N EMC measured from a green roof is 3.8 mg/L. The desired total N discharge mass load from the roof is 12 kg/ha year. For an annual rainfall of 1.1 m, find the fraction of rainfall that must be evapotranspired from the roof.

# 14

# Rainwater Harvesting

## 14.1 Introduction

Opportunities exist to capture rainfall and utilize it in a beneficial manner. This can bestow advantages from a stormwater management perspective but also may provide for a water source that can reduce the demand for potable water, possibly creating an economic incentive. The water for harvesting is normally captured from rooftops, though other sources are possible depending on need. Necessary components include a rainwater capture system, a storage tank, and a drainage/delivery system. Depending on the type of system used, pumps may be necessary and some type of water treatment. In some cases, the most beneficial use of stormwater may simply be to recharge natural surface or groundwater systems.

Rainwater harvesting has a number of advantages over other urban stormwater control measures (SCMs) and is a promising technology for future utilization. It can be employed in ultraurban environments, water can be captured at relatively high elevations, providing a source of energy, harvested water can be employed for beneficial use, even indoor use when public health safety is ensured, and water harvesting can have potential for near-zero runoff applications. Many countries use rainwater as either the primary or secondary water supply (Figure 14.1).

Collecting runoff from other surfaces, such as parking lots, sidewalks, and landscaped areas can also be referred to as stormwater harvesting, though the lower quality of runoff leaving these surfaces may require greater treatment than runoff from roof surfaces. This chapter specifically focuses on designing systems principally capturing runoff from rooftops.

Typically, water harvesting storage systems will be one of two types. Larger systems that would serve a commercial, institutional, or other large building would employ a large cistern for storage. Water capture at the individual downspout is generally accomplished using rain barrels.

*Green Stormwater Infrastructure Fundamentals and Design*, First Edition. Allen P. Davis, William F. Hunt, and Robert G. Traver.

Companion Website: www.wiley.com/go/davis/greenstormwater

Figure 14.1 Rainwater Harvesting Systems in (A) New Zealand and (B) California (Photo by Authors).

## 14.2 Potential as a Water Resource

The translation of rainfall to a usable water volume represents an opportunity for sustainable urban water management, reducing volumes of runoff and concurrently reducing energy and capacity stresses on potable water systems. Assuming that 95% of annual rainfall will be large enough to produce rooftop runoff, 1 m of rainfall corresponds to 950 L/m$^2$ (23.3 gal of water per ft$^2$) of roof area. A moderate-sized institutional building (60 ft × 200 ft, 18.3 × 61 m) will have the capacity to produce 10$^5$ L (280,000 gal) of water per year, an average of 2900 L (765 gal) per day. This water can be used beneficially outside or within the building to decrease stormwater impacts, reduce potable water needs (and energy associated with the water purification and delivery), and reduce corresponding costs to the user.

Nonetheless, performance and benefits may not be applicable during winter months where freezing temperatures are common. Stored water cannot be practically used for irrigation during times of the year when temperatures are frequently below freezing. As well, the economics of water harvesting may not be favorable due to the relatively low cost of potable water in many areas.

From a municipal perspective, a number of uses of harvested rainfall are possible. These can include various aspects of irrigation, including grassed areas, street trees, and even vegetated SCMs such as green roofs and bioretention facilities. Other uses can include washing of municipal vehicles, streets, and buildings and for firefighting. Harvested water can be used as makeup water in evaporative cooling tower systems and for fountains and other decorative water features.

## 14.3 Harvested Roof Water Quality

Discussions on water quality must consider the water source, the storage, and also the designated end use of the harvested water. Beneficial use of harvested water can include subsurface and surface irrigation, vehicle washing, toilet flushing, laundry, and even drinking. All of these uses have different water-quality needs. Some type of advanced treatment system will be needed for harvested water when high-quality water is required. The regulatory arena for use of harvested water is complex and evolving. In the United States, many decisions on water-quality requirements for harvesting systems are delegated to local government agencies and guidelines can vary greatly from place to place.

An overview of rooftop water quality emphasizes local conditions in influencing microbial loads and the need for molecular-based methods in characterizing pathogen types and levels in rooftop runoff (Lye 2009). Many factors affect the quality of roof runoff. These include the roof material and characteristics (i.e., slope), the type of the rainfall event, and other weather- and climate-related parameters. It is hard to find many simple correlations between water quality and conditions. Metal roofs and roofs that contain metal components (such as flashing) will have higher metals concentrations. Copper flashing and galvanized materials will lead to high levels of copper and zinc, respectively. Metal roofs, however, have also been found to generate better microbial quality than other roofing materials (Mendez et al. 2011) which may be due to higher temperatures (in Texas) or toxicity due to leached metals. Another possibility is that metal roofs are more easily washed and the majority of the contamination is removed with the diversion of the first flush.

As discussed in Chapter 5, urban runoff pollutant concentrations generally follow a first flush pattern. With the first flush, the highest concentrations of pollutants occur in the first few volume increments of the runoff. This is true with roof systems as well. The first rainfall washes off materials that have accumulated on the roof. As a result, many water harvesting systems have (or require) some type of diversion of the first flush from the runoff, leaving subsequent cleaner water for harvesting. The recommended diversion in Texas is 1–2 gal per 1000 $ft^2$ (4–8 $L/100\ m^2$) of roof area (Texas Water Development Board 2005).

Bringing harvested rainwater into human or near-human contact opens a health-risk pathway. Several studies have detected microbial pathogens in rooftop runoff with possible sources including insects, small animals, and birds that come into contact with the roof, as well as atmospheric deposition (Evans et al. 2006). As a result, the use of rooftop runoff for vehicle washing, toilet flushing, or other higher level uses may introduce pathogen exposure routes to humans. Currently, several states have published guidelines that specify the level of water quality necessary for various end uses including toilet flushing, fire protection, construction purposes, landscaping, and street cleaning (USEPA 2004). The extent of the guidelines varies from state to state; however, most guidelines include standard values for $BOD_5$, TSS, turbidity, and coliform bacteria.

Specific study on the quality of small residential rain barrels indicated that bacterial contamination was common (Shuster 2013). Concentrations of total coliform and enterococci were higher than the United States Environmental Protection Agency limits for recreational contact water quality. A recent review documents the microbiological quality of harvested rainwater (Hamilton et al. 2019). Worldwide *Escherichia coli* was detected in 24–92% of the rainwater tanks in documented studies. For drinking, no *E. coli* should be in the water. The presence of *E. coli* suggests fecal contamination. The lowest concentrations were noted in winter, with the highest in fall. Many different pathogenic microorganisms have been detected in rainwater harvesting systems, including *Acinetobacter, Aeromonas, Campylobacter, Legionella, Salmonella, Shigella,* and *Staphylococcus* (Hamilton et al. 2019). Microbial source tracking has implicated birds as a possible source of bacterial contamination. Length of storage will also impact microbiological water quality. Without advanced treatment, harvested rainwater should not be used for drinking or even in-home use. However, health risks should be small for using harvested water for outdoor irrigation, especially if the water is not used for spray irrigation.

## 14.4 Rain Barrels

Rain barrels generally hold about 200 L, but are available in different sizes. They are installed off-line on individual downspouts. The barrels can be simple, made of plastic and available from a typical home improvement store. Other, more expensive wooden barrels can be used as well. A cut is made in the downspout where a diverter is installed that collects water around the edges of the downspout. The diverter is kept level to the water level in the rain barrel so that when the barrel becomes full, no water will enter it and will continue down the downspout. A residential rain barrel is shown in Figure 14.2.

Rain barrels have a spigot tap about 30 cm from the bottom of the barrel. A garden hose or a soaker hose can be attached to the spigot. The water in the barrel can be used to water flower and vegetable gardens, and the lawn. However, the water pressure in the barrel will be low, since the water head will be less than 1 m even when completely full. A soaker hose or some type of semipermanent installation will work best with the rain barrel, allowing the water to drain slowly from the barrel into the garden. Currently, there are low head soaker hoses designed specifically for rain barrels.

Rain barrels will accumulate pollen, leaves, and other material and biological growth may occur. At least once per year they should be rinsed out and cleaned. In colder

Figure 14.2 Residential Rain Barrel Installed with Collection off Downspout (Photo by Authors).

climates, rain barrels and cisterns that are aboveground will be susceptible to freezing. In freezing months, the rain barrels should be unhooked from the downspout systems or the valve should be kept wide open so the little storage will occur and that a solid "ice barrel" will not be formed that could damage the rain barrel. Filters may be attached to the outflow to protect the pore size of the soaker hose.

## 14.5 Rainwater Harvesting Regulations

In some regions, such as in the western United States, rainwater harvesting can be complicated due to issues related to water rights and water ownership. Some states such as Colorado have changed their water rights law to allow for rain barrels and other types of harvesting.

When referring to regulatory documents, it is essential to recognize the definitions relating to different types of water, as some regulations may apply to only one water type.

- *Rainwater*: Water collected from the roof of a building or other catchment surface during a rainfall event and stored in a reservoir for nonpotable use.
- *Greywater*: Waste discharge of bathtubs, showers, lavatories, clothes washers and laundry sinks.
- *Reclaimed water*: Treated wastewater (also referred to as recycled water, reuse water).
- *Nonpotable water*: Water from any source not used for drinking.

### 14.5.1 Non-stormwater Regulations

In many states, the plumbing code applies only if the harvested rainwater is brought inside a building. Plumbing codes are uniform throughout a state and are implemented by local jurisdictions via plan reviews, building inspections and the issuance of a certificates and/or permits.

All local municipalities and/or water suppliers (cities, counties, utility/water supply companies, etc.) have their own regulations that serve to protect citizens and the public water supply system. These regulations may pertain to backflow prevention, metering, and/or billing, among other things. Additionally, many neighborhoods/homeowners associations have rules dictating aesthetic requirements for rainwater harvesting systems. Finally, some municipalities administer incentive or rebate programs to promote the implementation of rainwater harvesting. In some cases, rainwater harvesting systems must comply with program requirements to be eligible for program benefits. All rainwater harvesting systems should comply with the regulations pertaining to the area in which the system is installed to prevent monetary or judicial penalties.

Most codes require the implementation of backflow prevention measures to protect potable public water supplies from contamination via rainwater harvesting systems. This code is applicable when rainwater harvesting systems are equipped with a municipal water backup supply.

### 14.5.2 Stormwater Regulations

Collecting and storing rooftop runoff, and providing a consistent, dedicated, and reliable end use, will reduce the volume of runoff and enable the reduction in size of other required stormwater treatment systems on the site. In watersheds requiring nutrient removal from stormwater, dedicated uses of the collected rainwater or proper treatment/infiltration can reduce stormwater nutrient removal requirements.

To optimize treatment, some states require that the rooftop runoff must be captured and either (1) used on site; (2) treated, filtered, and released; (3) infiltrated; or (4) some combination of 1–3. For example, the captured rainwater can be used for irrigation or vehicle washing and then infiltrate into the ground or evaporate. The captured rainwater can also be used for toilet flushing, which then becomes a wastewater and is then directed into the wastewater system. The captured rainwater can also be retained in the tank/cistern for a period where it is partially treated and slowly discharged to a practice like a sand filter or infiltration system/area. The footprint of these devices should be smaller because of the capability of the cistern to store and slowly release the rainwater. When properly designed, rainwater harvesting systems can conserve water, save money by reducing the size of other on-site SCMs, and save money by reducing the amount of potable water consumed by irrigation, vehicle washing, toilet flushing, or other end uses. Collected rainwater shall not be directly sent to the wastewater system without first being used in the abovementioned uses. If for any reason the designed dedicated end use becomes unavailable because of some change, it will be required that an approved alternative end use or a properly designed SCM treatment system be installed on site to treat the roof runoff.

1) If for any reason the designed dedicated end use becomes unavailable because of some change, it will be required that an approved alternative end use or a properly designed SCM treatment system be installed on site to treat the roof runoff.

2) The harvesting system shall be labeled and identified as nonpotable water, where appropriate.
3) The harvesting system shall meet all local and state building and plumbing codes.

## 14.6 Designing Rainwater Harvesting Systems

### 14.6.1 General Characteristics and Purpose

Larger scale rainwater harvesting systems provide the dual benefits of (1) acting as alternate water supply sources, and (2) providing detention/retention of roof runoff that would otherwise become stormwater runoff (Basinger et al. 2010; Zhang et al. 2009; Ahmed et al. 2011; Fewkes and Warm 2000; Guo and Baetz 2007; Kim and Yoo 2009; Kim et al. 2012). However, storage for water supply and storage for runoff detention are sometimes opposing goals and require designers and operators to make tradeoffs between the two. That is, for a rainwater harvesting system to detain stormwater, there must be room available in the cistern for runoff. Despite a full cistern being ideal for water conservation (providing water when it is needed), a full cistern cannot provide the stormwater management benefit of detention.

Rainwater harvesting systems are a compilation of many components and processes (Figure 14.3), including:

- *Cistern*—The cistern is the aboveground or buried tank that holds the rainwater until it is used. It can be made from various materials; typically plastic or metal. This is also called a rainwater tank.
- *Filtration/Screens*—Screens help filter out debris such as leaves and other tree debris, roof shingle grit, and coarse dirt. They can also help prevent mosquito breeding in the cistern. Screens and filters may be added in the gutters and after the pump to provide additional filtration of sediments.
- *First flush diverter*—This optional device bypasses the initial roof runoff (approximately 1 mm, equal to 0.04 in) that may contain shingle grit, dirt, leaves, and pollen. This is not the same as the water-quality storm, or "first flush," used for other SCMs.
- *Gutter system*—The gutter system conveys the rainwater from the roof to the cistern. It includes the gutters along the roof edge and the gutter downspouts.
- *Overflow*—The overflow allows rainfall in excess of the designed storage volume of the cistern to discharge.
- *Pump*—The pump is needed in order to get the water to its designated use. An elevated tank or bladder tank with pump will minimize pressure fluctuations and wear on the pump.
- *Secondary water supply*—During drought conditions, an automated secondary water supply may be provided to supplement the rainwater typically captured and used for dedicated water reuse. Local plumbing codes shall be followed to prevent cross-contamination of potable water supplies.

For the majority of systems, precipitation falls onto the catchment surface (rooftop), is collected via gutters, and conveyed to the tank/cistern by a pipe network. Prestorage gross and coarse solids filtration is often utilized to prevent sediment, leaves, and debris from entering the storage. A piping network then conveys the water to the storage tank. A first flush diverter may be incorporated into the conveyance piping to

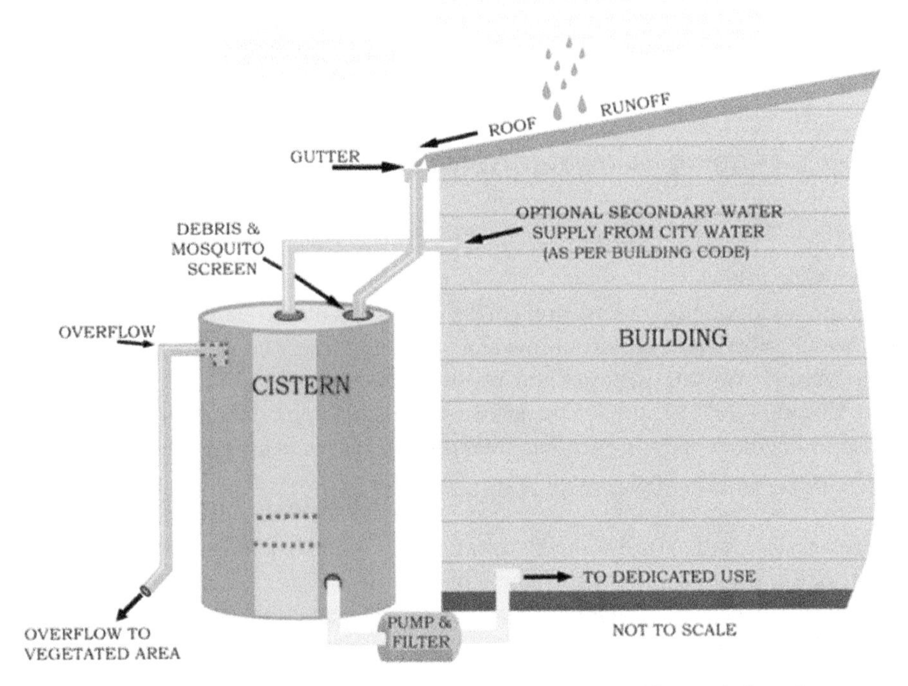


Figure 14.3 Typical Rainwater Harvesting System (Credit: Authors' Research Group).

divert the dirtiest of the runoff water (usually the first 1–3 mm) away from the storage container, thereby preserving the quality of water collected for later use (Kus et al. 2010). From the tank/cistern, the water either drains via gravity or is pumped to the point(s) of use. Poststorage treatment is sometimes included in the rainwater harvesting system, depending upon the quality of water in the storage tank and the quality needed for designated uses (e.g., toilet flushing or laundry).

### 14.6.2 Rainwater Storage Sizing Techniques

In general, the cistern must be sized to treat the design rainfall designated by the appropriate regulatory agency (e.g., 1-in, 1.5-in, 1-year 24-h storm) from the roof area directed to the water harvesting system. A minimum factor of safety equal to 1.2 must be applied to the calculated cistern volume required.

Therefore, the basis for the calculation required volume for a rainwater harvesting cistern is Equation 14.1. Note that the $SA_{roof}$ variable is the area of the birds-eye view plane of a roof surface (Figure 14.4).

$$V_{mReq'd} = (SA_{roof})(R_D)(FoS) \tag{14.1}$$

$V_{mReq'd}$—minimum required volume for the cistern;
$SA_{roof}$—horizontal surface area of the roof captured;
$R_D$—design storm rainfall depth;
FoS—factor of safety.

If a passive release component is incorporated as part of the rainwater harvesting system, then the $V_{mReq'd}$ is the volume to which the detention volume should be set.

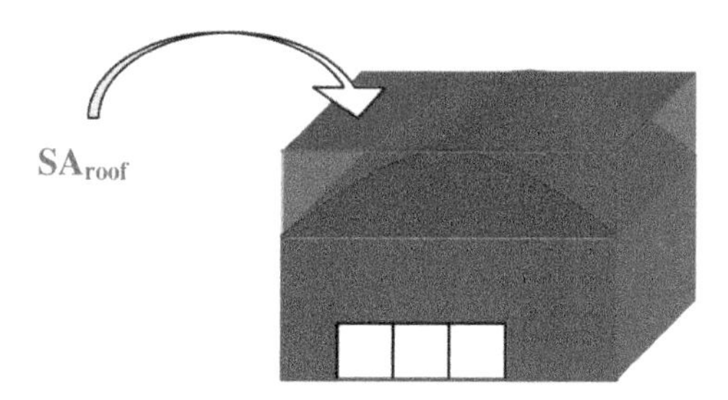


Figure 14.4 Finding Rainwater Volumes from Rooftops.

**Example 14.1** Find the minimum cistern volume to capture water from a building 15 m by 20 m for a 3-cm (1.2 in) rainfall.

*Solution* Using Equation 14.1 and a 1.2 (20%) factor of safety.

The roof surface area is 15 m × 20 m = 300 m$^2$.

$$V_{mReq'd} = (300\text{m}^2)(3\text{cm})(1.2) = 1080\text{m}^2\,\text{cm} = 10.8\text{m}^2 = 10{,}800\text{L}$$

For larger scale usage, typical reservoir sizing techniques can be used to determine what storage is required to maintain a set outflow. For an area where rainfall occurs fairly regularly year-round, storage equal to 4–5 days of demand may represent a reasonable compromise between inadequate storage and cistern size.

Routing a continuous rainfall record through a cistern of a fixed volume can provide more detailed information on the balance between cistern size and makeup water. Figure 14.5 demonstrates the results of technique with two different size cisterns. In this example, the roof is 4000 ft$^2$ (372 m$^2$) with 90% rainfall becoming runoff. The daily water demand for the building is 200 gal/day (757 L/day). A 90-day rainfall record from Baltimore-Washington International Airport is used. Overflow occurs when the rainfall exceeds the demand and available storage. If no rainfall or stored water are available, makeup water is required to satisfy the demand. Figure 14.5A shows the results for a 1000-gal (3785 L) cistern (equal to 5 days water demand). For each rain event, the cistern fills and frequently overflow occurs. The water balances for the 90-day data set are shown in Table 14.1. In this case, rainfall satisfied about 69% of the water demand, with 43% of the available rainfall bypassed to overflow.

Increasing the cistern size to 1500 gal had a major impact (Figure 14.5B, Table 14.1). In this case, the 81% of the demand was supplied by the rainfall and only 33% of the rainfall was lost. Again, a balance must be struck between cistern size/cost and rainfall utilization.

### 14.6.3 Design

The following design criteria are provided for rainwater harvesting systems in North Carolina (USA). Exact requirements will vary among climates and jurisdictions, e.g., for North Carolina: https://content.ces.ncsu.edu/rainwater-harvesting-guidance-for-homeowners:

1) All stormwater collected must either have (1) a dedicated, year-round, use to assure no overflow of the system during a design rainfall or (2) a release mechanism that

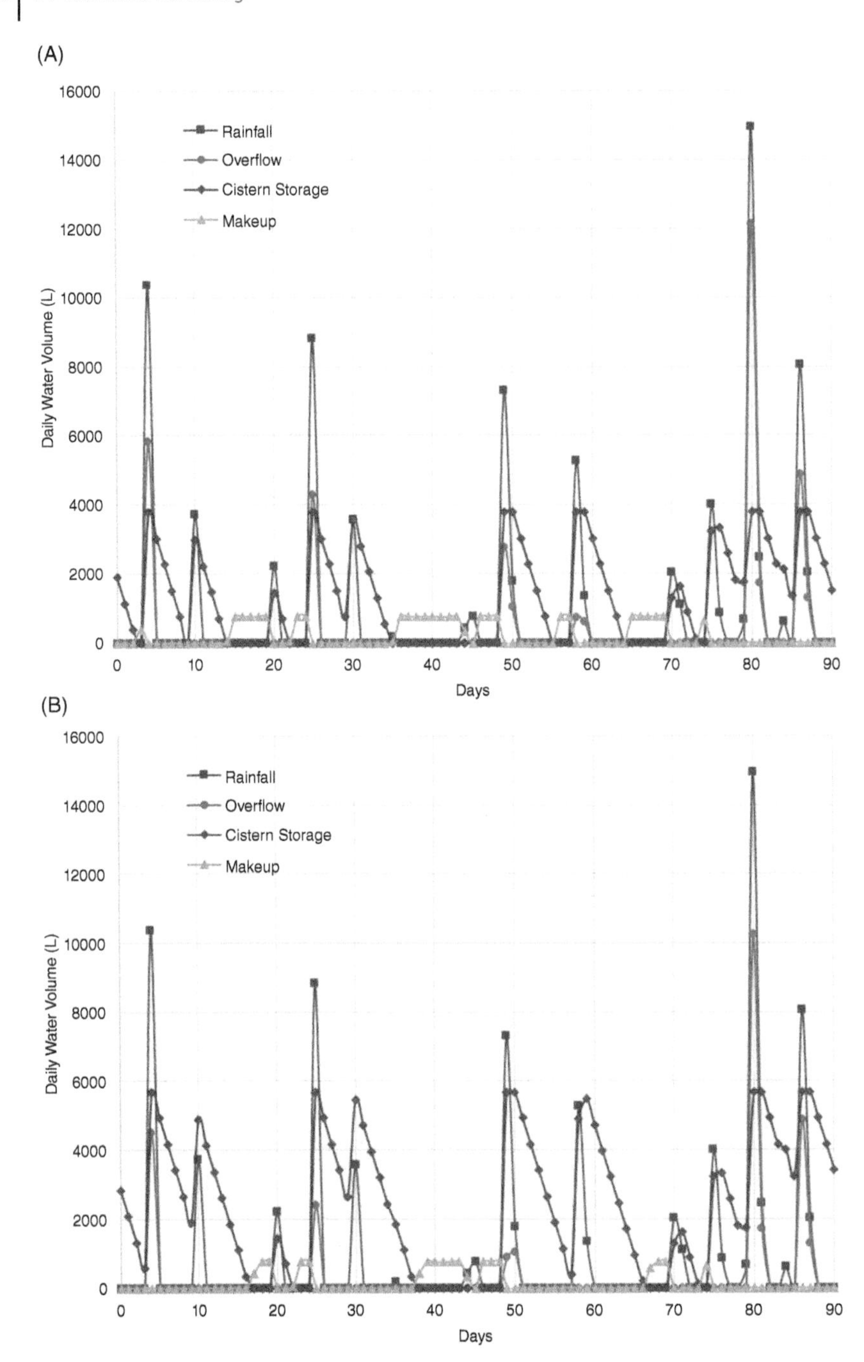


**Figure 14.5** Daily Volumetric Balances for Rainwater Harvesting System Using 90-day Rainfall Record from Baltimore-Washington International Airport: (A) 1000 gal (3785 L) Cistern and (B) 1500 gal (5678 L) Cistern.

**Table 14.1** Water balances for cistern simulations of Figure 14.5. All volumes in L.

| Rainfall | Overflow | Makeup | Rainfall used | Total use |
|---|---|---|---|---|
| 82,557 | 35,235 | 20,807 | 47,322 | 68,129 |
| 82,557 | 27,475 | 13,047 | 55,082 | 68,129 |

discharges detained water to a filtration or infiltration SCM. A water balance calculation must be used to establish the dedicated use of volumes and rates. The water balance calculation must demonstrate that the water-quality volume can:

a) be drawn down (used) within a 2–5-day period to allow for available volume in the system for the next rain event to be captured and stored, or
b) have an overflow of no more than 10–20% of the annual average historic rainfall. Eighty-six percent volume capture was determined by Smolek et al. (2015) to be the typical volume of water treated annually by a wet pond and other noninfiltrating SCMs in North Carolina. These SCMs were designed to capture the water-quality volume and detain the water volume for slightly more than 2 days. These leads to a value of 14% overflow, or
c) be drawn down within 5 days and discharged to a properly designed SCM.

The dedicated water use system must be established and reliable (typically automated) to ensure that the water will be used at the rate and volume designed. A permitting agency may require a meter on the cistern outlet to determine if the water is actually being used at the design-dedicated volume.

2) Overflows or bypass water shall be discharged in a nonerosive manner for the 10-year rainfall event. These flows can be directly discharged to the storm sewer network or to a natural water body. Overflows may be discharged to another properly designed SCM, but it is not required.
3) The elevation of the overflow pipe from the cistern shall be at or above the design volume elevation.
4) A first flush diverter should be used. If the first flush includes a slow leak that is discharged interevent, this slowly discharged water must discharge to a properly designed SCM. The first flush is typically the most polluted water leaving the roof and associated with the 0.02–0.04 in (0.5–1 mm) of rainfall. The first flush can be directed to the same SCM or infiltration zone that a passive release tank discharges to.
5) At a minimum, a 1-mm or smaller screen at the entrance to the cistern from the gutter system shall be provided to filter out debris. Screens in the gutters may also be provided on the gutters to prevent large debris from entering the gutter system.
6) If a secondary or makeup water supply is used, it shall be designed to place a minimal amount of volume in the cistern at any one time. The design must allow for adequate storage for the full volume of the next design storm. The secondary water supply must not be used to completely fill up the cistern. Local plumbing codes shall be followed to prevent cross-contamination of potable water supplies.
7) A properly designed footing for the cistern must be designed if the load of the cistern at full capacity is greater than that which the soils will support. If it is buried, buoyancy calculations must be provided to show the cistern will not float when empty.
8) An appropriate pump shall be selected to provide adequate pressure for its designated uses at the required pressures.
9) The gutter system and overflow shall be sized appropriately to handle the design rainfall.
10) Aboveground cisterns should be made of a material or color, or other screening is provided, that prevents light from entering the cistern, which helps prevent algae growth within the cistern.

11) Metal cisterns may require the use of an internal waterproof bladder to minimize leakage.
12) If the only use is irrigation, then a passive-release design should be implemented, unless a detailed plan is provided that assures year-round usage, which probably will require the inclusion of automatic mechanisms for rainfall detection and soil moisture monitoring. Irrigation water from a cistern shall be applied so that the water infiltrates into the ground; this probably will allow some degree of "over-irrigation" relative to standard agronomic rates.

## 14.7 Designing for Enhanced Stormwater Performance

### 14.7.1 Passive Release Mechanism

A dual purpose system may be employed to increase the stormwater mitigation potential (maximize capture efficiency and minimize overflow) of rainwater systems (Figure 14.6). This design feature will almost always be used when irrigation is the main intended use of captured rainwater. A dual-purpose rainwater harvesting tank can be created by dividing a storage tank into two portions, designated as the detention storage volume and retention storage volume. The retention storage volume comprises the bottom portion of the storage tank and water is extracted from this section to meet user demands. The detention storage volume comprises the top portion of the storage tank and serves as a temporary holding space for runoff. The volume dedicated to the detention storage can be estimated using Equation 14.1. The two storage volumes are delineated by a passive release orifice that allows the water in the detention portion to slowly drain between rain events. The intended purpose of this design feature is so that the detention storage is emptied prior to the subsequent rain event, allowing the rainwater harvesting system to reliably capture runoff from a large percentage of rain events while still serving as an alternative source of water for users (Brodie 2008; Hermann and Schmida 1999; Reidy 2010).

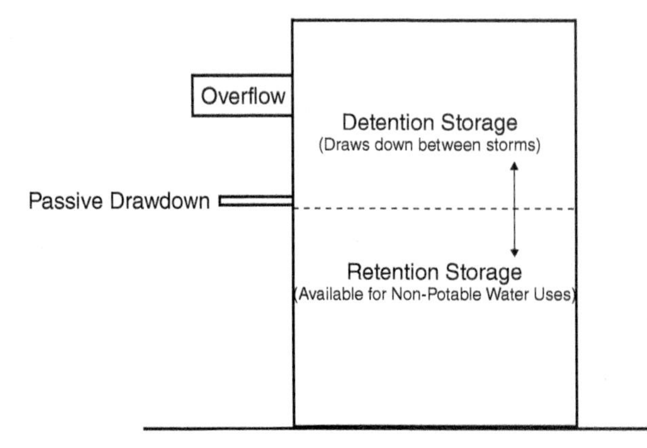


**Figure 14.6** Rainwater Harvesting System with Passive Release Design (Credit: Authors' Research Group).

This design method can often be retrofitted to existing rainwater harvesting systems (in addition to being part of newly installed systems. As a retrofit, it is employed with existing systems for which usage does not fully utilize the full capacity of the tank (i.e., the tank never empties). The steps for designing a passive release mechanism are as follows:

1) Determine the design rainfall associated with the contributing drainage area for the system. The passive release orifice should be placed such that the detention storage volume is equivalent to or greater than the design rainfall volume (Equation 14.1). In select cases where the passive release system is paired with a tank that has reliable and frequent withdrawals for demand, it is possible that long-term simulation modeling will show the detention volume required will be less than the water-quality volume.
2) Choose the drawdown time for the detention storage volume. Commonly, to receive stormwater credit, the design rainfall/detention storage volume typically must drain from the system within 2–5 days. A drawdown time of 2.5–3.0 days is recommended, as this is slightly greater than the minimum but maximizes the capture of the water-quality volume by preparing the system for the next rain event quickly.
3) Calculate the tank cross-sectional area and storage height based on the storage volume. The ratio of the height and area may be fixed by the commercial availability of storage tanks.

$$V_d = A_S h_0 \tag{14.2}$$

$V_d$ is the detention storage volume;
$A_S$ is the cross-sectional area of storage tank;
$h_0$ is the vertical distance between the center of the release orifice and the bottom of the overflow pipe.

4) Calculate the cross-sectional area of the passive release orifice. Since the flow rate will change and the water level falls, Equation 14.3 is used:

$$A = \frac{2A_S\left(h_0^{1/2} - h_f^{1/2}\right)}{C_d t\left(2g\right)^{1/2}} \tag{14.3}$$

$A$ is the cross-sectional area of the passive release orifice and
$t$ is the chosen drawdown time.

5) Determine the diameter, $d$, of the passive release orifice:

$$d = \sqrt{\frac{4 \cdot A}{\pi}} \tag{14.4}$$

**Example 14.2** A cistern storage will have a passive release of stored water. The detention storage volume is 8000 L, to be released over 3 days. The passive release outlet will be 90 cm below the cistern overflow. Determine the diameter of the passive release orifice.

***Solution*** The cross-sectional area of the storage tank is given by Equation 14.2

$$A_s = \frac{8000\ \text{L}}{90\ \text{cm}} = \frac{8\,\text{m}^3}{0.9\ \text{m}} = 8.89\ \text{m}^2$$

The time to drain the stored water is assumed as 3 days, equal to 2.592 × $10^5$ s. From Equation 14.3, the orifice area is given by

$$A=\frac{2\left(8.89\text{m}^2\right)\left(0.9\text{m}^{1/2}-0^{1/2}\right)}{0.6\left(2.592\times10^5\text{s}\right)\left(2\left(9.8\text{m}/\text{s}^2\right)\right)^{1/2}}=2.45\times10^{-5}\text{m}^2$$

And the diameter is found from Equation 14.4:

$$d=\sqrt{\frac{4\cdot2.45\times10^{-5}\text{m}^2}{\pi}}=5.59\times10^{-3}\text{m}=5.59\text{mm}$$

The flow rate when the tank is full is given by the orifice equation (Chapter 7):

$$Q=C_dA\left(2gh\right)^{1/2}=0.6\left(2.45\times10^{-5}\text{m}^2\right)\left(2\left(9.8\text{m}/\text{s}^2\right)0.9\text{m}\right)^{1/2}$$

$$=6.17\times10^{-5}\text{m}^3/\text{s}=0.0617\text{L}/\text{s}=3.70\text{L}/\text{min}$$

The average flow rate is

$$Q=\frac{8000\text{ L}}{2.592\times10^5\text{s}}=0.0609\text{L/s}=1.852\text{L/min}$$

Design considerations:

- The passive release mechanism should be equipped with some type of filter, located on the inside of the tank, to prevent clogging.
- The passive release orifice should drain to (1) an infiltrative area, or (2) an infiltration-based SCM (such as an infiltration trench, or a filtration-based SCM [e.g., a sand filter]).
- While typically not recommended, it may be acceptable to equip the passive release orifice with a valve; however, the handle of the valve should be removed/locked to prevent undesirable hampering of the release mechanism.
- Use caution if a secondary water supply is incorporated in the design of the rainwater harvesting system. The trigger point for the secondary water supply must not be at a volume greater than the retention storage volume and, ideally, should be at a substantially lower level.

The design of the passive release mechanism inherently ensures an automatic release of the detention storage volume. This design approach alleviates the need for human intervention and greatly improves the reliability with which the system meets stormwater management objectives.

### 14.7.2 Active Release Mechanism

An alternative to the passive release rainwater harvesting system is one that employs an active release mechanism. This design approach incorporates computer logic into the workings of the system. A release valve is placed at the outlet of the rainwater harvesting system and is controlled by computer programming logic. The program retrieves the precipitation forecast from the National Oceanic Atmospheric Administration (NOAA), which predicts the probability and amount of rainfall that will occur. The program determines if there is enough storage space within the rainwater harvesting system to fully capture the stormwater runoff volume generated by the expected precipitation (indicated by the readings of a pressure transducer located within the storage tanks). If there is sufficient space, the release valve stays shut. If not enough storage space is available, the release valve is automatically opened and stored water is released from the system until sufficient storage space is created by the release of water from the tank to capture the anticipated runoff. An active release program should include logic to prevent the release of stored water during a rainfall event. Computer simulations incorporating active release are not generally publicly available. If an active release system is chosen, then a proprietary model and private entity will need to demonstrate the long-term effectiveness of the system.

The active release mechanism addresses many of the disadvantages of the passive release mechanism. The use of weather forecasting ensures that water is only released from the system if it will be replaced by an imminent precipitation event. This eliminates the "wasting" of needed water, as well as the decrease in the volume of usable storage within the rainwater harvesting system. Thus, when applied to an existing rainwater harvesting system, the full system volume is still available to the user to meet water demands. When applied to the design of a new system, extra volume (and cost) to compensate for a loss in storage is not needed. Finally, the fail-safe clause in the program logic prevents the release of stored water during a precipitation event, which is a primary goal of stormwater management. Like passive release rainwater harvesting systems, active release devices must have their pre-event discharged water infiltrate or be filtered.

Drawbacks to the active release mechanism include cost, complexity, and resource requirements. The installation of the equipment and program logic is typically expensive. Furthermore, the complexity of the programming logic and the potential for technical glitches requires oversight by a highly knowledgeable company or individual, which can also be expensive. Finally, as the program is based upon the ability to access NOAA forecasts, there must be uninterrupted power and internet accessibility at the site.

### 14.7.3 Alternative Approaches for Irrigation-based Systems

Many rainwater harvesting systems are used for irrigation, as irrigation can require a substantial amount of water during the growing season. However, during the non-growing season, there is no use for harvested rainwater and systems often remain full, thus providing no stormwater management benefit (DeBusk et al. 2013).

Excess irrigation is the application of irrigation water to turfgrass or other vegetation at rates higher than the *minimum* needed to maintain health and quality. The conventional agronomic irrigation recommendation is based upon applying minimum volumes of water to conserve water (which is appropriate when using potable water); however, this does not benefit stormwater management. Irrigating at rates higher than these agronomic minimums will draw down cistern-stored water more frequently, making more storage available for roof runoff during subsequent rainfall events. Often, irrigation can continue throughout the nongrowing season. In addition, if done properly such that overirrigating does not produce runoff, water application on turf can serve as a de facto SCM by filtering, infiltrating, and evaporating applied water.

The type of soil and vegetation at the proposed location of filtration will dictate the maximum amount of water that can be applied per unit area without jeopardizing the health of the vegetation and/or causing runoff due to excess irrigation. Sandy soils are ideal for employing the excess irrigation approach. Studies conducted in North Carolina revealed that the application of up to 2 in (5 cm) per week, applied year-round, did not harm Bermuda turfgass or create moisture problems within the soil. Systems located in regions with less permeable soils can still employ excess irrigation; however, the amount that may be applied without damaging turf or creating runoff will most likely be lower than in the sandier regions. Preliminary information obtained per consultation with turfgrass specialists recommend that a 1-in/week (2.5-cm/week) irrigation rate may be appropriate outside sandier soil landscapes.

Higher irrigation rates result in the depletion of harvested rainwater faster than lower application rates. While this is ideal for stormwater management, if the cistern volume is not replenished by rainfall soon enough, turf health may decline due to lack of water during drought periods. Consequently, designers/owners may choose to incorporate a secondary water supply or tiered approach. If a secondary supply is used, it must be designed such that a minimal amount of irrigation is applied to maintain turf health; the excess irrigation depth is *not* to be applied when using potable water. A tiered approach may be used in which the excess irrigation depth is applied until the cistern volume reaches a certain threshold capacity, at which time the irrigation scheme switches to a conservation mode (the minimal amount of irrigation water is applied to maintain turf health). Whichever approach is used, the irrigation controls must be fully automated and not rely on human intervention to switch between rainwater and potable water or between irrigation schemes.

**Design and operation considerations**

- Studies on the excess irrigation approach have only been conducted on turfgrass; however, this approach may also apply to certain woody vegetation without adverse consequences. Designers/owners should consult a horticultural expert to confirm.
- A rain sensor must be part of an irrigation system to prevent irrigation during precipitation events.
- When irrigating year-round, care should be taken to insulate all exposed piping to prevent freezing. Irrigation water must *never* be applied to frozen ground. During a frost period, irrigation should be applied during the middle of the day (if at all) to prevent damage to turf.
- The excess irrigation rate shall never be applied when utilizing a secondary potable water supply.

### 14.7.4 Designing an Infiltration or Filtration Area

As noted previously, any runoff that is captured and subsequently drained from the detention portion of a dual-purpose tank must be directed to either a filter SCM or an infiltration SCM or area. Because of the detention outlet design, release rates are relatively consistent and minimal (e.g., a tank serving a 2000-ft$^2$ (186-m$^2$) rooftop may have a drawdown rate <0.01 cfs [<0.3 L/s]); this often leads to small footprints being dedicated to infiltration areas or filtration SCMs of these rainwater harvesting systems. Note that all overflow need not be directed to a filtration or infiltration SCM, only drainage from the detention zone. If overflow is directed to the filtration or infiltration SCM, the device footprint could become substantially larger.

The footprint of the filtration/infiltration device is dependent upon three principal factors: (1) the water-quality volume, (2) the intended duration of drawdown (typically

2.5–3.0 days), and (3) the infiltration rate of the media or soil to which the water is discharged. Equation 14.5 is a simplified equation to calculate this footprint.

$$SA = \frac{Q}{I} \tag{14.5}$$

where SA is the filtration/infiltration surface area, $Q$ is the drawdown flow rate, and $I$ is the media/soil infiltration rate.

Figures 14.7 and 14.8 show the footprint of the filter/infiltration area depending upon surface infiltration rate. The figures are based upon a drawdown rate of 2.5 days for 95% of the water-quality volume and a 90% capture factor. If a filtration SCM is used (recommended when the infiltration rate of the receiving soil is less than 0.5 in/h [1.3 cm/h]), a media depth (e.g., sand or a bioretention mix) of at least 2 ft (61 cm) is required.

**Example 14.3** A cistern storage will have a passive volume of 8000 L, to be released over 3 days. The infiltration rate of the infiltration area is 0.5 in/h (1.4 cm/h). Determine the infiltration area needed.

*Solution* As calculated in Example 14.2, the passive release flow rate is 30.9 $cm^3$/s.

$$Q = \frac{8000\ \text{L}}{3\ \text{days}} = 30.9\ \text{cm}^3/\text{s} = 111\ \text{L/h}$$

From Equation 14.5, the infiltration area is given by

$$\text{SA} = \frac{111\text{L/h}}{1.4\text{cm/h}} = 7.95 \times 10^4\ \text{cm}^2 = 7.95\text{m}^2$$

This is an area about 2.65 m wide and 3 m long.

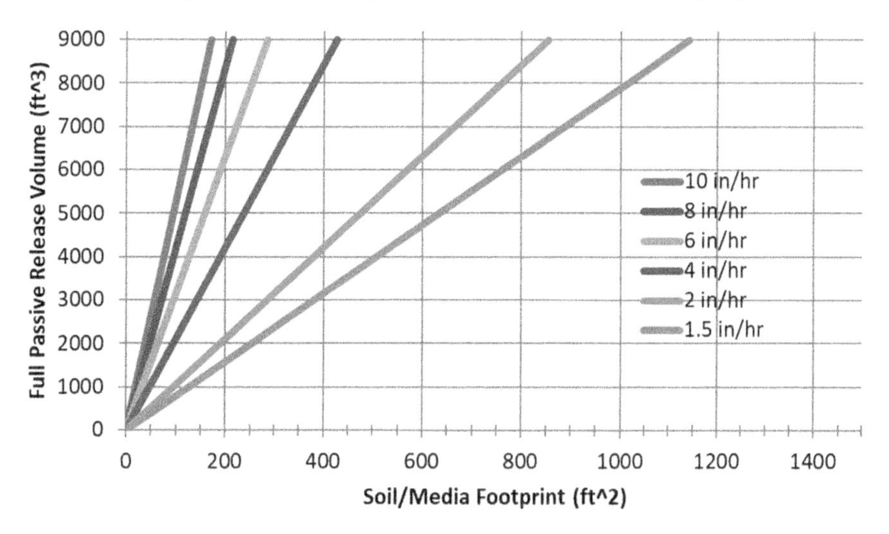


Figure 14.7 Soil/Media Area Needed for Infiltrating Stored Water over 2.5 days for High Infiltration Rate Media (Credit: Authors' Research Group).

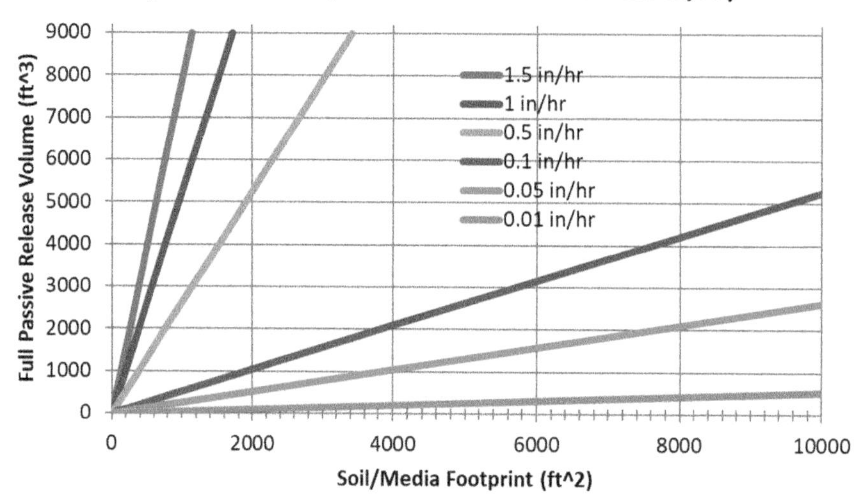


Figure 14.8 Soil/Media Area Needed for Infiltrating Stored Water over 2.5 days for Low Infiltration Rate Media (Credit: Authors' Research Group).

## 14.8 Treatment for High-quality Use

If harvested water is to be brought into a building for use, some type of treatment would almost always be required. This is to reduce the number of possible microbial pathogens that are washed into the storage cistern and also pathogens that may grow in the cistern during storage. The latter can occur since carbon and other nutrients can be washed into the cistern during the water harvesting process.

Treatment systems usually have at least three components. The first is some type of mixing system that will cycle on in the storage tank. This keeps the water well mixed and provides some aeration, keeping the water from becoming stagnant. The other two components are specifically used to reduce risks from microbial contaminants. The first is a filtration system. This is likely some type of replaceable cartridge filter. The filter may be paper or fabric with openings in the 3–5-μm ranges to filter out particulate matter (PM) and microorganisms attached to PM. An activated carbon filter may also be added for additional filtration as well as adsorption of compounds that may contribute tastes and odors to the water.

Finally, some type of disinfection process may be included. Small UV systems are common that can provide direct disinfection of the water as it is used, or as it is cycled in the storage tank. Alternatively, ozone or chlorine can be added for disinfection, but these systems require chemical handling.

An alternative treatment process is reverse osmosis (RO). RO systems use pressure and a semipermeable membrane to create very high-quality water. Nonetheless, RO systems can be expensive to operate and produce a water stream that must be wasted.

## 14.9 Inspection and Maintenance

Rain barrels and cisterns need regular inspections and periodic maintenance to ensure that they function as intended. Inspections should evaluate the entire flow pathways, to include the roof, downspouts, screens, overflow, discharge, and diversion piping. Screens should be cleaned to remove clogged material. Holes, leaks, and breaks in plumbing should be repaired. The rain barrel/cistern should be drained and rinsed with clean water at least once per year to remove any accumulated sediment and biofilm within the storage system.

Any water treatment equipment used in the water harvesting system will have its own inspection and maintenance requirements, usually specified by the manufacturer. Filters have to be periodically cleaned and possibly replaced. Pumps and more advanced treatment systems will have more complex requirements.

## References


Ahmed, W., Gardner, T., and Toze, S. (2011). Microbiological quality of roof-harvested rainwater and health risks: A review. *Journal of Environmental Quality* 40 (1): 13–21.

Basinger, M., Montalto, F., and Lall, U. (2010). A rainwater harvesting system reliability model based on nonparametric stochastic rainfall generator. *Journal of Hydrology* 392 (3-4): 105–118.

Brodie, I.M. (2008). Hydrological analysis of single and dual storage systems for stormwater harvesting. *Water Science and Technology* 58 (5): 1039–1046.

DeBusk, K.M., Hunt, W.F., and Wright, J.D. (2013). Characterization of rainwater harvesting performance and demonstrating stormwater management benefits in the humid southeast USA. *The Journal of American Water Resources Association* 49 (6): 1398–1411.

Evans, C.A., Coombes, P.J., and Dunstan, R.H. (2006). Wind, rain and bacteria: The effect of weather on the microbial composition of roof-harvested rainwater. *Water Research* 40 (1): p. 37–44.

Fewkes, A. and Warm, P. (2000). Method of modelling the performance of rainwater collection systems in the United Kingdom. *Building Services Engineering Research and Technology* 21 (4): 257–265.

Guo, Y. and Baetz, B.W. (2007). Sizing of rainwater storage units for green building applications. *Journal of Hydrologic Engineering* 12 (2): 197–205.

Hamilton, K., Reyneke, B., Waso, M., Clements, T., Ndlovu, T., Khan, W., DiGiovanni, K., Rakestraw, E., Montalto, F., Haas, C.N., and Ahmed, W. (2019). A global review of the microbiological quality and potential health risks associated with roof-harvested rainwater tanks. *Npj Clean Water* 2 (1): 7.

Hermann, T. and Schmida, U. (1999). Rainwater utilisation in Germany: Efficiency, dimensioning, hydraulic and environmental aspects. *Urban Water Journal* 1 (4): 1307–1316.

Kim, K. and Yoo, C. (2009). Hydrological modeling and evaluation of rainwater harvesting facilities: Case study on several rainwater harvesting facilities in Korea. *Journal of Hydrologic Engineering* 14 (6): 545–561.

Kim, H., Han, M., and Lee, J. Y. (2012). The application of an analytical probabilistic model for estimating the rainfall–runoff reductions achieved using a rainwater harvesting system. *Science of The Total Environment* 424: 213–218.

Kus, B., Kandasamy, J., Vigneswaran, S., and Shon, H.K. (2010). Analysis of first flush to improve the water quality in rainwater tanks. *Water Science and Technology* 61 (2): 421–428.

Lye, D.J. (2009). Rooftop runoff as a source of contamination: A review. *Science Total Environment* 407 (21): p. 5429–5434.

Mendez, C.B., Klenzendorf, J.B., Afshar, B.R., Simmons, M.T., Barrett, M.E., Kinney, K.A., and Kirisits, M.J. (2011). The effect of roofing material on the quality of harvested rainwater. *Water Research* 45 (5): 2049–2059.

Reidy, P.C. (2010). Integrating rainwater harvesting for innovative stormwater control. *In proc. World Environmental and Water Resources Congress*, 448–454. Providence, RI, USA: American Society of Civil Engineers.

Shuster, W.D., Lye, D., De La Cruz, A., Rhea, L.K., O'Connell, K., and Kelty, A. (2013). Assessment of Residential Rain Barrel Water Quality and use in Cincinnati, Ohio. *JAWRA Journal of the American Water Resources Association* 49 (4): 753-765.

Smolek, A.P., Hunt, W.F., and Grabow, G.L. (2015). Influence of drawdown period on overflow volume and pollutant treatment for detention-based stormwater control measures in Raleigh, North Carolina. *Journal of Sustainable Water in the Built Environment* 1 (2): 05015001.

Texas Water Development Board (2005). *The Texas Manual on Rainwater Harvesting*, 3rd Ed. Austin, Texas.

USEPA (2004). *Guidelines for Water Reuse, U.S.E.P.* Washington, D.C: Agency, Editor.

Zhang, D., Gersberg, R.M., Wilhelm, C., and Voigt, M. (2009). Decentralized water management: Rainwater harvesting and greywater reuse in an urban area of Beijing, China. *Urban Water Journal* 6 (5): 375–385.

## Problems

**14.1** A rooftop of 10 m × 30 m will have a water harvesting system installed. How much first flush water should be wasted?

**14.2** A rooftop of 15 m × 40 m will have a water harvesting system installed. How much first flush water should be wasted?

**14.3** Size a cistern for water harvesting for a 1.2-in rainfall over a 10-m × 40-m rooftop.

**14.4** Size a cistern for water harvesting for 3-cm rainfall over a 20-m × 40-m rooftop.

**14.5** Size a cistern for water harvesting for a 1-in rainfall over a 50-ft × 50-ft rooftop.

**14.6** Size a cistern for water harvesting for a 1-year, 24-h storm in your area, over a 40-ft × 75-ft rooftop.

**14.7** Size a cistern for water harvesting for a 1-year, 24-h storm in your area, over a 12-m × 46-m rooftop.

**14.8** Find the orifice diameter needed to drain a cistern in 4 days. The cylindrical cistern contains 8000 L with a 1.72-m diameter. Therefore, the full water height is 3.44 m.

**14.9** Find the orifice diameter needed to drain a cistern in 5 days. The cylindrical cistern contains 1800 gal with a 64-in diameter. Therefore, the full water height is 129 in.

**14.10** A cistern system is being installed to collect rainfall from a building roof. This water is then slowly released to an infiltration area. The roof area is 130 ft by 160 ft. The first 0.01 in of rainfall is held on the roof and does not run off. The first 0.02 in of runoff are diverted from the cistern as a first flush management process to maintain relatively high water quality.

a) *Find the cistern volume necessary to hold one inch of runoff (from 1.03 in of rainfall) from the roof (gal).*

b) The cistern overflow rectangular weir/spillway must handle a rainfall of 2 in/h on the roof. The weir width will be 2× the height. Find the dimensions of the weir (in). Assume that all rainfall becomes runoff at this intensity. (The weir coefficient is 3.3 to produce $Q$ in $ft^3/s$, for $H$ and $W$ in ft.)

c) A 2.8-in rainfall occurs over a 24-h period. How much water will overflow the cistern (gal)?

d) The cistern is an upright cylinder with a 3.5-m diameter. The passive release outlet is a circular orifice 5.2 m below the cistern overflow and must drain the storage volume over 3 days. Find the diameter of the orifice necessary to meet this requirement (in). Use an orifice coefficient of 0.62. (Note that the head is not constant.)

e) *The cistern will drain into an infiltration area. The infiltration rate of this area is 0.5 in/h. Determine the infiltration area needed ($ft^2$).*

**14.11** A cistern system is being installed to collect rainfall from a building roof. This water is then slowly released to an infiltration area. The roof area is 30 m × 60 m. The first 0.3 mm of rainfall is held on the roof and does not run off. The first 0.6 mm of runoff are diverted from the cistern as a first flush management process to maintain relatively high water quality.

a) Find the cistern volume necessary to hold 3 cm of runoff (from 3.09 cm of rainfall) from the roof (L).

b) The cistern overflow rectangular weir/spillway must handle a rainfall of 6 cm/h on the roof. The weir width will be 2× the height. Find the dimensions of the weir (in). Assume that all rainfall becomes runoff at this intensity.

c) An 8-cm rainfall occurs over a 24-h period. How much water will overflow the cistern (gal)?

d) The cistern is an upright cylinder with a 3.5-m diameter. The passive release outlet is a circular orifice 5.2 m below the cistern overflow and must drain the storage volume over 3 days. Find the diameter of the orifice necessary to meet this requirement (in). Use an orifice coefficient of 0.62. (Note that the head is not constant.)

e) The cistern will drain into an infiltration area. The infiltration rate of this area is 1 cm/h. Determine the infiltration area needed ($m^2$).

**14.12** A cistern storage will have a passive volume of 6000 L, to be released over 4 days. The infiltration rate of the infiltration area is 2 in/h. Determine the infiltration area needed.

**14.13** A cistern storage will have a passive volume of 1800 gal to be released over 4 days. The infiltration rate of the infiltration area is 0.2 in/h. Determine the infiltration area needed.

**14.14** A cistern storage will have a passive volume of 7500 L, to be released over 4 days. The infiltration rate of the infiltration area is 4 cm/h. Determine the infiltration area needed.

**14.15** Discuss the water-quality requirements for different uses of harvested rainfall:

a) Vehicle washing
b) Irrigation
c) Indoor toilet flushing
d) Laundry
e) Drinking

# 15

# Permeable Pavement

## 15.1 Introduction

Permeable pavement is exactly what it sounds like, pavement that allows water to pass through it rather than shed off. It is normally composed of a gravel storage layer, typically ranging in depth from 10 to 45 cm (4 to 18 in.), and as much as 120 cm (4 ft) when sized for extreme events. The surface is overlain by a permeable surface such as permeable concrete or asphalt, or permeable interlocking concrete pavers. Increasingly, some designers are turning to either large perforated pipes, or modular crate systems instead of gravel; these options allow for a smaller bed design (thus, less excavation and material) due to larger void space. Depending upon the pavement's underlying soil, an under-drainage system may be used in locations where 100% infiltration is not a goal. Should an underdrain be used, recommended techniques include use of an elevated or restricted outfall. Simple designs include upturned elbows or underdrain PVC caps that can be drilled and converted into a slow release system.

All permeable pavement types operate essentially the same way. Flow is generally one dimensional. Rainfall passes vertically through the permeable pavement (or travels a short horizontal distance in the case of pavers) and partially fills the gravel storage layer, where it is stored and later exfiltrates (Figure 15.1). Depending upon the rainfall intensity, rainfall volume, and existing soil permeability, water then either exits the bottom of the permeable paver system via soil exfiltration or underdrain, if present. When a storage volume is created, water will build up in this space until it is full and then either flows over the top of the pavement or through a raised underdrain. Although infiltration rates through the permeable pavements are high, very intense rainfall rates can produce runoff from permeable pavement, particularly on concrete grid paver systems filled with sand. After the event is over, stored water will exfiltrate into the surrounding soils, with the possibility of small amounts of evaporation. Some permeable pavement designs are able to capture and infiltrate nearly 100% of events, even extreme events, when designed to meet this requirement.

In early designs, permeable pavement SCMs were not meant to manage runoff from other surfaces, but to reduce the generation of excess runoff. Even today, applications with no (or incidental) run-on generally function longer than those with a fair amount of run-on. Designs that manage the runoff from surrounding buildings and standard pavements (run-on) must be protected (e.g., clogging prevention) so that the filtration

*Green Stormwater Infrastructure Fundamentals and Design*, First Edition. Allen P. Davis, William F. Hunt, and Robert G. Traver.

Companion Website: www.wiley.com/go/davis/greenstormwater

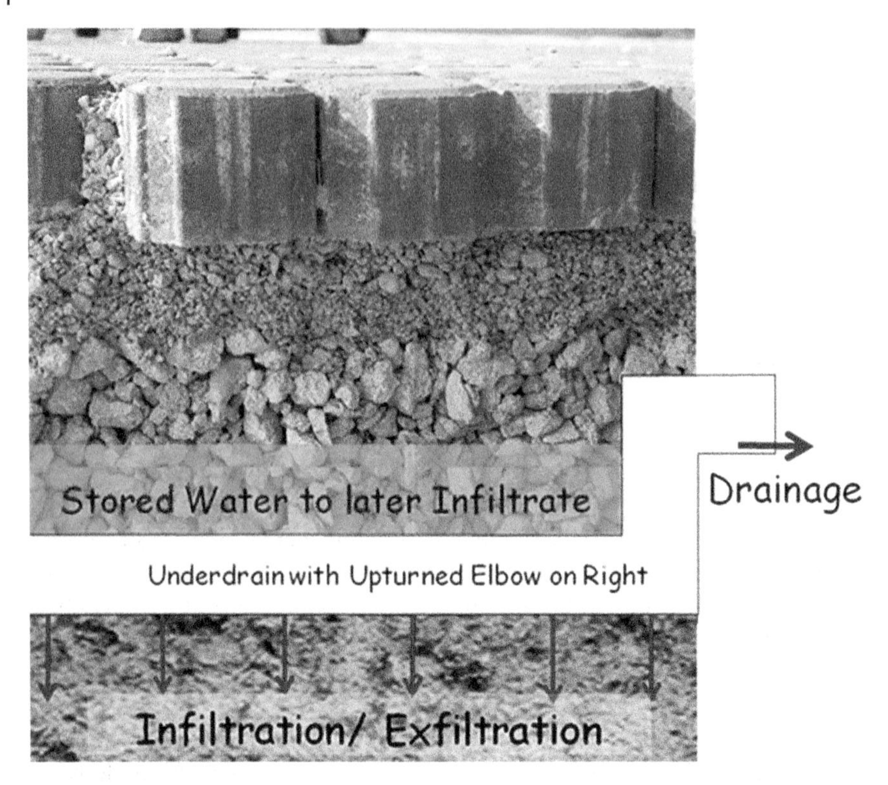


Figure 15.1 Permeable Pavement Cross-Section with Underdrain. Water that Collects in the Drainage Layer Will Infiltrate, or Outflow When Underdrains are Used (Credit: Authors' Research Group).

and infiltration properties of the permeable pavement are maintained. Most larger sites incorporate an overflow system to prevent excess water from expressing itself through the pavement surface. As water passes through the pavement, many pollutants can be trapped either inside the pavement, at the interface with the underlying soil, or in the underlying soil proper (Welker et al. 2013).

While some highways have been constructed of pervious material, generally permeable pavements should not be expected to withstand the structural loads and wear and tear common to heavy vehicular use. Permeable pavements are thus designated for areas with lower structural loadings (vehicular traffic), such as parking spaces or low-speed travel ways. Ideal applications include cul-de-sacs, road shoulders, bicycle lanes and sidewalks. Limited-use parking lots (such as event or overflow parking) are ideal for permeable pavements. Steeply sloped surfaces provide greater challenges for permeable pavement use. The subsurface layers must be built as terraces (essentially subsurface detention) so that water which flows through the top of the pavement does not seep out of the pavements at the lower slopes, resulting in surface flow.

The use of permeable pavements in regions that routinely freeze is often a concern; as liquid water stored in the pavement expands, it potentially breaks apart or heaves permeable pavements. However, because permeable pavements are designed to quickly infiltrate into the gravel storage lavers below the pavement surface, stored water is very rarely (if ever) exposed to freezing conditions. Proof of this can be observed by the extensive use of this SCM in cold weather climates such as Chicago and the Toronto region. Another benefit of permeable pavement surface infiltration is that these pavements are safer than traditional pavements, preventing the formation of "black ice."

## 15.2 Types of Permeable Pavements

Five types of permeable pavements are commonly used, Permeable Asphalt (PA), Permeable Concrete (PC), Permeable Interlocking Concrete Pavers (PICP), Concrete Grid Pavers (CGP), and Plastic Grid Pavers (PG). Photos of the various pavements are shown in Figure 15.2. General structural design considerations are discussed for each of the pavements below.

Permeable concrete (PC) is a mixture of Portland cement, fly ash, washed gravel, and water. The water to cementitious material ratio is typically 0.35–0.45. A fine, washed gravel, less than 1.3 cm (0.5 in.) in size (No. 8 or 89 stone), is added to the concrete mixture to increase the void space. Unlike traditional installations of concrete, permeable concrete usually contains a void content of 15–25% which allows water to infiltrate the pavement surface, percolate directly through the pavement, and (often) exfiltrate to the subsurface. An admixture improves the bonding and strength of the pavements. These pavements are typically poured to be 10–20 cm (4–8 in.) thick and commonly contain a gravel base course for additional storage or infiltration. Compressive strength can range from 2.8 to 28 MPa (400 to 4000 psi). Updated PC standards are available through the National Ready-mix Concrete Association (nrmca.org). A photo comparing PC to traditional concrete sidewalk is shown in Figure 15.3. The lack of fines is readily apparent in the PC.

Figure 15.2 Types of Permeable Pavement. Top L-R: Permeable Concrete (PC), Permeable Asphalt (PA), Permeable Interlocking Concrete Pavers (PICP). Bottom L-R: Concrete Grid Pavers (CGP), Plastic Reinforcing Grids (PG) Filled with Gravel, and PG with Grass (Photos by Authors).

Figure 15.3 Transition between Traditional (Left) and Permeable (Right) Concrete on a Sidewalk in Virginia (Photo by Authors).

Because of this lack of fines and high porosity, PC is highly susceptible to damage by chlorides. PC should not be used in areas that are expected to be heavily salted as part of deicing.

Permeable asphalt (PA) consists of fine and course aggregate stone bound by a bituminous-based binder. The amount of fine aggregate is reduced to allow for a larger void space (typically 15–20%) (Figure 15.4). Thickness of the asphalt depends on the traffic load, but usually ranges from 7.5 to 18 cm (3 to 7 in.). Current standards are available through the National Asphalt Pavement Association (asphaltpavement.org). A required underlying base course increases storage and adds necessary strength (Ferguson 2005).

Permeable interlocking concrete pavements (PICP) are available in many different shapes and sizes. When lain, the blocks form patterns that create openings through which rainfall can infiltrate. These openings, generally 5–20% of the surface area, are typically filled with pea gravel aggregate, but can also contain top soil and grass. *ASTM C936* specifications state that the pavers be at least 6.0 cm (2.36 in.) thick with a compressive strength of 55 MPa (8,000 psi) or greater. Typical installations consist of the pavers and gravel fill, a 3.8 to 7.6 cm (1.5 to 3.0 in.) fine gravel bedding layer, and a gravel base course storage layer (ICPI 2004). Bricks can also be used instead of concrete blocks. Aesthetically pleasing pavers used as a walkway are shown in Figure 15.5. The bricks have small tabs along their edges to aid in installation, preventing them from being placed too close together.

Concrete grid pavers (CGP). ASTM C 1319, *Standard Specification for Concrete Grid Paving Units* (2021) describes properties and specifications for concrete grid pavers (CGP). CGP are typically 9.0 cm (3.5 in.) thick with a maximum 60 × 60 cm

Figure 15.4 Permeable Asphalt Maintenance Road near Hangzhou, China (Photos by Authors).

(24 × 24 in.) dimension. The percentage open area ranges from 20% to 50%; the void space is filled with topsoil and grass, sand, or aggregate (Figure 15.6). The minimum average compressive strength of CGP can be no less than 35 MPa (5,000 psi). A typical installation consists of grid pavers with fill media, 2.5 to 3.8 cm (1 to 1.5 in.) of bedding sand, gravel base course, and a compacted soil subgrade (ICPI 2004). Care must be taken in CGP applications because of its propensity to "rock" or differentially settle.

Plastic reinforcement grid (PG) pavers, also called geocells, consist of flexible plastic interlocking units that allow for infiltration through large gaps filled with gravel or topsoil planted with turfgrass. A sand bedding layer and gravel base course are often added to increase infiltration and storage. The empty grids are typically 90–98% open space, so void space is dependent on the fill media (Ferguson 2005). To date, no uniform standards exist; however, one product specification defines the typical load-bearing capacity of empty grids at approximately 13.8 MPa (2000 psi). This value increases up to 38 MPa (5500 psi) when filled with various materials.

Figure 15.5 Permeable Brick Pavers at the Villanova University Campus. Note the Tabs on the Sides of the Bricks (Photos by Authors).

## 15.3 Permeable Pavement Installation

For all permeable pavements, proper installation is critical for effective performance and to ensure long-term life. Permeable pavements, especially PC, require experienced installers. Numerous examples of failure have been noted due to improper installation. Installation of PC and PA are shown in Figures 15.7 and 15.8, respectively.

## 15.4 Designing for Infiltration and Percolation

Permeable pavements can be specifically designed to optimize infiltration and percolation into the surrounding soils. Designers can adjust the following parameters:

- Ratio of run-on drainage area to pavement surface area
- Surface Infiltration (dependent on pavement type and media selection)
- Depth/Volume of storage layer
- Underdrain need/configuration
- Siting based on in situ soil

Figure 15.6 Two Installations of CGP with Various Fill (Soil on Top for Grass Growth, and Sand/Aggregate on Bottom) (Photos by Authors).

### 15.4.1 Surface Infiltration

The type of pavement used has a minor effect on surface infiltration; all pavement types exhibit similar strong performance. Pavers employing a sand or sandy soil fill have lower surface infiltration rates than pavers designed with pea gravel fill, permeable concrete, or permeable asphalt. The difference, however, among pavement types over long-term operation is not substantial (Bean et al. 2007a; Collins et al. 2008). High rainfall intensities of 2.5–5 cm/h (1–2 in./h) may cause runoff from CGP filled with sand and PG filled with sand. Rainfall intensities of 10 cm/h (4 in./h) may cause runoff from the other pavement types (Bean et al. 2007a).

### 15.4.2 Run-on Ratio

Run-on ratio relates the surface area of what drains to the permeable pavement to the surface area of the permeable pavement. Herein, the surface area of the permeable pavement is included in the area of what drains to the pavement. Run-on ratios

Figure 15.7 Installing Permeable Concrete on the Villanova University Campus (Photo by Authors).

Figure 15.8 Installing Permeable Asphalt on the Villanova University Campus (Photo by Authors).

will vary from 1:1 (where no run-on occurs) to 6:1 (an area 500% larger than the permeable pavement surface drains to the permeable pavement surface). To minimize risk of clogging a 1:1 ratio is typically best, but that is often not an economically-wise choice. Surface clogging of permeable pavement has been shown to be rapid once the run-on ratio exceeds 2:1, especially when impervious asphalt drains to permeable pavement. When watersheds are clean (such as concrete pavement or a polycarbonate roof), proportionally larger areas of drainage can be routed toward permeable pavement. The higher the run-on ratio the deeper the volume of the storage layer required.

### 15.4.3 Depth/Volume of Storage Layer

On average, each cm of gravel storage can store 1/3 cm of rainfall/ runoff. So, a 20-cm (8-in.) gravel storage layer can hold up to 6.7 cm (2.7 in.) of water at a given moment, assuming the drainage layer is flat. The gravel storage layer also provides structural support, which is required for most permeable pavement types (with the notable exception of permeable concrete). Provided the underlying (or in situ) soils allow infiltration to occur, deeper storage layers therefore allow more water to infiltrate during large storm events. Gravel thickness is, thus, one of the most important design elements for permeable pavement.

### 15.4.4 Underdrain Need

The need for an underdrain is dependent upon the characteristics of the in situ soil and the design intent. The storage layer must be able to drain within 2–3 days to recover the storage for the next storm event. If the soil infiltration rate is below minimal thresholds, an underdrain is required, and may be desired for construction or maintenance needs. Highly pervious sandy soils do not need an underdrain. Soil testing is key. Estimations of hydraulic conductivity based on infiltration tests or soil particle size provide the most useful information. Percolation or similar tests may vary from actual *in situ* values by factors ranging from 10 to 20, especially if construction compaction protocols are not followed. For example, compaction can change the infiltration rate for a sandy clay loam from 1.3 cm/h (0.5 in./h) to 0.06 to 0.13 cm/h (0.025 to 0.05 in./h), an order of magnitude lower (Bean et al. 2007b).

Even soils with low permeability will exhibit some infiltration due to the extended time available (between storm events) and the increased water head present during the storage. Some evapotranspiration may also occur during inter-event dry periods. Significant reductions in peak flows and discharge volumes have been documented using permeable pavements over clayey soils (Braswell et al. 2018; Fassman and Blackbourn 2010).

### 15.4.5 Underdrain Configuration

Research indicates that if an underdrain is needed, an elevated or upturned underdrain is more effective than a horizontal underdrain in reducing outflow volumes from permeable pavement systems (Figure 15.9) and is conceptually similar to that of a non-underdrained system (Collins et al. 2008; Horst et al. 2011; Wardynski et al. 2013). With an upturned underdrain, water that initially pools internally in the pavement (1) does not immediately drain and (2) can infiltrate the subbase, increasing times to peak, reducing runoff volumes, and lowering peak outflow rates. Similar results can be obtained through outflow restrictions; both options limit outflow rates. This is done using underdrains with a small diameter or by capping the underdrains with a cap that has a restrictive orifice or hole. The effectiveness of either technique depends entirely on the infiltration capacity of the underlying soil. An elevated underdrain is generally accepted to be modestly better for mitigating temperature and reducing nutrients (to be discussed).

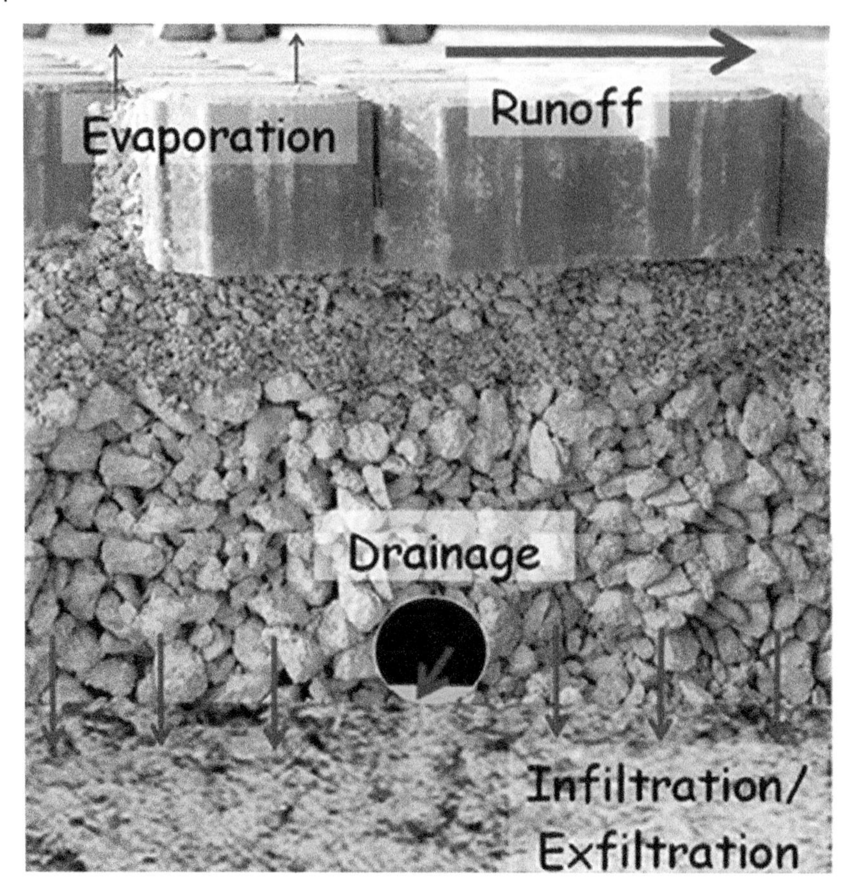


Figure 15.9 An Upturned Underdrain Elbow Creates an Internal Storage Zone for Water. This enhances the Ability of the Pavement to Exfiltrate into in Situ Soils (Credit: Authors' Research Group).

### 15.4.6 In Situ Soils

A developed site may have surprisingly varied underlying soils. Depending on the region, some soils may be borderline impermeable while others will have reasonable permeability. As with the majority of GI SCMs, if the designer is able to identify locations with more permeable underlying soils, the permeable pavements will likely infiltrate a substantially larger volume of water. Some jurisdictions now require soil tests per every 485 $m^2$ (10,000 $ft^2$) of permeable pavement to better understand underlying soils, which facilities more accurate designs (NCDEQ 2020).

## 15.5 Permeable Pavement Hydrologic Design Strategies

The common design procedure for permeable pavements is to size the pavement to capture a specified design event. The following represents aspects of the design.

- 100% capture of the specified event is assumed. The design volume will include all direct rainfall, plus any run-on from surrounding impervious surfaces.

- The gravel layer below the pavement provides the water storage. The depth of this layer is varied so that the design storm is captured without producing outflow.
- Gravel porosity ranges from 0.3 to 0.4; the default recommended value is 0.33.
- Any surface slope should be managed by varying the thickness of the gravel storage base. The bottom of the gravel storage is ideally flat. On sloped terrain, the permeable pavement can be built as terraces (mini-detention basins filled with gravel).
- If surrounding soils necessitate the use of an underdrain, it should be sized to convey the incoming flow. The underdrain may have an upturned elbow, if some infiltration is possible.
- Seasonal high water table (SHWT) should be at least 30 cm (1 ft) and preferably 60 cm (2 ft) from the bottom of the pavement base. If the SHWT is within 30 cm (1 ft) of the permeable pavement base, the permeable pavement should either be located somewhere else on the site where the SHWT is not restrictive or the practice should be eliminated from consideration.
- Run-on from impervious asphalt is ideally kept at a ratio no greater than 1:1 (1 part impervious asphalt to 1 part permeable pavement). Greater ratios lead to rapid surface clogging/ blinding of the permeable pavement.

Alternatively, hydrologic design can be evaluated through either a continuous or design storm simulation using a hydrologic model.

It is common to size the storage space to completely capture a storm event of a specific size, such as 1 in. (25 mm), which assumes no intra-event infiltration. However, it should be noted that infiltration occurring during events may greatly increase the hydrologic performance of the permeable pavement SCM, depending on the surrounding soils. As reviewed in Chapter 7, infiltration rate is a function of temperature (Braga et al. 2007) and will vary seasonally. This is clearly seen in the permeable concrete infiltration data shown in Figure 15.10 (Horst et al. 2011)

The drawdown of permeable pavement storage via an underdrain at the bottom of the storage layer is often controlled by orifice flow. The underdrain system can be set up in a manifold arrangement, with the final discharge controlled by an orifice.

**Example 15.1** A permeable pavement is required to store 5 cm of rainfall falling on perfectly flat pavement. There is run-on from a rooftop which has the exact same surface area as the pavement. Find the required thickness of the gravel storage layer

***Solution*** The gravel storage layer is assumed to have a porosity for water storage equal to 0.33. Therefore, the gravel layer thickness is

$$Thickness = \frac{Rainfall\ Depth}{Gravel\ Porosity} \times Runon\ Ratio = \frac{5\ cm}{0.33} \times 2 = 30\ cm$$

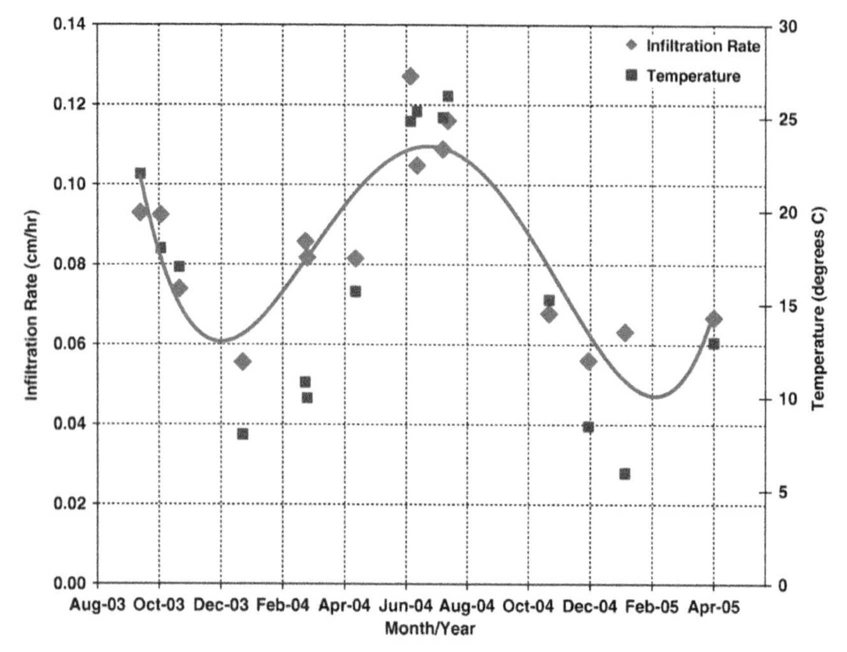


**Figure 15.10** Seasonal Infiltration Rate from a Permeable Concrete Site at Villanova University (Horst et al. 2011, with Permission from ASCE).

**Example 15.2** A permeable pavement covers an area 15 ft by 300 ft. Its rock storage layer is 10 in. deep, with a porosity of 0.4. The storage must be drained from full to a 1-in. depth within 24 h through the underdrain. Find the required orifice diameter

***Solution*** The orifice equation for describing flow with a decreasing head and constant cross-sectional area was presented in Chapter 7. This equation can be rearranged to solve for the orifice area:

$$A_o = \left( h_0^{\frac{1}{2}} - h^{\frac{1}{2}} \right) \left( \frac{2}{g} \right)^{\frac{1}{2}} \frac{A_s \varepsilon}{C_d t}$$

where:

$A_o$ is the orifice area
$h_0$ is the initial water height over the orifice centerline
$h$ is the final water height over the orifice centerline
$g$ is gravity
$A_s$ is the area of the permeable pavement being drained
$\varepsilon$ is the storage zone porosity
$C_d$ is the orifice discharge coefficient (0.6)
$t$ is the drain time

The pavement area is 15 × 300 = 4500 ft$^2$. The 24-h drainage time corresponds to 86,400 s.

$$A_o = \left( \frac{10^{\frac{1}{2}}}{12} - \frac{1^{\frac{1}{2}}}{12} \right) ft^{\frac{1}{2}} \left( \frac{2}{32.2 \frac{ft^2}{s}} \right)^{\frac{1}{2}} \frac{\left(4500\,ft^2\right)0.4}{(0.6)(86400s)}$$

$$= 5.4 \times 10^{-3}\,ft^2 = 0.778\,in^2 = \pi D^2 / 4$$

$$D = 1.0\,in$$

## 15.6 Permeable Pavement Hydrology

Permeable pavements act as a stormwater control measure primarily by reducing the creation of runoff. Would-be runoff is instead stored below the pavement. The stored water is mostly infiltrated into the surrounding soils, although studies show that some evaporation occurs. Three factors can contribute to highly effective hydrologic performance of permeable pavements. (1) From a design perspective, constructing a large storage volume below the pavement can allow for significant rainfall storage. (2) Pervious underlying soils promote exfiltration to the surrounding subsurface, reducing surface runoff. (3) Siting the pavements in areas with low particulate loadings, or following a strict maintenance program, will keep the pavement system from clogging (Bean et al. 2007a). This latter factor is the reason to restrict run-on from asphalt surfaces to permeable pavement at a 1:1 surface area ratio, as mentioned previously.

Generally stated, permeable pavement systems can be modeled as an infiltration site through routing of the stormwater through the SCM (Chapter 7). Inflow rates to the storage/infiltration zone would include rainfall falling directly on the permeable pavement and any run-on from other surfaces (like rooftops, nonpermeable pavements, and surrounding landscapes). Infiltration and potential underdrain flows act as an outfall. Evaporation through the pavements can occur, although it requires specific design criteria (Nemirovsky et al. 2013) and is not expected to be substantial.

### 15.6.1 Hydrographs

Two examples of storm event hydrographs and permeable pavement discharges are shown in Figure 15.11. The study system consisted of a concrete paver section incorporated into a roadway in New Zealand (Fassman and Blackbourn 2011). The surrounding soils were clayey with very low estimated permeability ($10^{-3}$ cm/day). Figure 15.11a shows results from a 63 mm (2.5 in.), 2-year storm event. A delay in the discharge of underdrain flow from the paver system is obvious compared to the control impervious pavement. Peak flows are significantly reduced by the paver system. Even under the 15.23 cm (6 in.), 10-year average return period extreme event shown in Figure 15.11b, the peaks are dampened and the total volume is reduced. Overall, the

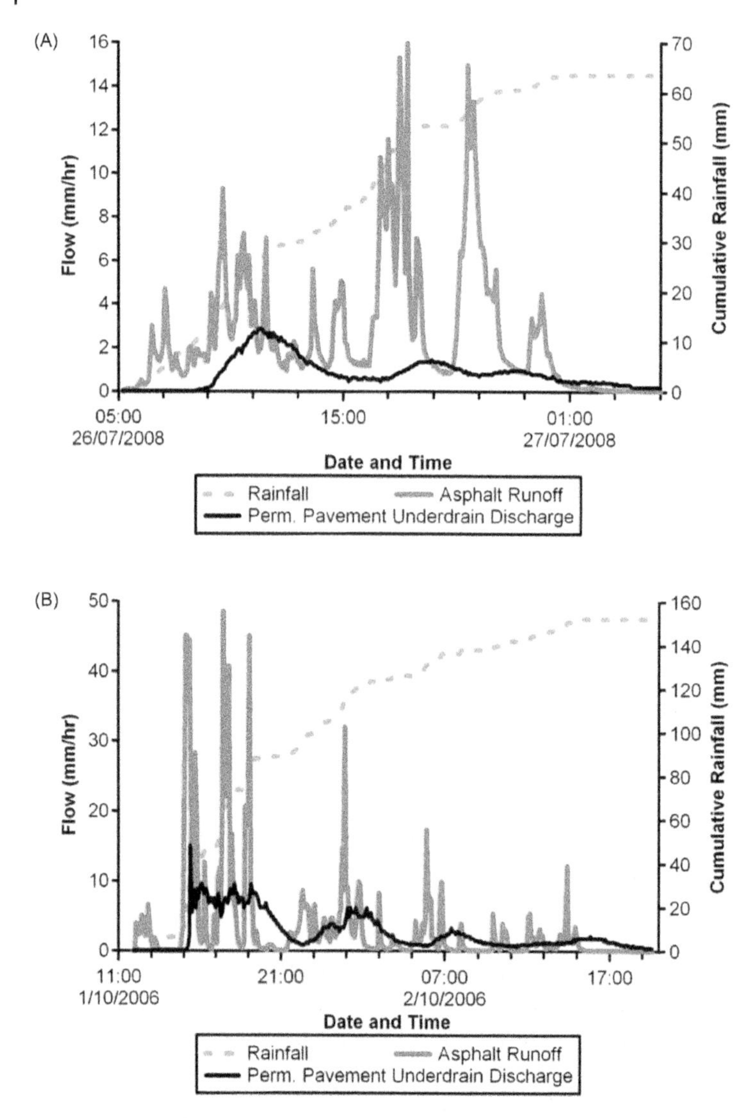


Figure 15.11 Flows from Asphalt (Impervious) and Permeable Pavement from Large Storms in New Zealand. (A) 63 mm Event, (B) 152.3 mm Event. (Fassman and Blackbourn 2010, with Permission from ASCE).

permeable pavement section proved highly successful in reducing runoff volumes and reducing peak flows.

### 15.6.2 Curve Numbers and Storage

Chapter 6 describes the use of curve numbers for determining the depth and volume of runoff generated from a surface cover. Using curve number reductions to represent SCM hydrologic performance can be problematic. The statistical basis for the curve numbers does not relate well to SCMs, and should be considered as a simplification for smaller or less complex sites. Curve numbers were calculated for four permeable pavements in eastern North Carolina by Bean et al. (2007b). The average curve number of different types of permeable pavements was found to range from a low of 45 to a high of 89. (The curve number for standard impermeable pavement is 98.) The reason for the variation in the North Carolina study was due to two factors: base (or storage) depth and underlying soil composition. The less the water storage and the more clayey the underlying soil, the higher the curve number.

Figure 15.12 shows the storage/overflow characteristic response from a permeable pavement in North Carolina. As with most SCMs, an initial storage capacity is noted, followed by discharge, as discussed in Chapter 11. Rainfall less than 35 mm (1.4 in.) is completely stored by the pavement; rainfalls larger than this were discharged off the pavement without any additional removal or storage (Bean et al. 2007b).

### 15.6.3 Evaporation

While permeable pavements typically do not have a substantial amount of evaporation loss, a few pavement types may hold water near the surface long enough for minor evaporation losses to occur. A system that captures and stores water near the surface of the pavement, such as CGP and PG filled with sand, have been estimated to temporarily store at least 0.6 cm (0.24 in.) of most storms and presumably "release" this water to the atmosphere by evaporation (Collins et al. 2008). On an annual basis, up to 33% of all precipitation events could be "captured" and all water returned to the atmosphere in this way by these pavements. Limited evaporation has been found for permeable concrete when the water level is close to the surface (Nemirovsky et al. 2013).

## 15.7 Water Quality Design

Permeable pavements frequently show improvement to stormwater runoff quality, but not always. Limited mechanisms exist for pollutant removals in permeable pavements. Filtration will occur as the water moves down through the pavement or the sand/

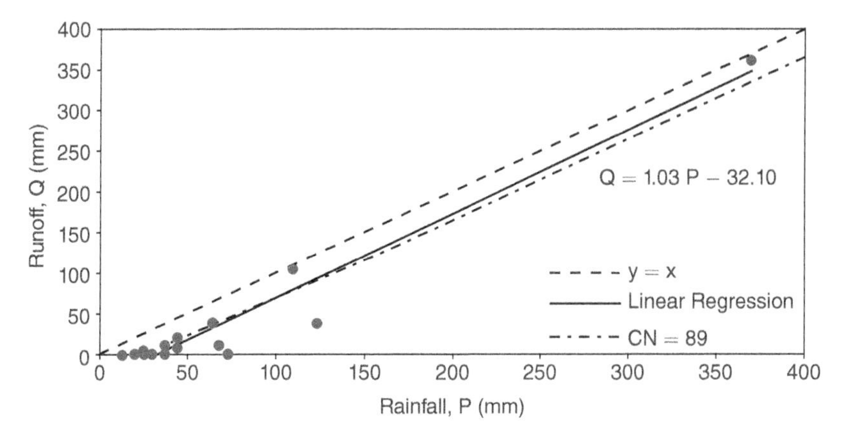


Figure 15.12 Runoff Response as a Function of Rainfall from Concrete Grid Pavements in North Carolina (Bean et al. 2007b, with Permission from ASCE).

gravel interlayers allowing removal of particulate matter (Welker et al. 2013). The extent of the filtration will depend on the fill media used. Minimal removal of dissolved pollutants occurs.

### 15.7.1 Particulate Matter

Sediments and other larger particulate material will be filtered by the pavement or the sand and gravel between pavers. Lower TSS concentrations have been measured in concrete paver underdrainage (39 mg/L) compared to nearby traditional asphalt pavement (84 mg/L; Fassman and Blackbourn 2011). This will reduce sediment concentrations, but can also lead to clogging in the permeable pavement systems. Vacuum sweeping of PA and PC removed particulate matter at a rate of 64 to 500 g/m$^2$ (Welker et al. 2012). Most of this material resulted from attrition of the pavements themselves, with $d_{50}$ values ranging from about 0.9 to 4.3 mm. If sediment or other particulate loadings become excessive (such as after several years of neglect), removing the schmützdecke (clogged surface) can become very problematic (Winston et al. 2016a).

Winston et al. (2016b) noted excess discharge of suspended solids from a permeable pavement system in Ohio (USA). The release of TSS correlated with winter season and the use of salts for roadway deicing. TSS discharge continued for several months after the winter season. The TSS release was attributed to the mobilization of fine materials from the bottom soil layer due to disaggregation and breakdown of soil structure resulting from the excess sodium present.

### 15.7.2 Metals

Copper, and especially zinc, concentrations in the drainage from concrete pavers (total Cu = 5.7 μg/L; total Zn = 15 μg/L) were lower than those from asphalt pavement (total Cu = 14.7 μg/L; total Zn = 250 μg/L; Fassman and Blackbourn 2011). Reductions in dissolved Cu and Zn were noted also. Most heavy metals are captured in the top layers (2.5 to 5 cm) of material in permeable pavement void space (Dierkes et al. 2002; Welker et al. 2012). Design features can improve permeable pavement performance; e.g., adding fly ash to permeable concrete can increase the removal of dissolved heavy metals (Pb, Zn,Cd) via adsorption onto the fly ash (Holmes et al. 2017).

Metals levels in vacuumed particulate matter from permeable pavements averaged 65 mg/kg for Cu, 6.0 mg/kg for Pb, and 48 mg/kg for Zn, indicating considerable accumulation of these metals, with the conclusion that significant removal of runoff metals occurs (Welker et al. 2012). For PICP, CGP, and PG that are filled with sand, this implies that standard street sweeping will remove the majority of heavy metals collected in the pavement fill material, provided this street sweeping is conducted with sufficient frequency and the proper street sweeping type is used (Winston et al. 2016a). Exact recommendations for disposal of collected material have yet to be made.

### 15.7.3 Nutrients

The fate of nitrogen in permeable pavements depends on the hydraulic design. Particulate N may be captured via filtration (Pagotto et al. 2000). Limited capture of dissolved N species has been documented, but the mechanism of removal is not clear.

The removal may be related to adsorption onto captured particulate matter or to the pavement itself.

Several studies have suggested that aerobic conditions, which result as permeable pavements drain, can result in nitrification of ammonium ($NH_4$-N) to nitrate ($NO_3$-N). This can lead to *increased* levels of nitrate discharging from the permeable pavement system for underdrained systems. Compared to conventional asphalt, substantially lower $NH_4$-N and total Kjeldhal nitrogen (TKN) concentrations, and higher $NO_3$-N concentrations in permeable pavement drainage have been measured in multiple studies (Bean et al. 2007b, Collins et al. 2010; Horst et al. 2011). It also appears that CGP and PG filled with sand have greater ability to reduce TN compared to other permeable pavement designs. This occurs because CGP filled with sand very much resembles a low head (ponded water), limited media depth sand filter. Sand filters have repeatedly been shown to reduce TN concentrations (Barrett 2003). Employing pollutant removal processes that reduce nitrogen may not be possible within many permeable pavements, which suggests the need to use a treatment train approach to water quality improvement (see Chapter 21). Incorporating an internal water storage zone at the bottom of permeable pavement (even if slight, e.g., 15 cm) can be a design element to enhance denitrification (Braswell et al. 2018). Care must be taken, however, to avoid storing water above plastic soils.

Significant removal of Total P through permeable pavement has been documented (Bean et al. 2007). Particulate P can be filtered through the pavement/pavement filler. Phosphate may adsorb onto the pavement materials or onto captured particulate matter, or may be precipitated with concrete at high pH and calcium concentrations (Ramsey et al. 2018). If phosphorus removal is an important design target, soil tests of underlying soil should be conducted to verify that the present soils are not high in phosphorus (as measured, for example, using the P-Index). It has been observed that locating permeable pavement over soils laden with P resulted in flushing of phosphorus (Collins 2007).

### 15.7.4 Hydrocarbons

A few studies have indicated that hydrocarbons are captured in permeable pavement systems (Pagotto et al. 2000; Scholz and Grabowiecki 2007). Some of this capture should also result from capture of hydrocarbons affiliated with particulate matter. Captured hydrocarbons can support a wide range of microorganisms that then degrade these hydrocarbons, indicating that the pavements can behave as an aerobic biodegradation chamber (Coupe et al. 2003).

### 15.7.5 pH

Concrete-based permeable pavements can buffer acidic rainfall pH (Dierkes et al. 2002; Collins et al. 2010; Kwiatkowski et al. 2007), likely due to the presence of calcium carbonate and magnesium carbonate in the pavement and aggregate materials. These pavements provide a greater pH buffering capacity than asphalt, due to the greater surface area provided by contours in the pavement geometry and the additional coarse aggregate layer through which water migrates. Of all pavement types, PC offered the most buffering capacity because it provided influent water the greatest contact time with cementitious materials (Collins et al. 2010).

### 15.7.6 Thermal Pollution (Temperature)

Permeable pavements can reduce thermal pollution (Karasawa et al. 2006; Wardynski et al. 2013) relative to standard pavements. Relative to conventional asphalt, the measured decrease was 10 to 25°F (6.7 to 13.9°C), in great part due to the light color of the pavement compared to the dark asphalt. Barbis and Welker (2010) evaluated pavement and runoff temperatures for 12 storm events over permeable asphalt and concrete. Average summer temperatures were reduced by about 2°C by allowing flow into the subsurface rock bed storage. Maximum temperature reductions approached 5°C. Negligible differences were found in runoff temperatures from permeable asphalt compared to permeable concrete. A design feature that might enhance thermal load reduction is internal water storage. Wardynski et al. (2013) demonstrated that permeable pavement's greatest thermal load reduction factor was infiltration.

### 15.7.7 Pollutant Loads

Because most permeable pavements substantially reduce the volume of runoff/outflow, they will also reduce pollutant loads. Several studies confirm that permeable pavements demonstrate lower total pollution discharge loadings than standard pavements (Bean et al. 2007b; Fassman and Blackbourn 2010; Horst et al. 2011; Wardynski et al. 2013). However, some of this lower load may have resulted due to their siting in less traveled areas, and the higher frequency of vacuuming.

Table 15.1 presents research results describing removal of sediment, metals, and nutrients by a permeable concrete pavement, and the impact on pH. Generally stated, the primary removal mechanisms are the filtration of stormwater by the surface layer, and infiltration. Importantly, the water discharge was only 3.2 cm (1.25 in.) out of an inflow of 45.06 cm (17.74 in.), demonstrating how important water storage is in permeable pavement's gravel layer. This alone reduces runoff and pollutant loadings by 92.9%. Some additional removal of N and P were found. The most significant removals were found with suspended solids and Cu (which is expected to be affiliated with TSS). Note the large episodic effect of snow melt chemicals, manifested by high discharge loadings of chloride and dissolved solids. Permeable pavement appears to be susceptible to pollutant discharge when located in heavily salted watersheds (Horst et al. 2011;

**Table 15.1** Water Quality Performance (based on mass loads) at the Villanova University Permeable Concrete Site (Horst et al. 2011, with permission from ASCE).

| Constituent | Inflow | Outflow | Removal Efficiency |
|---|---|---|---|
| Water quantity (cm) | 45.06 | 3.20 | 92.9 |
| pH | 4.17 < pH < 8.42 | 6.65 < pH < 9.75 | |
| Conductivity (μS/cm) | 2.96 < Cond < 89.2 | 9.0 < Cond < 2860 | |
| Copper (kg) | 1.58 | 0.02 | 98.5 |
| Total nitrogen (kg) | 2.21 | 0.10 | 95.3 |
| Total phosphorous (kg) | 0.44 | 0.03 | 94.3 |
| Chloride (kg) | 1.28 | 11.59 | −806.8 |
| Suspended solids (kg) | 30.29 | 0.17 | 99.4 |
| Dissolved solids (kg) | 24.24 | 27.61 | −13.9 |

Winston et al. 2016b). Designers will need to recognize that and perhaps locate snow disposal areas in locations that do not drain to permeable pavement.

Vadose zone water samples underneath the permeable concrete storage zone and outside the PC bed in a lawn area were compared. Conductivity, copper, Total N, Total P, and chloride were higher in the grass area than underneath the PC storage zone suggesting that either these pollutants were not transferred to subsurface water via the PC (Horst et al. 2011) or that the application of salt throughout the upslope landscape was so high that the permeable pavement actually was diluting the groundwater (Dietz et al. 2016).

### 15.7.8 Long-term Pollutant Fate

Pollutants that are removed from stormwater by the permeable pavement system will either remain in the upper (surface) layer of the pavements or in the subsurface pavement storage space. A major concern regarding long-term pollutant control is the threat of these pollutants leaching and causing groundwater pollution. Long-term studies and simulations of permeable pavement pollutant distributions have revealed low risks of subsoil pollutant accumulation and groundwater contamination (Dierkes et al. 2002; Horst et al. 2011; Kwiatkowski et al. 2007). Pollutants are likely to migrate as far as 0.6 m (2 feet) from the gravel-underlying soil interface. Thus, it is important that seasonally high water tables do not encroach the interface of the pavement base and the subbase. This is the reason for a 0.3 m–0.6 m SWHT separation from the bottom of the permeable pavement cut, as a high water table would cause soil saturation, then leach disassociated pollutants from the soil into the groundwater.

**Example 15.3** A designer would like to use permeable pavement to treat runoff from a 2,000 sf roof surface. She has allocated a 1,000 sf surface for permeable pavement. The seasonally high water table is 30 in. from the surface. She must capture the water quality storm of 1.5 in. Does she have enough vertical capacity for the pavement so that water quality concerns are addressed?

***Solution*** The run-on ratio is 2:1, so that for every inch of rainfall on the site, three inches of storage are needed. Assuming the pavement area is perfectly flat, the depth of the gravel storage layer is:

$$Gravel\ Thickness = \frac{Rainfall\ Depth}{Gravel\ Porosity} \times Runon\ Ratio = \frac{1.5\ in.}{0.33} \times 3 = 13.5\ in.$$

If PICP is used, a typical thickness is 3.25 in. This is added to the gravel thickness to produce a total depth of 16.75 in. The difference between the SHWT depth and the required pavement system thickness is:

$$Separation = SHWT\ Depth - Depth\ Pavement\ System = 30\ in. - 16.75\ in. = 23\ in.$$

Thus, this is a borderline case. An entire 2 feet of separation is just barely not available, but it is very close. The decision is, ultimately, left to the plan reviewer; however, the authors would – all things being equal – recommend installing permeable pavement here, provided great care is maintained in all other phases of the design and construction processes.

## 15.8 Maintenance

Because the filtration of the water takes place right at the surface of the pavement, build-up of collected materials will result in reduction of the infiltration rate through the pavement layer (Al-Rubaei et al. 2013). This necessitates a maintenance step for the removal of the accumulated materials. Vacuum, most likely, or standard street sweeping is required to break apart a schmützdecke and suck up the captured particles. Permeable pavements do not magically clog. There are 4 principal causes for clogging once a landscape is established: (1) run-on and leaf-fall from permeable landscapes, (2) run-on from impermeable asphalt and shingled roofs, (3) passage/use of dirty vehicles (such as garbage trucks), and (4) snow/ice disposal. A cost of sweeping, however, is the removal of sand or gravel in between PICP pavers. A typical recommendation is to sweep permeable pavements at least every six months depending on usage. Ideally, though, simple checks (such as the simple infiltration test, Winston et al. 2016c, Figure 15.13) will be conducted in areas susceptible to clogging.

In locations that are hard to reach via a street sweeper or a vacuum truck, permeable pavements have been shown to be unclogged using pressure washers and shop vacuums (Dougherty et al. 2011; Winston et al. 2016b).

## 15.9 Design Summary

A summary of permeable pavement design guidance is found in Table 15.2.

## 15.10 Permeable Pavement Cost Factors

Cost Factors unique to permeable pavement include the following:

1) The type of pavement used. Permeable pavement types do not cost the same. In general the permeable version of a pavement (e.g., permeable concrete as

Figure 15.13 The Simple Infiltration Test (Described in Detail by Winston et al. 2016c) Is Used to Determine Clogging. A Frame Is Made from 8 ft 2 × 4 Lumber. Plumbers Putty Is Placed along the Bottom Edge, the Frame Is Pressed onto the Pavement Surface, and 18-L (5-gal) of Water Is Dumped onto the Surface. If the Contents Dewater within 1.5 min, the Pavement Is Deemed to Work Sufficiently. Otherwise, that Section of the Permeable Pavement Is Subject to Renewal.

Table 15.2 Permeable Pavement Design Guidance Summary.

| Design Parameter | Guidance | Rationale |
|---|---|---|
| Optimal Pavement Types for Runoff Reduction | All are excellent. PC, PA, PICP, and CGP filled with gravel are best. | Research shows that all types of pavement types reduce runoff substantially. CGP and PG filled with sand have slightly higher runoff rates. |
| *Design* Surface Infiltration Rate | 2.5 to 7.5 cm/h (1 to 3 in./h) | Studies show that 90% of all study sites had at least 2.5 cm/h (1 in./h) surface infiltration rates, with 5 to 7.5 cm/h (2 to 3 in./h) being a median range for partially clogged permeable pavement. |
| Design Base Exfiltration Rate | 0.025 to 2.5 cm/h (0.01 to 1 in./h) 0.25 cm/h (0.1 in./h) is default for loamy sand. | Even in somewhat sandy soils, exfiltration rates from the base were impacted by compaction that occurred during construction. It is recommended that the $k_{sat}$ of the underlying soil be divided by 10 to provide an approximate exfiltration rate. |
| Slope | Flat ideal, or terraced | Flat applications of permeable pavement will utilize all of its storage capacity. Systems installed on a slope can be terraced, or effectively, little detention basins filled with gravel. |
| Underdrain Flow Rate | Release water so that the design event is emptied in 2–7 day depending on regional weather patterns. | Include underdrain IF infiltration is not sufficient to empty the site within the required time period. Allows a mimicking of predevelopment hydrology stream recharge post event. Mitigates peak flow. |
| Optimal Pavement Types for Metal Removal | All are excellent. CGP and PG filled with sand are easiest to maintain. | It is easier for street sweepers to remove the *schmützdecke* (or clogged layer) from the top of the sand column associated with CGP and PG. |
| Optimal Pavement Types for Nutrient Removal | CGP and PG filled with sand. | These pavements act as if they are low head, shallow depth sand filters. More research is needed to confirm this interim finding. |
| Optimal Pavement Design for Thermal Load Mitigation | Infiltration best. Multiday retention discharged through internal water storage next best. | Any water that infiltrates will approach subsurface ground temperature. Retaining discharge water for sufficient periods allows the cooler deeper depths of pavement to act as a heat exchanger. |
| Run-on Ratio | 1:1 Ideal, but 5:1 possible under certain conditions | The more run-on, the more likely a pavement is to clog. Impermeable asphalt, shingled roofs, and landscapes are most likely to cause clogging, so their run-on ratio would be limited to 1-part run-on surface to 1-part permeable pavement. A clean concrete pavement, however, could have up to a 5:1 run-on ratio and still function. |
| Seasonally High Water Table (SHWT) | 30 cm (1 ft), preferably 60 cm (2 ft) from the bottom of the pavement base | SHWT closer to the base will (1) impede exfiltration from the pavement and (2) lead to pollutant leaching from the pavement. |

compared to standard concrete) is approximately 15% more expensive to the standard, impermeable version.

2) The depth of the gravel bedding layer. A deeper gravel base costs more, but does allow for larger storm capture from the system, which can also support larger loads.
3) The gravel bedding layer needs to be washed. This increases the cost over a standard crusher run, which contains fines.
4) Slope of the lot. A flat lot is easier to construct as a permeable pavement. Pavement applications built on a slope will need internal berms to create underground ponding zones.
5) Other costs not unique to permeable pavements include surface area of the practice, impact of utility lines, and the ability to keep the permeable pavement free of off-site fine sediment.
6) Annual maintenance costs.

## 15.11 Permeable Friction Course

Permeable friction course (PFC) is an overlay of permeable asphalt on a layer of standard asphalt. The PFC is about 5 cm (2 in.) thick and has about 18–22% void space (Barrett and Shaw 2007). PFC was first employed in Texas to address safety issues related to heavy road spray during large storms. Later research indicated that runoff from PFC has much better quality than runoff from traditional asphalt pavements. Pollutants that are commonly affiliated with particulate matter, such as metals, were also much lower (Barrett 2008; Barrett and Shaw 2007).

The mechanism of treatment by PFC appears to be filtration of particulate matter by the pervious asphalt top layer, capturing particles. Dissolved pollutant concentrations are not significantly affected. This leads to the conclusion that, in order to maintain its effectiveness, PFC must be cleaned to removal accumulated particulate matter so that it will stay effective as a SCM. However, a long-term study in NC and TX demonstrated a minimal degradation of performance over the course of one decade (Eck et al. 2012).

Flows through PFC can be described as one-dimensional, steady-state Darcy-type flow. The flow fills the PFC layer, with the depth depending on the rainfall intensity. Possibly because of the decreased splashing/spray that transports some rainwater from the pavement, runoff volumes from PFC are somewhat higher than equivalent size conventional pavements (Pagotto et al. 2000).

The time of concentration through the PFC can be estimated by

$$T_c = h_{max} n / r \tag{15.2}$$

where $h_{max}$ is the maximum drainage depth in the PFC, n is the pavement porosity, and r is the PFC infiltration rate (Charbeneau and Barrett 2008). The time of concentration can be used as an estimate of hydraulic retention time in the PFC to address water quality concerns.

## References


Al-Rubaei, A., Stenglein, A., Viklander, M., and Blecken, G. (2013). Long-term hydraulic performance of porous asphalt pavements in Northern Sweden. *Journal of Irrigation and Drainage Engineering* 139 (6): 499–505.

ASTM Standard C1319 (2021). Standard Specification for Concrete Grid Paver Units. ASTM International, West Conshohocken, PA, 2015, doi: 10.1520/C1319-21, www.astm.org.

Barbis, J. and Welker, A.L. (2010). Stormwater temperature mitigation beneath porous pavements. World Environmental and Water Resources Congress 2010, 3971–3979. doi:10.1061/41114(371)404.

Barrett, M.E. (2003). Performance, cost, and maintenance requirements of Austin Sand filters. *Journal of Water Resources Planning and Management* 129 (3): 234–242.

Barrett, M.E. (2008). Effects of a permeable friction course on highway runoff. *Journal of Irrigation and Drainage Engineering-Asce* 134 (5): 646–651.

Barrett, M.E. and Shaw, C.B. (2007). Benefits of porous asphalt overlay on storm water quality. *Transportation Research Record* 2025: 127–134.

Bean, E.Z., Hunt, W.F., and Bidelspach, D.A. (2007a). Field survey of permeable pavement surface infiltration rates. *Journal of Irrigation and Drainage Engineering* 133 (3): 247–255.

Bean, E.Z., Hunt, W.F., and Bidelspach, D.A. (2007b). Evaluation of four permeable pavement sites in eastern North Carolina for runoff reduction and water quality impacts. *Journal of Irrigation and Drainage Engineering* 133 (6): 583–592.

Braga, A., Horst, M., and Traver, R. G. (2007). Temperature Effects on the Infiltration Rate through an Infiltration Basin BMP. *Journal of Irrigation and Drainage Engineering* 133 (6): 593–601.

Braswell, A.S., Winston, R.J., and Hunt, W.F. (2018). Hydrologic and water quality performance of permeable pavement with internal water storage over a clay soil in Durham, North Carolina. *Journal of Environmental Management* 224: 277–287.

Charbeneau, R.J. and Barrett, M.E. (2008). Drainage hydraulics of permeable friction courses. *Water Resources Research* 44 (4): W04417.

Collins, K.A. (2007). *A field evaluation of four types of permeable pavement with respect to water quality improvement and flood control.* M.S. Thesis. North Carolina State University, Raleigh, NC, USA.

Collins, K.A., Hunt, W.F., and Hathaway, J.M. (2008). Hydrologic comparison of four types of permeable pavement and standard asphalt in Eastern North Carolina. *Journal of Hydrologic Engineering* 13 (12): 1146–1157.

Collins, K. A., Hunt, W. F., and Hathaway, J. M. (2010). Side-by-Side Comparison of Nitrogen Species Removal for Four Types of Permeable Pavement and Standard Asphalt in Eastern North Carolina. *Journal of Hydrologic Engineering, American Society of Civil Engineers*, 15 (6), 512–521.

Coupe, S.J., Smith, H.G., Newman, A.P., and Puehmeier, T. (2003). Biodegradation and microbial diversity within permeable pavements. *European Journal of Protistology* 39 (4): 495–498.

Dierkes, C., Kuhlmann, L., Kandasamy, J., and Angelis, G. (2002). Pollution retention capability and maintenance of permeable pavements. Proc. 9$^{th}$ Int. Conf. on Urban Drainage, Global Solutions for Urban Drainage. ASCE, Portland, Oregon, USA.

Dietz, M.E., Angel, D.R., Robbins, G.A., and McNaboe, L.A. (2016). Permeable asphalt: A new tool to reduce road salt contamination of groundwater in urban areas. *Groundwater*. doi:10.1111/gwat.12454.

Dougherty, M., Hein, M., Martina, B.A., and Ferguson, B.K. (2011). Quick surface infiltration test to assess maintenance needs on a small pervious concrete sites. *Journal of Irrigation and Drainage Engineering* 132 (8): 553–563.

Eck, B.J., Winston, R.J., Hunt, W.F., and Barrett, M.E. (2012). Water quality of drainage from permeable friction course. *Journal of Environmental Engineering* 138 (2): 174–181.

Fassman, E.A. and Blackbourn, S. (2010). Urban runoff mitigation by a permeable pavement system over impermeable soils. *Journal of Hydrologic Engineering* 15 (6): 475–485.

Fassman, E.A. and Blackbourn, S.D. (2011). Road runoff water-quality mitigation by permeable modular concrete pavers. *Journal of Irrigation and Drainage Engineering-Asce* 137 (11): 720–729.

Ferguson, B.K. (2005). *Porous Pavements*. Boca Raton, Florida: CRC Press.

Holmes, R.R., Hart, M.L., and Kevern, J.T. (2017). Enhancing the ability of pervious concrete to remove heavy metals from Stormwater. *Journal of Sustainable Water in the Built Environment* 3 (2): 04017004.

Horst, M., Welker, A., and Traver, R. (2011). Multiyear performance of a pervious concrete infiltration basin BMP. *Journal of Irrigation and Drainage Engineering* 137 (6): 352–358.

Interlocking Concrete Pavement Institute (ICPI). (2004). Tech Spec 8, *Concrete Grid Pavements*. Washington, DC.

Karasawa, A., Toriiminami, K., Ezumi, N., and Kamaya, K. (2006). Evaluation of performance of water-retentive concrete block pavements. Proc. 8th Int. Conf. on Concrete Block Paving, *Sustainable Paving for Our Future*. ICPI, San Francisco, CA, USA.

Kwiatkowski, M., Welker, A.L., Traver, R.G., Vanacore, M., and Ladd, T. (2007). Evaluationi of an infiltration best management practice utilizing pervious concrete. *Journal of the American Water Resources Association* 43 (5): 1–15.

Nemirovsky, E., Welker, A., and Lee, R. (2013). Quantifying evaporation from pervious concrete systems: Methodology and hydrologic perspective. *Journal of Irrigation and Drainage Engineering* 139 (4): 271–277.

NCDEQ (2020) North Carolina Department of Environmental Quality Stormwater Design Manual, C-5 Permeable Pavement. https://deq.nc.gov/media/17539/download (accessed December 2021).

Pagotto, C., Legret, M., and Le Cloirec, P. (2000). Comparison of the hydraulic behaviour and the quality of highway runoff water according to the type of pavement. *Water Research* 34 (18): 4446–4454.

Ramsey, A.J., Hart, M.L., and Kevern, J.T. (2018). Nutrient removal rates of permeable reactive concrete. *Journal of Sustainable Water in the Built Environment* 4 (2): 04018004.

Scholz, M. and Grabowiecki, P. (2007). Review of permeable pavement systems. *Building and Environment* 42 (11): 3830–3836.

Wardynski, B.J., Winston, R.J., and Hunt, W.F. (2013). Internal water storage enhances exfiltration and thermal load reduction from permeable pavement in the North Carolina mountains. *Journal of Environmental Engineering* 139 (2): 187–195.

Welker, A., Barbis, J., and Jeffers, P. (2012). A side-by-side comparison of Pervious Concrete and Porous Asphalt. *Journal of the American Water Resources Association* 148 (4): 809–819.

Welker, A., McCarthy, L., Gilbert Jenkins, J.K., and Nemirovsky, E. (2013). Examination of the material found in the pore spaces of two permeable pavements. *Journal of Irrigation and Drainage Engineering* 139 (4): 278–284.

Winston, R.J., Al-Rubaei, A.M., Blecken, G.T., Viklander, M., and Hunt, W.F. (2016a). Maintenance measures for preservation and recovery of permeable pavement surface infiltration rate – The effects of street sweeping, vacuum cleaning, high pressure washing, and milling. *Journal of Environmental Management* 169: 132–144.

Winston, R.J., Al-Rubaei, A.M., Blecken, G.T., Viklander, M., and Hunt, W.F. (2016c). "A simple infiltration test for determination of permeable pavement maintenance needs. *Journal of Environmental Engineering* 142 (10): 06016005.

Winston, R.J., Davidson-Bennett, K.M., Buccier, K.M., and Hunt, W.F. (2016b). Seasonal variability in stormwater quality treatment of permeable pavements situated over heavy clay and in a cold climate. *Water, Air, and Soil Pollution* 227 (5): 140.

## Problems

**15.1** Find the required thickness of the gravel storage layer beneath a permeable pavement to store a 1.2 in. rainfall. Use a porosity of 0.35.

**15.2** Find the required thickness of the gravel storage layer beneath a permeable pavement to store a 1.8 in. rainfall. Use a porosity of 0.32.

**15.3** Compare the required thickness of a permeable pavement gravel storage layer (porosity = 0.33) to a manufactured material layer (porosity = 0.85) to store a 1.2 in. rainfall.

**15.4** A permeable pavement covers 3 acres. The gravel storage layer is 4 in. thick, with a porosity of 0.35. Find the time for the system to drain through an underdrain orifice with a 1.25 in. diameter.

**15.5** A permeable pavement covers 2 ha. The gravel storage layer is 15 cm thick, with a porosity of 0.35. Find the time for the system to drain through an underdrain orifice with a 3 cm diameter.

**15.6** Find the runoff volume from a 3 acre parking lot, 50% permeable pavement (CN = 74) from a 1 in. storm and compare to a 100% impervious lot.

**15.7** Find the runoff volume from a 2 ha parking lot, 50% permeable pavement (CN = 72) from a 2.5 cm storm and compare to a 100% impervious lot.

**15.8** Create a volume in/volume out plot similar to Figure 15.12 for a permeable pavement that covers 3 acres. The gravel storage layer is 4 in. thick, with a porosity of 0.35. The underlying soils will infiltrate 20% of the stored water.

**15.9** Create a volume in/volume out plot similar to Figure 15.12 for a permeable pavement that covers 2 ha. The gravel storage layer is 15 cm thick, with a porosity of 0.35. The underlying soils will infiltrate 15% of the stored water.

# 16

# Infiltration Trenches and Infiltration Basins

## 16.1 Introduction

Infiltration trenches and infiltration basins are specifically designed and constructed as surface or subsurface areas of storage, with a permeable bottom and sides to promote infiltration of the stored water into the surrounding soils (Figure 16.1). Trenches are linear and smaller to fit in the urban environment, for example, under a sidewalk or along a road as in Figure 16.1c. Basins are typically larger and can be thought of as a wider infiltration trench. Trenches and basins can be differentiated by the following: the depth of a trench is of greater distance than its width. For a basin, both length and width are greater than that of the depth.

What makes them attractive is that the larger volumes can also double as detention, to help reduce peak flows, especially in combined sewer areas. Surface systems combine storage with infiltration in areas that have sufficiently permeable soils. Subsurface systems are usually deep, with 1.2–2.4 m (4–8 ft) being typical, and filled with gravel and/or rock (Figure 16.1b). In some cases, large perforated pipe or modular crate systems are used instead of gravel, allowing for a smaller bed design due to larger void space (Figure 16.2). Several hybrid systems exist, such as a system that utilizes pervious pavements as one of the entry paths, with either rooftop or other sources of runoff. Typically, infiltration systems are built with an outlet structure, and a good design would incorporate a cleanout for maintenance or possible use as an underdrain if needed. Pretreatment for these systems is critical due to the higher runoff loading and concerns for clogging. The best designs will include these SCMs as part of a treatment train (e.g., using permeable pavement to serve as pretreatment for a subsurface filtration system) to reduce exposure to clogging sediments.

## 16.2 Types of Basins

Infiltration trench and basin SCMs can be designed in several configurations. Infiltration trenches are generally more linear in nature to fit the urban environment (Figure 16.1c) and are simply storage areas that are built into the ground. These areas are filled with rock or stone for structural support. Water is directed into the rock beds for storage of a few hours and infiltration to surrounding soils. The sides and

*Green Stormwater Infrastructure Fundamentals and Design*, First Edition. Allen P. Davis, William F. Hunt, and Robert G. Traver.

Companion Website: www.wiley.com/go/davis/greenstormwater

Figure 16.1 Infiltration Basins and Trenches. (A) Surface Infiltration Basin in Wilmington, NC (Photo by Authors); (B) Subsurface Infiltration Basin in Philadelphia PA (Credit: Philadelphia Water Department); and (C) Infiltration Trench in Wilmington, NC (Photo by Authors).

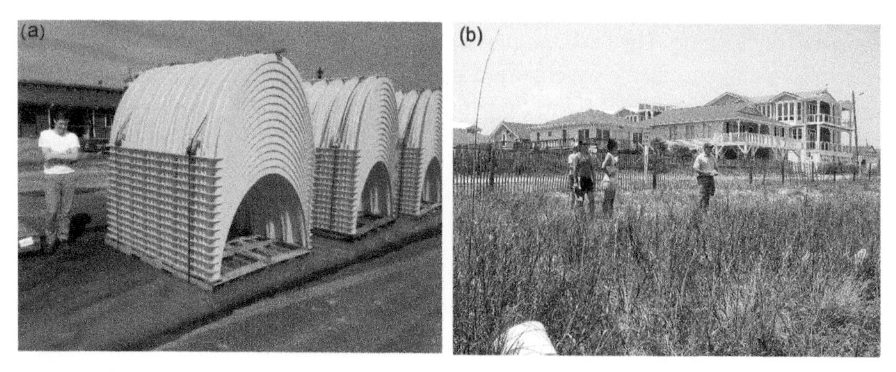

Figure 16.2 Subsurface Infiltration. (a) Plastic Chambers to Be Placed Underground as an Infiltration Basin in Tryon, NC; and (b) Clean Out Portals for a Dune Infiltration System in Kure Beach, NC. (Credit: Authors' Research Group).

bottoms of the trench are usually covered with a permeable geotextile to prevent soil from back-filling up the void space and then filled with clean aggregate. In some areas, they can be combined as part of a tree trench system (Figure 16.3).

Infiltration basins appear similar to detention ponds or bioretention (Figure 16.1a) but are located above permeable soils and thus do not retain runoff. Generally, they are used where surface land is available and where soils are permeable enough to allow infiltration of stored water within an acceptable length of time (usually within 72 h, but some jurisdictions allow up to 120 h for a water-quality event to completely infiltrate).

Subsurface infiltration basins are similar to trenches but have a much larger storage capacity to provide detention benefits mimicking detention basins (Figures 16.2 and 16.4). A number of proprietary systems are also available that are constructed subsurface for storage and infiltration. Most are made of a strong heavy plastic and can be placed under parking areas.

Pretreatment is critical to the operation of these SCMs, particularly if the infiltration component occurs subsurface. Pollutant loads carried by proportionally large volumes of runoff can pass through these devices, so pretreatment needs to account for this (Figure 16.4a). Studies have shown that lack of pretreatment has caused sealing (Emerson et al. 2010) of the bottom, while other sites with clear runoff have operated for potentially multiple decades (Horst et al. 2011; Price et al. 2013). Site and maintenance considerations are critical in deciding to employ these SCMs.

## 16.3 Mechanisms of Treatment

The primary mechanism in infiltration systems is, as its name implies, infiltration of stored stormwater into the surrounding soils. Successful infiltration will lead to groundwater recharge and reduction of surface runoff. The use of these systems for stormwater management is limited to areas where the local soils will allow some reasonable degree of infiltration. A typical infiltration threshold is 13.2 mm/h (0.52 in/h) or greater. This will include soils of hydrologic soil group A (sand, loamy sand, sandy

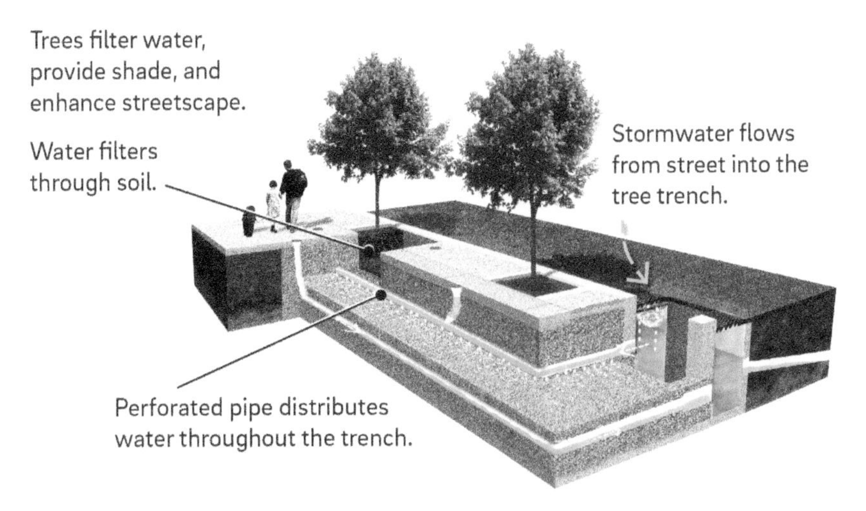


Figure 16.3 Stormwater Infiltration Tree Trench. (Credit: Philadelphia Water Department).

Figure 16.4 Subsurface Infiltration System. (a) Large Underground Detention Chamber Provides Pre-Treatment When Used in Series with (b) Infiltration Gallery in Raleigh, NC (with Permission, Patrick Smith, P.E., Soil & Environmental Consultants, PA).

loam) and loam, which is group B. However, designers can employ infiltration basins under tighter soils if (1) the SCM footprint is sufficiently large and (2) if the storage area is sufficiently shallow. The infiltration rate of the soils must be confirmed at the site through approved testing procedures. Infiltration basins and trenches may be prohibited in karst areas or when high water tables are present.

A second hydrologic mechanism is detention. Similar to surface basins, when designed to meet these goals, they reduce peak flows or delay flows until combined sewers regain capacity. They can be designed utilizing routing as discussed in Chapter 7, modeling the multiple outflow mechanisms.

Infiltration systems are not specifically designed for water-quality improvement but do utilize infiltration as a mechanism of removal. Plus, as water passes through underlying soil, pollutants undergo filtration. When infiltration basins provide large volume and flow rate mitigation, they reduce downstream erosion. Because infiltration devices will remove and accumulate sediment via sedimentation and filtration, they run the risk of clogging, which renders the most important function, infiltration, ineffective. As stated earlier, pretreatment for sediment removal or placing these SCMs downstream of particulate-free runoff is critical for long-term performance of these SCMs. Generally, mechanisms for adsorptive removal or biological transformations are limited in infiltration trenches and basins.

## 16.4 Infiltration

As discussed in Chapter 7, infiltration is a complex physical process. Infiltration rates will differ depending on the characteristics of the media through which the infiltration is occurring, the degree of saturation, the water head driving the infiltration, and fluid properties. As a result, advanced mathematical descriptions may be necessary to explain infiltration phenomena. Even with simplifying assumptions, infiltration rates can change during the infiltration process, leading to time-dependent performance of infiltration systems.

For many designs, simple infiltration models are used, such as some form of Darcy's law to describe saturated flow. Even simpler designs assume a constant infiltration rate, independent of the hydraulic gradient. Infiltration through side walls can be a challenge to describe.

## 16.5 Surface Infiltration Basins

Surface infiltration basins are designed using similar criteria as retention ponds but add the dynamic infiltration capability. Simple design of infiltration basins can assume a constant infiltration rate. This rate is found through geotechnical investigation of the soils at the site. For design purposes, the native rate is commonly reduced by one-half to provide a conservative rate that accounts for reductions in infiltration that may occur over time due to soil compaction (likely during construction), clogging by accumulated sediment and biological growth. Multiplying the native infiltration rate by reducing factors accounts for clogging and biological growth.

For design, the basin depth, $H$, is based on the design infiltration rate ($f$) and the maximum allowable ponding time, $t_m$.

$$H = f\,t_m \tag{16.1}$$

For surface infiltration systems, the water pooling depth is often set to 3 ft (0.9 m). The volume of the infiltration basin is calculated as $V_w$ minus the volume of water that infiltrates intra-event. The volume of rain that falls on the basin is usually small compared to the runoff volume and is ignored in this calculation.

$$V = V_w - ftA_b \tag{16.2}$$

where $t$ is the time duration of the inflow and $A_b$ is the bottom area of the basin.

Using a trapezoidal basin, the volume is calculated as

$$V = \frac{\left(A_t + A_b\right)}{2}H \tag{16.3}$$

where $A_t$ is the top basin surface area.

Combining Equations 16.2 and 16.3 results in an expression for the basin bottom area as a function of filling time:

$$A_b = \frac{2V_w - A_t H}{(H + 2f\,t_m)} \tag{16.4}$$

For a rectangular basin and sideslopes $Z$, the following relationships for length ($L$) and width ($W$) are found:

$$L_t = L_b + 2ZH \tag{16.5a}$$

$$W_t = W_b + 2ZH \tag{16.5b}$$

and

$$A_t = A_b + 2ZH(L_b + W_b) + 4Z^2H^2 \tag{16.6}$$

In Maryland, a value of 2 h is commonly used as the filling time of the basin. The length and width are selected to fit the site. If the infiltration is located at the surface, side slopes are typically set at 4:1 and must have extensive coverage of vegetation to prevent erosion, which therefore limits in-SCM footprint erosion and sedimentation. In non-sandy locations, the bottom of the basin is excavated an additional 0.3 m (1 ft) below the permanent bottom and may be scoured during construction to promote vertical percolation. This area is then filled with 0.3 m of coarse sand to promote infiltration into the native soils. When constructed over HSG A soils, the 0.3 m of course sand layer just described is usually omitted. Surface infiltration basins in Maryland (with a 0.3-m course sand layer) and North Carolina (without said layer) are presented in Figure 16.5.

**Example 16.1** A surface infiltration basin will be constructed to manage 1 in (25 mm) of runoff over a 0.7-ha drainage area. The surrounding soils are sandy loam with an infiltration rate of 1.8 in/h (4.6 cm/h). The basin filling time is designed as 2 h and the maximum holding time is 24 h. Assume that sediment and biogrowth clogging factor is 0.49.

A. For sideslopes of 4:1 and a *bottom* length-to-width ratio of 4:1, find the dimensions of the infiltration basin.

B. Find the volume of sand necessary for the bottom fill of the infiltration basin.

***Solution*** The design runoff volume is calculated as the 1-in depth over the 0.7-ha drainage area: $V_w = dA = 0.025\text{ m}(0.7 \times 10{,}000\text{ m}^2) = 178\text{ m}^3$

The infiltration rate is adjusted to account for sediment and biogrowth:

$$f = 1.8\text{in/h}(0.49) = 0.88\text{in/h}$$

The basin depth is found using Equation 16.1:

$$H = (0.88\text{in/h})(24\text{ h}) = 21.2\text{in} = 1.76\text{ ft} = 0.54\text{ m}$$

For $L_b = 4W_b$, $A_b = 4W_b^2$: $A_t = 4W_b^2 + 2ZHW_b(5) + 4Z^2H^2$

Equation 16.5 gives $4W_b^2 = \dfrac{2V_w - (4W_b^2 + 2ZHW_b(5) + 4Z^2H^2)H}{(H + 2ft)}$

with $V_w$ = 178 m$^3$, $Z$ = 4, $H$ = 0.54 m, $f$ = 0.88 in/h = 0.0224 m/h, and $t$ = 2 h:

$$W_b = 7.44 \text{ m};\ L_b = 29.8 \text{ m};\ A_b = 221 \text{ m}^2$$

$$W_t = 11.8 \text{ m};\ L_t = 34.1 \text{ m};\ A_t = 401 \text{ m}^2$$

From Equation 16.3: $V = 168 \text{ m}^3$

The sand depth will be 0.3 m at the basin bottom:

$$\text{Sand volume} = 221 \text{ m}^2 (0.3 \text{ m}) = 66.3 \text{ m}^3$$

Figure 16.5 Surface Infiltration Basins Immediately Following Storm Events: (a) in Maryland and (b) in North Carolina. (Photos by Authors).

## 16.6 Infiltration Trench and Subsurface Infiltration Basin Design

Subsurface infiltration basins are almost identical to those of infiltration trenches, except for geometry, and the possible use of volume-enhancing components. The infiltration trench is 0.9–2.4 m (3–8 ft) deep, filled with clean stone of diameter 1.5–2.5 in (38–64 mm). Since inflow to the SCM ideally will first go through pretreatment for sediment removal, generally flow is piped in. The surface of the SCM is often paved or left as an open rock bed. Another design component is access to the underground area for maintenance operations. An underdrain or relief valve would aid in this work. An infiltration trench covered with permeable pavers in Pennsylvania is presented in Figure 16.6.

The design steps for an infiltration trench are similar to those of a detention basin with a few exceptions. First, the infiltration trench usually has vertical side walls, simplifying the volume calculations. As the infiltration basins are filled with rock, the rock porosity

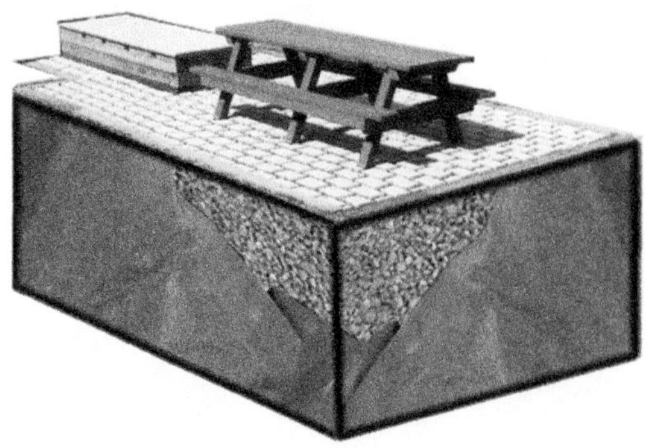

Figure 16.6 Infiltration Trench in Pennsylvania. (Credit: Authors' Research Group).

**Example 16.2** An infiltration trench will be constructed to manage 1 in (25 mm) of runoff over a 0.1-ha drainage area. The surrounding soils are sandy loam with an infiltration rate of 1.8 in/h. The trench filling time is designed as 2 h and the maximum holding time is 48 h. For a depth of 2 m and a bottom length-to-width ratio of 4:1, find the dimensions of the infiltration basin. (Infiltration from the sides of the basin may be ignored in this example.)

*Solution* Using the same equation as in Example 16.1, the design runoff volume is calculated as $V_w = dA = 0.025\ \text{m}\left(0.1 \times 10{,}000\ \text{m}^2\right) = 25\ \text{m}^3$
Using the default rock porosity of 0.4 and Equation 16.6, the trench area is found. The infiltration rate of 1.8 in/h is equal to 0.0458 m/h.

$$At = A_b = \frac{25\ \text{m}^3}{(0.4)(2\ \text{m}) + (0.0458\ \text{m/h})(2\ \text{h})} = 28\ \text{m}^2$$

At $L_b = 4W_b$, this gives $W_b = 2.64$ m; $L_b = 10.6$ m.

defines the available storage volume. Finally, the overall infiltration rate must consider depth due to infiltration through the side walls and the head driving the process. An assumption can be made to estimate the sidewall infiltration, such as to assume that side infiltration will occur using an area equal to one-third of the sidewall face.

## 16.7 Infiltration Trench and Basin Performance

As noted in Chapter 7, infiltration under saturated conditions is described by Darcy's law. The driving force is the hydraulic gradient, which is related to the depth of water in the basin or trench. The hydraulic conductivity of the soil below the infiltration SCM will control the infiltration rate, as discussed above. Also, infiltration of water is affected by the viscosity of the water, which is strongly dependent on temperature. As a result, infiltration areas are expected to show seasonal dependencies in infiltration rate that could vary by a factor of at least two (Emerson and Traver 2008). As a designer, this concept can be incorporated by using another (reducing) factor to multiply to the infiltration rate. Alternatively, infiltration rates in the winter, when infiltration is the slowest, could be used for the design.

Pretreatment removal of sediment is critical for proper long-term operation of infiltration facilities. Sediment that enters the upper layers of the infiltration trench will be removed by the gravel. This will result in water-quality improvement through reduction in TSS and associated pollutants (Hatt et al. 2007; Price et al. 2013). No mechanisms exist inside the SCM for dissolved pollutant removal and the removal is minimal, except through infiltration into the surrounding soil. Sediment that enters the infiltration chamber will be filtered at the chamber's interface with the underlying soil, as infiltration occurs.

## 16.8 Inspection and Maintenance

The function of an infiltration basin is straightforward: it needs to effectively infiltrate stored stormwater. Inspection of infiltration basins needs to confirm this function. The inspection can consist of observing the water levels in the basin for few days after a water-quality storm and documenting that the water level is adequately falling (Figure 16.7). Other inspection items may include ensuring that inlets are not blocked by debris and vegetation.

Should problems in infiltration be noted, maintenance of infiltration basins is not usually simple. If clogging is noted at the surface of the infiltration bed material, perhaps a thin layer of material can be removed and replaced with fresh media. In cases where the infiltration bed consists of rock, it is likely that sediment will migrate to the bottom of the rock layer. Once these infiltration facilities become clogged with sediment, maintenance to restore the infiltration capacity can be very difficult and may require complete reconstruction of the facility.

Examples exist where various types of infiltration basins have clogged but have maintained some degree of stormwater management performance. For subsurface systems, should the bottom become clogged, infiltration can take place through the sides of the SCM (Emerson and Traver 2008), as sediment builds. ET can be an important water removal mechanism in warm weather, if the basin or trench is not buried. Over time, clogged infiltration basin systems may take on wetland-like characteristics and still manage stormwater flows through storage and ET (Natarajan and Davis 2015a) and provide improvements in water quality via sedimentation, some adsorption, and nutrient processing (Natarajan and Davis 2015b, 2016).

Figure 16.7 Observing a Properly Functioning Infiltration Basin in Wilmington, NC; (Top) During a 75-mm Event and (Bottom) the Day After. (Photos by Authors).

## References

Emerson, C.H. and Traver, R.G. (2008). Multiyear and seasonal variation of infiltration from storm-water best management practices. *Journal of Irrigation and Drainage Engineering* 134 (5): 598–605.

Emerson, C.H., Wadzuk, B.M., and Traver, R.G. (2010). Hydraulic evolution and total suspended solids capture of an infiltration trench. *Hydrological Processes* 24 (8): 1008–1014.

Hatt, B.E., Fletcher, T.D., and Deletic, A. (2007). Treatment performance of gravel filter media: Implications for design and application of stormwater infiltration systems. *Water Research* 41 (12): 2513–2524.

Horst, M., Welker, A.L., and Traver, R.G. (2011). Multiyear performance of a pervious concrete infiltration basin BMP. *Journal of Irrigation and Drainage Engineering* 137 (6): 352–358.

Natarajan, P. and Davis, A.P. (2015a). Hydrologic performance of a transitioned infiltration basin managing highway runoff. *Journal of Sustainable Built Environment* 1: 3) 04015002.

Natarajan, P. and Davis, A.P. (2015b). Water quality performance of a transitioned' infiltration basin, Part 1: TSS, metals, and chloride removal. *Water Environment Research* 87 (9): 823–834.

Natarajan, P. and Davis, A.P. (2016). Performance of a 'Transitioned' infiltration basin, Part 2: Nitrogen and Phosphorus removal. *Water Environment Research* 88 (4): 291–302.

Price, W.D., Burchell II, M.R., Hunt, W.F., and Chescheir, G.M. (2013). Long-term study of dune infiltration systems to treat coastal stormwater runoff for fecal bacteria. *Ecological Engineering* 52: 1–11.

## Problems

**16.1** Find the dimension for an infiltration basin to treat 25 mm of runoff over 1 ha. The surrounding soils are classified as loamy sand. The filling time is 2 h and the storage time is 24 h. The basin should be trapezoidal with 3:1 side slopes and the *L*:*W* ratio should be 4:1.

**16.2** An infiltration basin in a loamy sand soil has a bottom width of 5 m and bottom length of 18 m. The side slopes are 4:1. For a filling time of 2 h and storage time of 24 h, find the design water-quality volume at a basin depth of 1.2 m.

**16.3** Find the dimension for an infiltration trench to treat 25 mm of runoff over 1 ha. The surrounding soils are classified as loamy sand. The filling time is 2 h. The depth should be 1.8 m and the L:W ratio should be 2:1.

**16.4** Repeat Problem 10.3, but assume that some clogging of the soils occurs, rendering the infiltration rate value only 75% of that listed for a loamy sand.

**16.5** An infiltration trench in a sandy soil has a depth of 2 m, width of 4 m, and length of 12 m. For a filling time of 2 h, find the design water-quality volume to be managed by this trench.

# 17

# Sand Filters

## 17.1 Introduction

Sand filters are a relatively simple stormwater control measures (SCMs) with a simple purpose: to filter runoff to remove particulate matter. Nonetheless, there are a wide a variety of design configurations. Their performance can vary widely depending on configuration, but most importantly, also on maintenance frequency. While sand filters do not have vegetative component, they do share a number of attributes with other GI SCMs. They represent a class of advanced water-quality SCMs and, being unvegetated, can be employed below the surface of highly urbanized areas (saving valuable space), or in combination with other SCMs. Sand filters can often be found in karst regions or near large bodies of water where surface volume reduction may not be critical.

## 17.2 Basic Sand Filter Operation

The basic mechanism of operation of a sand filter is particulate matter filtration, as fundamentally discussed in Chapter 8. Runoff is directed to a specific SCM that contains a layer of sand (or a similar inert filter media). The runoff is allowed to pool above the sand and infiltrates through the sand. An underdrain is usually placed at the bottom of the sand layer to collect the runoff after it passes through the media. Particulate matter is removed by the sand media during the treatment process. Captured particulate matter will accumulate at and near the surface of the sand media, so frequent maintenance is required, which entails scraping off or vacuuming away a thin surface layer of the sand (with the captured particulate matter).

## 17.3 Sand Filter Options and Configurations

Figures 17.1–17.3 show design schematics and photos of several sand filter systems. The common feature is the sand layer through which the stormwater flows for filtration and particulate matter removal. All three of these filters are encased in concrete structures. All also have a sedimentation basin separated by a concrete wall. The filter

*Green Stormwater Infrastructure Fundamentals and Design*, First Edition. Allen P. Davis, William F. Hunt, and Robert G. Traver.

Companion Website: www.wiley.com/go/davis/greenstormwater

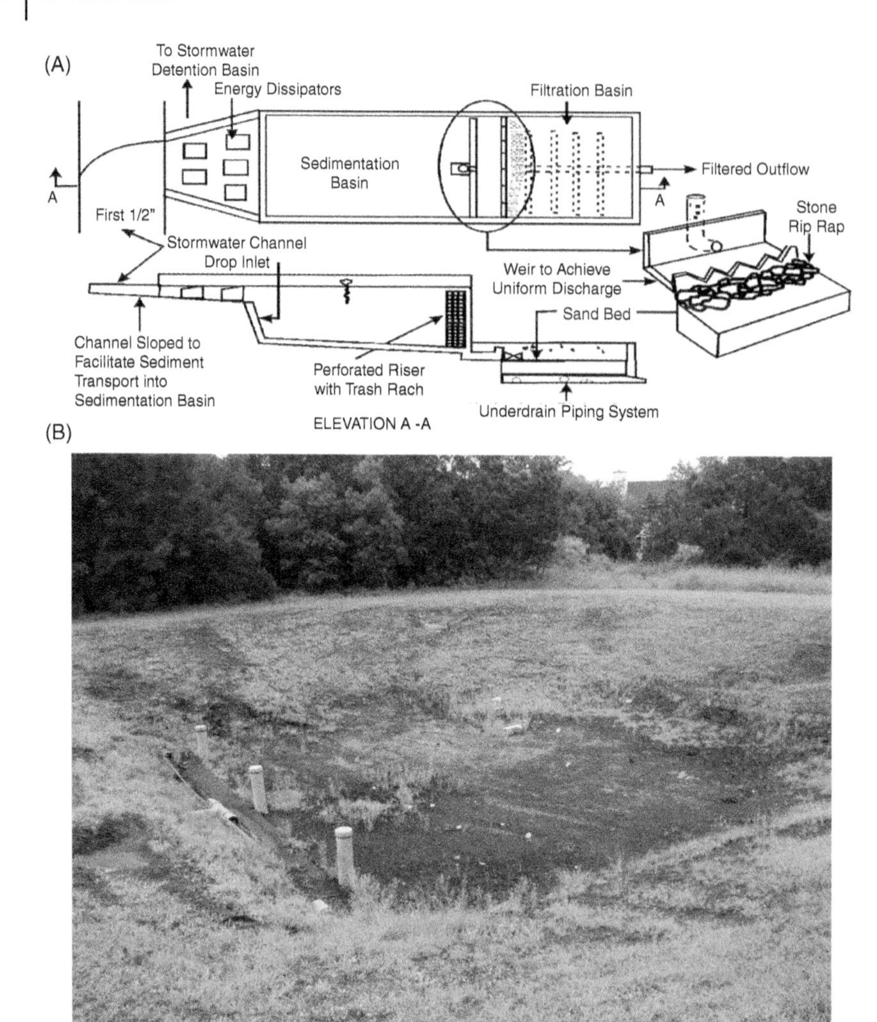


Figure 17.1 Austin Sand Filter: (A) Schematic Drawing (Credit: USEPA 1999) and (B) Photograph (Photo by Authors).

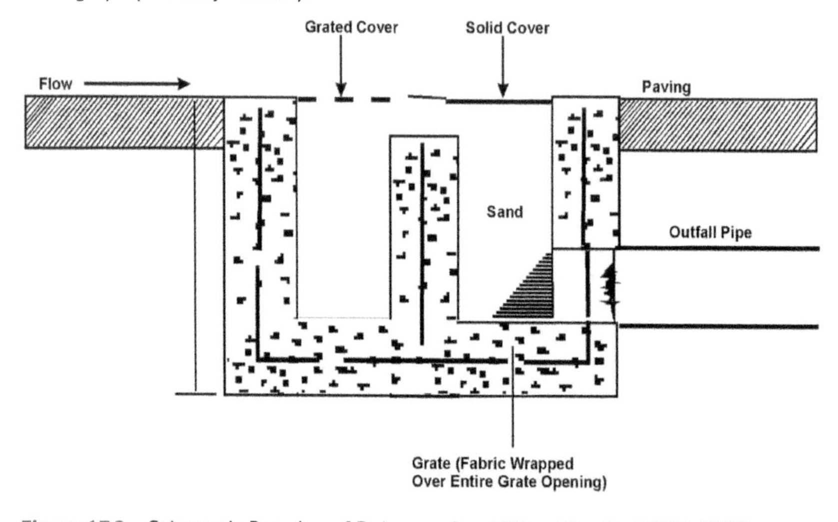


Figure 17.2 Schematic Drawing of Delaware Sand Filter (Credit: USEPA 1999).

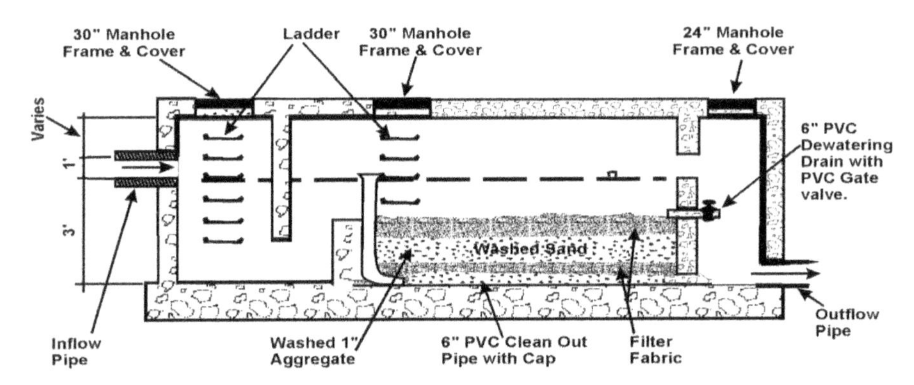


Figure 17.3 Schematic Drawing of DC Sand Filter (Credit: USEPA 1999).

is drained by a PVC underdrain or a grate wrapped with a geotextile fabric. The Austin sand filter (Figure 17.1) is open air and the sedimentation basin is sized to capture the entire water quality volume. The sedimentation basin is designed to discharge the captured runoff to the filter basin in 24 h.

The Delaware (Figure 17.2) and District of Columbia (DC) (Figure 17.3) sand filters are installed completely subsurface, allowing them to be used in ultraurban areas, being placed under parking lots or light-use highways.

Sand filters may also be installed directly into the soil, with a geotextile liner on the sides and bottom to isolate the filter from the surroundings (Figure 17.4). Otherwise, the surface sand filter is similar to the Austin filter.

In the designs presented in Figures 17.1–17.4, the sand filter is established as a device with an impervious bottom that does not permit water percolation into surrounding soils. Therefore, hydrologic impacts from sand filters are minor, as no volume reduction will occur, except through evaporation from the media between storm events. Also, reductions in pollutant loads must occur by treatment only and not due to runoff volume reduction.

## 17.4 Sand Filter Design

Sand filters should have a sand layer 45-cm (18 in) deep, with a minimum of 30 cm (12 in). The bed surface should be level. The sand should be AASHTO-M-6 or ASTM C 33 concrete sand, size range of 0.5–1 mm. A 10- or 15-cm perforated PVC pipe, with 9.5 mm (3/8 in) holes, is used for the underdrain. The underdrain should be surrounded with 9.5–19 mm (3/8–3/4 in) diameter gravel. Depending on the filter type and design, the water pool above the filter media can be a few cm to more than a meter.

The sand filter area, $A_f$, is computed by

$$A_f = \frac{Vd}{k(h+d)t} \tag{17.1}$$

where $d$ is the sand filter depth, typically 45 cm; $k$ is the sand saturated hydraulic conductivity; $h$ is the average height of water above filter bed; V is volume treated and $t$ is the design filter bed drain time.

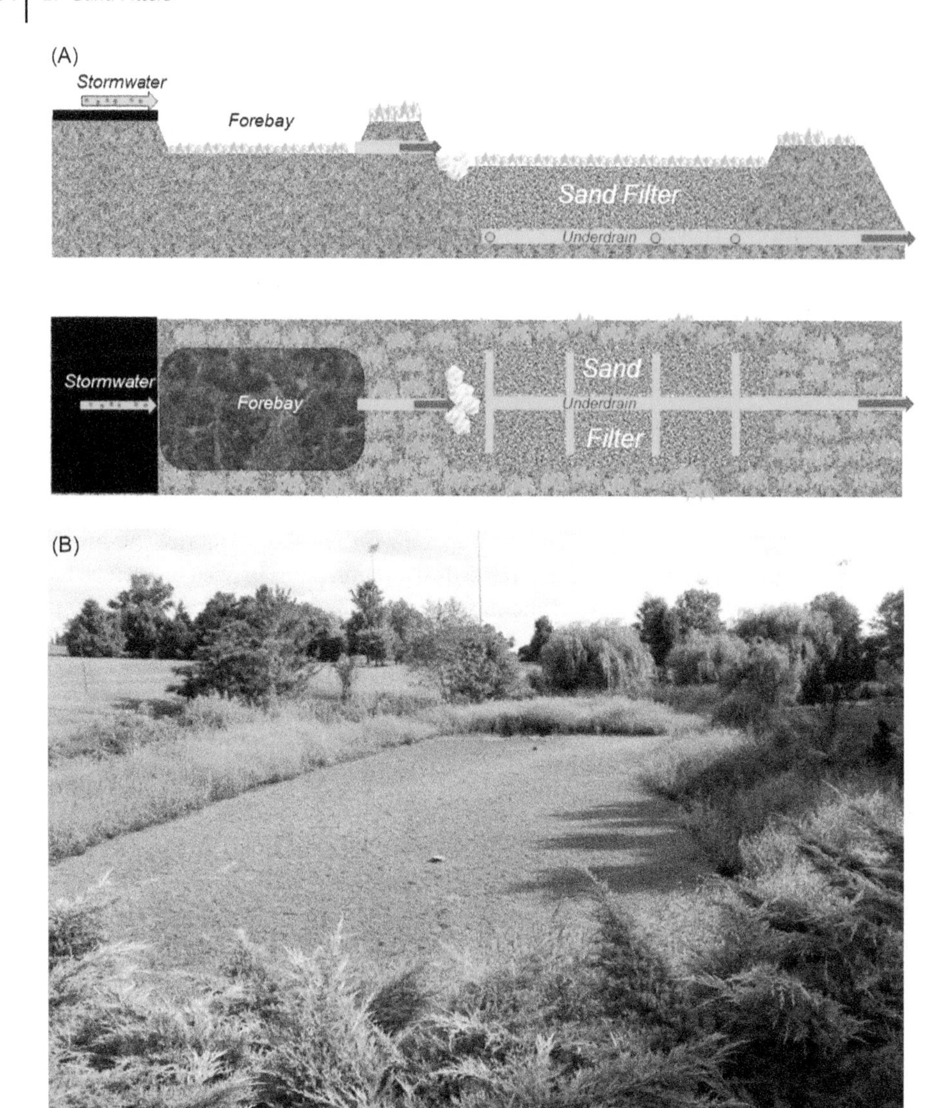


Figure 17.4 Surface Sand Filter: (A) Configurations and (B) Photograph (Photo by Authors).

**Example 17.1** The water quality volume from a parking lot is calculated as 82,000 L (2900 ft$^3$). A sand filter is designed with a depth of 45 cm and a hydraulic conductivity of 107 cm/day. Find the filter area that would drain in 1.67 days at an average water height of 60 cm (2 ft).

***Solution*** Equation 17.1 is used, along with the default design parameters.

$$A_f = \frac{(82{,}000 \text{ L})(45 \text{ cm})}{(107 \text{ cm/day})(45 + 60 \text{ cm})(1.67 \text{ day})} = 197 \text{ L/cm}$$

$$= 197{,}000 \text{ cm}^2 = 19.7 \text{ m}^2 = 212 \text{ ft}^2$$

## 17.5 Water Quality Performance

Typically, sand filters are very efficient in the removal of particulate matter. However, dissolved pollutant removal is generally poor.

### 17.5.1 Particulate Matter Removal

Barrett (2003) found that the effluent concentration of total suspended solids (TSS) through sand filters was approximately constant at 7.8 mg/L (with uncertainty at the 90% confidence level of 1.2 mg/L), regardless of the influent concentration. The consistent effluent concentration suggests that there is little difference in the total mass of the smallest sized particles in the runoff regardless of influent TSS concentration. This implies that differences among influent concentrations are generally caused by larger sized fractions that will not pass through the filter, while the smallest size fractions are not removed by the filter. This agrees with filtration theory as discussed in Chapter 8 and the concept of "irreducible" particulate matter concentrations is discussed in Chapter 11. In a similar manner, removal of particulate-affiliated metals in sand filters will be parallel that of particulate matter removal. In the same 2003 study, regardless of influent particulate copper concentration, the effluent concentration was relatively constant at about 2 μg/L (Barrett 2003). Reductions of fecal coliform were noted also.

**Example 17.2** Estimate the filtration removal efficiency for a 100-μm particle (specific gravity of 2.0) through a 0.6-m sand filter. Assume an infiltration rate of 100 cm/day. The sand media $d_{10}$ is 0.7 mm and the porosity is 0.43. The particle attachment efficiency is 0.05 and the transport efficiency is dominated by the sedimentation mechanism. The temperature is 15°C.

*Solution* The infiltration rate of 100 cm/day corresponds to $1.157 \times 10^{-5}$ m/s. Therefore, the sedimentation transport efficiency is given as

$$\eta_s = \frac{(\rho_p - \rho_w) g d_p^{\,2}}{18 \mu v} = \frac{\left(2.0(1000) - 1000 \text{ kg/m}^3\right) 9.8 \text{ m/s}^2 \left(100 \times 10^{-6} \text{ m}\right)^2}{18\left(1.14 \times 10^{-3} \text{ kg/(m s)}\right)\left(1.157 \times 10^{-5} \text{ m/s}\right)} > 1$$

The maximum value for this term is 1.0. Therefore, the particle removal fraction is

$$\frac{C}{C_0} = \exp\left(-1.5 \frac{(1-0.43)(0.05)(1.0)(0.6 \text{ m})}{\left(0.7 \times 10^{-3} \text{ m}\right)}\right) = 10^{-16}$$

This result indicates that filtration of 100-μm particles is highly efficient and that, based on filtration theory, the concentration leaving the filter is very small.

### 17.5.2 Dissolved Pollutant Removal

The primary treatment mechanism for sand filters is filtration. No major mechanism exists for removal of dissolved pollutants and the impact on dissolved pollutant concentrations is expected to be minimal.

#### 17.5.2.1 Nitrogen and Phosphorus

Austin sand filters typically show a negative removal of nitrate. Barrett (2003) noted an average nitrate effluent event mean concentration (1.10 mg/L) significantly higher than the influent (0.63 mg/L), likely due to nitrification reactions taking place in the sand filter media. It is expected that particulate N species are filtered at the media surface and subsequently mineralized/nitrified to nitrate. While N transformations in sand filters are expected, total nitrogen removal is expected to be minimal.

Particulate phosphorus was reduced to a constant value of 0.07 mg/L by Austin sand filters (Barrett 2003). Phosphate removal was noted, with the removals dependent on the influent concentration.

#### 17.5.2.2 Metals and Hydrocarbons

Removal of high concentrations of dissolved metals has been noted in an Austin sand filter (Barrett 2003). The reasons for this removal are unclear but suggests adsorption on the sand grains or possibly accumulated sediment to be the responsible mechanisms. Small removal of total petroleum hydrocarbons was noted.

## 17.6 Sand Filter Headloss

Important to understanding sand filter performance, maintenance, and clogging is the concept of headloss development. The sand media provides resistance to infiltrating runoff. Subsequent accumulation of collected particulate matter can greatly increase this resistance, leading to decreased infiltration rates, problems with filter performance, and ultimately, failure.

The Carmen–Kozeny equation is commonly used to describe filter headloss in drinking water treatment systems:

$$\frac{h}{L} = \frac{5\mu(1-\varepsilon)^2 S^2 v}{g\rho\varepsilon^3 {d_m}^2} \tag{17.2}$$

where $h$ is the headloss; $L$ is the media travel length; $\mu$ is the viscosity; $\varepsilon$ is the media porosity; $v$ is the approach (superficial) water velocity (infiltration rate); $S$ is the specific surface = 6 (spheres) to 8 (irregular shapes); $d_m$ is the mean media grain diameter; and $\rho$ is the water density.

This equation can be used to estimate infiltration rate (the velocity) for stormwater sand filters based on the water head in the filter. As solids are captured and build up in the filter the infiltration rate will slow.

**Example 17.3** A sand filter has a 0.6-m layer of 0.6-mm media. Specific surface, S=8, for irregular shaped media. The porosity is 0.43. Find the clean media infiltration rate when the water level is 20 cm above the media surface. The temperature is 15°C.

*Solution* Equation 17.2 is solved for $v$, the infiltration rate. The headloss is the total water head through the media.

$$v = \frac{g\rho\varepsilon^3 d_m{}^2 h}{5\mu(1-\varepsilon)^2 S^2 L} = \frac{9.8\ \text{m/s}^2\left(1000\ \text{kg/m}^3\right)(0.43)^3\left(0.6\times10^{-3}\ \text{m}\right)^2(0.6+0.2\ \text{m})}{5\left(1.14\times10^{-3}\ \text{kg/(m s)}\right)(1-0.43)^2\, 8^2\,(0.6\ \text{m})} =$$

$$= 3.16\times10^{-3}\ \text{m/s} = 11.54\ \text{m/h}$$

This is clearly a very high flow rate that will decrease rapidly with solids buildup.

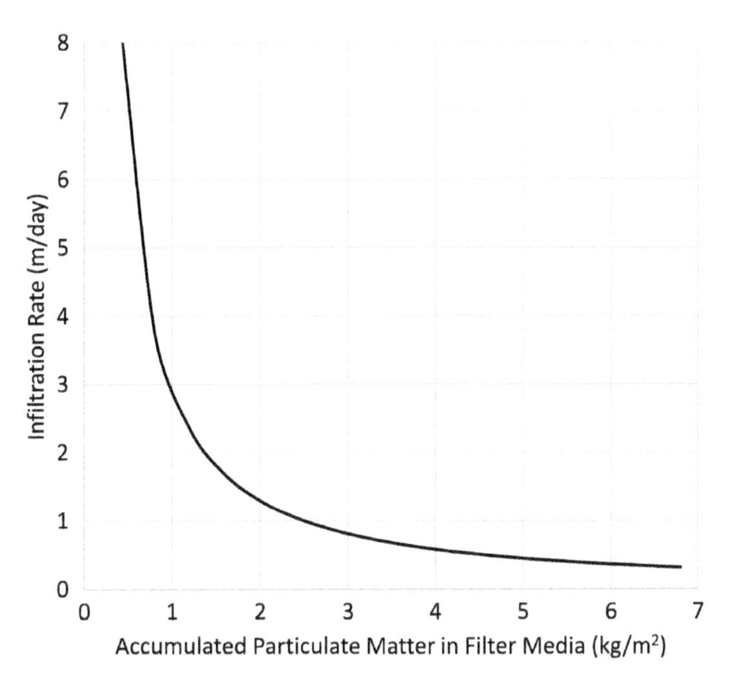


Figure 17.5 Impact of Accumulated Particulate Matter on Sand Filter Infiltration Rate (Adopted from Urbonas 1999).

## 17.7 Solids Accumulation and Clogging

Consideration of sand filter design and clogging was evaluated by Urbonas (1999). Sand filters clog quickly, based on the accumulated particulate matter. A design equation based on sand filter performance data is shown in Figure 17.5. The infiltration rate falls quickly based on cumulative particulate matter removed (accumulated), with the design equation:

$$v = 2.91 L_m^{-1.165} \tag{17.3}$$

where $v$ is the infiltration rate in m/day and $L_m$ is the solids accumulation (kg/m$^2$).

As seen in Figure 17.5, the infiltration rate falls very quickly with accumulated sediment. Therefore, the design infiltration value should not be the initial rate, but an average value or one near the clogged rate. With knowledge of flow rates, particulate matter concentrations, and filtration efficiency, Figure 17.5 can be used for design and analysis.

As seen in Figure 17.5, sand filter infiltration rates decrease sharply after accumulating about 2 kg/m$^2$ of particulate matter. Barrett (2003) noted clogging of Austin sand filters after *loading* of 5–7.5 kg/m$^2$ of particulate matter; recognizing that not all particulate matter is captured by the filter, values of about 2–3.8 kg/m$^2$ of accumulated particulate matter resulted in need for maintenance.

Sand filter maintenance usually requires the removal of the top 8 cm of sand and accumulated material. This can be done by hand or with a vacuum truck. At some point, the sand layer becomes too shallow and additional sand must be added to increase the media depth.

Without regular maintenance, the sand media becomes completely clogged with accumulated sediment. This will prevent the passage of any water and result in water pooling up to the level of the filter bypass. This SCM will no longer provide stormwater treatment and all water will be bypassed. This standing water can be a health hazard.

**Example 17.4** The sand filter of Example 17.1 requires a minimum infiltration rate of 1.07 m/day sized at 19.7 m$^2$. At an average input TSS concentration of 80 mg/L and output concentration of 10 mg/L, find (a) the volume of runoff treated before reaching the infiltration rate threshold, and (b) assuming that the average storm produces 20% of the water quality volume and that a storm occurs every 5 days, the time the filter will operate before needed maintenance.

***Solution*** At 1.07 m/day, Equation 17.3 is solved for $L_m$.

$$L_m = (q/2.91)^{-1/1.165} = (1.07/2.91)^{-1/1.165} = 2.36 \text{ kg/m}^2$$

The total filter area is 19.7 m$^2$, therefore, the total accumulated particulate matter is $2.36 \text{ kg/m}^2 \times 19.7 \text{ m}^2 = 46.5 \text{ kg} = 46.5 \times 10^6$ mg.

The runoff treated is therefore

$$V = 46.5 \times 10^6 \text{ mg}/(80 - 10 \text{ mg/L}) = 6.64 \times 10^5 \text{ L}$$

Twenty percent of the water quality volume (82,000 L) is 16,400 L. Therefore, the amount of runoff treated corresponds to 40.5 average storm events. At 5 days each, the useful filter life is 202 days before the filter requires maintenance.

## 17.8 Sorptive and Reactive Media

While traditional sand filters are highly effective at removal of particulate and particulate-bound pollutants, sand filter performance is lacking for dissolved pollutants. A number of studies have evaluated the use of various granular media or media coatings for enhanced removal of various stormwater dissolved pollutants of interest. Effective media amendments can increase the duration and capacity of the media for dissolved pollutant removal as compared to simple silica sand. Various geomedia, including calcite/limestone, zeolite, iron-based media, aluminum-based media, manganese-based media, fly ash, olivine, and several commercial proprietary media have been found to be effective to some degree for a specific pollutant or group of pollutants.

Mechanisms for the removal of dissolved phosphorus species (phosphate and organic phosphorus) include precipitation by calcium, aluminum, or iron, and surface adsorption to iron oxide or aluminum oxide. Addition of iron materials (steel wool and iron fillings) to sand filter media (Erickson et al. 2007, 2012; Reddy et al. 2014), coating/using media with aluminum oxide (Wu and Sansalone 2013a; 2013b; Duranceau and Biscardi 2015), or use of fly ash (Zhang et al. 2008) has shown to be effective in enhanced removal of dissolved phosphate. Steel–wool-enhanced media was predicted to only raise construction costs by approximately 3–5% (Erickson et al. 2007).

Manganese oxide-coated sand and iron-based materials can be effective for heavy metals removal (Liu et al. 2005; Reddy et al. 2014). Removal of multiple pollutants can be accomplished by layering of appropriate media types (Prabhukumar et al. 2015).

Reactive media can promote beneficial transformations of stormwater pollutants. A manganese oxide-coated sand was able to oxidize several toxic organic pollutants (Grebel et al. 2016). Zero valent iron may be effective in nitrate removal (Reddy et al. 2014). Contact with copper-modified zeolite and granular activated carbon can promote the inactivation of *Escherichia coli* (Li et al. 2014).

## 17.9 Geotextile Filters

Geotextile filters may be considered as a simple alternative filtration SCM to remove particulate matter from urban stormwater runoff. A geotextile filtration media could be a better choice than sand because it is light and easily transportable (Franks et al. 2013). A laboratory column study determined that a geotextile with an opening size of 150 μm can remove TSS below a target concentration of 30 mg/L from a synthetic urban runoff via a filtration mechanism (Franks et al. 2013, 2014). The results also showed that the change in hydraulic conductivity of the filter system can be related to the concentration of captured TSS, which can also be used to predict the flow rate through the filter throughout its life cycle. The clogging results indicated that a geotextile filter lasts more than 50% longer than a sand filter under urban stormwater conditions. Geotextile filters may also provide a substrate for biofilm formation, allowing for some organic pollutant removal (Paul and Tota-Maharaj 2015).

## 17.10 Inspection and Maintenance

Sand filters have a long history of performance of TSS removal and are considered an important component of green stormwater infrastructure. They are most useful in areas where volume reduction is not important or advantageous, or as part of a treatment train. Regular inspection and maintenance is critical for successful sand filter operation. Inspection is relatively simple and will document the presence of standing water above the filter. Sand filters are designed for relatively rapid treatment rates and should not have ponding water for long periods. Research has shown that sand filters will require maintenance after accumulation of approximately 2–3.8 kg/m$^2$ (0.41–0.78 lb/ft$^2$) of particulate matter. Filters that treat drainage areas with higher sediment loads can be expected to require more frequent maintenance.

Maintenance of sand filters requires the removal of the top layer of clogged sand. This layer may be up to 4 in (10 cm) deep. The remaining sand layer thickness may still be adequate for further stormwater treatment after the removal of this upper layer. New sand must be added after several maintenance cycles.

An advantage of sand filters is that they can be constructed completely subsurface in highly urbanized areas where space is at a premium; this, however, can complicate maintenance. For subsurface filters, the clogged sand will have to be vacuumed out through a manhole.

Sand filters have been documented to provide excellent removal of particulate matter from stormwater. However, this removal will decrease to zero if a proper maintenance schedule is not followed, the filter clogs, and all incoming stormwater is bypassed.

## References

Barrett, M.E. (2003). Performance, cost, and maintenance requirements of Austin sand filters. *Journal of Water Resources Planning and Management* 129 (3): 234–242.

Duranceau, S.J. and Biscardi, P.G. (2015). Comparing adsorptive media use for the direct treatment of phosphorous-impaired surface water. *Journal of Environmental Engineering* 141 (8): 04015012.

Erickson, A.J., Gulliver, J.S., and Weiss, P.T. (2007). Enhanced sand filtration for storm water phosphorus removal. *Journal of Environmental Engineering* 133 (5): 485–497.

Erickson, A.J., Gulliver, J.S., and Weiss, P.T. (2012). Capturing phosphates with iron enhanced sand filtration. *Water Research* 46: 3032–3042.

Franks, C.A., Aydilek, A.H., and Davis, A.P. (2013). Modeling hydraulic conductivity of a geotextile filter during suspended solids accumulation. *Geosynthetics International* 20 (5): 332–343.

Franks, C.A., Davis, A.P., and Aydilek, A.H. (2014). Effects of runoff characteristics and filter type on geotextile storm water treatment. *The Journal of Irrigation and Drainage Engineering* 140 (2): 04013014.

Grebel, J.E., Charbonnet, J.A., and Sedlak, D.L. (2016). Oxidation of organic contaminants by manganese oxide geomedia for passive urban stormwater treatment systems. *Water Research* 88: 481–491.

Li, Y.L., Deletic, A., and McCarthy, D.T. (2014). Removal of E.coli from urban stormwater using antimicrobial-modified filter media. *Journal of Hazardous Materials* 271: 73–81.

Liu, D.F., Sansalone, J.J., and Cartledge, F.K. (2005). Comparison of sorptive filter media for treatment of metals in runoff. *Journal of Environmental Engineering-Asce* 131 (8): 1178–1186.

Paul, P. and Tota-Maharaj, K. (2015). Laboratory studies on granular filters and their relationship to geotextiles for stormwater pollutant reduction. *Water* 7 (4): 1595–1609.

Prabhukumar, G., Bhupal, G.S., and Pagilla, K.R. (2015). Laboratory evaluation of sorptive filtration media mixtures for targeted pollutant removals from simulated stormwater. *Water Environment Research* 87 (9): 789–795.

Reddy, K.R., Xie, T., and Dastgheibi, S. (2014). Adsorption of mixtures of nutrients and heavy metals in simulated urban stormwater by different filter materials. *Journal of Environmental Science and Health Part a-Toxic/Hazardous Substances & Environmental Engineering* 49 (5): 524–539.

Urbonas, B.R. (1999). Design of a sand filter for stormwater quality enhancement. *Water Environ. Research* 71 (1): 102–113.

USEPA (1999). Storm water technology fact sheet. EPA 832-F-99-007, Washington DC.

Wu, T. and Sansalone, J. (2013a). Phosphorus equilibrium. II: Comparing filter media, models, and leaching. *Journal of Environmental Engineering* 139 (11): 1325–1335.

Wu, T. and Sansalone, J. (2013b). Phosphorus equilibrium. I: Impact of AlOx media substrates and aqueous matrices. *Journal of Environmental Engineering* 139 (11): 1315–1324.

Zhang, W., Brown, G.O., Storm, D.E., and Zhang, H. (2008). Fly-ash-amended sand as filter media in bioretention cells to improve phosphorus removal. *Water Environment Research* 80 (6): 507–516.

## Problems

**17.1** The water quality volume from parking lot is calculated as 100,000 L. Find the filter area using default design values and an average water height of 90 cm.

**17.2** The water quality volume from parking lot is calculated as 120,000 L. Find the filter area using default design values and an average water height of 100 cm.

**17.3** A sand filter has dimensions 5 m × 10 m. Find the water quality volume treated by the filter using default design values and an average water height of 100 cm.

**17.4** A sand filter is loaded at an average TSS concentration of 96 mg/L. The expected output TSS concentration is 8 mg/L. The minimum sand filter infiltration rate is 0.9 m/day. What should be the surface area of the filter if yearly filter maintenance is planned?

**17.5** The minimum flow rate through a sand filter is 33 $m^3$/day. For a 1-year maintenance cycle and an expected TSS discharge of 12 mg/L TSS, what should be the runoff input TSS concentration?

**17.6** A sand filter for stormwater runoff treatment has been installed. The width is 10 ft and the length is 20 ft. The sand media characteristics are given in the following table:

| Depth | $d_{10}$ | Porosity | Specific gravity | Specific shape | Sticking factor |
|---|---|---|---|---|---|
| 2.0 ft | 0.8 mm | 0.42 | 2.65 | 7.2 | 0.3 |
| 610 mm | $2.62 \times 10^{-3}$ ft | | | | |

The filter performance will vary during a storm event because pollutant concentrations and flow rates will change during the storm duration.

a) During a 1.0-in/h rainfall (producing a flow rate of 1600 gpm), the TSS concentration is 240 mg/L, dominated by 50 μm particles. Find the TSS concentration in the filter effluent under these conditions (mg/L).

b) Find the particle size that is 90% removed when the filter is treating a 0.5-in/h rainfall, producing a flow rate of 800 gpm (μm).

# 18

# Bioretention

## 18.1 Introduction

Bioretention is a vegetated infiltration/filtration stormwater control measure (SCM) (Figure 18.1). Because it utilizes vegetated media, when designed, built, and maintained correctly it is expected to provide significant impacts in reducing volume and erosive peak flow rates, restoring evapotranspiration and recharge, and improving water quality. The basic configuration of bioretention is a shallow bowl for runoff storage and a vegetated media layer. An increasing number of design options are available depending on location and treatment goals. Some modifications of bioretention exist under different names.

Figure 18.1 Bioretention SCM in Dare County, North Carolina. (Photo by Authors).

*Green Stormwater Infrastructure Fundamentals and Design*, First Edition. Allen P. Davis, William F. Hunt, and Robert G. Traver.

Companion Website: www.wiley.com/go/davis/greenstormwater

## 18.2 Bioretention Classifications

The simplest class of bioretention is commonly termed as a *rain garden*. A rain garden is a small facility, usually at the scale that can be built by a homeowner on a residential lot. A professional engineer would not be needed for this design. The rain garden would be placed near a downspout to collect, store, and infiltrate runoff. It is sized based upon the contributing impervious surface (Figure 18.2).

A larger bioretention facility employed to meet specific stormwater management requirements would have a more complex design (Figure 18.3). Bioretention may or may not contain an underdrain; the later configuration has been termed *bioinfiltration*. A bioretention cell used for conveyance has been termed a *bioswale*. A number of other options are possible related to media depth and characteristics, and the underdrain configuration. Bioretention can also be used as part of a *treatment train* with other SCMs. All of these configurations have the basic characteristics of bioretention and share the same unit processes. They are discussed under the umbrella terminology of bioretention, with emphasis placed on design options and configurations.

Various surface planters and tree boxes are also used in highly urbanized areas. They typically manage downspout runoff from buildings. These systems are best described as a class of bioretention systems (Figure 18.4).

Figure 18.2 Home Rain Garden in Burnsville, MN, after and before Construction. (Courtesy of Fred Rozumalski, Barr Engineering, and Minnesota Pollution Control Agency).

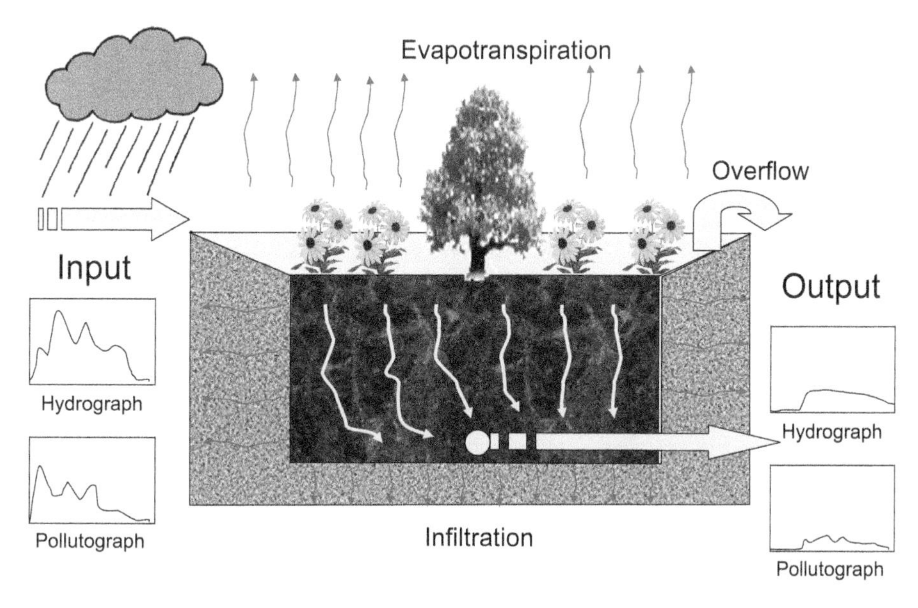


Figure 18.3 Diagram of Bioretention Facility with Underdrain.

Figure 18.4 Urban Stormwater Planter Box on a 100% Impervious Cover Residential Mid-rise Project in Old Town Alexandria, VA. Roof-drains Are within Building Façade.
(Photo by Authors).

## 18.3 Bioretention Components

The basic components of bioretention facilities are the (1) flow entrance, (2) pretreatment, (3) storage bowl and overflow, (4) media, (5) vegetation, (6) in situ soils, and (7) underdrain, although some of these components are optional. Urban hydrology and space limitations may add an eighth component, a storage or rock bed, to increase runoff detention.

## 18.4 Siting and Configuration

Bioretention is a highly flexible SCM and that has led to its popularity in many areas. A number of design options and configurations are available to tailor its use to meet different stormwater goals. It can be narrow or wide, shallow or deep, and densely or sparsely vegetated. Exfiltration to surrounding soils can be encouraged or prevented.

The first factor that needs to be considered is simply the contributing drainage area that would drain to the bioretention SCM. As shown in Figure 18.5, the drainage area contributing to the SCM is outlined in blue. The directly connected impervious fraction is of interest as it generates the majority of runoff, especially for smaller events. Use of ratios of the impervious drainage area to that of the infiltration foot print area are in common usage but should only be considered a planning guide. Ratios of 5:1 to 10:1 are common, but as they do not consider weather patterns and soil types, they can be limiting; a more detailed engineering analysis may permit larger ratios.

The source of inflow is also of interest. Care must be taken to prevent clogging and erosion of the infiltrating surface. Rooftop runoff, roads with minimal traffic, and runoff from mature vegetated areas is desirable due to limited sediment content. Where sediment loads are of concern, loading to the bioretention cell can be minimized through pretreatment processes (e.g., forebay) and ensuring that the drainage catchment area is properly stabilized.

The infiltration rate of the existing soil, high ground water tables, karst topography, and underground utilities may dictate the need to use bioretention facilities with slow release underdrains that are also designed to maximize evapotranspiration. In this way, stormwater volume is still reduced, and the underdrain flows are slowed to maximize water quality and downstream erosion. DeBusk et al. (2011) demonstrated that bioretention underdrain flow, if properly designed, can mimic base flow.

Urban settings can also be found for bioretention implementation, though space is generally at a premium (Figure 18.6). Areas with wider sidewalks or unused spaces can be repurposed. Suburban or less-urbanized areas are usually much simpler in finding space for bioretention (Figure 18.7) and require less use of concrete.

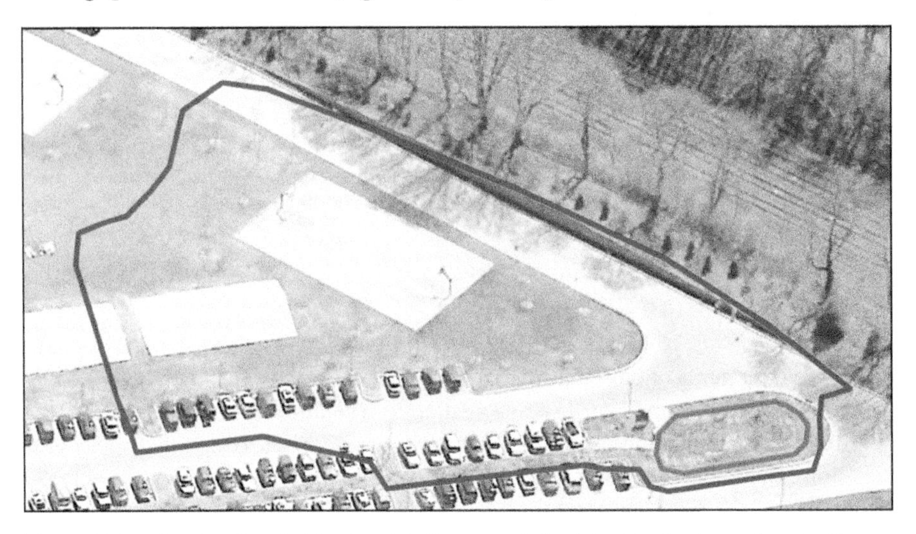

Figure 18.5 Bioretention Cell Outlined in Green, with Contributing Drainage Area in Blue. (Credit: Authors' Research Groups).

Figure 18.6 Urban Rain Gardens–Sidewalk Planters: (A) St Paul Minnesota; (B) Philadelphia (Winter); and (C) New York City. (Photos by Authors).

Figure 18.7 Suburban Rain Gardens at the Gettysburg Battlefield Visitor Center. (Photo by Authors).

Figure 18.8 Bioretention Retrofit Sites: (a) Maryland; (b) North Carolina and (c) Villanova University. (Photos by Authors).

The bioretention designer should also consider retrofit areas. Figure 18.8A and B shows, respectively, a Maryland and North Carolina bioretention facility built around existing drainage structures in parking lots. Figure 18.8C depicts a bioinfiltration site at Villanova University constructed in a traffic island. Retrofit sites can reduce existing problems, not just mitigate new construction. These photos also demonstrate the importance of the SCM becoming an attractive addition to the community. Note the signage at the North Carolina site promoting community education.

## 18.5 Bioretention Flow Entrances, Inlets, and Forebays

The first bioretention design component is simply to capture the water and introduce it to the bioretention site. Inlets are discussed in Chapter 12. Figure 18.9A and B provides examples of inadequate inlet design. In Figure 18.9A, the curb cut is parallel to the flow of water such that the water passes by the entrance, bypassing the raingarden. Figure 18.9B shows that though the curb cut captures the water, it is a source of erosion. Figure 18.9C shows a curb-cut with a depressed forebay to capture the water, which then forces a turn and weir overflow to reduce energy and capture grit.

Bioretention can be very effective for treatment of particulate matter. Because of this, the particulate matter loading on the surface of the bioretention media can be large, possibly leading to maintenance challenges. For areas where high sediment loads are expected, constructing a small forebay at the inlet can allow for the majority

Figure 18.9 Various Ineffective Bioretention Inlet Configurations: (A) Curb Cut That Does Not Allow Inflow; (B) Erosion at Inlet; (C) Inlet with Forced Turn to Enter Bioretention. (Photos by Authors).

of the particulate matter to be removed before the runoff flow enters the bioretention cell (Figure 18.9C). The forebay will capture larger sediment in an area where it can be easily removed during routine maintenance. Details on forebay designs are presented in Chapter 12.

If a forebay is not employed in a bioretention cell, accumulation of sediment in the cell near the inlet is likely over time. This area can become a zone of captured sediment that will produce a crusted layer several millimeters thick. Over time, this area will not infiltrate, pushing the burden of flow further into the cell away from the inlet. Simple maintenance of scraping off this thin crust should restore normal bioretention function.

Depending upon the design, forebays can be located either “internal” or “external” to the bioretention cells. “Internal” to the cell implies that bioretention fill media lies directly beneath the forebay. Forebays “external” of the cell are hydraulically separated from the bioretention cell and do not have fill media under them. This subtle difference impacts both the design and the maintenance of both forebay types.

Because forebays are concave, "internal" bioretention forebays, if unlined, provide a bypass mechanism for incoming water, subjecting runoff to less media depth for treatment. For example, if bioretention media is typically 0.75 m (2.5 ft) deep and the forebay storage depth is 0.45 m (18 in) deep, only 0.3 m (1 ft) of media would lie underneath the forebay. For many pollutants, this is an insufficient depth needed for treatment (Hunt et al. 2012). To prevent short-circuiting of water provided by the forebay, bioretention cells with "internal" forebays should have those forebays lined with an impermeable membrane. Rip-rap may then be placed over the membrane. The result of this design, of course, is standing water for a potentially long period. As such, the forebay can become a mosquito breeding ground. To prevent mosquito infestation, it is recommended that products that contain Bti (a bacterium that kills mosquito larvae, e.g., "mosquito dunks") be placed in "internal" forebays on a monthly basis during the mosquito season.

"External" forebays, which are hydraulically separated from the primary SCM, should not be lined. If the external forebay is located in clayey soils, they may still pond. In these cases, it is recommended to drain forebays with a simple underdrain, which can be connected to a nearby drainage conveyance (such as the outlet of the bioretention). Slowly draining an external forebay from a bioretention cell can prevent this forebay from becoming a mosquito breeding ground. Many external forebays will dewater within 24 h without the aid of an underdrain, particularly in HSG A and B underlying soils.

While no hard and fast rules exist for sizing forebays for bioretention cells (and level spreader-vegetated filter strip systems), forebays should be sufficiently large to still water so that the entry velocity into the bioretention cell or level spreader does not exceed 1–3 ft/s (0.3–0.9 m/s).

## 18.6 Storage Bowl

The storage bowl represents the concave surface storage for the bioretention facility. Collected runoff will pool in the bowl, allowing it to infiltrate into the media. Some of this water may also be evapotranspirated as it infiltrates into the upper media levels. The bowl depth can be as shallow as a few inches in a residential rain garden application or as deep as 18 in (46 cm) in a larger bioretention facility that has relatively clean runoff such as from a roof, a lightly traveled paved area, or an effective pretreatment configuration. The maximum depth in the bowl is limited by some type of overflow berm or weir built into the facility. This may just be the edge of the bowl in a rain garden. In a larger bioretention cell, the bowl storage may be limited by overflow into a manhole grate (Figure 18.10), or a weir at the outlet (Figure 18.11).

The depth of the bowl storage will control volume reduction and help to manage flows. The bowl storage depth must be matched by media characteristics so the ponded water does not remain for more than 48–72 h. This restriction results from concern due to plant health and in some areas mosquito breeding. Additionally, safety should always be a consideration. Local jurisdictions may require the facility to be fenced if the pond depth poses a drowning hazard to children.

Figure 18.10 Raised Grate for Control of Bowl Storage Depth in Bioretention Cell. (Photo by Authors).

Figure 18.11 Outlet Weir for Control of Bowl Storage Depth in Bioretention Cell. (Photo by Authors).

## 18.7 Bioretention Design: Static Storage and Hydrologic Performance

The bioretention footprint in relation to its drainage area will play a large role in dictating its hydrologic performance. Larger facilities will have greater capacity for storing and infiltrating runoff, along with providing surface areas for evapotranspiration.

From a static design perspective, the storage volume of a bioretention cell can be estimated based on its size, media characteristics, and drainage configuration. The volume of water that a bioretention cell can hold before it exhibits underdrain flow or bowl overflow is defined as the *bioretention abstraction volume (BAV)* and can be set as the volume of runoff that can be managed by a bioretention facility. The BAV is based upon storage in the bowl and media layers, with the media storage defined by specific points related to the water holding capacity in the media pores.

For a bioretention cell with no underdrain and cells with internal water storage, the BAV is given by (Davis et al. 2012):

$$\mathrm{BAV} = \mathrm{Bowl\,Vol.} + \mathrm{RZMS} * (\mathrm{SAT} - \mathrm{WP}) \tag{18.1}$$

where Bowl Vol. is the storage volume in the bowl. RZMS is the root zone media storage, equal to the total bulk volume of the media from the surface through a defined root zone depth. SAT and WP represent the saturated fraction and wilting point, respectively, for the type of media used in the cell. Values of SAT and WP are presented for various soil textures in Chapter 7. Without additional information, the depth of the root zone as a default is assumed as 0.3 m (1 ft).

For an underdrained bioretention cell, underdrain flow generally occurs before bowl storage. Also, the full media depth is utilized. The BAV in this case is given by

$$\mathrm{BAV} = \mathrm{RZMS} * (\mathrm{SAT} - \mathrm{WP}) + \mathrm{LMS} * (\mathrm{SAT} - \mathrm{FC}) \tag{18.2}$$

The LMS is the lower media storage, equal to the media volume below the root zone. FC is the field capacity of the media, also discussed in Chapter 7.

Once the storage capacity of the bioretention facility is exceeded, the cell will begin to discharge through the underdrain, or excess water will overflow if an underdrain is not provided. The relationship between volume in and out of the bioretention cell will depend on the exfiltration characteristics of the water from the bioretention media into the surrounding soils. If the surrounding soils are tight, most of the water that enters beyond the BAV will subsequently exist. If the surrounding soils have a moderate-to-high infiltration rate, the infiltration will reduce the volume of water leaving the facility. An example of the volumetric relationship found for the Villanova bioinfiltration facility is given in Figure 18.12 (Davis et al. 2012).

**Example 18.1** An underdrained bioretention cell has a length of 32 ft, width of 22 ft, and a media depth of 3 ft. The media is classified as a sandy loam. Assuming a vegetated root zone of 1 ft, what is the BAV for this cell?

***Solution*** From Chapter 7, SAT, FC, and WP for a sandy loam soil are estimated at 0.45, 0.179, and 0.081, respectively. The area of the bioretention cell is 32 ft × 22 ft = 704 $\mathrm{ft}^2$. At a 1-ft depth, the volume of the root zone is 704 $\mathrm{ft}^3$; from 1 to 3 ft, the lower media storage volume is 1408 $\mathrm{ft}^3$. Therefore, using Equation 18.2,

$$\mathrm{BAV} = 704\ \mathrm{ft}^3 * (0.45 - 0.081) + 1408\ \mathrm{ft}^3 * (0.45 - 0.179) = 641\ \mathrm{ft}^3$$

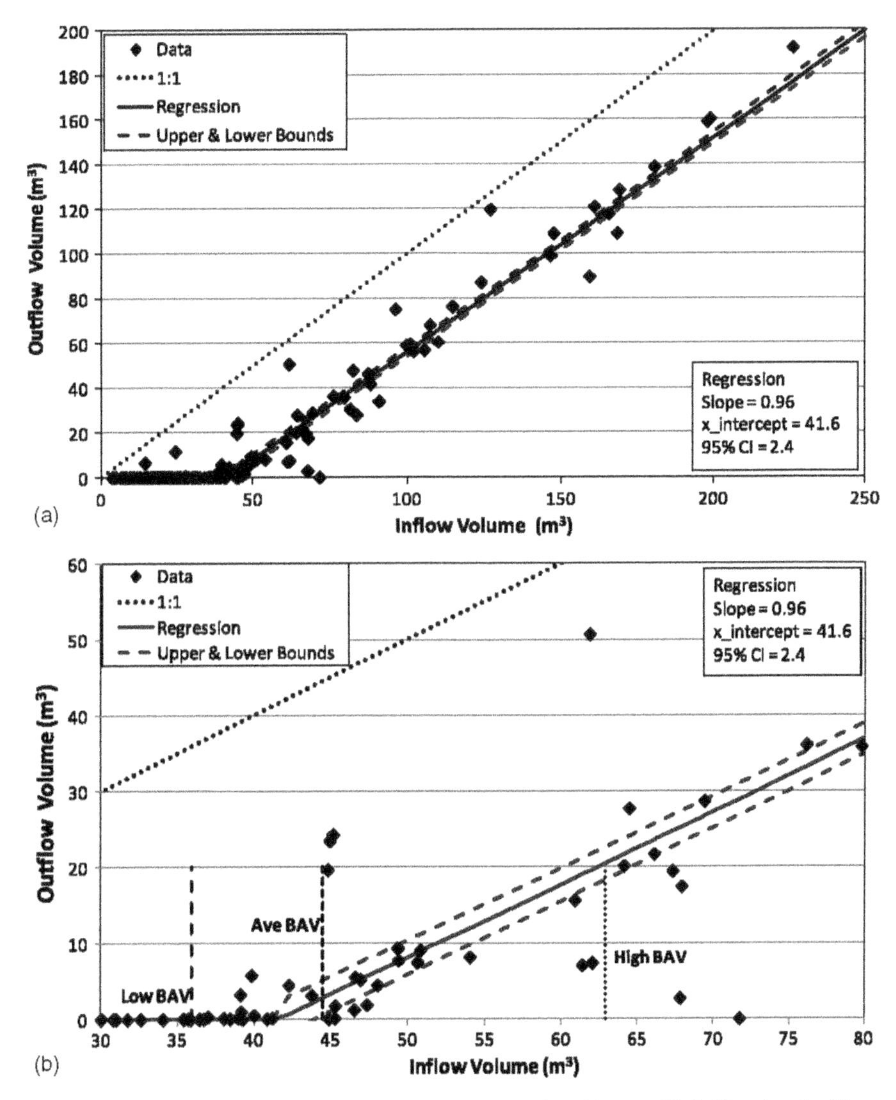


Figure 18.12 Inflow/Outflow Volumetric Relationship for Villanova Bioinfiltration Facility; (a) All Data; (b) Data near BAV Determination (Davis et al. 2012 with Permission from ASCE).

## 18.8 Dynamic Storage

Considerations of dynamic conditions can allow designs that can handle greater stormwater volumes than calculated using static design assumptions. With dynamic design continuous infiltration into the surrounding soils is considered in addition to static storage. The infiltration continuously regenerates some portion of the SCM storage, allowing effective management of larger events, especially those with low intensity, which allow a greater fraction of the stormwater to be infiltrated.

## 18.9 The Media

The physical, chemical, and biological characteristics of the media are key to the overall performance of bioretention. It must have a high hydraulic conductivity so that it can infiltrate a large runoff volume efficiently. However, it must also have some chemical reactivity so that it can remove dissolved pollutants; this usually means that fines are needed. Also, some degree of fines/water holding capacity can assist in promoting evaporation and infiltration from the bioretention media. The media must be able to support the vegetation and promote evapotranspiration. Therefore, a balanced media texture is necessary. It may have different layers to provide different treatments and different depths to balance costs and performance.

### 18.9.1 Rain Gardens

With a homeowner rain garden, the media should be pervious enough so that pooled water will not remain after 24 h. If the existing soil is sandy, nothing needs to be done. Otherwise, after excavating, the existing soil can be mixed with sand and placed back into the hole. The media should not be too sandy that it will not support vegetation growth.

### 18.9.2 Standard Media

Most jurisdictions have a standard specification that they require for bioretention media. Several of these recipes have evolved from practice. Generally, all bioretention media are some mix of coarse sand, low-fines soil, and some organic matter, such as compost, mulch, or peat. A high media sand content is necessary so that the media will not clog, though some designs are appearing with higher silt content to promote plant health and evapotranspiration. The overall sand content is about 85–95% of the media, with fines (silts + clay) making up the rest. The media is generally characterized as a sandy loam or loamy sand soil. The organic matter is required to add water holding capacity and chemical reactivity but should be limited to the needs of the plants; compost will leach nutrients and may be problematic. These media are purchased as specified from a supplier or may be mixed on site by the contractor. Figure 18.13 shows the media being mixed onsite using a mechanical sieve at the Villanova bioinfiltration site in 2001. Note the rough graded bowl in the foreground and piles of sand left and native material waiting to be mixed.

The typical bioretention media depth is 0.75–1 m (2.5–3 ft). This provides enough media to provide significant improvements in both hydrology and water quality, yet still not generate large construction costs. Shallower media may provide some water quality benefits but limited volume and flow reduction. Deeper layers provide diminishing benefits but at exponentially higher costs.

### 18.9.3 Surface Mulch Layer

Many bioretention designs require a surface layer of mulch, about 5 cm (2 in) thickness. Typically, a triple-shredded hardwood mulch is used that is resistant to floating.

Figure 18.13 Media Being Mixed During Construction of Villanova Bioinfiltration Facility in 2001. (Photo by Authors).

The mulch layer helps to keep some moisture for the plants in the media during prolonged dry spells. It also provides capacity for removing and holding heavy metals and possibly other pollutants of interest. Care should be taken to prevent build-up of mulch that can clog the surface infiltration.

## 18.10 Evapotranspiration

Evapotranspiration is a major mechanism for removal of captured water from bioretention and related facilities. Between storms, evapotranspiration (ET) (along with infiltration/percolation) will remove water that has been stored within the facility during the preceding rainfall event. These processes are important in order to reclaim available storage volume before the next rainfall event.

Average ET determined from experimental rain garden lysimeters in Pennsylvania ranged from 2.7 to 4.3 mm/day (Hess et al. 2017) representing 43–70% of the water budget. As expected, values were highest in the summer and lowest in the winter. The greatest ET occurred for a system with internal water storage. Both the Hargreaves and ASCE Penman–Monteith equations (see Chapter 7) can effectively describe the rain garden ET (Figure 18.14, Hess et al. 2019). The use of crop coefficients in the ET models improved the model predictive capabilities. Crop coefficients for early-, middle-, and ending-stage vegetation ranged from 0.45 to 0.84, 1.02 to 1.51, and 0.35 to 1.2, respectively. Crop coefficients for IWS vegetation were higher, at 0.95 to 1.71 for all stages. Trees can play an important role in bioretention ET (Tirpak et al. 2019). Growth rate and canopy size may be important to the overall rate.

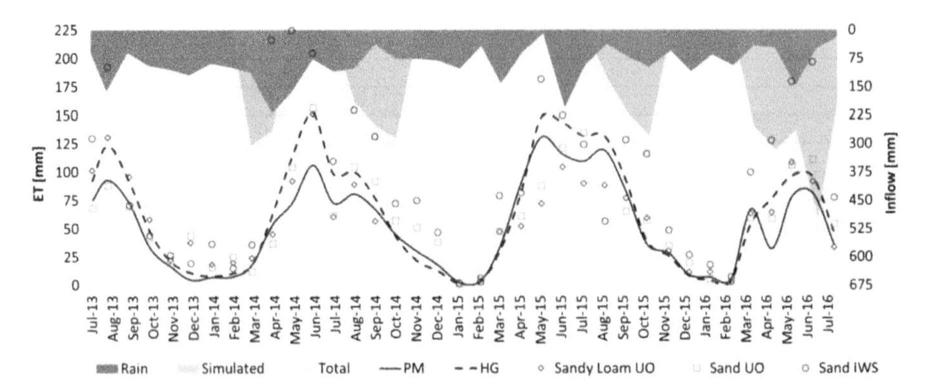


Figure 18.14 Monthly Summations of Daily Added Water (Rain, Simulated, and Total), ET (Sandy Loam UO, Sand UO, and Sand IWS), and ET Models [ASCE Penman–Monteith Equation (PM) and Hargreaves Equation (HG)] for the Three-year Period (Hess et al. 2019, with Permission from ASCE).

**Example 18.2** After a rainfall event in August, the media of a bioretention cell at 39° N latitude is at field capacity (17.9%). Find the time for the top 1 ft of the media to return to a moisture content of 8.1% (the average wilting point for a sandy load soil–Table 7.1). Use the Hargreaves equation; the August average maximum and minimum temperatures are 29.5 and 18.4°C, respectively.

*Solution* The depth of water is found from the media depth and difference in media moisture contents:

$$D = 12\ \text{in}(0.179 - 0.081) = 1.18\ \text{in} = 29.9\ \text{mm}$$

The Hargreaves equation is used to find the ET. From Table 7.4, the August extraterrestrial radiation ($R_a$) measured in terms of evaporation rate is 15.3 mm/day at 38° and 15.2 mm/day at 40°, so use 15.25 mm at 39°. The average temperature is 24°C.

$$\begin{aligned} ET_0 &= 0.0023\ R_a\left(\text{TC} + 17.8\right)\left(T_{\max} - T_{\min}\right)^{0.5} \\ &= 0.0023(15.25\ \text{mm/day})(24 + 17.8)(29.5 - 18.4)^{0.5} \\ &= 4.88\ \text{mm/day} \end{aligned}$$

Therefore, the time to evapotranspire 112.5 mm of water is

$$\text{Time} = \frac{29.9\ \text{mm}}{4.88\ \text{mm/day}} = 6.1\ \text{days}$$

## 18.11 The Media and Particulate Matter Removal

Bioretention has been found to be very effective in the removal of particulate matter. A number of studies have documented TSS concentrations in underdrain and overflow streams that are much lower than that of the inputs (Davis 2007; Hatt et al. 2009; Li and Davis 2009).

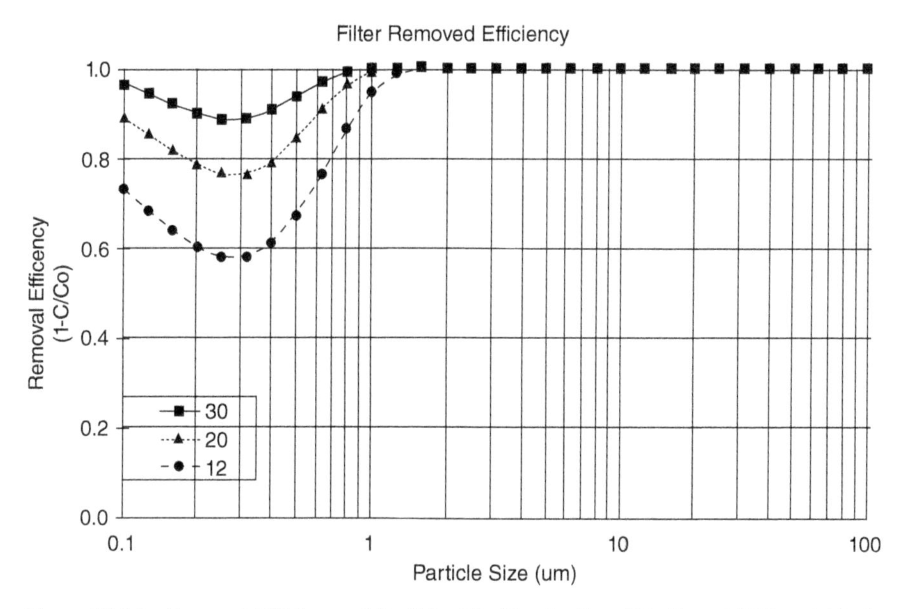


Figure 18.15 Removal Efficiency of Particles Via Bioretention Filtration at Different Media Depths (12, 20, and 30 in; 30, 51, and 76 cm), Based on Filtration Theory (Chapter 8). Flow Rate = 2 in/h (5 cm/h), Media Diameter = 0.5 mm, 40% Media Porosity, $\alpha = 0.1$.

Particulate matter removal occurs via sedimentation and filtration, both of which are active in bioretention. Sedimentation will occur doing storage and pooling of runoff in the bowl and will follow Stoke's law.

As water infiltrates into and through the media, filtration of suspended solids will occur. As noted in Chapter 8, the mechanism of filtration depends on media size, the size of the particulate being captured, and the contact time with the media, or media depth.

A simulation using filtration theory can be completed using operational filtration values typical for bioretention. The results, shown in Figure 18.15 for three bioretention depths, indicate that filtration should be very effective for all but the smallest particles and the complete removal is predicted for particles greater than 1.5 $\mu$m in diameter.

Filtration theory also indicates that the particle removal will be concentrated at the media surface, where the polluted water will first contact the media. Laboratory studies support this surface accumulation, as do field studies of metals attached to particulate matter (Jones and Davis 2013; Li and Davis 2008a, 2008b). Theory, laboratory, and field studies suggest that only a shallow media is necessary for effective PM reduction and that PM removal performance does not vary significantly with bioretention design (Hatt et al. 2009).

**Example 18.3** The media in a bioretention cell has a $d_{10}$ of 0.1 mm and a porosity of 0.38. The concentration of 0.015 mm runoff particles is 45 mg/L. At an infiltration rate of 2 in/h, find the concentrations of 0.015 mm particles at media depths of 3, 6, 12, and 24 cm. Assume that the filtration capture mechanism is dominated by interception (see Chapter 8) and that the capture efficiency is 10%.

***Solution*** From Chapter 8, the interception transport term is calculated as

$$\eta_I = \frac{3}{2}\left(\frac{d_p}{d_c}\right)^2 = \frac{3}{2}\left(\frac{0.015 \text{ mm}}{0.1 \text{ mm}}\right)^2 = 0.0338$$

The equation for estimating filtration efficiency is

$$C = C_0 \exp\left(-1.5\frac{(1-\epsilon)\alpha\eta L}{d_c}\right)$$

Therefore,

$$C = 45 \text{ mg/L} \exp\left(-1.5\frac{(1-0.38)(0.1)(0.0338)(L \text{ cm})}{(0.01 \text{ cm})}\right)$$

$$C = 45 \text{ mg/L} \exp\left(-0.314\, L(\text{cm})\right)$$

For $L$ values of 3, 6, 12, and 24 cm, $C$ is equal to 17.5, 6.8, 1.0, and ≪1 mg/L, respectively.

## 18.12 The Media and Heavy Metals Removal

Since heavy metals (Cu, Pb, Zn) exist in runoff in both the particulate and dissolved phases, several unit processes are active in their removal via bioretention. Removal of particulate metals follows particulate matter removal, which, as noted in Section 18.11, is very good.

Dissolved metals are removed via adsorption processes on the organic matter and fine inorganic particles in the media (Chapter 8). Metals will strongly adsorb onto these media components and metals removal via bioretention is overall very effective. A probability plot showing Zn concentrations in and out of two Maryland bioretention cells is presented in Figure 18.16 (Li and Davis 2009). Overall reductions were approximately an order of magnitude and all bioretention Zn discharges were below the aquatic toxicity water quality target of 120 μg/L.

Most adsorption will occur at the first point at which the water reaches the media, so the majority of the metal removal will occur in the surface media and this is where the metals will accumulate in the media. Samples of bioretention media have shown that metals accumulate at the very surface, especially for Pb (Li and Davis 2008b). Metals will also accumulate most near the entrance point to the cell, which receives the first flush runoff and runoff from most all storms (Jones and Davis 2013). Media furthest from the entrances may not be utilized for smaller events.

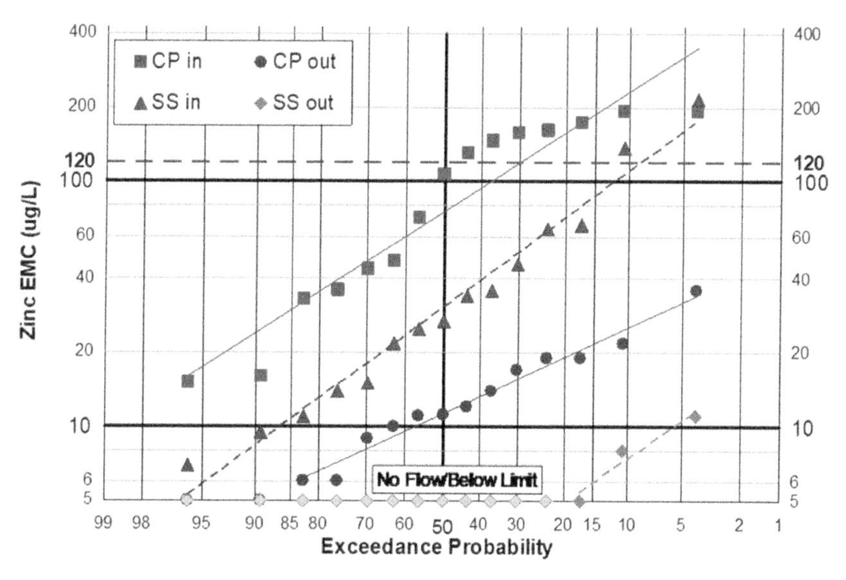


Figure 18.16 Concentrations of Zn Entering and Discharged from Two Maryland Bioretention Cells (Li and Davis 2009, with Permission from ASCE).

Bioretention capacity for heavy metals is an ongoing concern. Because of the high capacity of the media and the relatively low metal concentrations in runoff, a long media lifetime for metals is expected. Lifetime estimates of greater than 15–20 years have been made based on metals capture and metal accumulations after 4 years have indicated great excess in capacity (Jones and Davis 2013). Metal uptake into vegetation is possible, but a laboratory study has suggested that this pathway is very minor compare to accumulation in the media (Sun and Davis 2007).

## 18.13 The Media and Organic Pollutants Removal

Hydrocarbons in water will sorb to organic matter in bioretention media. Effective removal of various hydrocarbons, such as toluene, naphthalene, and motor oil onto a leaf compost bioretention surface media has been noted in laboratory studies (Hong et al. 2006). A shredded cedar mulch was found to be effective for the sorptive removal of several herbicides (Huang et al. 2006).

A bioretention field study found significant removal of both dissolved and particulate polycyclic aromatic hydrocarbon (PAH) compounds (DiBlasi et al. 2009). Runoff particulate-bound total PAH concentrations were about 0.2–4.7 μg/L, while the dissolved concentrations were about an order of magnitude less. Effluent total PAH concentrations were always less than 0.37 μg/L. Because the majority of the PAH was particulate-bound, removal correlated with TSS removal. PAH accumulated in bioretention media at the surface, as expected based on filtration of particulate-bound PAH and adsorption of dissolved compounds (DiBlasi et al. 2009).

Biodegradation of hydrocarbons has been noted in both laboratory and field studies. Toluene, naphthalene, and used motor oil that was sorbed onto a leaf mulch compost was biodegraded after a few days, showing little evidence of the parent compounds (Hong et al. 2006). Numbers of hydrocarbon-degrading bacteria increased during the biodegradation process. Column studies on the fate of naphthalene in bioretention indicate that the majority of the hydrocarbon becomes adsorbed, with subsequent biological mineralization and limited uptake into vegetation (LeFevre et al. 2012a, 2012b). Vegetation in the column systems increased overall naphthalene removal efficiency.

In field testing, total petroleum hydrocarbon (TPH) levels found in bioretention soils were low (<3 μg/L), suggesting biodegradation with no buildup of TPH. Media sampled from several sites showed mineralization of naphthalene, with the rate of mineralization being a function of the number of copies of bacteria genes related to hydrocarbon degradation found in the media. Facilities with deep-rooted plants contained higher levels of these genes than systems with turf grass or mulch (LeFevre et al. 2012a).

Overall, evidence to date suggests that the fate of hydrocarbons is first via capture onto bioretention media, followed by biological mineralization. Hydrocarbon-degrading bacteria exist naturally in the media and will flourish when exposed to the hydrocarbons. No evidence of significant hydrocarbon accumulation is noted. These mechanisms appear to be enhanced by thick-rooted vegetation. The capture and biodegradation of hydrocarbons appear to follow a sustainable process in bioretention.

## 18.14 The Media and Phosphorus Removal

Phosphorus (P) treatment and removal via bioretention is complex. Phosphorus concentrations targeted for excellent water quality in fresh waters can be very low, on the order of 0.008–0.04 mg/L. Strict attention to the media and its characteristics must be made in order to meet stringent P targets.

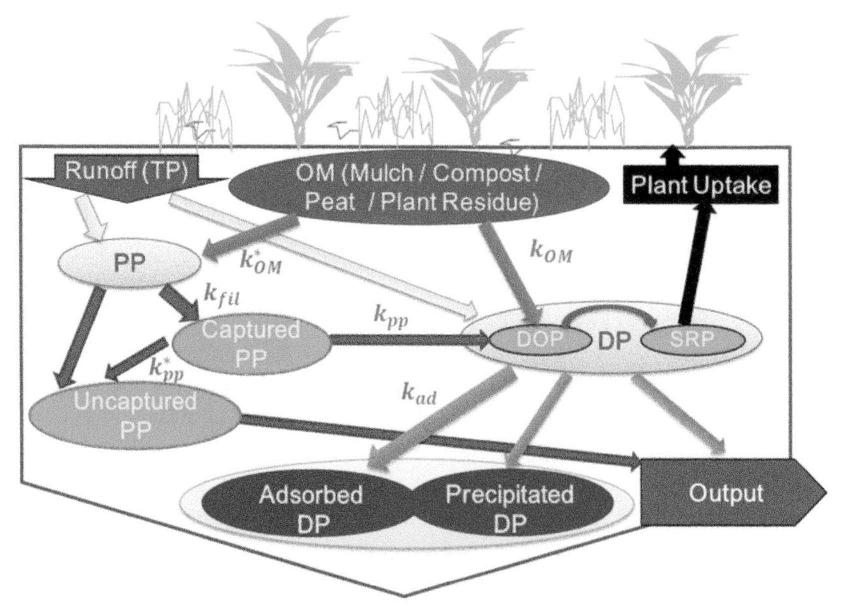


Figure 18.17 Diagram Showing the Sources and Fates of Various Forms of Phosphorus in Bioretention (Li and Davis 2016).

As discussed in Chapter 3, phosphorus has a complex chemical speciation, existing in particulate and several dissolved forms. Additionally, because P is a macronutrient, it is ubiquitous in the natural environment. A diagram showing the fates of P in bioretention is presented in Figure 18.17 (Li and Davis 2016).

### 18.14.1 Phosphorus Removal in Bioretention

Like heavy metals, runoff P includes both particulate and dissolved P. Therefore, the removal of these species are independent, although release of captured P from one phase may end up in the other. Particulate P will be removed via filtration and dissolved P via adsorption mechanisms. Particulate P removal via bioretention can be very effective, as particulate matter is filtered from the runoff (Li and Davis 2016), collecting at the media surface, and correlations between particulate matter and particulate P removal have been found. However, over time, P that had been affiliated with the particulate matter may be released and manifest as dissolved P during a later storm event.

In addition to incoming runoff as a source of P, internal sources are possible as well. Decaying vegetation in the bioretention cell can release phosphorus. Very important to the control of P in bioretention is the makeup and characteristics of the media. Most bioretention designs call for some amount of organic matter, to assist with plant growth, to hold water, to assist in heavy metals removal, and for other benefits. However, as this organic matter breaks down, it will release P. Several older studies have demonstrated bioretention cells exporting concentrations of P higher in the underdrain discharge than found in the incoming runoff (Dietz and Clausen 2005; Hunt et al. 2006; Li and Davis 2009). The source of this P can be the organic matter employed, or the soil-based media may have already been exposed to high concentrations of P, which leaches when submitted to high flow rates in the bioretention. In one specific case noted by Hunt et al. (2006), the bioretention media was synthesized by onsite soil. The previous land use included animal agriculture and the soil used had a high P content. When employed in the bioretention cell, this excess P was leached away.

In a similar manner, several recent design documents propose high amounts of compost to create bioretention media. Compost, especially fresh compost, will leach high levels of dissolved P (Hurley et al. 2017); that is one reason it is effective in promoting plant growth. Phosphorus concentrations exceeding 1 mg/L have been reported in bioretention and other SCMs that employ compost media (Li et al. 2011; Mullane et al. 2015; Palmer et al. 2013; Toland et al. 2011). Compost should be avoided in bioretention, especially those with an underdrain.

As discussed in Chapter 8, the adsorption of dissolved P in soils can be correlated with the amount of amorphous iron and aluminum in the soil. Based on these parameters, or related parameters, the *Phosphorus Index*, or *P Index* can be defined. In North Carolina, the P Index is based on the amount of P extracted from the media, which is then normalized to a value indicative of the amount of P which typically causes leaching. A value of 10–30 is required for bioretention media (NCDENR 2009).

Long-term accumulation of ortho-P in a bioretention cell showed P accumulation in the top 10 cm of the media (Komlos and Traver 2012). The extracted media ortho-P concentration at the surface was about twice that deeper in the cell. Results indicate excess P sorption capacity in the media for >20 years.

**Example 18.4** Estimate the adsorption capacity lifetime of a bioretention system for the adsorption of phosphate. The annual rainfall is 100 in/year, with 90% becoming runoff. The runoff concentration contains 0.5 mg/L phosphate, as P. The bioretention cell area is 5% that of the drainage area and is 3 ft deep. The media has a bulk density of 90 lb/ft$^3$ and contains 0.05 mg/g adsorbed phosphate when it is placed into the facility. The adsorption of the phosphate can be described by a Langmuir isotherm with $Q$ = 1.5 mg/g as P and $K_L$ = 0.3 L/mg.

*Solution* Adsorption isotherm information from Chapter 8 will be used to solve this problem. Several of the parameters for the adsorption calculation need to be determined from the given information. Since areas are not given, they must cancel in the equation.
The media mass is the volume times the bulk density.

$$M = A_B H \rho_b = 0.05A(3\text{ ft})(100\text{ lb/ft}^3) = 15\,A\text{ lb}$$

where media area $A_B$ is 5% of the drainage area, in ft$^2$.
The annual volume is the runoff depth times the drainage area.

$$V = DA = 0.9\frac{100}{12}\text{ ft } A = 7.5A\text{ ft}^3\text{/year} = 212A\text{ L/year}$$

The saturated phosphate amount is given from the isotherm, as presented in Chapter 8.

$$q = \frac{QK_LC}{1+K_LC} = \frac{(1.5\text{ mg/g})(0.3\text{ L/mg})(0.5\text{ mg/L})}{1+(0.3\text{ L/mg})(0.5\text{ mg/L})} = 0.196\text{ mg/g} = q_e$$

Therefore, rearranging the isotherm equation, with $q_0$ = 0.05 mg/g:

$$t = \frac{M(q_e - q_0)}{VC_0} = \frac{15A(\text{lb})(0.196-0.05\text{ mg/g})(454\text{ g/lb})}{212A(\text{L/year})(0.5\text{ mg/L})} = 12.7\text{ years}$$

### 18.14.2 Quantifying Phosphorus Removal

Dissolved P removal via bioretention appears to behave based on an adsorption equilibrium type of relationship. During the flow of runoff through the media, a local equilibrium occurs between the runoff and the media. The concentration of P in the runoff in equilibrium with the P in the media is given by $C^*$ (Li and Davis 2016). $C^*$ is related to the amount of P adsorbed or otherwise affiliated with the media, such as by a Freundlich isotherm relationship, as introduced in Chapter 8.

$$q = K_F C^{*1/n} \tag{18.3}$$

where $q$ is the amount of P adsorbed or otherwise affiliated with the media. When the media contains a large amount of adsorbed P, $q$ is relatively large. This leads to a relatively large value of $C^*$.

When the incoming runoff has a concentration of P that is greater than $C^*$, then adsorption of P onto the media will occur and removal of P from the stormwater takes place. However, if the incoming P concentration is low, specifically less than $C^*$ for the media, then no P removal will occur and in fact, release of P from the media is expected (Figure 18.18).

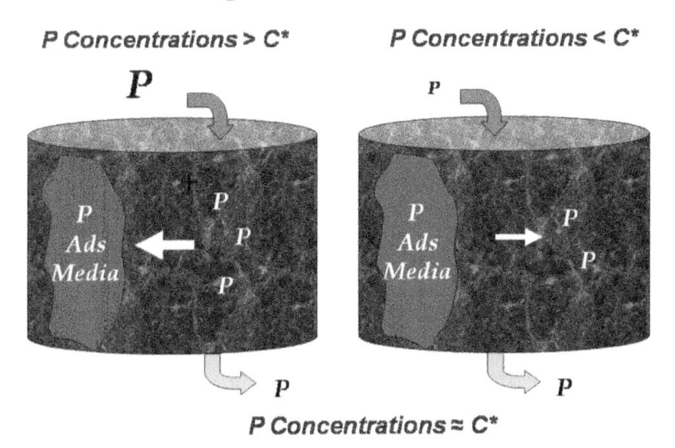


Figure 18.18 Diagram Showing the $C^*$ Equilibrium Model for Phosphorus Fate and Removal in Bioretention Media.

This $C^*$ model explains the P performance results described in the previous section. For compost media or media with a high P Index, the $q$ is high and the corresponding $C^*$ is large. Therefore, when the runoff enters the bioretention media, since $C^*$ is larger than the runoff $C$, leaching of P from the media occurs, resulting in higher effluent concentrations than influent (Li and Davis 2016).

The $C^*$ model can explain P performance for standard bioretention media. McNett et al. 2011) collected influent and effluent P data from bioretention cells in North Carolina and Maryland. The data were found to group into two categories. For influent P concentrations greater than 0.19 mg/L, P removal was found. However, for influent P concentrations less than 0.19 mg/L, essentially no removal was noted. These data suggest that typical bioretention media specifications will specify a media that leads to a $C^*$ of about 0.19 mg/L. Influent concentrations above this value will receive treatment, as the P will adsorb onto the media. However, lower concentrations will not be reduced and may see an increase in P concentration as the media approaches equilibration with the runoff water, at about 0.19 mg/L.

### 18.14.3 Media Enhancements for Phosphorus Removal

Because the value of $C^*$ is controlled by the P adsorption characteristics of the media, modifications of the media have potential to create media with higher P adsorption capacity, which may lead to lower values of $C^*$ and correspondingly lower effluent P concentrations. The adsorption of P in soil is controlled by the amount of amorphous iron and aluminum (hydr)oxide minerals in the soil ($(Fe + Al)_{am}$). Also, high calcium concentrations can lead to precipitation/coprecipitation of calcium phosphate minerals. Therefore, amending media with $(Fe + Al)_{am}$ and/or a calcium source can lead to enhanced P removal.

$(Fe + Al)_{am}$ amended bioretention media has been invested in several laboratory studies (Lucas and Greenway 2011c; O'Neill and Davis 2012a, 2012b; Yan et al. 2016, 2017a, 2017b). The $(Fe + Al)_{am}$ was added as "red mud," which is a byproduct of aluminum production, or as water treatment residual (WTR). The latter is a waste product from the treatment of potable water during the coagulation process. During coagulation, alum or ferric chloride is employed as an inorganic coagulant to promote particle agglomeration. The residual from this process is an amorphous aluminum or iron sludge, mixed with the particulate matter from the raw water supply.

Additionally, another waste product, steel slag, has been evaluated as a bioretention media amendment. Steel slag is a byproduct of steel production with a very high free lime content. Because of this lime, calcium is released, which can bind P. Addition of small amounts of fly ash to sand resulted in significant increases in P uptake capacity, again reducing effluent P concentrations to values acceptable for sensitive water discharge (Zhang et al. 2008).

Adding WTR or fly ash to bioretention media can lead to improved P uptake and overall lower effluent P concentrations in runoff. Laboratory studies have indicated that dissolved P concentrations consistently as low as 0.025 mg/L may be obtained using WTR-enhanced P media (O'Neill and Davis 2012b). The oxalate ratio (OR) can be employed to quantify the P adsorption capacity for the media (O'Neill and Davis 2012a; Yan et al. 2017b):

$$OR = \frac{\left(Al_{ox} + Fe_{ox}\right)}{P_{ox}} \tag{18.4}$$

where $Al_{ox}$, $Fe_{ox}$, and $P_{ox}$ represent the oxalate-extractable Al, Fe, and P concentrations, respectively, in mmol/kg. Increasing the OR increases the media P sorption capacity (Figure 18.19). Values of OR between 20 and 40 appear to provide for large P adsorption capacity and small P runoff effluent concentrations. Additional modifications to increase the OR > 100 using alum or partially hydrolyzed aluminum amendments showed even greater P adsorption capacity (Figure 18.20).

Field evaluation of the addition of WTR to bioretention media has indicated enhanced P removal, with no impact on volume management or particulate matter removal (Liu and Davis 2014). Significant removal of particulate P occurred via filtration and adsorption of inorganic phosphate was noted. Concentrations of organic P were essentially unchanged, suggesting input of organic P, possibly due to release from captured particulate P.

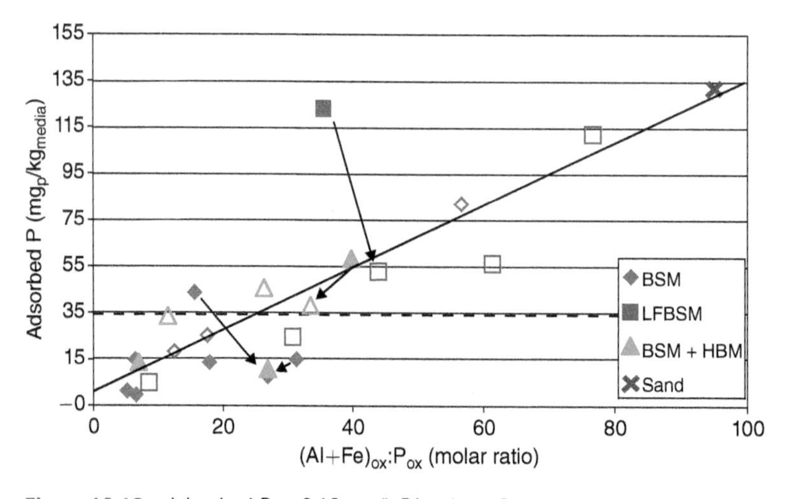


Figure 18.19 Adsorbed P at 0.12 mg/L Dissolved Ortho-phosphorus at Different Oxalate Ratios (OR). OR Was Increased by Addition of Aluminum-based Water Treatment Residual. BSM = Bioretention Soil Media, LFBSM = Low Fines Bioretention Soil Media, HBM = Hardwood Bark Mulch (O'Neill and Davis 2012a, with Permission from ASCE).

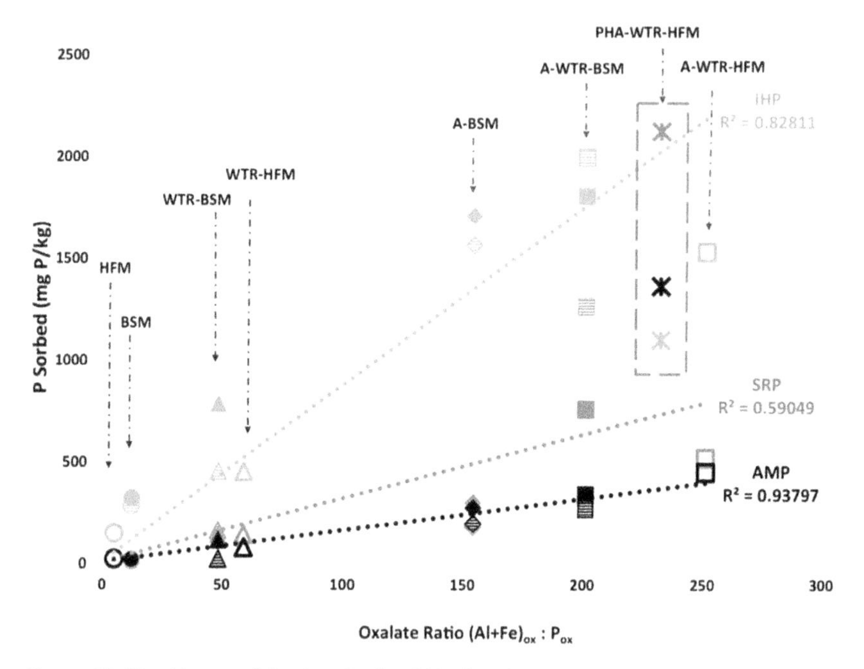


Figure 18.20 Measured Oxalate Ratio of Media Mixtures and Their Phosphorus Sorption Capacities (at Ce = 0.2 mg P/L); Bioretention Soil Media (BSM) Batch Data: Hatched Symbols; BSM Column Data: Solid Symbols; High-flow Media (HFM, Column Data: Open Symbols) (BSM Data from Yan et al. 2016). Partially Hydrolized Al (PHA)–WTR–HFM Data Not Included in Linear Regression (Yan et al. 2017b, with Permission from ASCE).

IHP = Inositolhexaphosphate, SRP = Soluble reactive phosphorus, AMP = Adenosine monophosphate

**Example 18.5** The bioretention media of Example 18.4 has been modified with an aluminum amendment. This amendment increases the media adsorption capacity for phosphate so that the phosphate Langmuir isotherm parameters are $Q$ = 4.1 mg/g as P and $K_L$ = 0.68 L/mg. All other parameters are the same as given in Example 18.4.

*Solution* Following the solution of Example 18.4, media mass is 15$A$ lbs (where $A$ is the bioretention footprint area) and the runoff flow rate is 212$A$ L/year. The saturated phosphate amount is given by the Langmuir isotherm:

$$q = \frac{QK_LC}{1+K_LC} = \frac{(4.1\text{ mg/g})(0.68\text{ L/mg})(0.5\text{ mg/L})}{1+(0.68\text{ L/mg})(0.5\text{ mg/L})} = 1.04\text{ mg/g} = q_e$$

As in Example 18.4, rearranging the isotherm equation, with $q_0$ = 0.05 mg/g:

$$t = \frac{M(q_e - q_0)}{QC_0} = \frac{15A(\text{lb})(1.04 - 0.05\text{ mg/g})(454\text{ g/lb})}{212A(\text{L/year})(0.5\text{ mg/L})} = 63.6\text{ years}$$

This lifetime is five times longer than the 12.7 years predicted for the unamended media. This lifetime should be greater than the useful life of a bioretention cell.

## 18.15 The Media and Nitrogen Removal

As discussed in Chapters 3 and 8, the urban nitrogen cycle is complex. Nitrogen (N) will exist in stormwater in various forms and transformations will occur during storage and flow through bioretention. Nitrogen performance data for bioretention tend to show variable reductions in total concentrations ranging from N leaching to some degree of removal (Hatt et al. 2009; Li and Davis 2009, 2014). Overall removal of nitrogen from the bioretention facility will require significant design modifications to promote denitification, or intensive management and removal of vegetation biomass.

### 18.15.1 Nitrogen Processing in Standard Bioretention Systems

Nitrogen in runoff exists primarily as organic N, both particulate and dissolved. A fraction is present as nitrate, along with low concentrations of ammonium and nitrite. A schematic of nitrogen fate in bioretention is provided in Figure 18.21.

In addition to N in the runoff, other sources of N must be considered. These include decay of plant litter and residue, release of N from decomposing organic media amendments, and the possible covert fertilizer application to supplement vegetation nutrients.

Some amount of the total nitrogen is particulate in form and will be effectively filtered out at the bioretention surface. Some fraction of the dissolved organic nitrogen and ammonium will be held by various components of the bioretention media, likely soil fines and organic matter. However, nitrate is not held by the media and will directly wash through. In underdrained bioretention systems, as the captured water

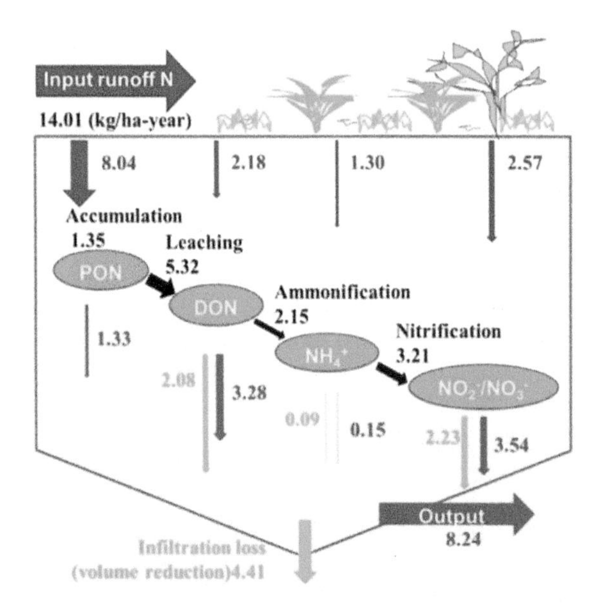


Figure 18.21 Simplified Pathway of Nitrogen Fate in Bioretention. Values of N Loadings are in kg/ha year. PON = Particulate Organic N, DON = Dissolved Organic N. Reprinted with Permission from Li, L., and Davis, A. P. (2014). "Urban Stormwater Runoff Nitrogen Composition and Fate in Bioretention Systems." Environmental Science & Technology, 48(6), 3403–3410. Copyright 2014 American Chemical Society.

drains and is evapotranspirated, the media remains aerobic between rainfall events. During this time, the captured nitrogen species become biotransformed. Some organic N can be mineralized to ammonium, while the ammonium can be nitrified to nitrate. Therefore, the net bioretention performance result for nitrogen is little accumulation by the media, as all the nitrogen species follow the nitrogen cycle toward nitrate (Li and Davis 2014). When the next rainfall occurs, nitrogen is washed out as organic N and nitrate.

The overall picture for N fate in bioretention is summarized by the data presented in Figure 18.22. Particulate N is decreased due to filtration by the media. Dissolved organic N (DON) increases due to leaching from various organic sources, including captured particulate N, vegetation litter, and possibly organic matter in the media. Ammonium is held by the fines in the media, so the effluent concentration is small.

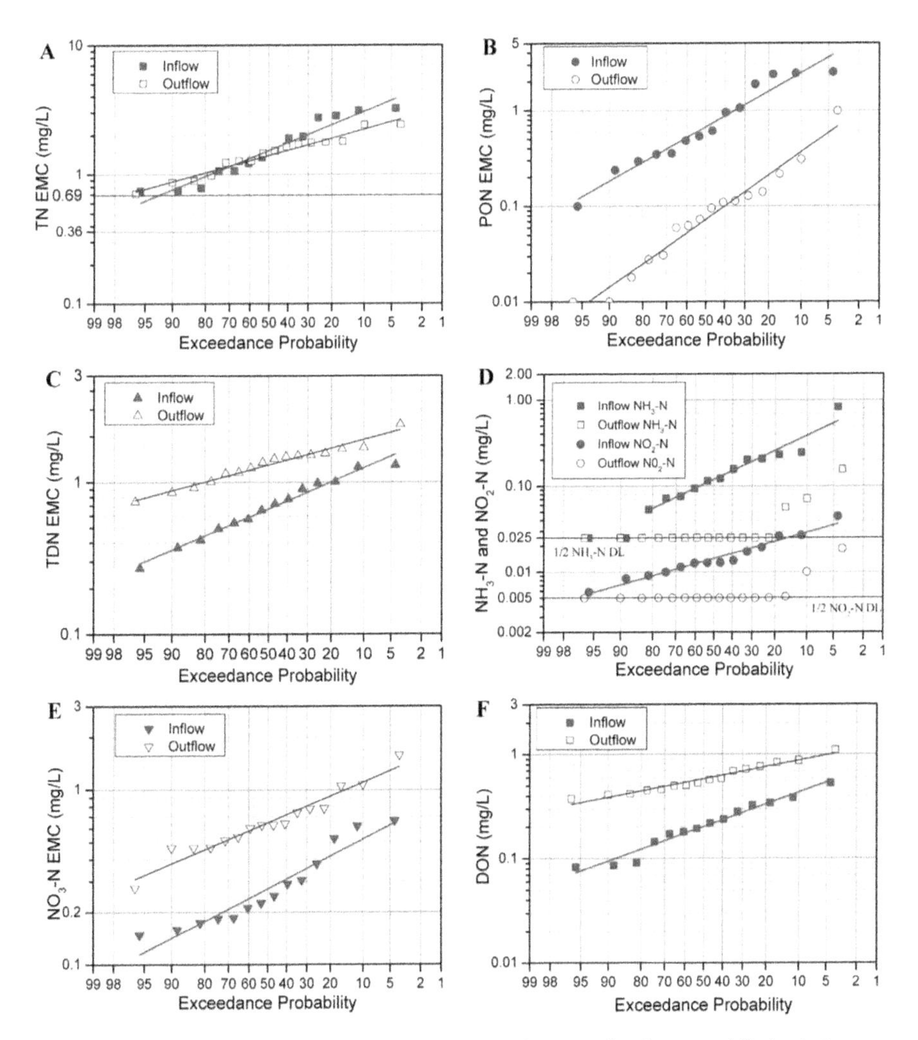


Figure 18.22 Summary Data of N Concentrations in Bioretention Input and Underdrain Flows. Reprinted with Permission from Li, L., and Davis, A. P. (2014). "Urban Stormwater Runoff Nitrogen Composition and Fate in Bioretention Systems." Environmental Science & Technology, 48(6), 3403–3410. Copyright 2014 American Chemical Society.

$NO_x$ is increased because held ammonium becomes nitrified between storm events (Chen et al. 2013). Some degree of removal may be expected due to vegetative uptake and pockets of denitrification in the media, but these pathways are generally minor. The net result is no measurable change in effluent N concentration.

In calculations of N mass loads, reductions of total N are typically recorded. The N mass reduction, however, is dominated by the reduction of surface runoff volume via exfiltration to the surrounding soils (Figure 18.21). For most pollutants, such as heavy metals and P, exfiltration losses are not of great concern because additional sorption is expected to continue to lower pollutant concentrations in the surrounding soils. This is not the case with nitrate, which will not be attenuated by the natural soils, but can continue to be transported in the subsurface.

### 18.15.2 Enhanced Nitrogen Removal

Like phosphorus, compost will also leach N (Hurley et al. 2017). High N concentrations have been found leaching from compost-containing SCMs (Li et al. 2011; Mullane et al. 2015; Palmer et al. 2013; Toland et al. 2011). Biosolids-based compost flushes extremely high N concentrations, much higher than other compost types (Gardina 2016).

To remove nitrogen from runoff using bioretention, two pathways are available, biological denitrification and plant uptake. However, in order for denitrification to be effective in the overall N discharge, all of the N must be converted to nitrate. Layering of media can impart a "treatment train" approach to bioretention. Specific media can be tailored to address treatment of specific pollutants. This has been evaluated in laboratory studies, primarily in attempts to improve N removal (Cho et al. 2009; Hsieh et al. 2007). Upper layers can capture and hold organic N and ammonium, while lower layers restrict water flow and promote denitrification.

Amendments to bioretention media can be added to sorb and hold the organic N that leaches from the system. Activated carbon has been noted as an effective media enhancement for the removal of various organic N compounds that may be expected in urban runoff (Mohtadi et al. 2017). The carbon can be mixed with the upper media layers, allowing adsorption of the organic N and allowing time for the minerization of the organic N compounds to ammonium.

Ammonium can be adsorbed onto the fines in bioretention media. Clinoptilolite, a zeolite mineral, has been shown to effectively capture ammonium from runoff solutions (Khorsha and Davis 2017a, 2017b). The zeolite can hold the ammonium during rainfall events and buffer high incoming ammonium concentrations. Sorbed ammonium is biotransformed to nitrate (Figure 18.23). The nitrate must be managed using an alternative design, as will be discussed in Section 18.19.

### 18.15.3 Biological Nitrogen Transformations

Biotransformation of N appears complex in bioretention systems. Research methods that employ quantitative determination of nitrification and denitrification genes spatially in bioretention media, along with isotopic studies, have provided some information on these processes. Nitrogen cycling, with specific functional genes, is shown in Figure 18.24. Nitrification (*amoA*) genes have been measured in bioretention media,

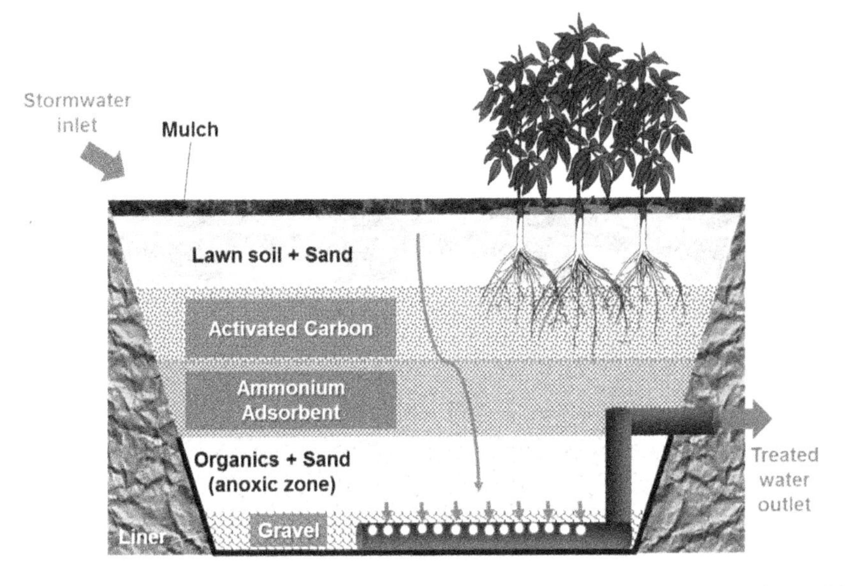


Figure 18.23 Layering of Bioretention Media for Removal of Nitrogen (Mohtadi et al. 2017).

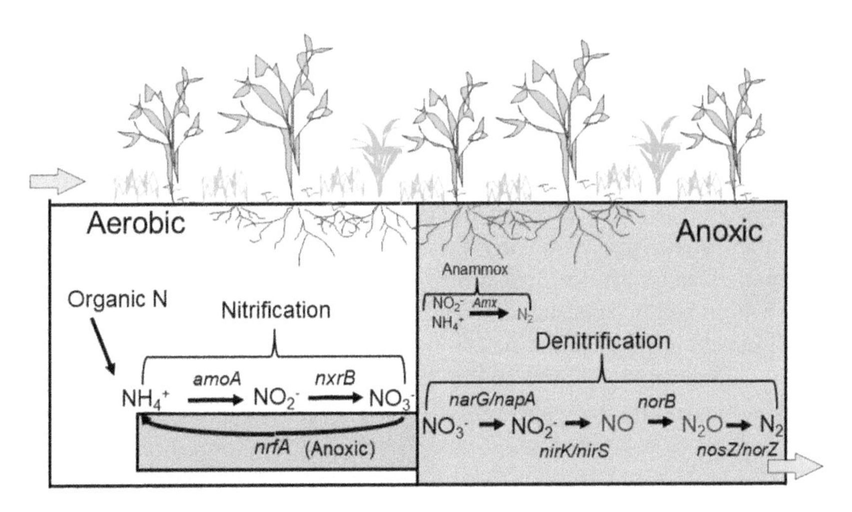


Figure 18.24 Redox Layering of Bioretention Media for Removal of Nitrogen (Modified from Chen et al. 2019).

generally at decreasing concentrations with increasing depths for multiple media cores from a single bioretention cell (Chen et al. 2013) and vegetated bioretention mesocosms (Chen et al. 2019).

Several studies have quantified various denitrification genes (*nirS*, *nirK*, *norB/cNor/qNor*, and *nosZ*) in field bioretention media (Chen et al. 2013; Waller et al. 2018; Willard 2017) and laboratory studies (Chen et al. 2019; Morse et al. 2018). In essentially all cases, higher concentrations are found near the surface of the media as opposed to deeper, even in systems that had internal water storage that would be expected to promote microbial denitrification. This is somewhat counterintuitive. It has been suggested that the higher organic matter content in the upper layers, and possibly root extrudates, promote the higher concentrations of denitrifiers.

The type of plant used in bioretention appears to affect the denitrification microbial consortia (Morse et al. 2018); bioretention mesocosms with certain plant species had higher concentrations of *Nar*, *nirK*, *qNor*, and *nosZ* than others. Data suggested a competition between plant N uptake and microbial denitrification processes (Morse et al. 2018). In mesocosms with greater root volume, more N was assimilated by the plants and less was denitrified. This could be related to the oxic local environment very near the roots (Chen et al. 2019).

## 18.16 The Media and Bacteria Removal

The behavior of bacteria in bioretention is complex. First, important to the stormwater treatment process is the capture of bacteria and removal from the stormwater. However, as important is the fate of captured bacteria. Bacteria are living organisms and this complicates understanding their fate in bioretention.

Initial capture of bacteria follows mechanisms similar to the capture of colloids and particulate matter. Porous media filtration should be the operative capture mechanism. Based on standard filtration theory, moderate removal of bacteria, with sizes near 1 μm, should be expected. Laboratory column studies with an *Escherichia coli* species found the removal to be moderate (40–60% removal) during the initial studies with fresh media, but *t* increased to essentially 100% removal after about 6 months of runoff application (Zhang et al. 2011). Field studies have shown moderate but also somewhat mixed results for bacteria removal. *E. coli* and fecal coliform have been found to be reduced by about an order of magnitude by a bioretention facilities in Maryland (Zhang et al. 2012) and North Carolina (Hathaway 2009; Hathaway et al. 2011). As expected from filtration theory, media depth is important. A shallow 25-cm media bioretention cell did not demonstrate significant *E. coli* removal (Hathaway et al. 2011). Field studies are complicated by the high variability of various bacteria in runoff. The seasonal variation can be several orders of magnitude and concentrations at freezing/near temperatures may be negligible.

The fate of capture *E. coli* cells was evaluated in laboratory column studies. At temperatures of 25 and 37°C, the bacteria died off quickly, within a few days. Die-off was much slower at 15 and 5°C, lasting 15–22 days. The die-off appeared to be caused by growth of protozoa in the media, with higher growth at higher temperatures (Zhang et al. 2012).

## 18.17 Vegetation

Vegetation plays an important role in the effective operation of bioretention. Private rain gardens can be planted with a mix of flowers and other ornamental plants that will survive. Grasses, flowers, shrubs, and small trees are commonly specified as the bioretention vegetation for larger facilities, although different planting plans have been used in areas to meet specific needs. Research has shown that vegetation plays an

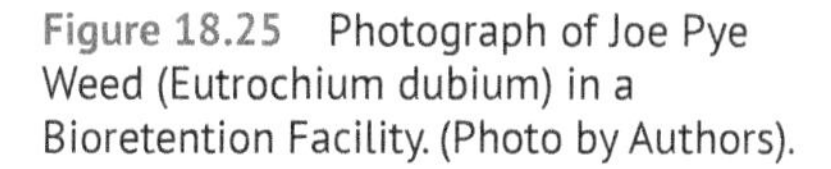

Figure 18.25 Photograph of Joe Pye Weed (Eutrochium dubium) in a Bioretention Facility. (Photo by Authors).

important role in reducing discharge levels of nutrients, especially N (Barrett et al. 2013; Lucas and Greenway 2008; Read et al. 2010).

Plant survival is important in bioretention systems. The fairly harsh bioretention environment can pose many challenges to plant survival. The media is very sandy to allow high infiltration rates. This produces very dry media during prolonged times of drought and many types of vegetation cannot tolerate such dry conditions. On the other hand, during large rainfall events, runoff will pool in the bioretention bowl to a depth of 20–30 cm (8–12 in). This pooling and possible slow drainage also create conditions that some plants cannot adapt to. Bioretention facilities are placed in urban areas where temperatures in summer can be very high and in some cases little maintenance for managing the media and competing volunteer vegetation may occur. Finally, the plants are exposed to myriad pollutants that can exert toxic effects, including heavy metals, oils and other hydrocarbons, and occasional high concentrations of salts from roadway deicing. The Joe Pye weed (*Eutrochium dubium*, Figure 18.25) is a flowering plant that has been found to be highly successful in Mid-Atlantic bioretention systems (Muerdter et al. 2016). *E. dubium* can have aboveground biomass height of around 90 cm (3 ft), and below ground penetration of 30 cm (1 ft).

Nonetheless, because the runoff is concentrated into the bioretention cell by a factor of approximately 20, the bioretention vegetation receives a large annual water volume and nutrient load. Some plant species thrive in the bioretention environment. Recommended bioretention plantings have been developed by a number of organizations and jurisdictions. In addition to stormwater management, the vegetation has obvious aesthetic functions. For bioretention to be accepted by the community, it must be attractive and blend with the area landscape. Some bioretention facilities have been designed with minimal vegetation and a large amount of exposed mulch (Figure 18.26). Generally, in these bioretention cells, the vegetation is well managed, with regular maintenance performed. Volunteers are removed frequently and existing vegetation is trimmed. In other bioretention facilities, little management of the vegetation occurs. In this case, the cell tends to be dominated by very dense vegetation of various sizes and very little of the mulch layer is exposed after a few years (Figure 18.27). In some cases, the situation may dictate that turf is the best vegetated

Figure 18.26 Maryland Bioretention Facility with Minimal Vegetation. (Photo by Authors).

Figure 18.27 Maryland Bioretention Facility with Dense Vegetation. (Photo by Authors).

cover. A bioretention cell covered completely in turf was found to demonstrate no difference in performance as compared to one with a more traditional vegetation mix (Passeport et al. 2009). Deeper rooted vegetation can access more plant available water, creating more storage space. Growth and death of roots create macropores maintaining infiltration capacity.

The bioretention vegetation appears to play an important role in N removal from runoff. Vegetated column and mesocosom studies consistently demonstrate that vegetated systems will remove nitrogen, while unvegetated columns show little nitrogen removal (Barrett et al. 2013; Bratieres et al. 2008; Lucas and Greenway 2011a, 2011b). It appears that thick-rooted systems are most effective at N removal. Different plant species will provide for different impacts on N removal. It is not clear how much of the impact may be seasonal, dependent on the growing season. Also, while plant uptake may provide a removal and sequestration mechanism from runoff to remove the N from the facility, some degree of vegetation harvesting and removal must be initiated so that N is not returned to the media as litter and dying vegetation.

The role of vegetation in P bioretention is still not well defined. Pilot-scale (Lucas and Greenway 2008) and column studies (Barrett et al. 2013) have noted clear improvements in P removal in planted bioretention systems compared to unplanted. Contributions from plant uptake and rizosphere microbial populations are likely.

## 18.18 The Underdrain and Subsurface Storage

The presence or lack of an underdrain, and its configuration can have a major impact on the performance of a bioretention cell. Rain gardens, as simple systems to be built by homeowners, will not have underdrains. Additionally, if the native soils are sandy and have adequate native infiltration rates, an underdrain should not be necessary. Even in more moderately draining soils, underdrains may be avoided if the design prevents long-term standing water.

The traditional underdrain is a perforated pipe that runs the length of the bioretention cell. If the cell is wide, it may have several underdrain laterals that come together in a manifold configuration. The underdrain is usually established in a layer of stone beneath the media layer to ensure that the entire media layer is drained. The underdrain will discharge into the storm sewer network or may directly discharge to a natural water body.

A modification of the underdrain configuration is the use of an up-turned elbow to create an internal water storage (IWS) layer (Figure 18.28), as described in Chapter 11. The IWS can assist in promoting infiltration, nitrogen removal, and allowing bioretention to be used in areas where traditional deep underdrains would not work due to topography challenges.

The IWS is created with an upturned elbow in the underdrain, usually raised from 12 to 24 in (30–60 cm). This allows infiltrating water to pool in the bioretention media before any underdrainage occurs. This water remains in the media and will drain down based on the hydraulic conductivity of the surrounding soils. Hydrologically, this promotes greater exfiltration and groundwater recharge than the traditional underdrain. The longer time and 12–24 in (30–60 cm) of head can produce recharge even in very clayey soils.

Figure 18.29 shows water performance results for shallow and deep IWS in bioretention cells in a sandy soil and sand clay loam (SCL) soil, highlighting the hydrologic benefit of the IWS. The sandy soil bioretention system drained quickly and only no surface discharge was measured over the 2-year study, with >94% of the water exfiltrated (from the bioretention media to the surrounding soils). In the case of the SCL, 70–82% of the runoff was exfiltrated (Brown and Hunt 2011).

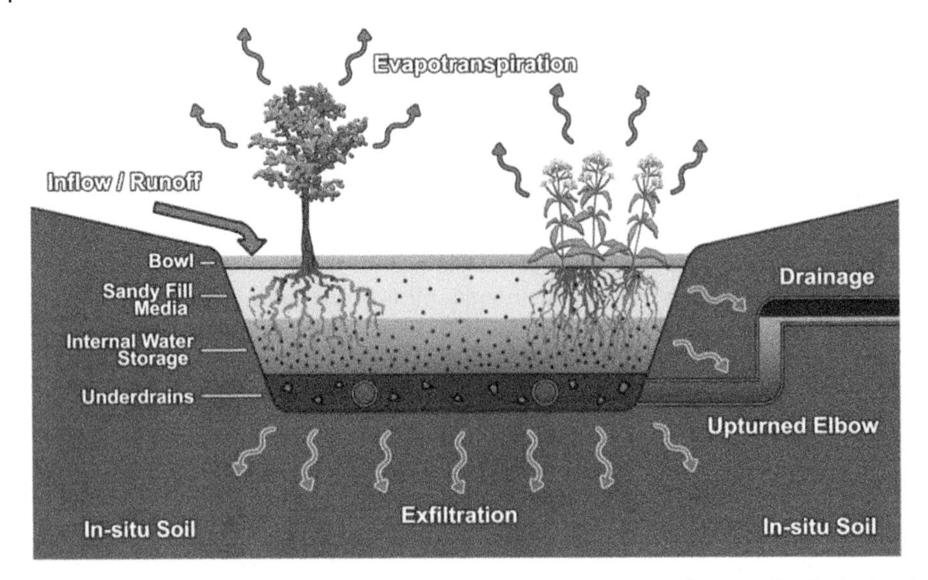


Figure 18.28 Diagram of a Bioretention System with Internal Water Storage. (Credit: Authors' Research Groups).

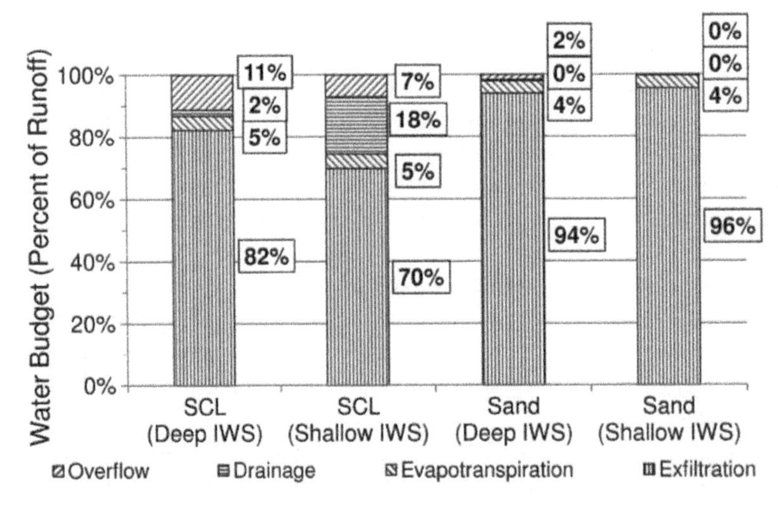


Figure 18.29 Water Balance for Bioretention Systems with Internal Water Storage Showing Fates of Incoming Runoff (Brown and Hunt 2011, with Permission from ASCE).

Additional subsurface storage may be included with bioretention systems if needed for greater volume control. This storage can be a rock bed or proprietary storage system below or adjacent to the bioretention media. These systems are receiving great interest in cities that have combined sewers and need additional stormwater volumetric capacity. Examples of these designs are shown in Figure 18.30.

Mounding of groundwater beneath bioretention facilities is of interest as infiltration is encouraged. Machusick et al. (2011) found that a rainfall of 8.03 cm caused localized mounding of over a meter under a bioinfiltration site. Most storms though produced much less of an effect.

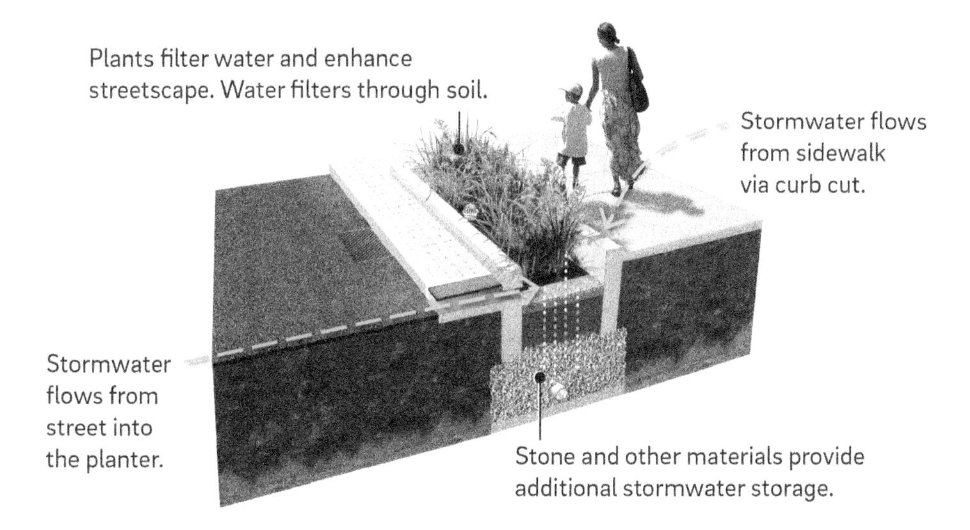


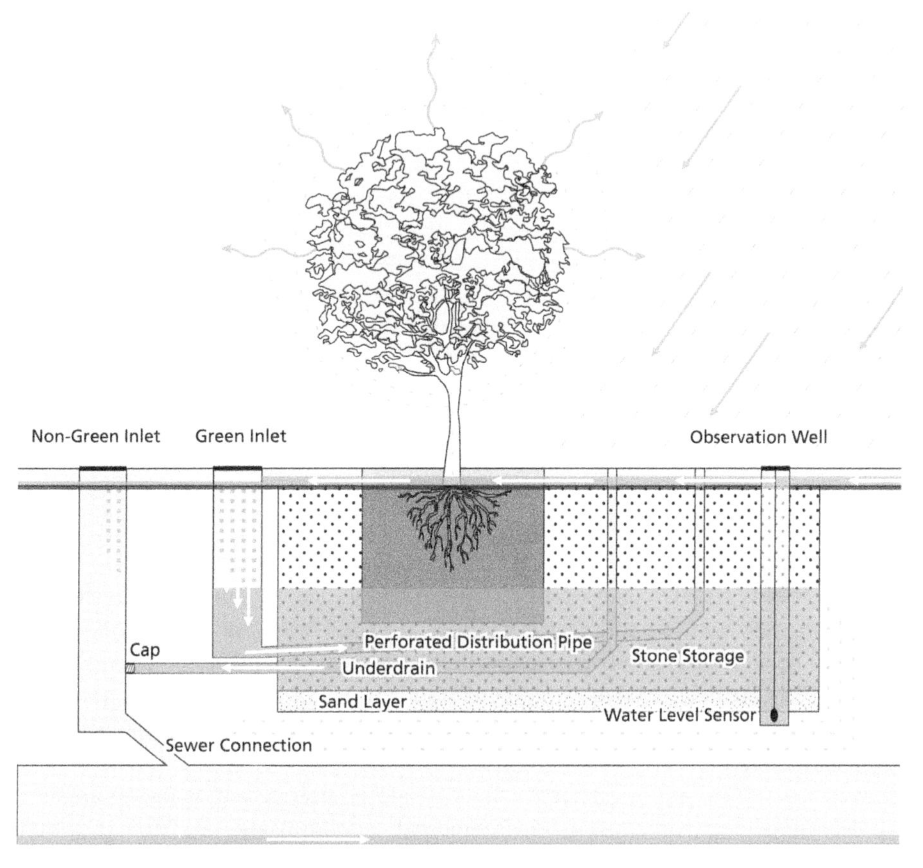


Figure 18.30 Conceptual Designs of Water Storage Added Below and Adjacent to Bioretention Facilities in Philadelphia.

## 18.19 Internal Water Storage and Nitrogen Removal

Nitrate removal is a challenge. As noted in Chapter 8, requirements for denitrification include (1) an electron donor, typically organic carbon; (2) the absence of oxygen so that nitrate will be the electron acceptor; (3) denitrifying microorganisms; (4) moisture; and (5) adequate time for the denitrification reaction to occur.

Several research studies have evaluated bioretention denitrification zones in laboratory settings. Columns containing materials such as wood chips, newspaper, straw and other materials were used as the carbon source/electron donor (Kim et al. 2003; Peterson et al. 2015). Denitrification readily occurred in these laboratory systems, indicating that microorganisms were naturally available without the need for any culturing. The challenges with the denitrification are with organic N leaching from the carbon source and providing adequate time for denitrification.

Peterson et al. (2015) worked to optimize the denitrification process and also reduce overall organic N leaching with oak wood chips. They found a first-order denitrification rate constant of $0.17 \pm 0.19\ h^{-1}$ at 22°C. In follow-up work, Igielski et al. (2019) determined a somewhat lower first-order rate constant equivalent to $0.066\ h^{-1}$. Halaburka et al. (2017) estimated the wood chip denitrification process to be zero order with a value of approximately 0.13 mg N/L h at 22°C, which is equivalent to first-order constants ranging from 0.012 to $0.065\ h^{-1}$ using the nitrate concentrations employed by the authors. Data from He et al. (2020) indicate a first- and zero-order rate constants of approximately $0.06\ h^{-1}$ and 0.11 mg N/L h for an internal water storage zone containing newspaper scraps. Altogether, these studies present fairly consistent data considering the variability inherent in biological processes, the different wood/paper characteristics, and the different reactor systems used. The implication of a zero order denitrification rate found by Halaburka et al. (2017) is that the rate is limited primarily by retention time. With the first-order denitrification rate constant range of $0.012–0.17\ h^{-1}$, 9.5–134 h would be needed to reduce 3 mg/L $NO_3$–N to 0.6 mg/L. All of these studies indicate the slow kinetics of the denitrification reaction and the long retention times needed for effective denitrification; the implication from this is that little denitrification will take place during an actual storm event and storage between events is necessary for measurable denitrification.

Igielski et al. (2019) found microorganisms capable of denitrifying in their wood chip IWS. Waller et al. (2018) detected more denitrifying microorganisms in upper bioretention layers than in lower layers, even in IWS layers. Denitrifier concentrations correlated with organic matter content in upper layers. Grassed bioretention cells had lower concentrations of denitrifiers than did landscaped and overgrown facilities. Similar results were noted by Willard et al. (2017). The concentrations of denitrifying genes *nirK* and *nosZ* were highest in the upper layers of a bioretention cell with an IWS.

Internal water storage (IWS) has the ability to store water subsurface between rainfall events. As long as the proper environmental conditions are met (anoxic, carbon source, etc.) denitrification is expected. Field studies have demonstrated some reductions in nitrate concentration in bioretention systems that are lined or emplaced in

**Example 18.6** A bioretention cell is designed with an internal water storage layer 1 ft deep. The bioretention cell footprint is 20 ft × 40 ft. The porosity of the IWS layer is 0.45. The nitrate concentration entering the IWS is 3.2 mg/L $NO_3$-N at a flow rate of 5.3 $ft^3$/min. Using a first-order dentification rate of 0.1 $h^{-1}$ and assuming plug flow conditions, find the nitrate concentration exiting the IWS.

*Solution* The IWS (water) volume is found as

$$V = (1\ \text{ft})(20\ \text{ft})(40\ \text{ft})(0.45) = 360\ \text{ft}^3$$

the hydraulic retention time is found:

$$t_R = V/Q = 360\,\text{ft}^3/\left(5.3\ \text{ft}^3/\text{min}\right) = 67.9\ \text{min} = 1.132\ \text{h}$$

For a plug flow reactor, from Equation 8.3, the discharge concentration of nitrogen is found:

$$C = 3.2\ \text{mg/L}\exp\left[\left(-0.1\ \text{h}^{-1}\right)(1.132\ \text{h})\right] = 2.86\ \text{mg/L}\,NO_3 - N$$

The N removal is only about 11%.

high-clay surrounding soils (Brown and Hunt 2011; Davis 2007), but not in sandy soils where the retention time is inadequate. Brown and Hunt (2011) noted no statistical difference in total N concentrations (~1.31 mg/L) between input runoff and discharge effluent in a bioretention system installed in a sandy soil. Conversely the EMC dropped from an average of 1.02–0.43 mg/L Total N for a bioretention system in a sandy clay loam, which exhibited water storage for up to 7 days. Denitrification may also be occurring in bioinfiltration, where the infiltrating water can readily become anoxic (Lord 2013). Various types of media layering may also be effective in managing flows and the nitrogen cycle (Cho et al. 2009; Hsieh et al. 2007). Increasing retention time through increased anoxic storage appears to be the key design variable for nitrogen removal.

## 18.20 Bioretention Pollutant Load Reductions

With bioretention, both hydrologic runoff volume reduction and pollutant treatment is expected for most water quality parameters. As a result, the overall pollutant load reduction can be significant. Load into and out of two bioretention cells in Maryland for a suite of pollutants is provided in Table 18.1. For pollutants that undergo significant treatment by the bioretention media, such as TSS and metals, the load differences in Table 18.1 are substantial.

**Example 18.7** The TSS load for a 65-ha (160 ac) subdivision must be reduced by 3400 kg/year. Bioretention cells will be built in the subdivision, each treating approximately 0.2 ha. Find the number of bioretention cells needed to meet the TSS load reduction.

*Solution* In Table 18.1, several data sets for TSS load reductions are listed. In this example, use the mean input and output from the three studies on the College Park bioretention cell. The mean input is 1047 kg/(ha year) and the mean output is 41 kg/(ha year), for a difference of 1006 kg/(ha year)—let's just round to 1000. To reduce the TSS load by 3400 kg/year:

$$\text{Area} = \frac{3400\ \text{kg/year}}{1000\ \text{kg/(ha yr)}} = 3.4\ \text{ha}$$

Therefore, 3.5 ha of impervious area must be treated. Since bioretention cell will treat a 0.2-ha drainage area, the number of cells is

$$\text{Bioretention cells} = \frac{3.4\ \text{ha}}{0.2\ \text{ha/cell}} = 17\ \text{cells}$$

**Table 18.1** Pollutant Loads, in kg/(ha year), Measured into and Out of (via Underdrain) Bioretention Cells in Maryland (Compiled from DiBlasi et al. 2009; Li and Davis 2009, 2014; Liu and Davis 2014.) Multiple data sets result from different studies.

| | College Park bioretention | | Silver Spring bioretention | |
|---|---|---|---|---|
| | In | Out | In | Out |
| TSS | 1190 | 37 | 570 | 38 |
| | 860 | 39 | | |
| | 1090 | 47 | | |
| Chromium | 0.09 | 0.015 | 0.02 | ~0.007 |
| Copper | 0.26 | 0.073 | 0.12 | 0.045 |
| Lead | 0.09 | 0.013 | 0.03 | ~0.005 |
| Zinc | 1.0 | 0.063 | 0.36 | 0.017 |
| Chloride | 6800 | 458 | 320 | 25 |
| TN | 27 | 7.2 | 9.6 | 3.6 |
| | 14 | 8.2 | | |
| Nitrate-N | 12 | 2.5 | 3.7 | ~0.19 |
| | 2.4 | 3.5 | | |
| TKN | 15 | 4.1 | 6.0 | 3.6 |
| Particulate organic N | 8.0 | 1.3 | | |
| Dissolved organic N | 2.2 | 1.3 | | |
| TP | 3.6 | 0.72 | 0.9 | 0.38 |
| | 3.0 | 0.48 | | |

(*Continued*)

Table 18.1 (Continued)

| | College Park bioretention | | Silver Spring bioretention | |
|---|---|---|---|---|
| | In | Out | In | Out |
| Particulate P | 2.3 | 0.20 | | |
| Phosphate P | 0.41 | 0.16 | | |
| Dissolved Organic P | 0.30 | 0.12 | | |
| TOC | 44 | 154 | 43 | 78 |
| Total PAH | 0.018 | 0.0025 | – | – |

**Example 18.8** A bioretention cell treating a 0.18-ha parking lot was studied for one year. The study found that 1280 $m^3$ of stormwater entered the cell and only 490 $m^3$ was discharged over the year. The stormwater overall volume-weighted total P concentration was 0.78 mg/L and the dissolved P was 0.35 mg/L. The bioretention discharge concentrations were 0.26 mg/L TP and 0.18 mg/L DP. Find the input and bioretention discharge annual loads of TP, PP, and DP, all in kg/(ha year).

*Solution* The mass loads are found as volume time concentration over the year-long study, divided by the land area treated. For the stormwater total P:

$$L=\frac{VC}{At}=\frac{\left(1280\ \text{m}^3\right)\left(0.78\ \text{mg/L}\right)\left(1000\ \text{L/m}^3\right)}{\left(0.18\ \text{ha}\right)\left(1\ \text{year}\right)}=5.5\times10^6\ \text{mg/ha year}=5.5\ \text{kg/ha}$$

The same process is used for DP:

$$L=\frac{\left(1280\ \text{m}^3\right)\left(0.35\ \text{mg/L}\right)\left(1000\ \text{m}^{-3}\right)}{\left(0.18\ \text{ha}\right)\left(1\ \text{year}\right)}=2.5\times10^6\ \text{mg/ha year}=2.5\ \text{kg/ha year}$$

The value for PP is the difference between total and dissolved phosphorus:

$$L=5.2-2.5=2.7\ \text{kg/ha year}$$

For the discharge loads, the same calculations are used:

$$L=\frac{\left(490\ \text{m}^3\right)\left(0.26\ \text{mg/L}\right)\left(1000\ \text{L/m}^3\right)}{\left(0.18\ \text{ha}\right)\left(1\ \text{year}\right)}=7.1\times10^5\ \text{mg/ha year}=0.71\ \text{kg/ha ye}$$

for TP:

$$L=\frac{\left(490\ \text{m}^3\right)\left(0.18\ \text{mg/L}\right)\left(1000\ \text{L/m}^3\right)}{\left(0.18\ \text{ha}\right)\left(1\ \text{year}\right)}=4.9\times10^5\ \text{mg/ha year}=0.49\ \text{kg/ha ye}$$

for DP

And for PP:

$$L=0.71-0.49=0.22\ \text{kg/ha year}$$

## 18.21 Bioretention Exfiltration and Groundwater

One of the primary mechanisms for water management in bioretention and all GSI is the infiltration of excess stormwater. This process helps to mitigate problems associated with high stormwater flows, increases groundwater recharge, and assists in the goal of GSI in helping to return the site to predevelopment hydrology.

Nonetheless, the promotion of infiltration to surrounding soils must be carefully considered. Infiltration will lead to mounding of groundwater near the SCM, even if temporary. This mounding can lead to groundwater seepage into nearby buildings and under roadways where it is not desired. Nearby structures and the impact of increased groundwater levels must be considered during the implementation of GSI, especially in retrofit situations.

In addition to mounding, water quality impacts from infiltrating stormwater to groundwaters must be carefully evaluated. The inaccessibility of groundwater supplies and the difficulties of remediating them make groundwater contamination a serious environmental concern. A number of common stormwater pollutants are highly mobile in soil systems and can be readily transported to groundwaters. These can include nitrate, chloride, soluble pesticides, several other organic pollutants, and zinc. Clark and Pitt (2007) provide an overview of stormwater pollutant impacts to groundwater.

Road salts used for deicing in areas impacted by snow and ice are a specific concern. The primary deicing chemical is sodium chloride (NaCl). Both sodium and chloride can be transported through bioretention media, the chloride with essentially no attenuation and sodium with little. NaCl can be found at high concentrations in groundwater beneath bioretention and bioswale SCMs after managing winter storm events (Burgis et al. 2020). Source reduction may be the most effective way to address these mobile stormwater contaminants.

## 18.22 Inspection and Maintenance

While bioretention is very effective for particulate matter removal, this can lead to problems if excess particulate matter loads are applied to the cell. Clogging of the surface media is possible under high loadings. If a bioretention cell is being placed in an area where high sediment loads are expected, some type of pretreatment rip-rap, swale, or stilling area should be included. Thick vegetation appears to inhibit clogging, but more research is needed on this phenomenon. Additionally, highly sloped or unstabilized banks can erode soils into the bioretention cell, leading to clogging.

Inspection of bioretention facilities focus on three concerns: infiltration rate, vegetation density and type, and inlet/outlet clogging. Inspections should take place at least once per year. Ensuring that the bioretention facility still adequately infiltrates is the most important requirement of the inspection. This type of inspection can be done several ways, all in relatively simple manners. Simply driving by a facility 24- to 48 h after the end of a significant rainfall can provide some degree of confidence that the facility is working. If no ponded water is noted, infiltration must be taking place.

More quantitative information can be provided by installing a simple staff gauge and taking two or three measurements of the ponded water at different times. The slope of the change in water level as a function of time is the operationally defined infiltration rate (it should be corrected for temperature to be most accurate). A

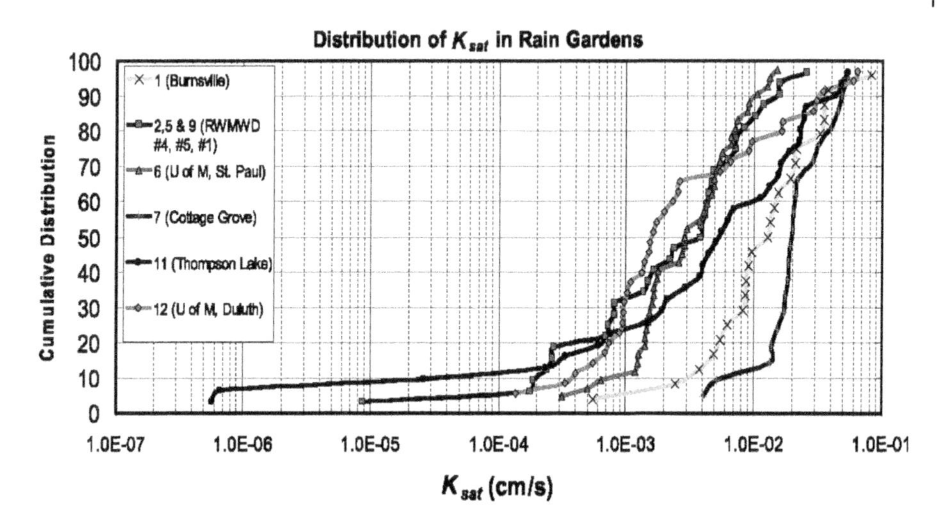


Figure 18.31 Distribution of Measured Saturated Hydraulic Conductivity ($K_{sat}$) for Eight Rain Gardens (Asleson et al. 2009).

rain event that produces at least 0.6 cm of rainfall in 8 h should be adequate for infiltration testing (Welker et al. 2013).

Visual inspections can provide a wealth of information. The presence of wetland plants or obvious loss of vegetation, along with indication of hydric soils (possibly gray/black in color with hydrogen sulfide odor) can indicate long-term saturation and compromised infiltration. Existing vegetation can be compared with the vegetation on original planting plans to determine changes. Invasive species can be evaluated, although this may or may not be an issue with respect to stormwater control performance. Obvious sediment accumulation in the drainage bowl and/or excessive erosion within or around the bioretention facility can indicate clogging. Clogged inlet and/or outlet structures should also be noted (Welker et al. 2013).

Direct infiltration testing can be done using a tanker truck or fire hydrant, rapidly filling the bowl with water, and measuring infiltration rate. Specific clogged areas can be noted. When calculating the overall infiltration rate or hydraulic conductivity of a bioretention cell, because of the high variability expected, the median or geometric mean is better than the mean to describe the overall infiltration rate (Asleson et al. 2009); see Figure 18.31, showing variations in saturated hydraulic conductivity within individual bioretention cells that had been operating for several years.

Should bioretention clogging be evident and some remedial measure be necessary, several steps should be applied. First a raking of the surface should be tried to see if a surface crust exists that can be broken up. If media replacement is necessary, only the surface layer should require removal. Clogging should be confined to the first 10 cm or so and removing this layer and replacing with fresh media should alleviate the clogging problem.

## References

Asleson, B.C., Nestingen, R.S., Gulliver, J.S., Hozalski, R.M., and Nieber, J.L. (2009). Performance assessment of rain gardens. *JAWRA Journal of the American Water Resources Association* 45 (4): 1019–1031.

Barrett, M.E., Limouzin, M., and Lawler, D.F. (2013). Effects of Media and Plant Selection on Biofiltration Performance. *Journal of Environmental Engineering, ASCE* 139 (4): 462–470.

Bratieres, K., Fletcher, T.D., Deletic, A., and Zinger, Y. (2008). Nutrient and sediment removal by stormwater biofilters: A large-scale design optimisation study. *Water Research* 42 (14): 3930–3940.

Brown, R.A. and Hunt, W.F. (2011). Underdrain configuration to enhance bioretention exfiltration to reduce pollutant loads. *Journal of Environmental Engineering, ASCE* 137 (11): 1082–1091.

Burgis, C.R., Hayes, G.M., Henderson, D.A., Zhang, W., and Smith, J.A. (2020). Green stormwater infrastructure redirects deicing salt from surface water to groundwater. *Science of the Total Environment* 729: 138736.

Chen, T., Liu, Y., Zhang, B., and Sun, L. (2019). Plant rhizosphere, soil microenvironment, and functional genes in the nitrogen removal process of bioretention. *Environmental Science: Processes & Impacts, the Royal Society of Chemistry* 21 (12): 2070–2079.

Chen, X., Peltier, E., Sturm, B.S.N., and Young, C.B. (2013). Nitrogen removal and nitrifying and denitrifying bacteria quantification in a stormwater bioretention system. *Water Research* 47: 1691–1700.

Cho, K.W., Song, K.G., Cho, J.W., Kim, T.G., and Ahn, K.H. (2009). Removal of nitrogen by a layered soil infiltration system during intermittent storm events. *Chemosphere* 76 (5): 690–696.

Clark, S.E. and Pitt, R. (2007). Influencing factors and a proposed evaluation methodology for predicting groundwater contamination potential from stormwater infiltration activities. *Water Environment Research* 79 (1): 29–36.

Davis, A.P. (2007). Field performance of bioretention: Water quality. *Environmental Engineering Science* 24 (8): 1048–1063.

Davis, A.P., Traver, R.G., Hunt, W.F., Brown, R.A., Lee, R., and Olszewski, J.M. (2012). Hydrologic performance of bioretention stormwater control measures. *Journal of Hydrologic Engineering - ASCE* 17 (5): 604–614.

DeBusk, K.M., Hunt, W.F., and Line, D.E. (2011). Bioretention outflow: Does it mimic nonurban watershed shallow interflow? *Journal of Hydrologic Engineering* 16 (3): 274–279.

DiBlasi, C.J., Li, H., Davis, A.P., and Ghosh, U. (2009). Removal and fate of polycyclic aromatic hydrocarbon pollutants in an urban stormwater bioretention facility. *Environmental Science & Technology* 43 (2): 494–502.

Dietz, M. and Clausen, J. (2005). A field evaluation of rain garden flow and pollutant treatment. *Water, Air, and Soil Pollution* 167 (1): 123–138.

Gardina, C. (2016). Water quality impacts due to the addition of biosolids-derived compost to bioretention. Master of Science Thesis, University of Maryland.

Halaburka, B.J., LeFevre, G.H., and Luthy, R.G. (2017). Evaluation of mechanistic models for nitrate removal in woodchip bioreactors. *Environmental Science & Technology* 51 (9): 5156–5164.

Hathaway, J. M., Hunt, W. F., Graves, A. K., and Wright, J. D. (2011). Field Evaluation of Bioretention Indicator Bacteria Sequestration in Wilmington, North Carolina. *Journal of Environmental Engineering, American Society of Civil Engineers*, 137(12), 1103–1113.

Hathaway, J. M., Hunt, W. F., and Jadlocki, S. (2009). "Indicator Bacteria Removal in Storm-Water Best Management Practices in Charlotte, North Carolina." *Journal of Environmental Engineering, American Society of Civil Engineers*, 135(12), 1275–1285.

Hatt, B.E., Fletcher, T.D., and Deletic, A. (2009). Hydrologic and pollutant removal performance of Stormwater Biofiltration Systems at the field scale. *Journal of Hydrology* 365: 310–321.

He, K., Qin, H., Wang, F., Ding, W., and Yin, Y. (2020). Importance of the submerged zone during dry periods to nitrogen removal in a bioretention system. *Water* 12 (3): 876.

Hess, A., Wadzuk, B., and Welker, A. (2017). Evapotranspiration in rain gardens using weighing lysimeters. *Journal of Irrigation and Drainage Engineering* 143 (6): 04017004.

Hess, A., Wadzuk, B., and Welker, A. (2019). Predictive evapotranspiration equations in rain gardens. *Journal of Irrigation and Drainage Engineering* 145 (7): 04019010.

Hong, E., Seagren, E.A., and Davis, A.P. (2006). Sustainable oil and grease removal from synthetic storm water runoff using bench-scale bioretention studies. *Water Environment Research* 78 (2): 141–155.

Hsieh, C.-H., Davis, A.P., and Needelman, B.A. (2007). Nitrogen removal from removal from urban stormwater runoff through layered bioretention columns. *Water Environment Research* 79 (12): 2404–2411.

Huang, X., Massoudieh, A., and Young, T.M. (2006). Measured and predicted herbicide removal by mulch. *Journal of Environmental Engineering*, 132 (8): 918–925.

Hunt, W.F., Davis, A.P., and Traver, R.G. (2012). Meeting hydrologic and water quality goals through targeted bioretention design. *Journal of Environmental Engineering, ASCE* 138 (6): 698–707.

Hunt, W.F., Jarrett, A.R., Smith, J.T., and Sharkey, L.J. (2006). Evaluating bioretention hydrology and nutrient removal at three field sites in North Carolina. *Journal of Irrigation & Drainage Engineering* 132 (6): 600–608.

Hurley, S., Shrestha, P., and Cording, A. (2017). Nutrient leaching from compost: Implications for bioretention and other green stormwater infrastructure. *Journal of Sustainable Water in the Built Environment* 3 (3): UNSP 04017006.

Igielski, S., Kjellerup, B.V., and Davis, A.P. (2019). Understanding urban stormwater denitrification in bioretention internal water storage zones. *Water Environment Research* 91 (1): 32–44.

Jones, P.S. and Davis, A.P. (2013). Spatial accumulation and strength of affiliation of heavy metals in bioretention media. *Journal of Environmental Engineering, ASCE* 139 (4): 479–487.

Khorsha, G. and Davis, A.P. (2017a). Characterizing clinoptilolite zeolite and hydroaluminosilicate aggregates for ammonium removal from stormwater runoff. *Journal of Environmental Engineering, ASCE* 143 (2): 04016082.

Khorsha, G. and Davis, A.P. (2017b). Ammonium removal from stormwater runoff using clinoptilolite zeolite and hydroaluminosilicate columns. *Water Environment Research* 89 (6): 564–575.

Kim, H., Seagren, E.A., and Davis, A.P. (2003). Engineered bioretention for removal of nitrate from stormwater runoff. *Water Environment Research* 75 (4): 355–367.

Komlos, J. and Traver, R.G. (2012). Long-term orthophosphate removal in a field-scale storm-water bioinfiltration rain garden. *Journal of Environmental Engineering, ASCE* 138 (10): 991–998.

LeFevre, G.H., Hozalski, R.M., and Novak, P.J. (2012a). The role of biodegradation in limiting the accumulation of petroleum hydrocarbons in raingarden soils. *Water Research* 46: 6753–6762.

LeFevre, G.H., Novak, P.J., and Hozalski, R.M. (2012b). Fate of naphthalene in laboratory-scale bioretention cells: Implications for sustainable stormwater management. *Environmental Science & Technology* 46: 995–1002.

Li, H. and Davis, A.P. (2008a). Urban particle capture in bioretention media I: Laboratory and field studies. *Journal of Environmental Engineering, ASCE* 134 (6): 409–418.

Li, H. and Davis, A.P. (2008b). Heavy metal capture and accumulation in bioretention media. *Environmental Science & Technology* 42 (14): 5247–5253.

Li, H. and Davis, A.P. (2009). Water quality improvement through reductions of pollutant loads using bioretention. *Journal of Environmental Engineering, ASCE* 135 (8): 567–576.

Li, J. and Davis, A.P. (2016). A unified look at phosphorus treatment using bioretention. *Water Research* 90: 141–155.

Li, L. and Davis, A.P. (2014). Urban stormwater runoff nitrogen composition and fate in bioretention systems. *Environmental Science & Technology* 48 (6): 3403–3410.

Li, M.-H., Sung, C.Y., Kim, M.H., and Chu, K.-H. (2011). Assessing performance of bioretention boxes in hot and semiarid regions; highway application pilot study. *Transportation Research Record* 2262: 155–163.

Liu, J. and Davis, A.P. (2014). Phosphorus Speciation and Treatment Using Enhanced Phosphorus Removal Bioretention. *Environmental Science & Technology* 48 (1): 607–614.

Lord, L.E. (2013). Evaluation of nitrogen removal and fate within a bioinfiltration stormwater control measure. Master of Science Thesis. Villanova University, Villanova, PA

Lucas, W.C. and Greenway, M. (2008). Nutrient retention in vegetated and nonvegetated bioretention mesocosms. *Journal of Irrigation and Drainage Engineering* 134 (5): 613–623.

Lucas, W.C. and Greenway, M. (2011a). Hydraulic Response and Nitrogen Retention in Bioretention Mesocosms with Regulated Outlets: Part I-Hydraulic Response. *Water Environment Research* 83 (8): 692–702.

Lucas, W.C. and Greenway, M. (2011b). Hydraulic response and nitrogen retention in bioretention mesocosms with regulated outlets: Part II-Nitrogen Retention. *Water Environment Research* 83 (8): 703–713.

Lucas, W.C. and Greenway, M. (2011c). Phosphorus retention by bioretention mesocosms using media formulated for phosphorus sorption: Response to accelerated loads. *Journal of Irrigation and Drainage Engineering* 137 (3): 144–153.

Machusick, M., Welker, A., and Traver, R. (2011). Groundwater mounding at a stormwater infiltration BMP. *Journal of Irrigation and Drainage Engineering-ASCE* 137 (3): 154–160.

McNett, J.K., Hunt, W.F., and Davis, A.P. (2011). Influent pollutant concentrations as predictors of effluent pollutant concentrations for Mid-Atlantic bioretention. *Journal of Environmental Engineering, ASCE* 137 (9): 790–799.

Mohtadi, M., James, B.R., and Davis, A.P. (2017). Adsorption of compounds that mimic urban stormwater dissolved organic nitrogen. *Water Environment Research* 89 (2): 105–116.

Morse, N., Payne, E., Henry, R., Hatt, B., Chandrasena, G., Shapleigh, J., Cook, P., Coutts, S., Hathaway, J., Walter, M.T., and McCarthy, D. (2018). Plant-microbe interactions drive denitrification rates, dissolved nitrogen removal, and the abundance of denitrification genes in stormwater control measures. *Environmental Science & Technology* 52 (16): 9320–9329.

Muerdter, C., Ozkok, E., Li, L., and Davis, A.P. (2016). Vegetation and media characteristics of an effective bioretention cell. *Journal of Sustainable Water in the Built Environment* 2 (1): UNSP 04015008.

Mullane, J.M., Flury, M., Iqbal, H., Freeze, P.M., Hinman, C., Cogger, C.G., and Shi, Z. (2015). Intermittent rainstorms cause pulses of nitrogen, phosphorus, and copper in leachate from compost in bioretention systems. *Science of the Total Environment* 537: 294–303.

North Carolina Department of Environmental and Natural Resources (2009). *Stormwater Best Management Practices*, Chapter 12, Bioretention. Raleigh NC.

O'Neill, S.W. and Davis, A.P. (2012a). Water treatment residual as a bioretention amendment for Phosphorus. I. Evaluation studies. *Journal of Environmental Engineering, ASCE* 138 (3): 318–327.

O'Neill, S.W. and Davis, A.P. (2012b). Water treatment residual as a bioretention amendment for Phosphorus. II. Long-term column studies. *Journal of Environmental Engineering, ASCE* 138 (3): 328–336.

Palmer, E.T., Poor, C.J., Hinman, C., and Stark, J.D. (2013). Nitrate and phosphate removal through enhanced bioretention media; Mesocosm study. *Water Environment Research* 85 (9): 823–832.

Passeport, E., Hunt, W.F., Line, D.E., Smith, R.A., and Brown, R.A. (2009). Field study of the ability of two grassed bioretention cells to reduce stormwater runoff pollution. *Journal of Irrigation and Drainage Engineering* 135 (4): 505–510.

Peterson, I.J., Igielski, S., and Davis, A.P. (2015). Enhanced denitrification in bioretention using woodchips as an organic carbon source. *Journal of Sustainable Water in the Built Environment* 1 (4): 04015004.

Read, J., Fletcher, T.D., Wevill, T., and Deletic, A. (2010). Plant traits that enhance pollutant removal from stormwater in biofiltration systems. *International Journal of Phytoremediation* 12 (1): 34–53.

Sun, X. and Davis, A.P. (2007). Heavy metal fates in laboratory bioretention systems. *Chemosphere* 66 (9): 1601–1609.

Tirpak, R.A., Hathaway, J.M., and Franklin, J.A. (2019). Investigating the hydrologic and water quality performance of trees in bioretention mesocosms. *Journal of Hydrology* 576: 65–71.

Toland, D.C., Haggard, B.E., and Boyer, M.E. (2011). Evaluation of nutrient concentrations in runoff water from green roofs, conventional roofs, and urban streams. *Transactions of the ASABE* 55 (1): 99–106.

Waller, L.J., Evanylo, G.K., Krometis, L.-A.H., Strickland, M.S., Wynn-Thompson, T., and Badgley, B.D. (2018). Engineered and environmental controls of microbial denitrification in established bioretention cells. *Environmental Science & Technology* 52 (9): 5358–5366.

Welker, A.L., Mandarano, L., Greising, K., and Mastrocola, K. (2013). Application of a monitoring plan for storm-water control measures in the Philadelphia Region. *Journal of Environmental Engineering* 139 (8): 1108–1118.

Willard, L.L., Wynn-Thompson, T., Krometis, L.H., Neher, T.P., and Badgley, B.D. (2017). Does it pay to be mature? Evaluation of bioretention cell performance seven years postconstruction. *Journal of Environmental Engineering, American Society of Civil Engineers* 143 (9): 04017041.

Yan, Q., Davis, A.P., and James, B.R. (2016). Enhanced organic phosphorus sorption from urban stormwater using modified bioretention media: Batch studies. *Journal of Environmental Engineering* 142 (4): 04016001.

Yan, Q., James, B.R., and Davis, A.P. (2017a). Lab-scale column studies for enhanced phosphorus sorption from synthetic urban stormwater using modified bioretention media. *Journal of Environmental Engineering* 143 (1): 04016073.

Yan, Q., James, B.R., and Davis, A.P. (2017b). Bioretention media for enhanced permeability and phosphorus sorption from synthetic urban stormwater. *Journal of Sustainable Water in the Built Environment* 4 (1): 04017013.

Zhang, W., Brown, G.O., Storm, D.E., and Zhang, H. (2008) Fly-Ash amended sand as filter media in bioretention cells to improve phosphorus removal. *Water Environment Research* 80 (6): 507–516.

Zhang, L., Seagren, E.A., Davis, A.P., and Karns, J.S. (2011). Long-term sustainability of Escherichia coli removal in conventional bioretention media. *Journal of Environmental Engineering, ASCE* 137 (8): 669–677.

Zhang, L., Seagren, E.A., Davis, A.P., and Karns, J.S. (2012). Effects of temperature on bacterial transport and destruction in bioretention media: Field and laboratory evaluations. *Water Environment Research* 84 (6): 485–496.

## Problems

**18.1** A bioretention cell is planned for the treatment of a 0.48-ac parking lot. Based on area ratio and typical media depth, find the dimensions of the bioretention cell. Use a length:width ratio of 4:1.

**18.2** An elliptical shaped bioretention cell has a major axis length of 8 m and a minor axis length of 3 m. At a 4.5% area ratio, how much area can this bioretention cell treat?

**18.3** Find the BAV for a bioinfiltration cell with an area of 1500 ft2 and a sandy loam media depth of 3 ft.

**18.4** Find the BAV for an underdrained bioretention cell with an area of 1500 ft2 and a sandy loam media depth of 3 ft.

**18.5** One inch of runoff from a 0.6-ac catchment must be completely stored by bioinfiltration. A loamy sand media will be used at a depth of 3.5 ft. What should be the bioinfiltration cell area?

**18.6** One inch of runoff from a 0.6-ac catchment must be completely stored by an underdrained bioretention cell. A loamy sand media will be used at a depth of 3.5 ft. What should be the cell area?

**18.7** A bioretention cell is being designed to manage stormwater runoff from a parking lot. The drainage area is 0.6 ac. Local regulations require that the bioretention cell store (in the surface pool) the volume equal to 0.5 in of rainfall over the drainage area. The area of the bioretention cell surface must be 5% of the drainage area. The media should be 3.5 ft deep with a 6-in diameter underdrain.

a) Find the bioretention surface area ($ft^2$).
b) Find the surface pool depth (in).
c) Find the volume of bioretention soil media (BSM) required ($ft^3$).
d) If a loamy sand BSM is used, estimate the bioretention abstraction volume (BAV) ($ft^3$).
e) If 95% of rainfall becomes runoff on the parking lot, find the rainfall depth that is completely managed by the BAV (in).
f) If the surrounding soils are sandy and 25% of the runoff volume is infiltrated into the surrounding soils during saturated bioretention conditions, find the bioretention discharge volume that will result from a 2.8-in rainfall ($ft^3$). Again, assume 95% rainfall to runoff.

**18.8** A bioretention facility treats stormwater runoff from a paved parking lot with a drainage area of 0.75 ac. Runoff input and output (from underdrain) data measured during a rain event are presented in the following table. The bioretention media dimensions are 50 ft × 30 ft × 2.5 ft ($L \times W \times H$).

| Input | | | Output | | |
|---|---|---|---|---|---|
| Time of day | Flow rate (L/s) | TSS (mg/L) | Time of day | Flow rate (L/s) | TSS (mg/L) |
| 05:10 | 1.5 | 348 | 08:20 | 0.06 | 22 |
| 05:30 | 0.52 | 98 | 08:50 | 0.11 | 21 |
| 05:50 | 2.1 | 118 | 09:20 | 0.15 | 9 |
| 06:10 | 0.54 | 52 | 09:50 | 0.16 | 7 |
| 06:40 | 0.40 | 63 | 10:20 | 0.14 | 6 |
| 07:10 | 0.38 | 87 | 11:20 | 0.14 | 5 |
| 07:40 | 0.58 | 161 | 12:20 | 0.13 | 4 |
| 08:10 | 1.9 | 282 | 13:20 | 0.11 | 5 |
| 09:10 | 1.4 | 188 | 14:20 | 0.09 | 6 |
| 10:40 | 0.34 | 61 | 16:20 | 0.03 | 4 |

a) Plot the hydrographs and pollutographs, input and output on same plot.
b) For the input, calculate the total TSS mass and the EMC.
c) Estimate the total rainfall for this event (in).
d) Determine if a TSS first flush occurs using the 50%-mass-in-25%-volume definition.
e) For the output, calculate the total TSS mass and the EMC.
f) Calculate the bioretention treatment efficiency for TSS, based both on EMC and on total mass.

Note 1: Integrate using trapezoids; extrapolate 1 $\Delta t$ in each direction to 0; multiply first, then integrate.
Note 2: If you are relying primarily on spreadsheets to complete this homework, ensure that all figures, tables, and explanations are unconditionally clear and can be easily followed.

**18.9** Several pollutants are required to be removed by SCMs, such as bioretention. For the following pollutants, please mention (1) in what part of the bioretention cross-section (bowl, mulch, top 8″ of soil, etc.) and (2) by what mechanism(s) (e.g., sedimentation), these pollutants are removed or sequestered:

TSS
TP
$NO_{2-3}$–N
Copper
*Escherichia coli*
Hydrocarbons
Thermal load

**18.10** Why is pH important in bioretention soils?

**18.11** List at least four advantages of using an IWS zone in bioretention? What are potential drawbacks of an IWS zone in bioretention (if any)? Are there any other practices that could benefit from a similar design feature?

**18.12** Consider that the bioretention media has a $d_{10}$ (collector diameter) of 0.1 mm and a porosity of 40%. For a TSS particle size of 0.08 mm and a sticking coefficient of 0.1, what is the predicted the TSS removal fraction $(1 - n/n_0)$ exiting the bioretention facility during the stormwater treatment (mg/L)? Briefly discuss your answer and compare with data in Problem 18.7.

**18.13** A bioretention cell is to be built to manage the stormwater runoff from a surface parking area. The parking area is 0.85 ac. The bioretention area should be underdrained and have storage capacity to hold 0.5-in of **rainfall** before any underdrain discharge occurs. Assume that 95% of rainfall becomes parking lot runoff. For the bioretention design, specify:

a) Media depth (ft)
b) Media type
c) Bioretention surface area ($ft^2$)

**18.14** A bioretention cell has a surface area of 1100 $ft^2$ and 3 ft deep loamy sand media. How much runoff volume ($ft^3$) will be discharged from the bioretention cell for a 3.3-in rainfall event (a 2-year, 24-h storm in Maryland)? Assume that your bioretention cell will infiltrate 75% of the runoff volume **beyond** the initial bioretention storage volume.

**18.15** Using the Hargreaves equation (Chapter 7), estimate the time it will take for 1.2 in of water to evapotranspire from a bioretention cell surface in May in Baltimore, MD (39°N). The average maximum and minimum temperatures are 24.4 and 14.4°C, respectively.

**18.16** Using the Hargreaves equation (Chapter 7), estimate the time it will take for 1.4 in of water to evapotranspire from a bioretention cell surface in September in Philadelphia, PA (40°N). The average maximum and minimum temperatures are 26.7 and 17.2°C, respectively.

**18.17** Using the Hargreaves equation (Chapter 7), estimate the time it will take for 1.0 in of water to evapotranspire from a bioretention cell surface in October in Portland, OR (45.5°N). The average maximum and minimum temperatures are 17.2 and 8.9°C, respectively.

**18.18** Using the Hargreaves equation (Chapter 7), estimate the time it will take for 2 cm of water to evapotranspire from a bioretention cell surface in this month in your city.

**18.19** A bioretention cell has a surface area of 1100 $ft^2$ and 3 ft deep loamy sand media. The phosphorus concentration in the incoming parking lot runoff averages 0.3 mg/L as P. The adsorption of P onto the bioretention media can be described using the Freundlich isotherm with $K_F = 0.085$ and $1/n = 0.62$ (for C in mg/L as P and $q$ in mg/g). Assuming 40 in of rainfall per year at 95% to runoff, how many years of P adsorption capacity does this bioretention cell have? The media bulk density is 90 $lbs/ft^3$.

**18.20** A bioretention cell is built to manage the stormwater runoff from a surface parking area. The parking area is 0.85 ac and the bioretention area is 25 ft × 60 ft. The phosphorus concentration in the parking lot runoff averages 0.3 mg/L as P. The adsorption of P onto the bioretention media can be described using the Freundlich isotherm with $K_F = 0.085$ and $1/n = 0.62$ (for C in mg/L as P and $q$ in mg/g). Assuming 40 in of rainfall per year at 90% to runoff, how many years of P adsorption capacity does your bioretention cell have? The media bulk density is 90 lbs/ft$^3$.

**18.21** A bioretention cell is built to manage the stormwater runoff from a surface parking area. The parking area is 0.85 ac and the bioretention area is 25 ft × 60 ft. The zinc concentration in the parking lot runoff averages 0.09 mg/L. The adsorption of P onto the bioretention media can be described using the Freundlich isotherm with $K_F = 0.75$ and $1/n = 0.52$ (for C in mg/L as P and $q$ in mg/g). Assuming 40 in of rainfall per year at 90% to runoff, how many years of Zn adsorption capacity does your bioretention cell have? The media bulk density is 90 lbs/ft$^3$.

**18.22** A bioretention facility treats stormwater runoff from a paved campus parking lot Runoff input and output data during a large rain event is presented in the following table:

| Input | | | Output | | |
|---|---|---|---|---|---|
| Time of day | Flow rate (L/s) | TSS (mg/L) | Time of day | Flow rate (L/s) | TSS (mg/L) |
| 11:50 | 2.8 | 175 | 13:00 | 0.8 | 22 |
| 12:10 | 3.3 | 85 | 13:30 | 1.6 | 25 |
| 12:30 | 4.9 | 45 | 14:00 | 2.3 | 13 |
| 12:50 | 3.6 | 45 | 14:30 | 2.4 | 28 |
| 13:10 | 5.7 | 75 | 15:00 | 2.3 | 15 |
| 13:30 | 5.6 | 80 | 16:00 | 2.5 | 18 |
| 13:50 | 20 | 280 | 17:00 | 2.2 | 22 |
| 14:20 | 15 | 145 | 18:00 | 1.7 | 29 |
| 14:50 | 5.4 | 115 | 20:00 | 1.4 | 21 |
| 15:50 | 1.2 | 21 | 22:00 | 0.6 | 17 |

a) Plot the hydrographs and pollutographs, input and output on same plot. Discuss.
b) For the input, calculate the total TSS mass and the EMC.
c) For the output, calculate the total TSS mass and the EMC.
d) Calculate the bioretention treatment efficiency for TSS, based both on EMC and on total mass.

Note: Integrate using trapezoids; extrapolate 1 $\Delta t$ in each direction to 0.

**18.23** Discuss the bioretention nitrogen performance data shown in Figure 18.22 with respect to nitrogen cycling. Why is TN unchanged during stormwater flow through bioretention while PON and $NH_4$–N decrease and TDN, DON, and $NO_3$–N increase?

**18.24** The internal water storage system of a bioretention facility is being designed for nitrate removal. About 2.8 mg/L $NO_3$–N is to be treated. Find the effluent $NO_3$–N concentration from a design flow of 5 $ft^3$/min through an IWS volume of 1200 $ft^3$ at 0.4 porosity. Use a first-order denitrification rate constant of 0.1 $h^{-1}$ and assume plug flow conditions.

**18.25** The internal water storage system of a bioretention facility is being designed for nitrate removal; 2.8 mg/L $NO_3$–N is to be treated. Find the effluent $NO_3$–N concentration from a design flow of 5 $ft^3$/min through an IWS volume of 1200 $ft^3$ at 0.4 porosity. Use a first-order denitrification rate constant of 0.08 $h^{-1}$ and assume plug flow conditions.

**18.26** The internal water storage system of a bioretention facility is being designed for nitrate removal; 2.8 mg/L $NO_3$–N is to be treated to 1.4 mg/L at a design flow of 5 $ft^3$/min. Find the IWS volume required. Use a porosity of 0.42, a first-order denitrification rate constant of 0.1 $h^{-1}$ and assume plug flow conditions.

**18.27** The internal water storage system of a bioretention facility volume has a volume of 1200 $ft^3$ at 0.4 porosity; 2.8 mg/L $NO_3$–N remains in IWS storage for 5 days between rainfall events. Find the $NO_3$–N concentration remaining in the IWS. Use a first-order denitrification rate constant of 0.08 $h^{-1}$.

**18.28** The internal water storage system of a bioretention facility volume has a volume of 1200 $ft^3$ at 0.4 porosity; 1.8 mg/L $NO_3$–N remains in IWS storage for 5 days between rainfall events. Find the $NO_3$–N concentration remaining in the IWS. Use a first-order denitrification rate constant of 0.08 $h^{-1}$.

**18.29** Consider a bioretention cell. Suggest possible designs that would have produced the following inflow and outflow volumes/pollutant concentrations. Some example design features include media type (including relative compost %), media depth, drainage configuration, watershed location, and so on; all concentrations in mg/L. If you would like to justify your answers, you may. Sometimes only one or two design feature(s) is/are possible to be ascertained per the information provided.

a) *Inflow*: [TKN] = 0.76, [TP] = 0.11.
*Outflow*: [TKN] = 4.10, [TP] = 0.56.

b) *Inflow*: [TKN] = 1.26, [$NH_4$–N] = 0.34, [$NO_{2\text{-}3}$–N] = 0.41, [TP] = 0.19.
*Outflow*: [TKN] = 0.70, [$NH_4$–N] = 0.10, [$NO_{2\text{-}3}$–N] = 0.43, [TP] = 0.13.

c) *Inflow*: [TKN] = 1.11, [$NH_4$–N] = 0.34, [$NO_{2\text{-}3}$–N] = 0.42, [TP] = 0.14.
*Outflow*: [TKN] = 0.57, [$NH_4$–N] = 0.10, [$NO_{2\text{-}3}$–N] = 0.28, [TP] = 0.05.

d) *Inflow*: 64 events produced inflow.
*Outflow*: 4 of the above events had outflow.

**18.30** To address a TMDL requirement (see Chapter 1), the stormwater load of total nitrogen must be reduced by 700 kg/year. Using the data of Table 18.1, how many ha of impervious area must be treated using bioretention. Clearly explain any assumptions you make.

18.31 To address a TMDL requirement (see Chapter 1), the stormwater load of total phosphorus must be reduced by 300 kg/year. Using the data of Table 18.1, how many ha of impervious area must be treated using bioretention. Clearly explain any assumptions you make.

18.32 To address a TMDL requirement (see Chapter 1), the stormwater load of total copper must be reduced by 24 kg/year. Using the data of Table 18.1, how many ha of impervious area must be treated using bioretention. Clearly explain any assumptions you make.

# 19

# Swales, Filter Strips, and Level Spreaders

## 19.1 Introduction

Swales and filter strips are vegetated stormwater control measures (SCMs) employed for conveyance and treatment of stormwater. Swales are linear systems, the flow is concentrated, and conveyance plays a large role in their application. To be effective, filter strips must treat sheet flow which is common along highways and in parking lots. In other applications, filter strips are coupled with a level spreader, which is a long, flat weir designed to deliver sheet flow to the filter strip. Both swales and filter strips are common in treating runoff from highways because of their compatibility with linear infrastructure.

## 19.2 Characteristics

### 19.2.1 Swales

Swales are channels that convey water from one place to another. Vegetated swales are one of the most commonly used SCMs worldwide (Figure 19.1). Most have been designed for safe conveyance of flows from events with infrequent return periods (e.g., average return intervals of 2 or 10 years). Stormwater swales are commonly constructed in the shape of a trapezoid, using a grass cover. Flow will enter one end of the swale from impervious areas, such as a parking lot or building downspout. Swales can be the primary SCM or they can be part of a treatment train, conveying flow to another SCM, providing treatment or pretreatment.

As the water flows along the swale, some infiltration can occur, depending on the underlying swale soils. Additionally, interactions with the vegetation can slow the water, allowing sedimentation, possibly filtration, and other unit processes to improve water quality. Swales may also receive runoff laterally along the side slopes.

### 19.2.2 Filter Strips and Level Spreaders

Filter strips are vegetated areas of flat land with a mild slope to allow sheet flow to infiltrate and be treated. They are used at sites where land area is available directly adjacent to the impervious area. The sheet flow is captured directly off of the

*Green Stormwater Infrastructure Fundamentals and Design*, First Edition. Allen P. Davis, William F. Hunt, and Robert G. Traver.

Companion Website: www.wiley.com/go/davis/greenstormwater

Figure 19.1 Swales Are Used World-wide, Such as Those: (A) in Raleigh, NC; (B) under a Bridge Deck in Knightdale, NC; and (C) Serving a Residential Road in Montgomery Co., Maryland (Photos by Authors).

impervious area and directed onto the filter strip. Filter strips can be placed adjacent to highways to treat the sheet flow coming directly off of the highway. Concentrated flow can converted to and applied as sheet flow to a filter strip using a level spreader. Filter strip discharge can lead to a stream or another SCM. A short filter strip receiving water directly from a highway, discharging laterally to a swale is shown in Figure 19.2. A filter strip supplied by a level spreader is presented in Figure 19.3.

## 19.3 Swale Design

Most swale guidance is based upon long-term experience with swale performance. However, these SCMs can also be explicitly designed to improve the quality of runoff, and design guidance can be based upon research conducted within the past 10–12 years. This guidance is for simple grass-lined dry swales, for example, those without underlying filter media or underdrains and does not cover either linear wetland-like swales or those with check dams. Swales designed specifically for infiltration, frequently known

Figure 19.2 Grass Filter Strip along Highway, Discharging into a Swale (Photo by Authors).

Figure 19.3 Grass Filter Strip Feed by Concrete Level Spreader (Photo by Authors).

as "bioswales" are discussed in Chapter 18. Simple grass swales are also known as "dry swales," since their common state is to be dry.

Water quality swale design is a two-step process, as is reviewed in the following sequence: first is the hydraulic design. During this phase, the swale depth of flow, cross-sectional geometry, and slope are determined. Second, the pollutant mitigation design is performed, which establishes the length of the swale.

### 19.3.1 Configurations

Stormwater swales are usually designed with trapezoidal cross-sections because of simplicity of construction. Additionally, trapezoidal cross-sections are recommended to maximize flow contact with water (USEPA 2004), prevent formation of erosive channels, and for ease of mowing. The swale bottom width is usually a minimum of 0.6 m (2 ft), with a maximum width of the swale often restricted by regulatory authorities to approximately 2 m (6 ft). The side slopes are set at 3:1 or 4:1 and the longitudinal slope can range from about 0.5% to 4%.

### 19.3.2 Hydraulic Design

#### 19.3.2.1 Setting Grass Height (and Depth of Flow)

For water quality purposes, the threshold depth for water quality flow is not to exceed the expected height of the grass. The maximum allowed grass height depends on grass species. It is in the designer's interest to choose somewhat tall, stiff grasses that grow uniformly, like a carpet. Clumping grasses are to be avoided. Table 19.1 presents a short list of relatively stiff grasses and their associated height. Knowing that maximum grass heights may not be maintained, the water quality flow design grass height should be slightly less than the maximum. So, in the case of tall fescue (*Festuca arundinacea*), a recommended design height of grass (and therefore design depth for flow) would be 150 mm (6 in)

#### 19.3.2.2 Channel Geometry

For flow conveyance, swales are designed using Manning's equation, introduced in Chapter 7.

$$v = \frac{1.49}{n} R_h^{\frac{2}{3}} s^{\frac{1}{2}} \tag{19.1}$$

where $v$ is the water velocity in the swale in ft/s, $n$ is the Manning roughness coefficient, $R_h$ is the hydraulic radius (in ft), and $s$ is the slope or hydraulic gradient. Equation 19.1 can be used with SI units without the 1.49 coefficient; in this case, $v$ is in m/s and $R_h$ in m.

Table 19.1 Suitably Stiff and Tall Grass Species Recommended for Use in Swales.

| Common name | Scientific name | Maximum height | Other notes |
|---|---|---|---|
| Tall fescue | *Festuca arundinacea* | 200 mm (8 in) | Can grow as tall as 1.2 m (48 in) with proper management. Cool season grass with US range into the southeast |
| St. Augustine | *Stenotaphrum secundatum* | 150 mm (6 in) | Warm season grass, found in southern tier of US states |
| Bluegrass | *Poa pratensis* | 200 mm (8 in) | Cool season grass with US range into the southeast |

The values of Manning's $n$ is 0.013 for concrete and smooth asphalt. Values for gravel channels or those with short grasses range from 0.025 to 0.027. Water quality flows through swales are meant to be at or below grass height, not overtopping the vegetation. Kirby et al. (2005) calculated flow impedance (or hydraulic resistance) provided by three different grass types for these types of flows. The data were presented using an "apparent" Manning's roughness coefficient, $n$. The range of values for each of the grasses is found in Table 19.2. The roughness coefficient is higher by almost a factor of 10 than what is typically used for grass liners, because when flow does not overtop grass (as is the case here), the grass imparts much more resistance to flow. A value of $n = 0.35$ may be a good estimate for swale design, although smaller values may be used also (Bäckström 2002).

The continuity equation may be employed for the calculation of swale flow:

$$Q = vA \tag{19.2}$$

Flow velocities must also be checked to ensure that they do not exceed erosional rates. Grass-tolerance velocity thresholds are about 1.2 m/s (4 ft/s) (Malcom 1993).

**Example 19.1** A grass swale has a longitudinal slope of 1.5%, 4:1 slide slopes, and a 3-ft bottom diameter, as diagrammed below. Find the velocity and flow rate at a water depth of 4 in and a Manning's $n$ value of 0.25.

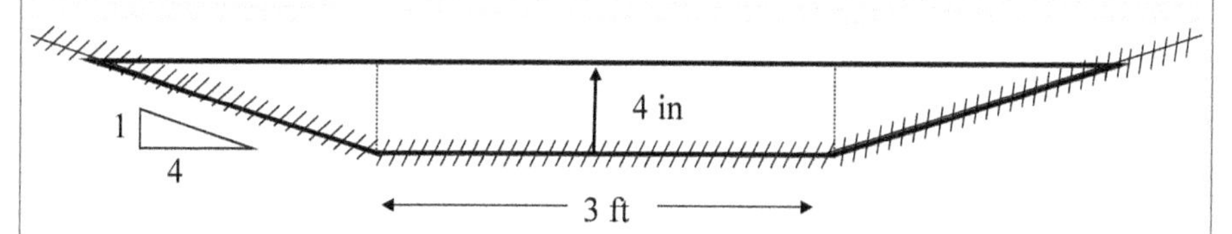


*Solution* The swale hydraulic radius must be calculated using the area and the wetted perimeter. The area is found as the sum of the rectangular area of the swale channel and the two triangular cross-sections of the slopes.

$$A = (0.33\ \text{ft})(3\ \text{ft}) + 2\left[\frac{1}{2}(0.33\ \text{ft})(4 \times 0.33\ \text{ft})\right] = 1.42\ \text{ft}^2$$

The wetted perimeter is the bottom width, plus the hypotenuse of each of the slide slopes.

$$P = (3\,\text{ft}) + 2\sqrt{(0.33\,\text{ft})^2 + (4 \times 0.33\,\text{ft})^2} = 5.74\,\text{ft}$$

Therefore, the hydraulic radius is found as $R = \frac{A}{P} = \frac{1.42\ \text{ft}^2}{5.74\ \text{ft}} = 0.247\ \text{ft}$

The velocity is calculated using Equation 19.1.

$$v = \frac{1.49}{0.25}(0.247\ \text{ft})^{\frac{2}{3}}(0.015)^{\frac{1}{2}} = 0.287\ \text{ft/s}$$

Flow rate is found from velocity and cross-sectional area:

$$Q = vA = (0.287\ \text{ft/s})(1.42\ \text{ft}^2) = 0.408\ \text{ft}^3/\text{s}$$

**Table 19.2** "Apparent" Manning's Roughness Coefficients for Grasses without Water Level Exceeding Grass Height (Adapted from Kirby et al. 2005).

| Common name | Scientific name | Blade length | Manning's $n$ |
|---|---|---|---|
| Centipede | *Ermochloa ophiuroides* | 50–80 mm | 0.27–0.95 |
| Bluegrass | *Poa pratensis* | 35–80 mm | 0.26–0.56 |
| Zoysia | *Zoysia* × *"Emerald"* | 40–80 mm | 0.28–1.35 |

In Example 19.1, the water depth was given and the flow rate was determined. Computationally, this is straightforward. However, for most swale designs, the channel geometry is determined by the water quality flow (determined from a runoff model, as discussed in Chapter 6), the roughness of the grass liner, and the slope. In practice, the channel geometry dimensions are adjusted until the carrying capacity of the channel matches or exceeds the target water quality flow rate. This requires a guess-and-check process.

**Example 19.2** A grass swale must treat a flow rate of 2 ft$^3$/s at a flow depth of 6 in. The swale has a longitudinal slope of 3% and 4:1 slide slopes. Find the swale bottom width. Use a Manning's $n$ value of 0.25.

***Solution*** The solution uses a guess-and-check process following the procedure of Example 19.1. A bottom width is assumed and the flow rate determined using Manning's equation and the continuity equation.

Assuming a width of 3 ft: $A = (0.5\text{ ft})(3\text{ ft}) + 2\left[\frac{1}{2}(0.5\text{ ft})(4\times 0.5\text{ ft})\right] = 2.5\text{ ft}^2$

$$P = (3\text{ ft}) + 2\sqrt{(0.5\text{ ft})^2 + (4\times 0.5\text{ ft})^2} = 7.12\text{ ft}$$

$$R = \frac{A}{P} = \frac{2.5\text{ ft}^2}{7.12\text{ ft}} = 0.351\text{ ft}$$

$$v = \frac{1.49}{0.25}(0.351\text{ ft})^{\frac{2}{3}}(0.03)^{\frac{1}{2}} = 0.514\text{ ft/s}$$

$$Q = vA = (0.514\text{ ft/s})(2.5\text{ ft}^2) = 1.28\text{ ft}^3\text{/s}$$

The calculated flow is too low. The process must be repeated with a larger bottom area until $Q$ = 2 ft$^3$/s. A value of 5.3 ft will suffice.

$$A = (0.5\text{ ft})(5.3\text{ ft}) + 2\left[\frac{1}{2}(0.5\text{ ft})(4\times 0.5\text{ ft})\right] = 3.65\text{ ft}^2$$

$$P = (5.3\text{ ft}) + 2\sqrt{(0.5\text{ ft})^2 + (4\times 0.5\text{ ft})^2} = 9.42\text{ ft}$$

$$R = \frac{3.65\text{ ft}^2}{9.42\text{ ft}} = 0.387\text{ ft}$$

$$v = \frac{1.49}{0.25}(0.387\text{ ft})^{\frac{2}{3}}(0.03)^{\frac{1}{2}} = 0.549\text{ ft/s}$$

$$Q = vA = (0.549\text{ ft/s})(5.3\text{ ft}^2) = 2.00\text{ ft}^3\text{/s}$$

For some cases of shallow slopes and low flow rates, a water quality dry swale cannot be constructed. The shallower slopes are simply unable to convey water at a small enough depth (less than or equal to grass height). In situations where this is the case, other practices, such as a wetland swale should be considered (USEPA 2004; Winston et al. 2012).

## 19.4 Filter Strip Design

Filter strips are wide, long vegetated areas designed to treat sheet flow runoff. It is critical that sheet flow be maintained; should the flow be channelized, it will become erosive, negating the treatment benefits of the filter strip and becoming destructive to the SCM.

### 19.4.1 Configurations

Filter strips must be completely level in the lateral direction, with an even, slight slope longitudinally. In order that sheet flow be maintained, an even homogeneous grass surface should be employed. No trees, shrubs, or other obstacles should be placed in the filter strip, which would encourage channeling. Filter strip design and analysis also is guided by Manning's Equation (19.1).

### 19.4.2 Flow Conveyance

In applications where the flow is concentrated, inclusion of the level spreader is critical to filter strip success. The level spreader is a long, narrow concrete (or similar smooth) berm to introduce sheet flow into the filter strip. Soil, stone, and wood should not be used as level spreaders, as they will not be completely level, or with wood, will rot, leaving an unlevel surface.

Runoff is directed into the upstream side of the level spreader. A small forebay may be installed to allow removal of large sediment and for easy maintenance of the SCM. The upstream channel must be sloped to encourage flow along the entire length. This channel is constructed of concrete, rock, or grass. This channel fills with water before the water reaches the height of the level spreader and spills over the top (Figures 19.3 and 19.4).

Over the lip of the level spreader, on the downstream side, about 1 m (3.3 ft) of stone should be placed over a strip of filter fabric. This will stabilize the area over the level spreader. The lip of the level spreader should be about 8 cm (3 in) above the top of the stone. Beyond this stabilization zone is the vegetated filter strip (Figure 19.5).

The level spreader width is designed based on conveying an extreme rainfall event, such as the 10-year storm. The width is designed at 13 ft of level spreader width per $ft^3/s$ runoff treated (140 $m/m^3/s$) (Hathaway and Hunt 2008).

**Example 19.3** A peak design flow from a roadway area is 2.5 $ft^3/s$. Find the level spreader/filter strip width.

*Solution* The level spreader design width is 13 ft per $ft^3/s$ runoff treated.

Therefore: $\text{Width} = \left(2.5\ \text{ft}^3/\text{s}\right)\left(\dfrac{13\ \text{ft}}{\text{ft}^3/\text{s}}\right) = 32.5\ \text{ft}$

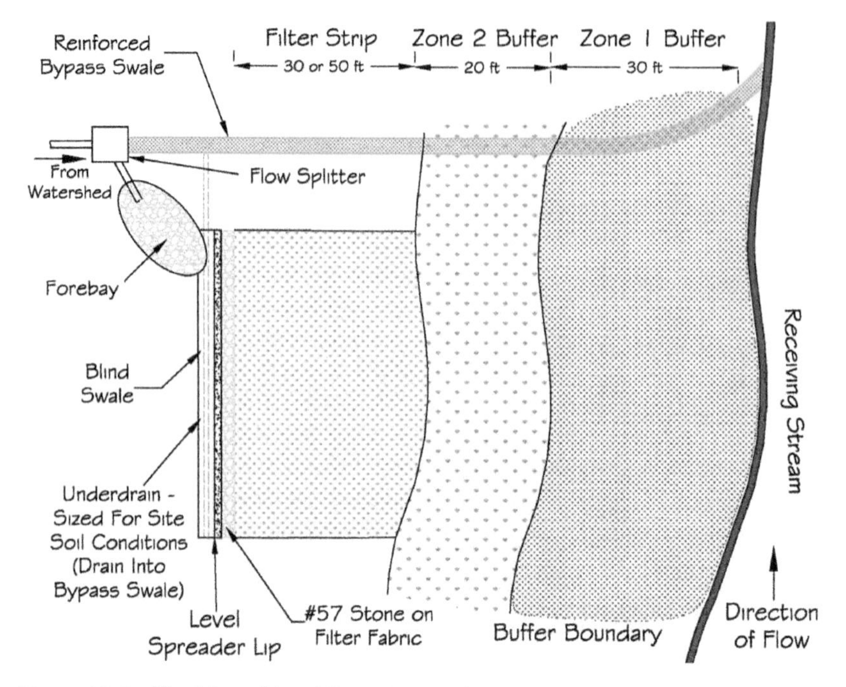


Figure 19.4 Plan View of Level Spreader and Grass Filter Strip System (Credit: Authors' Research Group).

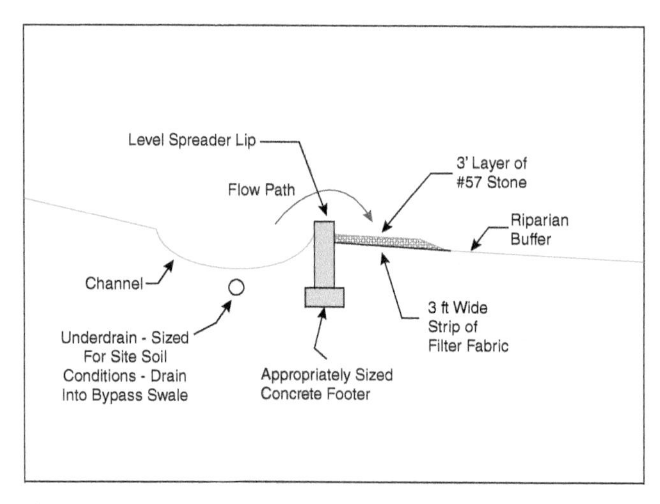


Figure 19.5 Cross-section of Level Spreader and Grass Filter Strip System (Credit: Authors' Research Group).

## 19.5 Filter Strips Conveying to Swales

As shown in Figure 19.2, filter strips are commonly used to convey water from a highway edge to a conveyance swale. The filter strip can play a very important role in reducing the stormwater volume from the pavement source as a significant fraction of the runoff is infiltrated during sheet flow over the filter strip. García-Serrana et al.

(2017a) derived a relationship to estimate the percentage of volume infiltrated during filter strip flow. The two controlling factors are a dimensionless saturated hydraulic conductivity ($K_s^*$) and the initial soil moisture deficit in the filter strip ($\Delta\theta$). The dimensionless hydraulic conductivity term is calculated as

$$K_s^* = \frac{K_{\text{sat}}}{I}\frac{W_S}{W_R} \tag{19.3}$$

where $K_{\text{sat}}$ is the saturated hydraulic conductivity of the filter strip soil, $I$ is the rainfall intensity, $W_S$ is the width of the filter strip parallel to the flow path and $W_R$ is the width of the roadway draining to the strip.

$V_i^*$, is defined as the ratio of infiltrated volume, $V_{\text{inf}}$ to the inflow volume to the filter strip, $V_{\text{in}}$:

$$V_i^* = \frac{V_{\text{inf}}}{V_{\text{in}}} \tag{19.4}$$

These dimensionless parameters are related by the expression (García-Serrana et al. 2017a):

$$V_i^* = 1.79 K_s^{*\,0.25} (\Delta\theta)^{0.68} \tag{19.5}$$

If a value greater than 1 is calculated for $V_i^*$, this parameter should be set equal to 1.0.

These expressions link together the important parameters for filter strip hydrologic performance, including filter strip soil hydraulic conductivity and soil moisture content, and the widths of both the highway supplying the stormwater and the filter strip that is infiltrating. García-Serrana et al. (2017b) employed a modified Green and Ampt model with a kinematic wave model to more fundamentally describe the relationships among these parameters.

**Example 19.4** A 6-m filter strip connects a 10-m roadway to a grassed swale. The saturated hydraulic conductivity of the filter strip soil is 2.0 cm/h. A rainfall with average intensity of 1.4 cm/h occurs when the filter strip soil moisture deficit is 30% (0.30). Find the fraction of runoff infiltrated by the filter strip.

***Solution*** The value of the dimensionless saturated hydraulic conductivity is calculated using Equation 19.3

$$K_s^* = \frac{(2.0 \text{ cm/h})}{(1.4 \text{ cm/h})}\frac{(6 \text{ m})}{(10 \text{ m})} = 0.857$$

The infiltration ratio is calculated using Equation 19.5.

$$V_i^* = 1.79\left(0.857^{0.25}\right)\left(0.30^{0.68}\right) = 0.76$$

Seventy-six percent of the incoming runoff should be infiltrated by the filter strip.

## 19.6 Water Quality Considerations

Swales and filter strips offer few opportunities for water quality improvement. Most of the stormwater management benefits result from infiltration and volume loss. Sedimentation of suspended solids is likely the dominant unit process for water treatment. Infiltration is important for pollutant mass removal, so designing for high infiltration rates will result in significant environmental benefits. As the water contacts the vegetation, limited opportunities for other treatment mechanisms may be generated, such as particle filtration and biological treatments.

Focusing on sedimentation, retention time and hydraulic loading (Q/A) become important design parameters. Tall grasses and check dams may help to slow flow and reduce turbulence. Wider swales can be more effective than narrow swales. Nonetheless, any accumulated sediment could be mobilized and washed away during a subsequent high-flow event and velocities should be kept low.

### 19.6.1 Designing for Pollutant Capture: Length of Swale

The mechanism for sediment-borne pollutant removal is settling or deposition. The settling rate of particulates is governed by Stokes' law, as described in Chapter 8. Particle removal by settling is a function of residence time in the swale, which depends on flow velocity, and the size and density of the particle; only particles with settling velocities, $v_s$, greater than or equal to the hydraulic loading rate will be removed. As noted in Chapter 8, this is given by

$$v_s > v\frac{H}{L} \tag{19.6}$$

where $v$ is the runoff velocity and $H$ and $L$ are the swale (or filter strip) flow depth and length, respectively. Thus, long swales and slow velocities will provide the greatest sedimentation of particulate matter.

Removal of other pollutants, like that of nitrogen or phosphorus, is only partially explained by particle settling. Biological processes, such as nitrification and denitrification, account for much of SCM nitrogen removal, and phosphorus removal is in part predicated upon adsorption. As retention time increases, more nutrients may be removed from the swale, provided conditions conducive to these mechanisms exist, as discussed in Chapter 8.

### 19.6.2 Designing for Particulate Matter Removal

Particle size distributions for highway runoff tend to be coarser than the soils of the surrounding catchment, which is in great part driven by the material composition of pavement. A thorough examination of urban roadway solids in Cincinnati, OH, for example, identified mean and median $d_{50}$ values of 555 and 570 μm, respectively (Sansalone et al. 1998).

With a focus on water quality improvement, swale length is calculated by establishing a target removal of particulate matter. To accomplish this, the particle size distribution (PSD) "typical" for that location needs to be known. Short swales can trap large particles (Barrett et al. 1998; Deletic 1999). Some fine particles are essentially untreated (Deletic and Fletcher 2006), even by long swales.

The particle "Fall Number," $N_f$, is a suggested tool for determining whether a particle is trapped (Deletic 2005; Hunt et al. 2020) during conveyance through a vegetated swale or filter strip. This model was specifically designed for urban flows and the low to moderate concentrations of particulate matter associated with postdevelopment (not active construction) conditions. The dimensionless Fall Number is calculated as

$$N_f = Lv_s / hV \tag{19.7}$$

where $L$ is the length of grass strip/swale, $v_s$ is the particle settling velocity (calculated using Stokes' Law), $h$ is flow depth, and $V$ is the flow velocity. $N_f$ is employed in the empirically based Aberdeen equation (Equation 19.5), used to predict particulate matter removal (Deletic 2005):

$$Tr_s = \frac{N_f^{0.69}}{N_f^{0.69} + 4.95} \tag{19.8}$$

where $Tr_s$ = particulate matter trapping efficiency (in decimal form).

The Aberdeen equation is used to set the length of the swale to meet a target removal rate for particulate matter. For example, if 50% of a particulate matter load is to settle in the swale, the designer would adjust the length of the swale (Equation 19.7) until at least a 50% removal has been achieved (Equation 19.8).

Particles of approximately 500 μm can easily be trapped within the first meter (3.3 ft) or so of vegetation, as has been shown through field data collection (Barrett et al. 2004) and the use of predictive models. Equation 19.8 predicts greater than a 90% removal rate for a 570-μm particle within the first 1.5 m (5 ft).

Equations 19.7 and 19.8 were used to predict removal efficiencies for two swales: (1) in the Albany suburb of Auckland, New Zealand (NZ) receiving parking lot runoff and (2) in Knightdale, NC, receiving bridge deck runoff (Hunt et al. 2020). One of those swales is pictured in Figure 19.1. The predicted results were then compared to actual field data collected for that swale. The Albany (NZ) swale was 73.6 m long and 1.0 m wide with nominally 150 mm tall grass. The Knightdale, NC, swale was 36.5 m long, 1.2 m wide with nominally 225 mm tall grass. Inflow runoff and particle size distribution were both available for three storms at each site. Predicted values were compared to actual field measurements for the events during which flow regimes were within their intended range (Table 19.3). The model was very accurate (within 8.4%) for four of the six events, while somewhat underpredicting removal efficiency for one event (Knightdale, December 13, 2010) and over-predicting efficiency for the final event (Albany, November 23, 2010). When viewed collectively, Equations 19.7 and 19.8 provide reasonably good accuracy for predicting particulate matter removal via swale flow.

Because the Fall Number model (Deletic 2005) did not consistently underpredict the actual removal efficiency of swales (though in four of the six cases the model was conservative), designers may want to consider adding a factor of safety (either added swale length or a reduced amount of assigned removal). An exact factor of safety may be a case-by-case example, but 1.3–1.5 is reasonable. A factor of safety is further needed due to potentially unreliable maintenance of swales. The grass may be cut too low or so infrequently that the grass liner loses stiffness.

A simple, but only approximate, way to estimate the removal efficiency of particulate matter by a swale is by calculating the removal efficiency for the $d_{50}$ particle. If, for example, 50% of particulate matter is meant to be removed by the swale, the length of the swale could be set such that at least 50% of the $d_{50}$ particle is trapped within that distance because all particles greater than $d_{50}$ will be completely removed, and those less than $d_{50}$ will be partially removed.

**Table 19.3** Actual and Theoretical TSS Removal Efficiencies for Monitored Swales in Auckland, New Zealand, and North Carolina, the United States (Hunt et al. 2020, with Permission from ASCE).

| Location | Event Date | $Q_p$[a], field measured (L/s) | $d_{50}$ (mm) | Depth of flow at $Q_p$, Calculated (mm) | Calculated removal efficiency (%) | Actual removal efficiency (%) | Difference (%) |
|---|---|---|---|---|---|---|---|
| Albany, Auckland, NZ | November 23, 2008 | 12.9 | 0.82 | 120 | 90 | 70 | −20 |
| Albany, Auckland, NZ | December 23, 2008 | 20.7 | 0.25 | 150 | 81 | 82 | 1 |
| Albany, Auckland, NZ | August 16, 2009 | 25.9 | 0.86 | 170 | 85 | 94 | 8 |
| Knightdale, NC, US | October 15, 2010 | 22.5 | 0.15 | 240 | 65 | 71 | 6 |
| Knightdale, NC, US | November 05, 2010 | 8.5 | 0.083 | 170 | 48 | 56 | 8 |
| Knightdale, NC, US | December 13, 2010 | 3.65 | 0.045 | 130 | 34 | 49 | 15 |

[a] Peak flow

**Example 19.5** Particulate matter removal of 50% is desired in a grass swale at flow conditions of Example 19.2. If the assumed particle matter diameter is 0.1 mm (specific gravity of 2.5, 50°F), find the length of the swale.

***Solution*** Because of the functional relationship of Equation 19.8, a guess-and-check process is necessary. A swale length is assumed and $N_f$ is calculated using Equation 19.7.

First, the particle settling velocity is found using the Stokes law (Chapter 8); 0.1 mm is equal to $3.28 \times 10^{-4}$ ft.

$$v_s = \frac{(\gamma_p - \gamma_w)d^2}{18\mu} = \frac{\left(2.5(62.4) - 62.4 \text{ lb/ft}^3\right)\left(3.28 \times 10^{-4} \text{ ft}\right)^2}{18\left(2.74 \times 10^{-5} \text{ lb s/ft}^3\right)} = 2.04 \times 10^{-2} \text{ ft/s}$$

A swale length of 150 ft is initially assumed; the velocity from Example 19.2 is 0.549 ft/s at a depth of 0.5 ft:

$$N_f = Lv_s / hV = (150\,\text{ft})(0.0204\,\text{ft/s}) / ((0.5\,\text{ft})(0.549\,\text{ft/s})) = 11.15$$

$$Tr_s = \frac{N_f^{0.69}}{N_f^{0.69} + 4.95} = \frac{11.15^{0.69}}{11.15^{0.69} + 4.95} = 0.516$$

The 150-ft swale assumption is good, with removal estimated at 51.6%. The exact solution is found with a 136-ft swale.

### 19.6.3 Designing for Particulate Matter Removal with Particle-size Distribution Available

The actual removal efficiency for the entire particulate matter load is determined by the PSD. PSDs with low $d_{90}/d_{50}$ ratios or high $d_{50}/d_{10}$ ratios will likely have a cumulative removal efficiency lower than that calculated for the $d_{50}$ particle alone. The ideal procedure is to determine performance to model a PSD-weighted removal efficiency to calculate the removal efficiency. Detailed information about the particle sizes of the inflow sediment are critical to the accuracy of this calculation.

**Example 19.6** Predict how well a 50-m long swale, receiving runoff from a roadway would treat the sediment particle-size distribution collected by Sansalone et al. (1998), given in the first two columns of the following table. The swale flow velocity is calculated as 0.135 m/s at a flow depth of 0.10 m.

Calculating a Weighted Trapping Efficiency for a 50-m Long Swale Using PSD.

| Particle size distribution, $d_x$ | Particle size, $d$ (μm) | Settling velocity, $v_s$ (m/s) | Fall Number, $N_f$ | Trap efficiency, $T_r$ | $T_r$ * Particle size fraction |
|---|---|---|---|---|---|
| 100 | 2200 | 4.22 | 15,640 | 99.4% | |
| 80 | 1300 | 1.47 | 5460 | 98.7% | 19.81 |
| 60 | 970 | 0.82 | 3040 | 98.1% | 19.68 |
| 50 | 820 | 0.59 | 2170 | 97.6% | 9.78 |
| 30 | 590 | 0.30 | 1120 | 96.3% | 19.39 |
| 10 | 130 | 0.015 | 54.6 | 76.1% | 17.24 |
| 0 | 16 | 0.00022 | 0.83 | 15.1% | 4.56 |
| | | | Cumulative removal efficiency | | 90.5% |

***Solution*** A procedure similar to that of Example 19.5 is used. The particle settling velocity is calculated from Stokes' law at 20°C and a particle specific gravity of 2.6. The results are presented in the third column, as $v_s$. The Fall Number is different for each particle; for the largest:

$$N_f = Lv_s / hV = (50\text{ m})(4.22\text{ m/s})/((0.1\text{ m})(0.135\text{ m/s})) = 15{,}630$$

The Fall Numbers are given in Column 4, with the trap efficiency given in Column 5 from Equation 19.8.

$$Tr_s = \frac{15{,}630^{0.69}}{15{,}630^{0.69} + 4.95} = 0.994$$

The last column uses the trapezoidal rule to find the cumulative removal for each particle size distribution interval (Column 1), times the average removal (Column 5). The total removal, 90.5%, is the sum of the removal of each particle size fraction. As expected, most of the largest particles are removed, while few of the smaller size particles are removed.

**Example 19.7** Evaluate the $d_{50}$ "short-cut" method and compare it to the PSD-based weighted trapping efficiency method in Example 19.6.

*Solution* Calculating the values of $T_r$ (removal efficiency) for solely the $d_{50}$ (820 μm), yields a 97.6% removal. When applying the more accurate estimation based upon weighting PSDs, a 90.5% removal rate is calculated as in Example 19.5. The 90.5% removal is found with approximately the $d_{24}$ particle. Use of the $d_{50}$ overestimates because of the difficulty in removing the smallest particles.

Winston et al. (2017) evaluated the use of Equations 19.7 and 19.8 for swale and filter strip particulate removal performance. Flow conditions for the SCMs were calculated based on Manning's equation, coupled with Fall Number calculations to predict particulate matter removal. Effects of swale geometry (triangular vs. trapezoidal), side slopes, longitudinal slopes, grass height, and swale/filter strip length were examined using different PSDs. Flows frequently exceeded grass heights for triangular cross-sections. As expected, particulate matter removal increased with increasing swale length (although with diminishing returns) and with flatter longitudinal slopes—both related to longer hydraulic retention times. Filter strips performance was superior to swales simply due to the lower flow velocities encountered.

### 19.6.4 Designing for Metals Removal

Metals follow similar, but not the exact, patterns of capture as sediment. As noted in Chapter 5, finer fractions of sediment carry a proportionally higher fraction of metals. Data from Zanders (2005) demonstrating this is shown in Chapter 5, repeated as Table 19.4. Therefore, 50% total suspended solids (TSS) removal would not equate to 50% Cu or Zn removal; the latter would be lower.

Table 19.4 Total Metal Concentrations as a Function of Particle-size Fraction (Adapted from Zanders 2005)

| Particle-size fraction (μm) | Total metal concentration (mg/kg) | | | Particle density (kg/m³) |
|---|---|---|---|---|
| | *Cu* | *Zn* | *Pb* | |
| 0–32 | 181 | 2080 | 316 | 2140 |
| 32–63 | 197 | 1695 | 322 | 2150 |
| 63–125 | 212 | 1628 | 334 | 2190 |
| 125–250 | 184 | 1073 | 251 | 2330 |
| 250–500 | 85 | 507 | 193 | 2530 |
| 500–1000 | 26 | 268 | 323 | 2540 |
| 1000–2000 | 21 | 226 | 36 | 2390 |

**Example 19.8** Using the metals data from Zanders (2005) and the Aberdeen model (Deletic 2005), find the percent removal of (particulate) copper through a 30-m long swale. The particle size distribution is given in the first two columns of the following table. The swale flow velocity is calculated as 0.14 m/s at a flow depth of 0.18 m.

*Solution* The first part of the solution requires finding the particulate matter removal, as in Example 19.6, shown in the following table.

Calculating a Weighted Trapping Efficiency for a 30-m Long Swale Using PSD.

| Particle size range (μm) | Average particle size (μm) | Size fraction | Particle specific gravity | Fall Number, $N_f$ | Trap efficiency, $T_r$ | $T_r$ * Size fraction |
|---|---|---|---|---|---|---|
| 0–32 | 16 | 16% | 2.14 | 0.19 | 6.0% | 0.96% |
| 32–63 | 17.5 | 14% | 2.15 | 1.68 | 22.4% | 3.1% |
| 63–125 | 94 | 21% | 2.19 | 6.82 | 43.2% | 9.1% |
| 125–250 | 188 | 15% | 2.33 | 30.3 | 68.0% | 10.2% |
| 250–500 | 375 | 14% | 2.53 | 139 | 85.9% | 12.0% |
| 500–1000 | 750 | 9% | 2.54 | 562 | 94.1% | 8.5% |
| 1000–2000 | 1500 | 11% | 2.39 | 2028 | 97.5% | 10.7% |
| | | | | Cumulative removal efficiency | | 54.6% |

The particle removal efficiency through the swale is 54.6%.

The example continues with the addition of the copper data from Zanders (2005).

Calculating Particulate Metals' Removal in a Swale Using PSD.

| Particle size range (μm) | Average particle size (μm) | Size fraction | Cu conc. (mg/kg) | Weighted Cu fraction | $T_r$ * Size fraction | Cu removed |
|---|---|---|---|---|---|---|
| 0–32 | 16 | 16% | 181 | 29.0 | 0.96% | 1.74 |
| 32–63 | 17.5 | 14% | 197 | 27.6 | 3.1% | 6.18 |
| 63–125 | 94 | 21% | 212 | 44.5 | 9.1% | 19.2 |
| 125–250 | 188 | 15% | 184 | 27.6 | 10.2% | 18.8 |
| 250–500 | 375 | 14% | 85 | 11.9 | 12.0% | 10.2 |
| 500–1000 | 750 | 9% | 26 | 2.34 | 8.5% | 2.20 |
| 1000–2000 | 1500 | 11% | 21 | 2.31 | 10.7% | 2.25 |
| | Average Cu concentration | | 145.2 | | | |
| | | | Cumulative removal | | | 60.6 |
| | | | Cumulative removal percent | | | 41.7% |

The weighted Cu fraction is found by multiplying the particle size fraction by the Cu concentration at that particle size. For the 16-μm particle size:

$0.16 \times 181 = 29.0$

The sum of these values gives the average Cu concentration in the sediment.

The removal fraction for each particle size is found from the previous table and transferred here. Multiplying the particle removal fraction by the Cu concentration gives the Cu removal.

$$0.0096 \times 181 = 1.74$$

These values are summed to give 60.6.

The fraction Cu removal is found by dividing the cumulative removal by the average Cu concentration.

$$60.6/145.2 = 0.417 = 41.7\%$$

Only 42% of Cu would be captured, compared to 55% of the total particulate matter. This occurs because a larger fraction of the Cu is affiliated with the smaller particles and the smaller particles are removed to a lesser extent than the larger ones.

### 19.6.5 Filtration through Swales and Filter Strips

A compilation of swale studies from the International BMP Database (http://www.bmpdatabase.org) (IBMPD, 2010) is presented in the probability plot shown in Figure 19.6. The majority of swale TSS effluent concentrations are less than the 25 mg/L target, suggested by Barrett et al. (2004). In fact, 80% of all swales included in the database had effluent concentrations less than 30 mg/L. The median observed effluent concentration was 8 mg/L. Perhaps the reason for this low-effluent TSS concentration was that most of the results in the IBMPD were for swales exceeding 30 m (100 ft) in length, providing adequate length for particle sedimentation and filtration. With more details available for these studies, namely, contributing watershed area, it may be possible to design swales such that specific drainage area to swale length ratios (DA:$L_{swale}$) can be selected to achieve a target effluent concentration. Knowledge about particle size distributions would also be helpful in understanding performance.

The concept of designing SCMs to discharge a target concentration as opposed to removing a percentage of a pollutant, has gained recent popularity, as discussed previously. Barrett et al. (2004) found that the grassed filter strips in California, USA, nearly uniformly reduced TSS concentrations to approximately 25 mg/L. The final concentration did not appear to be impacted by filter strip width. Winston et al. (2011) found similar results: increasing the travel path of the vegetated filter strips only modestly improved the effluent concentration of TSS; the effluent TSS concentrations were essentially 20–30 mg/L. While not examined in either Barrett et al. (2004) or Winston et al. (2011), the discharge particulate matter concentration may have been influenced by finer particle sizes within the stormwater. Large particles are effectively removed via filtration and sedimentation, while smaller particles are not, leaving behind a TSS concentration that becomes effectively independent of SCM size. This follows the "irreducible concentration" discussion in Chapter 11.

Some degree of filtration can occur as the runoff flows between the grass blades in a swale or filter strip. As discussed in Chapter 8, steady state filtration theory is given by

$$\frac{C}{C_0} = \exp\left(-1.5\frac{(1-\epsilon)\alpha\eta L}{d_c}\right) \tag{19.9}$$

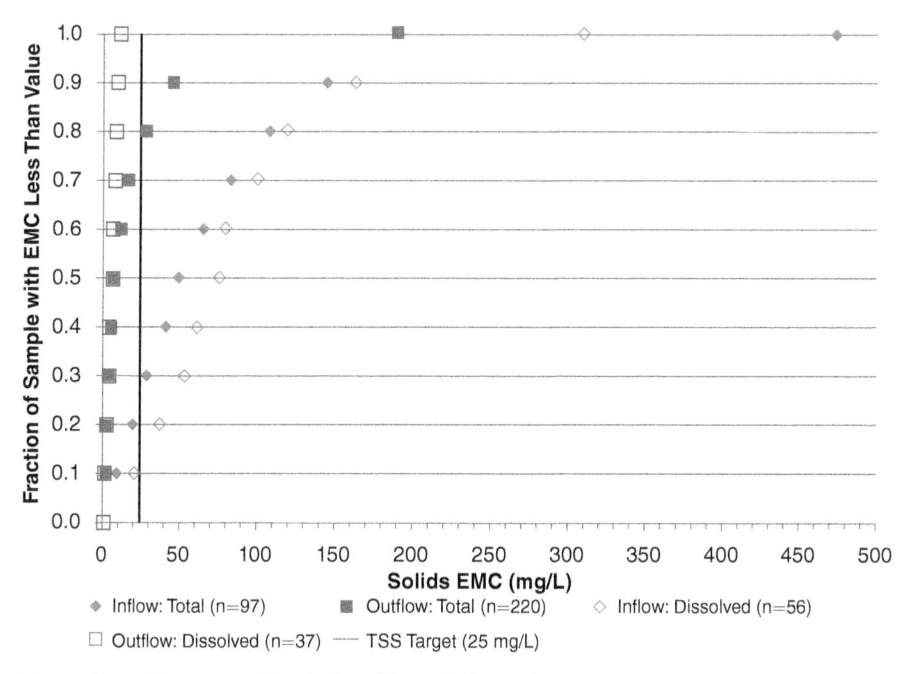


Figure 19.6 Summary of Particulate Matter Effluent Concentrations (Summarized from (IBMPD) 2010), Illustrating a Typically Low-effluent TSS Concentration. (Figure Modified from Fassman et al. 2010).

In the case of swales and filter strips, the flow/treatment direction is horizontal and the filter medium is the grass blades. Vertical sedimentation will also take place. The porosity is defined as the space between the grass blades, which may be a difficult parameter to estimate. Filtration theory does predict an exponential reduction of particulate matter (and affiliated pollutants) as a function of travel path length, which has been supported by some performance data.

A final note about particle sizes and discharge concentrations. As was demonstrated in Chapter 8, wide particle size distributions can lead to "irreversible" effluent TSS concentrations. Large particles can readily settle or be filtered via vegetated swales. The efficiencies of these treatment mechanisms are related to the square of the particle diameter and, in the case of filtration, is exponentially related to the travel path. The smallest particles are much more difficult to remove, resulting in residual amounts of TSS and other pollutants that can only be removed by resorting to very large (unreasonably large) SCMs.

### 19.6.6 Check Dams

Check dams are commonly placed in swales to retain some of the water as it flows through the channel. This will increase the water retention time in the swales, allowing for greater beneficial impacts to both hydrology and water quality. Ponding water behind check dams will allow for greater infiltration of the water and possibly greater ET after the storm. Ponded water will also allow for increased sedimentation of suspended solids.

Swale check dams are commonly made of wood planks that are placed across the channel. A weir may be cut in the check dam to manage flow. Check dams have also been constructed from rip rap and from thick, tall vegetation (Figure 19.7).

## 19.7 Swale Performance

Multiple field studies have shown that pollutant removal by swales is a function of many factors: cross-sectional geometry, slope, flow depth, grass type and height, pollutant type, particle size, and so on (Bäckström 2002, 2003; Barrett et al. 1998; Deletic 1999; Deletic and Fletcher 2006). Research findings are nearly uniform: swales and filter strips improve water quality for sediment and sediment-borne pollutants. The pollutant removal mechanisms dominant in swales are sedimentation and filtration/ straining by vegetation. These mechanisms are most effective when the vegetation is not overtopped, that is, when flow is spread out so that the full measure of water flows *through* the grass and not over it. Thus, a clear distinction exists between swales intended to convey flow for larger events and swales intended to improve water quality. For water quality swales, the water elevation during the design storm stays at or below the height of grass lining the swale. It is, of course, recommended that swales be designed for both functions.

### 19.7.1 Hydrologic Considerations

The performance of grass swales in volume management can be approximately divided into three regimes (Davis et al. 2012). For the smallest rainfall events, no swale discharge will occur. Runoff that is applied to the swale will be stored and infiltrated into

Figure 19.7 Grass Swale along Highway. Vegetated Check Dam Is Shown in Foreground (Photo by Authors).

the soil. The infiltration capacity of the swale will depend on the characteristics of the swale and native soils, which will impact the infiltration rate, and any surface storage that may be created with check dams. In the Davis et al. (2012) swales study, the storage capacity varied from about 0.4 to 2.2 cm of water depth over the vegetated area.

Once the volumetric capacity is reached, some degree of volume reduction is exhibited by the swales. Swales have a finite storage capacity that can be removed from the incoming volume, reducing volumetric discharge.

For the largest events, the volumetric storage in the swales is negligible. The flow is high and deep in the swales, so that the vegetation offers minimal resistance to flow. Under these conditions, the swales act primarily as conveyance devices only.

Figure 19.8 shows measured runoff volume discharges plotted against the input volumes for four swale design conditions, with and without a pretreatment filter strip and with/without vegetated check dams. The data are scattered, but it can be seen that most small events (and a few large ones) are completely captured by the swales. Larger events show various degrees of capture. Volumes exceeding input result as flow from surrounding areas enter the swale during large events.

Flow duration curves for two of the swales described above are shown in Figure 19.9, demonstrating similar phenomena. The highest flows are generally unaffected by the swales as they act as pure conveyance systems. Flow reduction occurs for the smallest flows and about half of the total volume input to the swale (the data in Figure 19.9 listed as HWY) is lost due to swale infiltration.

Peak flow reduction in swales result from the same design considerations as volume reduction. Increasing swale length, storage (through check dams), and infiltration will promote peak flow reduction.

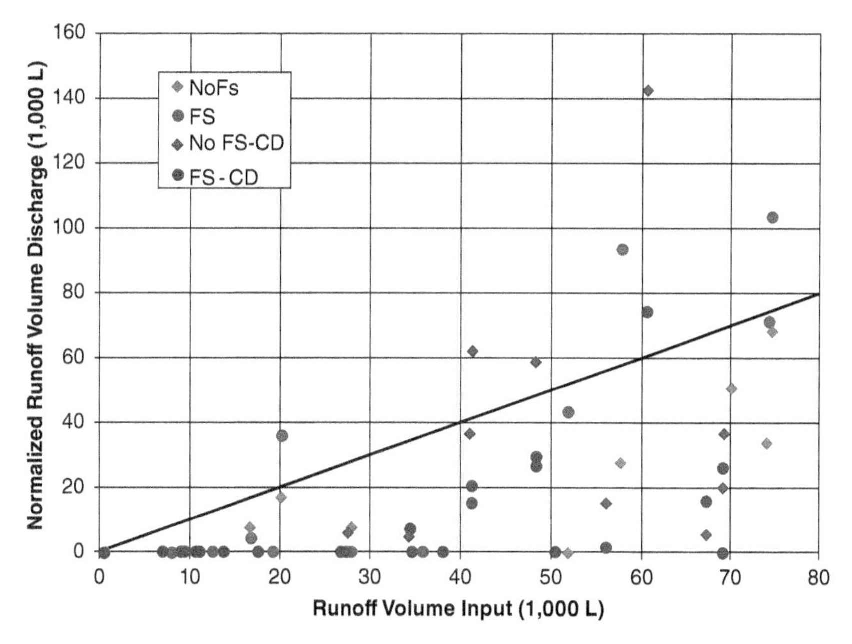


Figure 19.8 Volumetric Performance of Grass Swales in Managing Highway Runoff (Davis et al. 2012). Data Are from Four Different Swale Designs. FS Is a Swale with a Pretreatment Filter Strip. CD Is a Swale with a Vegetated Check Dam.

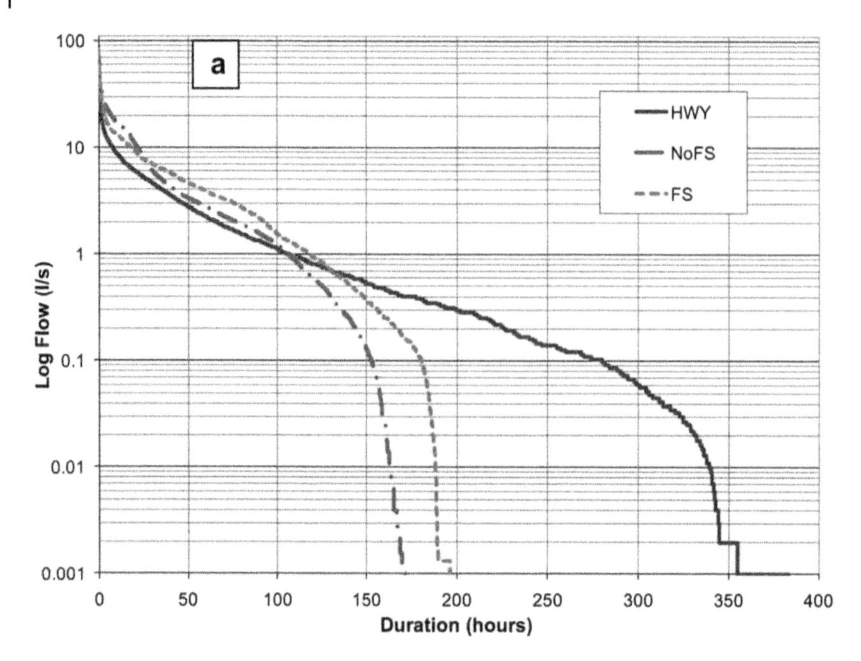


Figure 19.9 Flow Duration Performance of Grass Swales in Managing Highway Runoff (Davis et al. 2012). No-FS and FS Are Swales without and with a Pretreatment Filter Strip, Respectively. The Swale Input Flow Duration Is the HWY Designation.

### 19.7.2 Water Quality Considerations

As discussed above, TSS concentrations are reduced by grass swale treatment (e.g., Stagge et al. 2012; Fassman 2012; Winston et al. 2015), resulting from sedimentation/filtration of particulate matter. Reductions in concentrations of total Cu, Pb, and Zn are also noted, primarily because of their affiliation with the captured solids. Overall mass reduction is found for TSS and metals, resulting from the combination of treatment and water volumetric loss through infiltration. A summary of swale water quality performances is presented in Table 19.5. Data showing the range of concentrations measured (and normalized to account for rainfall directly on the swale) from two studies are provided. In the studies, the swales infiltrated all of the runoff in about half of the measured storm events, thus the NF (no flow) designation for the minimum concentration leaving the swales. Reductions are obvious for TSS and the metals; less so for the nutrients.

Overall, studies are inconclusive on TN and TP swale removal performance (Tables 19.5 and 19.6). Particulate or particulate-bound N and P are expected to be sequestered in a swale, but dissolved fractions would not, unless the swale is a wetland-like SCM with a high retention times. Studies with high nitrogen inlet concentrations (relative to Passeport and Hunt (2009) who found an average TN of 1.63 mg/L for impermeable surfaces) had nitrogen removal (Barrett et al. 1998; Deletic and Fletcher 2006; Winston et al. 2012). Sites with low nitrogen influent concentrations (Rushton 2001; Winston et al. 2012) saw an *increase* in effluent concentrations. Similar trends have been found for phosphorus.

Table 19.5 Range of Concentrations (Normalized to Account for Rainfall over the Swale) for Two Sets of Swale Studies (Davis et al. 2012). No-FS and FS Are Swales without and with a Pretreatment Filter Strip, Respectively.

| Water quality parameter | Swale input (highway runoff) | No check dams | | Swale input (highway runoff) | With check dams | |
|---|---|---|---|---|---|---|
| | | No-FS | FS | | No-FS | FS |
| TSS (mg/L) | 9.8–309 | NF–210 | NF–60 | 8–582 | NF–109 | NF–232 |
| TP (mg/L) | 0.1–2.3 | NF–1.2 | NF–1.0 | 0.1–1.5 | NF–1.3 | NF–0.6 |
| TKN (mg/L) | 0.8–10.2 | NF–3.0 | NF–3.2 | 0.3–12.1 | NF–14.7 | NF–62.4 |
| Nitrate-N (mg/L) | 0.8–3.8 | NF–10.4 | NF–8.2 | 0.27–16.3 | NF–6.4 | NF–2.63 |
| Zn (μg/L) | 54–1650 | NF–310 | NF–220 | 115–2320 | NF–440 | NF–210 |
| Cu (μg/L) | 12–195 | NF–42 | NF–53 | 15–182 | NF–161 | NF–117 |
| Pb (μg/L) | 2–70 | NF–30 | NF–45 | 6–960 | NF–63 | NF–150 |

Table 19.6 Influent and Effluent TSS Concentrations from Applied Field Studies.

| Study | Location | Influent TN | Effluent TN | Type change |
|---|---|---|---|---|
| Winston et al. (2012) | Site A | 1.48 | 1.65 | Increase |
| Winston et al. (2012) | Site D | 2.60 | 1.62 | Decrease |
| Rushton (2001) | F7 | 0.55[1] | 0.64 | Increase |
| Rushton (2001) | F8 | 0.55 | 0.72 | Increase |
| Barrett et al. (1998)[2] | US 183 | 3.08 | 1.92 | Decrease |
| Barrett et al. (1998) | MoPac | 3.88 | 2.42 | Decrease |
| Deletic and Fletcher (2006) | Brisbane Swale[3] | 2.6 | 1.12 to 1.46 | Decrease |

[1] Influent TN calculated as average of "Asphalt, no swale," F1 (0.556 mg/L) and F2 (0.548 mg/L);

[2] TN calculated as TKN + $NO_3$–N for Barrett et al. (1998); and

[3] experimental test with various flow rates.

Modest to moderate load reductions for TN were observed for swales located in sandy coastal plain soils of Florida, resulting from a high amount of infiltration (Rushton 2001); TP load results were mixed. Deletic and Fletcher (2006) examined a swale in Brisbane, Australia, and showed that TP and TN concentrations were lower than influent concentrations but did not appreciably change with flow rate.

It appears that seasonal effects that impact N and P cycling in the vegetation, including anthropogenic activities such as grass mowing, impact swale nutrient performance. Since a major fraction of stormwater nutrients is expected to be dissolved, sedimentation and filtration will not impact these species. While reductions in concentration may occur, mass removal has been noted in some swales, but overall export of mass was found for others for both total nitrogen and total phosphorus.

While dry swales did not clearly reduce the concentrations of nitrogen or phosphorus, Winston et al. (2012) found that TN and TP concentrations leaving a wetland-like swale were lower than those of dry swales, indicating that creation of an anaerobic condition in the swale, while maximizing retention time, may be a design tool used to improve nitrogen removal performance where conditions allow.

## 19.8 Construction, Inspection, and Maintenance

A major reason for the popularity of swales and filter strips is the ease of construction. Filter strips are flat, with a slight slope. Swales are (usually) trapezoidal channels. Both are easily constructed with standard construction equipment. However, level spreaders require specific consideration to ensure that they are perfectly level. This will require some construction skill. Any disturbances to the level of the level spreader will lead to concentrated flow at one part of the filter strip, eventually leading to failure of the strip.

The newly constructed swale channel and filter strip must be carefully stabilized to prevent soil erosion as the vegetation becomes established. Frequently this requires some type of matting to be installed to keep the soils in place. The matting may be biodegradable so that will not need to be removed once the site vegetation matures.

Inspection and maintenance of vegetated swales and filter strips are also usually simple. Grass is usually mowed for aesthetic reasons and for sight/safety concerns. However, the stormwater management benefits of mowing are debatable and grass clippings left to decay will release nutrients. Bare soil should be replanted immediately, as this soil can be eroded and will lead to high particulate matter in discharge waters; it is critical that complete vegetation coverage be established during construction. Erosion should be evident if flow is too concentrated and the source of the problem should be addressed.

Again, the level of the level spreader needs special consideration. The level needs to be checked on a regular basis and fixed if necessary. Check dams also need periodic inspection, and maintenance if necessary.

## 19.9 Summary

Vegetated swale design and performance can be separated into two general parts: hydraulic and pollutant capture. The hydraulic design is predicated upon several factors, with two of the most influential being the maximum depth of water allowed in the swale and the designed rainfall intensity (peak flow) the swale is expected to treat. The suggested maximum water height allowed in the swale mirrors that of the target grass height (perhaps 150 mm or 6 in, but exact depth is a function of grass species). The water quality design flow is based on having at least 95% of water pass through the swale at a height no greater than that of the grass.

Once the vegetative liner (a grass), slope, and cross-sectional dimensions have been determined, the length of the swale is calculated based on a desired pollutant capture. Hydraulic residence time is a key parameter for predicting pollutant removal; that is, runoff must be exposed to treatment mechanisms (sedimentation, grass filtration) for an adequate period. This establishes the length of the swale. Check dams can increase hydrologic and water quality benefits by increasing retention time.

For both sediment and sediment-borne pollutants, the PSD is the critical component for predicting pollutant capture. Coarser particles, such as those found in some highway runoff studies, need very little grass verge widths or swale lengths to be almost completely removed from stormwater. Conversely, extremely fine particles are not at all likely to collect in even a rather long swale.

Metals removal by a given swale will be similar to, but typically slightly less than, that of sediment, because metals like Cu and Zn are more closely associated with finer—and harder to remove—particle sizes. Nitrogen and phosphorus removal is best explained by (1) average hydraulic retention time and (2) the fraction of each nutrient that is in particulate form. In the limited studies conducted, nutrient removal is mixed.

An important implication of this guidance is that swales are appropriately maintained. A design factor-of-safety of 1.3–1.5 is recommended to account for potential maintenance lapses. Additionally, landscapers and other maintainers of SCMs should be made aware of the importance of grass height and cover to the success of a swale's water quality performance.

## References

Bäckström, M. (2002). Sediment transport in grassed swales during simulated runoff events. *Water Science and Technology* 45 (7): 41–49.

Bäckström, M. (2003). Grassed swales for stormwater pollution control during rain and snowmelt. *Water Science and Technology* 48 (9): 123–134.

Barrett, M.E., Lantin, A., and Austrheim-Smith, S. (2004). Stormwater pollutant removal in roadside vegetated buffer strips. *Transportation Research Record* 1890: 129–140.

Barrett, M.E., Walsh, P.M., Malina, J.F., and Charbeneau, R.J. (1998). Performance of vegetative controls for treating highway runoff. *Journal of Environmental Engineering* 124 (11): 1121–1128.

Davis, A.P., Stagge, J.H., Jamil, E., and Kim, H. (2012). Hydraulic performance of grass swales for managing highway runoff. *Water Research* 46 (20): 6775–6786.

Deletic, A. (1999). Sediment behaviour in grass filter strips. *Water Science and Technology* 39 (9): 129–136.

Deletic, A. (2005). Sediment transport in urban runoff over grassed areas. *Journal of Hydrology* 301 (1-4): 108–122.

Deletic, A. and Fletcher, T.D. (2006). Performance of grass filters used for stormwater treatment — A field and modelling study. *Journal of Hydrology* 317 (3-4): 261–275.

Fassman, E. (2012). Stormwater BMP treatment performance variability for sediment and heavy metals. *Separation and Purification Technology* 84: 95–103.

Fassman, E.A., Liao, M., Shadkam Torbati, S., and Greatrex, R. (2010). *Stormwater Mitigation through a Treatment Train*. Auckland, New Zealand: Prepared by Auckland UniServices, Ltd. for Auckland Regional Council. Auckland Regional Council Technical Report TR19/2010.

García-Serrana, M., Gulliver, J.S., and Nieber, J.L. (2017a). Infiltration capacity of roadside filter strips with non-uniform overland flow. *Journal of Hydrology* 545: 451–462.

García-Serrana, M., Gulliver, J.S., and Nieber, J.L. (2017b). Non-uniform overland flow-infiltration model for roadside swales. *Journal of Hydrology* 552: 586–599.

Hathaway, J.M. and Hunt, W.F. (2008). Field evaluation of level spreaders in the piedmont of North Carolina. *J. Irrigation Drainage Engineering* 134 (4): 538–542.

Hunt, W.F., Fassman-Beck, E.A., Ekka, S.A., Shaneyfelt, K.C., and Deletic, A. (2020). Designing dry swales for stormwater quality improvement using the Aberdeen equation. *Journal of Sustainable Water in the Built Environment* 6 (1): 05019004.

International BMP Database (IBMPD) (2010). Denver, CO. www.bmpdatabase.org

Kirby, J.T., Durrans, S.R., Pitt, R., and Johnson, P.D. (2005). Hydraulic resistance in grass swales designed for small flow conveyance. *Journal of Hydraulic Engineering* 131 (1): 65–68.

Malcom, H.R. (1993). *Fundamentals of Urban Hydrology*. Raleigh, NC: NCSU Industrial Extension.

Passeport, E. and Hunt, W.F. (2009). Asphalt parking lot runoff nutrient characterization for eight sites in North Carolina, USA. *Journal of Hydrologic Engineering* 14 (4): 352–361.

Rushton, B.T. (2001). Low-impact parking lot design reduces runoff and pollutant loads. *Journal of Water Resources Planning and Management* 127 (3): 172–179.

Sansalone, J.J., Koran, J.M., Smithson, J.A., and Buchberger, S.G. (1998). Physical characteristics of urban roadway solids transported during rain events. *Journal of Environmental Engineering* 124 (5): 427–440.

Stagge, J.H., Davis, A.P., Jamil, E., and Kim, H. (2012). Performance of grass swales for improving water quality from highway runoff. *Water Research* 46 (20): 6731–6742.

United States Environmental Protection Agency (USEPA) (2004). *Stormwater Best Management Practice Design Guide: Volume 2 Vegetative Biofilters*. EPA/600/R-04/121A. Washington DC: Office of Research and Development.

Winston, R.J., Anderson, A.R., and Hunt, W.F. (2017). Modeling sediment reduction in grass swales and vegetated filter strips using particle settling theory. *Journal of Environmental Engineering* 143 (1): 04016075.

Winston, R.J., Hunt, W.F., Kennedy, S.G., Wright, J.D., and Lauffer, M.S. (2012). Comparing dry and wet swale hydrology and water quality along Interstate 40 in coastal plain North Carolina. *Journal of Environmental Engineering* 138 (1): 101–111.

Winston, R.J., Hunt, W.F., Osmond, D.L., Lord, W.G., and Woodward, M.D. (2011). Field evaluation of four level spreader-vegetative filter strips to improve urban stormwater quality. *Journal of Irrigation and Drainage Engineering* 137 (3): 170–182.

Winston, R.J., Lauffer, M.S., Narayanaswamy, K., McDaniel, A.H., Lipscomb, B.S., Nice, A.J., and Hunt, W.F. (2015). Comparing bridge deck runoff and stormwater control measure quality in North Carolina. *Journal of Environmental Engineering* 141 (1): 04014045.

Zanders, J.M. (2005). Road sediment: Characterization and implications for the performance of vegetated strips for treating road run-off. *Science of the Total Environment* 339 (1): 41–47.

## Problems

**19.1** A flow rate of 1.2 $ft^3/s$ must be conveyed using a trapezoidal grass swale. The swale will have a longitudinal slope of 1% and 3:1 slide slopes. Find the swale bottom width necessary to provide a water depth of 8 in. Also, find the velocity at the design width. Use a Manning's $n$ value of 0.4.

**19.2** A flow rate of 2.1 $ft^3/s$ must be conveyed using a trapezoidal grass swale. The swale will have a longitudinal slope of 1.2% and 3:1 slide slopes. Find the swale bottom width necessary to provide a water depth of 8 in. Also, find the velocity at the design width. Use a Manning's $n$ value of 0.35.

**19.3** A grass swale has a bottom width of 2 ft 4:1 side slopes, and a longitudinal slope of 1.8%. The grass height is to be held at 6 in (Manning's $n = 0.28$). Find the allowable flow rate through this swale.

**19.4** A 28-ft highway drains to an 8-ft grass filter strip. The saturated hydraulic conductivity of the filter strip soil is 1.0 in/h. A rainfall with average intensity of 0.8 in/h occurs when the filter strip soil moisture deficit is 28%. Find the fraction of runoff infiltrated by the filter strip.

**19.5** A 28-ft highway will drain to a filter strip. The saturated hydraulic conductivity of the filter strip soil is 1.9 in/h. All of the highway runoff must be infiltrated from a 1.2-in/h rainfall intensity. Find the required width of the filter strip, assuming that the antecedent soil moisture deficit is 32%.

**19.6** A 12-m highway will drain to a filter strip. The saturated hydraulic conductivity of the filter strip soil is 2.6 cm/h. All of the highway runoff must be infiltrated from a 3.0-cm/h rainfall intensity. Find the required width of the filter strip, assuming that the antecedent soil moisture deficit is 26%.

**19.7** A 12-m highway drains to a 4-m grass filter strip. A rainfall with average intensity of 0.9 in/h must be completely infiltrated when the filter strip soil moisture deficit is 28%. Find the required saturated hydraulic conductivity of the filter strip soil.

**19.8** A 32-ft highway will drain to a 12-ft grass filter strip. The saturated hydraulic conductivity of the filter strip soil is 2.2 in/h. Make a plot of the fraction of infiltrated runoff as a function of rainfall intensity. Assume the filter strip soil moisture deficit is 30%. Discuss your results.

**19.9** A 32-ft highway will drain to a grass filter strip. The saturated hydraulic conductivity of the filter strip soil is 2.2 in/h. Make a plot of the fraction of infiltrated runoff as a function of filter strip width at 1 in/h rainfall intensity. Assume the filter strip soil moisture deficit is 30%. Discuss your results.

**19.10** Thousands of miles of interstate crisscross the United States, most of these miles are rural. A typical interstate cross-section is shown below. Note the typical cross-section is roadway/filter strip/swale.

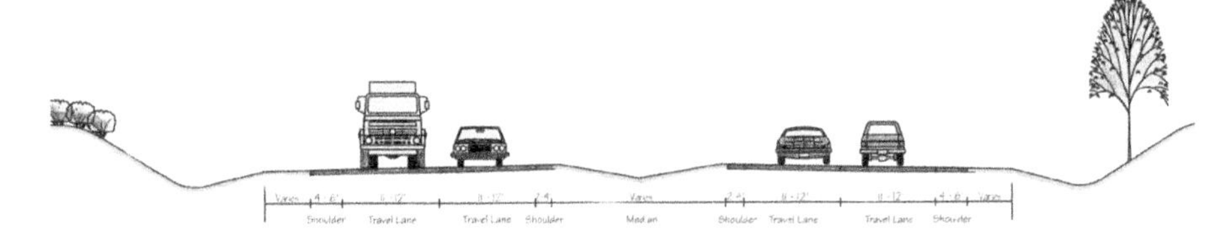


Four Lane Divided Roadway

Image provided by circeis.org.

Several studies have indicated particle size distributions (PSDs) for highway runoff. An example PSD for highway runoff is found below:

| PSD: $d_x$ | Size (m) |
|---|---|
| 100 | 0.0022 |
| 80 | 0.0013 |
| 60 | 0.00097 |
| 50 | 0.00082 |
| 30 | 0.00059 |
| 10 | 0.00013 |
| 0 | 0.000016 |

a) Estimate the percentage removal provided by filter strip adjacent to the roadway, before runoff collects in roadside swales.
b) How wide would the filter strip need to be to achieve 50% TSS capture?
c) Would copper removal be higher, lower, or the same as TSS capture?

In doing each, be sure to explicitly state all of your assumptions and provide reasoning or citations for each.

**19.11** Design a Splitter box for a level spreader in Greenville, NC. The watershed is 4 ac, residential with four dwelling units per acre. Include the invert, length (or diameter) for the "outlet" to the level spreader and the "outlet" to the high flow swale. State all assumptions. Show calculations, perhaps include a spreadsheet, if one is used. Provide a sketch.

**19.12** Three different steady state storm conditions for small, medium, and large storm events. Assume a design temperature of 50°F and that the particles have a density of 2.1 g/cm$^3$.

| | Small | Medium | Large |
|---|---|---|---|
| **Flow rate** | 5 cfs | 15 cfs | 50 cfs |
| **Water depth** | 2 ft | 3 ft | 4 ft |
| **Sediment conc.** | 75 mg/L | 180 mg/L | 440 mg/L |

**Sediment size distribution (by mass)**

| Dia. (μm) | % of total | % of total | % of total |
|---|---|---|---|
| <2 | 10 | 1 | 6 |
| 2–6 | 12 | 2 | 10 |
| 6–10 | 12 | 5 | 7 |
| 10–20 | 14 | 10 | 2 |
| 20–40 | 9 | 20 | 1 |
| 40–50 | 14 | 15 | 2 |
| 50–80 | 15 | 10 | 5 |
| 80–100 | 10 | 12 | 10 |
| 100–200 | 8 | 6 | 20 |
| 200–300 | 1 | 10 | 15 |
| 300–400 | 1 | 7 | 10 |
| 400–500 | 5 | 2 | 12 |

Find the expected sediment removal (%) and the discharge sediment concentration (mg/L) for the small storm traveling through a 30-m swale at a 0.1-m depth and velocity of 0.1 m/s.

**19.13** Find the expected sediment removal (%) and the discharge sediment concentration (mg/L) for the medium storm of Problem 19.6 traveling through a 30-m swale at a 0.12-m depth and velocity of 0.12 m/s.

**19.14** Find the expected sediment removal (%) and the discharge sediment concentration (mg/L) for the large storm of Problem 19.6 traveling through a 30-m swale at a 0.18-m depth and velocity of 0.15 m/s.

**19.15** Find the removal of lead from the swale, given the conditions in Example 19.6.

**19.16** Find the removal of zinc from the swale, given the conditions in Example 19.6.

**19.17** The particle size distribution for a runoff is given below. Find the removal of Cu traveling through a 30-m swale at a 0.12-m depth and velocity of 0.12 m/s.

| Average particle size (μm) | Size fraction |
|---|---|
| 16 | 9% |
| 17.5 | 15% |
| 94 | 11% |
| 188 | 22% |
| 375 | 18% |
| 750 | 14% |
| 1500 | 11% |

**19.18** The particle size distribution for a runoff is given in Problem 19.11. Find the removal of Pb traveling through a 30-m swale at a 0.12-m depth and velocity of 0.12 m/s.

**19.19** The particle size distribution for a runoff is given in Problem 19.11. Find the removal of Zn traveling through a 30-m swale at a 0.12-m depth and velocity of 0.12 m/s.

**19.20** A grass swale is to be designed to convey and treat runoff from a 2.4-ha parking lot. The swale is 120 m long, with a slope of 1.2%. It will have a trapezoidal cross-section with 1:1 side slopes. The grass produces a Manning's $n$ value of 0.04.

| Input | | |
|---|---|---|
| **Time of day** | **Flow rate (L/s)** | **TSS (mg/L)** |
| 11:50 | 26.6 | 145 |
| 12:10 | 46.9 | 280 |
| 12:30 | 126 | 175 |
| 12:50 | 175 | 115 |
| 13:10 | 41.3 | 85 |
| 13:30 | 46.2 | 75 |
| 13:50 | 32.2 | 80 |
| 14:20 | 30.1 | 45 |
| 14:50 | 44.8 | 45 |
| 15:50 | 15.4 | 21 |

Runoff entering the swale during a storm event is presented in the table above.

a) Calculate the total TSS mass (kg) and the EMC (mg/L) in the stormwater runoff. Note: Integrate using trapezoids; extrapolate Q and TSS concentration to 0 for one $\Delta t$ before and after the data set.

b) The swale design requires a flow depth of 28 cm at a flow rate of 175 L/s. Find the bottom width of the swale to meet the design requirement.

c) During the storm, sedimentation of particles will occur on the projected bottom area of the swale. Assume the swale bottom width is 0.4 m, the TSS particle size is 22 μm and specific gravity is 2.4, and the temperature is 15°C.

Although this is not a steady state system, use ideal sedimentation theory at each sample time in the table above to estimate TSS concentration leaving the swale. Calculate the total TSS mass (kg) and the EMC (mg/L) in the stormwater runoff leaving the swale; again, integrate using trapezoids; extrapolate Q and TSS concentration to 0 for one $\Delta t$ before and after the data set. For this exercise, disregard any infiltration into the swale soil.

19.21 A grass swale is to be designed to convey runoff from a parking lot to a river. The swale is 220 ft long, with a slope of 0.5%. It will have a trapezoidal cross-section with 1:1 side slopes. The grass produces a Manning's $n$ value of 0.06. For this exercise, disregard any infiltration into the swale soil.

a) The swale design requires a flow depth of 1 ft at the design flow rate. The design storm over the parking lot produces a flow rate of 5 $ft^3/s$. What should be the bottom width of the swale to meet the design requirements?
For parts a and b, use a trapezoidal swale with a bottom width of 2.5 ft. All other parameters as above.

b) The hydraulic retention time for the swale will affect the reactions and removal of several pollutants in the runoff.
   i. Make a plot of swale hydraulic retention time (min) as a function of swale flow rate up to 8 $ft^3/s$.
   ii. If the grass is permitted to grow taller, the Manning's $n$ value can be increased to 0.2. Repeat part (a) above with $n = 0.2$
   iii. Discuss your results from the plots.

c) Sedimentation of particles will also occur in the swale. Knowing the characteristics of the swale and the flow, the critical particle size for sedimentation can be calculated. Use a particle specific gravity of 2.5 and 50°F.
   i. Make a plot of the critical particle diameter (mm) as a function of swale flow rate up to 8 $ft^3/s$.
   ii. If the grass is permitted to grow taller, the Manning's $n$ value can be increased to 0.2. Repeat part (a) above with $n = 0.2$
   iii. Discuss your results from the plots.

# 20

# Stormwater Wetlands

## 20.1 Introduction

Due to nutrient reduction needs in large watersheds like the Chesapeake Bay in the USA and Moreton Bay in Australia, constructed stormwater wetlands have become a popular technique that merges "green engineering" with larger detention style devices. Stormwater wetlands come in a wide variety of sizes and operational characteristics. They have a number of advantages as a stormwater control measure (SCM), including very good management of both stormwater hydrology and water quality. They do, however, tend to have a large footprint and must maintain a permanent shallow water pool.

A stormwater wetland has permanent wet areas, commonly with variable depths and associated vegetation (Figure 20.1). They are designed to mimic natural wetlands

**Figure 20.1** Four Stormwater Wetlands That Have Been Part of SCM Research Studies: (A) Smithfield, NC; (B) Asheville, NC, Mountain Application; (C) Auckland Region, New Zealand; and (D) Singapore's Garden-by-the-Bay (Photos by Authors).

*Green Stormwater Infrastructure Fundamentals and Design*, First Edition. Allen P. Davis, William F. Hunt, and Robert G. Traver.

Companion Website: www.wiley.com/go/davis/greenstormwater

and the associated hydrologic, physical, chemical, and biological environmental benefits that they provide. A number of types of stormwater wetlands are defined, including shallow wetlands, extended detention shallow wetlands, pond/wetland systems, submerged gravel wetlands (SGWs), and pocket wetlands. Shallow wetlands provide high water quality improvements in a shallow pool that has a large surface area. An extended detention shallow wetland provides water quality improvement by a combination of shallow wetland and extended detention storage. A pond/wetland system differs from a shallow wetland due to its deep permanent pool that is placed before the shallow wetland. In a pocket wetland, the high water table or groundwater interception helps maintain a shallow wetland pool.

This chapter focuses on stormwater wetland operation and design points including (1) wetland sizing and definition of internal wetland features, (2) discussion of herbaceous plants that have been found to commonly thrive in *stormwater* wetlands, (3) a review of a proper growing soil medium, and (4) the importance for and design guidance on a flexible outlet structure. Preventing wetlands from becoming mosquito breeding grounds is also discussed.

Constructed wetlands are also used for municipal wastewater treatment and polishing. Many details on wetland design and function are provided by Kadlec and Knight (1996).

## 20.2 Sizing Stormwater Wetlands

Most constructed wetlands for small-to-medium-watershed sizes (up to 20 ha, 50 ac) are in-line structures, meaning they are located in ephemeral streams and all the runoff produced in the watershed flows through the wetland. The sizing for these systems is straightforward. In most jurisdictions, the water quality event (Chapter 9) is fully captured in the wetland and then slowly released over a 2–5-day period. Some regulations require a base volume of runoff per unit area to be treated, while others dictate that runoff from a certain design storm be captured and slowly released. The surface area $A$ of an in-line stormwater wetland is simply determined by dividing the water quality volume $V_{wq}$ by the average ponding depth $H_{ave}$ associated with that event. This depth commonly ranges from 22 to 38 cm (8.5–15 in.).

$$A = V_{wq} / H_{avg} \quad (20.1)$$

The temporary (or water quality volume) ponding depth is limited to this range because greater depths have been anecdotally shown to hurt plant growth. However, constructed stormwater wetlands can mitigate the peak flow of much larger events, such as those with 2–25-year return frequency (Chapter 2), provided the additional depth of water resides in the wetland over the course of hours, not days.

**Example 20.1** The water quality volume from an 8-ha lot has been calculated as 1625 m$^3$. Find the surface area of the in-line stormwater wetland and the fraction of drainage area that it occupies.

*Solution* From Equation 20.1, the wetland surface area is calculated. An average depth of 30 cm (0.3 m) of water storage is assumed.

$$A = V_{wq}/H_{avg} = 1625\ \text{m}^3/0.3\ \text{m} = 5420\ \text{m}^2$$

This represents 6.8% of the 80,000-$\text{m}^2$ drainage area.

## 20.3 Stormwater Wetland Features and Design

A stormwater wetland is designed to exploit physical, chemical, and biological processes of the system to enhance water quality (Imfeld et al. 2009). The ability of a stormwater wetland to improve water quality is dependent on particular vegetation, sediments and soil, microbial biomass, and an aqueous phase containing the pollutants. The use of stormwater wetlands is limited by specific constructs, including soil types, depth to groundwater, contributing drainage area, and available land area (Malaviya and Singh 2012).

Stormwater wetlands must either be sited on lands with very impervious soils, be lined with an impervious liner or clay layer, or intersect the groundwater table so that they will always remain wet. Infiltration rates should not exceed 0.025 cm/h (0.01 in./h) but could be much less. Medium-fine textured soils (e.g., loams and silt loams) are recommended for use in stormwater wetlands. These soils are optimal for culturing vegetation, retaining surface water, promoting groundwater recharge, and capturing pollutants.

The internal topography of a stormwater wetland can be divided into five features: (I) deep pool, (II) deep-to-shallow transition, (III) shallow water, (IV) temporary inundation, and (V) upper bank. An illustration of a stormwater wetland design is presented in Figure 20.2 while Figure 20.3 depicts a cross-section of the wetland topography by zone. The average depth of each zone is quite different, from 50 to 90 cm (20–30 in.) for deep pools, transitioning to 5–10 cm (2–4 in.) for shallow water. Each zone supports different vegetation and serves a specific purpose within the SCM.

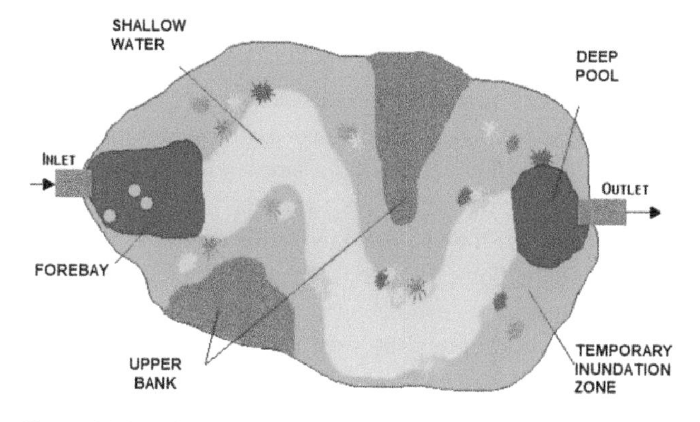


Figure 20.2 Plan View of Stormwater Wetland (Credit: Authors' Research Group).

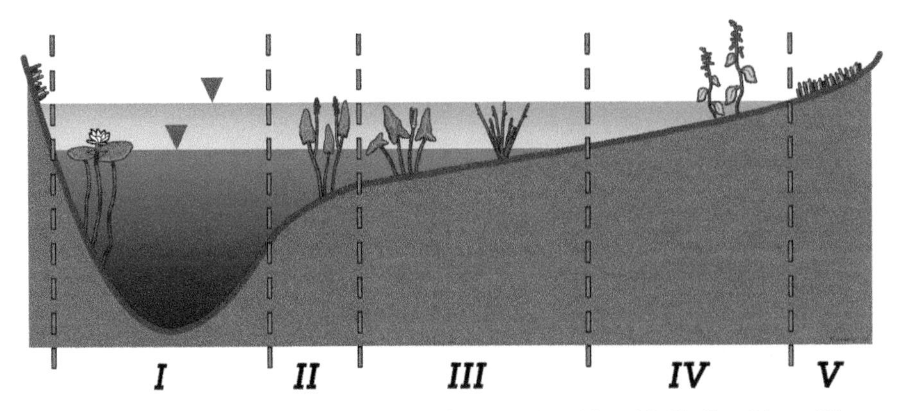


Figure 20.3 Interior Wetland Zones: (I) Deep Pool, (II) Transition, (III) Shallow Water, (IV) Temporary Inundation, and (V) Upper Bank (Credit: Authors' Research Group).

By incorporating all these zones in a single wetland, the designer creates (1) a system that dissipates stormwater flow energy by distributing that flow over the entire wetland, (2) multiple and unique zones for pollutant treatment, (3) a relatively diverse ecosystem for wetland plants and fauna (to limit mosquito populations), and (4) a more aesthetically pleasing addition to the local landscape.

Stormwater wetlands will not develop fully for 1–3 years postconstruction. Until the system is mature, its treatment will not be optimal.

### 20.3.1 Zone I—Deep Pools

Deep pools serve several functions in stormwater wetlands. They dissipate flow energy, trap sediment associated with incoming stormwater and provide an anaerobic environment for enhanced nitrate treatment. They also provide additional water storage that increases infiltration and evaporative losses which reduce outflow volumes (in locations where water tables are lower) and provide refuge for aquatic organisms during dry periods.

The deep pool is an important component of a "mosquito resistant" stormwater wetland, as it provides year-round habitat for mosquito predators that require water to thrive. As such, several deep pools should pockmark the stormwater wetland; this allows easier travel to/from the deep pools to all parts of the stormwater wetland (and mosquito habitats). An example of a small deep pool is shown in Figure 20.4.

The bottom elevation of the deep pool(s) should be at least 50 cm (20 in.) deeper than the designed water elevation at normal pool (defined as the water elevation within the wetland after complete drawdown following a storm event). The target depth of the pool should be 90 cm (30 in.), if possible. Because deep pools are a refuge for aquatic organisms, it is important that they be designed sufficiently deep so that they do not dry up during a long drought. Some local codes may require fencing along the perimeter of wetlands with deep pools.

The initial deep pool into which runoff enters the wetland is called the forebay. Forebays are discussed in more detail in Chapter 12. The forebay has two special purposes: (1) dissipate the energy of the entering runoff and (2) provide a storage zone for gross solids and large sediment to settle. The size of the forebay is nominally set to 10–15% of the total wetland surface area. A study conducted by North Carolina State

Figure 20.4 Small Deep Pool with Water Lilies in Stormwater Wetland (Photo by Authors).

University in 2004 and 2005 confirmed this to be a reasonable sizing standard. As gross solids and sediment accumulation in the forebay becomes significant, the collected material needs to be removed. The total wetland surface area occupied by deep pools (including the forebay(s)) should range between 20% and 25%.

Equation (20.2) incorporates a simple water balance that can be used to determine the minimum depth necessary for deep pools:

$$DP = RF_M * EF * (WS / WL) - ET - INF - RES \quad (20.2)$$

where DP is the depth of pool; $RF_M$ is the monthly rainfall during a drought period; EF is the fraction of rainfall that enters stormwater wetland from the watershed (0.20–0.25 estimate); WS/WL is the ratio of watershed area to wetland surface area; ET is the monthly evapotranspiration water loss; INF is the monthly infiltration water loss; RES is the reservoir of water for a factor of safety; DP, $RF_M$, ET, INF, and RES are water depths (in cm or in).

A projected worst-case condition may be assumed to evaluate pool depths. Deep pools should be designed to a depth to ensure year-round water. This condition will depend on the climate for the area.

As an example, a four-week drought period in the summer can be assumed as the worst-case condition; it must be noted that a one-month period without rainfall is extremely rare in many humid portions of the USA and world. Even a relatively small storm event (such as 1.2 cm, 0.5 in.) can contribute 8–16 cm (3–6 in.) depth of water to the stormwater wetland. Assuming this worst-case scenario, it is reasonable and conservative to expect up to 25 cm (10 in.) of water loss due to evapotranspiration by the wetland plants. If the wetland does not intersect shallow groundwater—and is therefore perched in areas that have very low natural infiltration rates or is lined—the wetland may still experience an infiltration loss, although it will generally be low. Assuming the maximum water loss rate of 0.025 cm/h, as discussed previously, approximately 0.65 cm/day will exit the wetland due to infiltration. In a week, about 4.5 cm (1.8 in.) would infiltrate the wetland, and during the course of a one-month drought—without any rain—20 cm (8 in.) of water loss is estimated.

**Example 20.2** During the month of July, a total of 2.5 cm of rainfall occurs. Of this rainfall, 20% enters the stormwater wetland. The watershed to wetland surface area ratio is 20. ET loss is estimated at 20 cm and the infiltration rate is 0.025 cm/h. The designer desires to keep 15 cm of water in reserve, as a factor of safety. Find the required depth of the wetland deep pool.

*Solution* First calculate the amount of water that will infiltrate into the soil below the wetland:

$$\text{INF} = \left(0.025\ \text{cm/h} * 24\,\text{h/d} * 31\,\text{d}\right) = 18.6\ \text{cm}$$

EF is given as 0.2 and WS/WL is 20.

With this information, the depth of the deep pool is calculated as

$$\text{DP} = 2.5\ \text{cm} * 0.20 * 20 - 18.6\ \text{cm} - 20\ \text{cm} - 15\ \text{cm} = -44\ \text{cm}$$

Therefore, the depth of the deep pool should be designed at about 44 cm (17 in.) below the depth of the normal pool.

### 20.3.2 Zone II—Deep to Shallow Water Transition Zone (Transition Zone)

The deep and shallow wetland pools should be connected with a maximum slope of 1.5:1 (1.5 horizontal per 1 vertical). Slopes steeper than this are not recommended inside the wetland for soil stability and safety reasons. The transition zone is essentially this gentle slope that serves to connect the pool and shallow water area. Only a few plants can tolerate the depth of the transition zone, but those that do survive are important to the function of the wetland, supporting both nitrification and denitrification. This zone tends to occupy the smallest amount of surface area in a stormwater wetland. The design water depth for this zone is between 15 and 23 cm (6 and 9 in.) when the stage within the wetland is at normal pool, but it is important to understand that this zone incorporates all depths between the deep pool (50 cm, 20 in.) and the shallow water zone (10 cm, 4 in.).

### 20.3.3 Zone III—Shallow Water Zone

Shallow water zones are also areas that retain water following drawdown of the wetland after a storm event. At lower flows, water entering the wetland should follow the course of the shallow water area. During extended drought periods, they will eventually become dry, but until they do, they form a connection between pools that allow passage for smaller fish, amphibians, and invertebrates. They are important in pollutant treatment because the shallow depths are better oxygenated to support aerobic nutrient transformations such as nitrification.

One of the most important departures from earlier stormwater wetland design guidance is the recommended depth of the shallow water zone. This depth *had formerly been* suggested to range from 15 to 30 cm (6 to 12 in.). Since the early 2000s, it has been observed that most wetland plants could not tolerate normal pool depths this deep for

Figure 20.5 A Stormwater Wetland in Durham North Carolina, during the Drought of 2002. Note the Pickerelweed (a Plant That Usually Flourishes in 3–6 in. (7.5–15 cm) of water) that is "High and Dry" (Photo by Authors).

extended periods. To promote a wider range of vegetation, thus avoiding a vegetation monoculture, average depths of 5–10 cm (2–4 in.) at normal pool are now recommended for the shallow water zone.

One common concern among designers is the ability of shallow water plants to survive during a drought. As Figure 20.5 shows, once established, shallow water plants can tolerate being dry (i.e., not inundated) during typical drought periods. Remember that naturally occurring wetlands also become dry occasionally. In fact, wetting and drying cycles enhance the ability of the wetland to effectively treat many pollutants. Even during droughts, soils within the wetland remain moist within a foot of the surface. As long as wetland plant roots are able to reach these moist soils, the wetland plants will survive during most droughts.

### 20.3.4 Zone IV—Temporary Inundation Zone

The temporary inundation zone acts as an internal floodplain, surrounding the shallow water channel and extending to the lower bank of the wetland. It is designed to be completely inundated when any storm larger than the design water quality event occurs in the watershed. This zone has no significant standing water remaining several days after a storm. The elevation of the ground surface is above the invert of the low-flow drawdown, to be discussed later. At normal pool, the elevation of land above the water line will range from 0 to (nominally) 30 cm (12 in.). The actual vertical extent of the temporary inundation zone is dependent upon the depth of water to which the designer wishes to store the water quality volume. Like the shallow water zone, the temporary inundation zone allows for a variety of vegetation to be grown, giving the wetland the potential to be a diverse ecosystem. The temporary inundation zone often includes a narrow strip of land which can be termed the lower bank. The lower bank is the part of a bank which is inundated when the water quality volume is captured.

### 20.3.5 Zone V—Upper Bank

The upper bank is essentially the upland portions surrounding the stormwater wetland. The surface area of the upper bank is not included as part of the wetland surface area but is necessary to tie the wetland topography back into the surrounding land. A wide variety of vegetation is able to survive in this zone, provided it can grow on slopes. The upper bank should not be sloped any steeper than 3:1, especially in sandy soils (although sometimes in retrofit applications, this maximum slope recommendation may need to be exceeded). This will minimize erosion and allow a reasonable grade for maintenance such as mowing or pruning.

## 20.4 Wetland Vegetation

Many more species of wetland vegetation can flourish in non-stormwater wetlands than in stormwater wetlands. Most initial lists of plants recommended for stormwater wetlands included many plants that were found in naturally occurring or constructed wastewater treatment wetlands. Much of this vegetation has been found to not tolerate the extreme conditions of a stormwater wetland. Unlike naturally occurring wetlands and wastewater treatment wetlands, stormwater wetlands have relatively high and frequent changes in water surface. The water depth measured from a point in the shallow area of a stormwater wetland can vary from 8 to 38 cm (3 to 15 in.) back to 8 cm in as few as 3 days. Over a long term, many plants cannot handle this hydroperiod unique to stormwater wetlands.

Based upon local observation in North Carolina, a select group of plants has been identified as being reliably able to survive in stormwater wetlands. This list has been divided into two tiers. The plants listed in Table 20.1, designated as Tier 1, are those that have been shown to develop extensive colonies inside the wetland. These are the most dominant species within a stormwater wetland. Tier 2 plants (Table 20.2) survive and can add color to the wetland but rarely have out-competed the plants listed as Tier 1 when establishing large colonies. Several of the species listed in Tables 20.1 and 20.2 are depicted in Figure 20.6.

Cattails (*Typha* spp.) are conspicuously absent from both lists. While native to many areas, cattails are well adapted to develop monocultures that shelter mosquitoes from their predators (Knight et al. 2003). In short, if a stormwater wetland is to be located near a population center, such as a commercial center parking lot, a schoolyard, or a residential neighborhood, it is advised to keep cattail populations under control. If more than 15% of a stormwater wetland (located near populated areas) is colonized by cattails, it is recommended to remove the majority—if not all—of the cattails present. However, if stormwater wetlands are to be constructed in rural areas, such as along highways, it is acceptable to allow cattail growth, as these plants are tolerant of relatively high pollutant loads and propagate easily.

The denser the initial planting, the more quickly the vegetation will establish, and the less likely invasive species of plants will dominate the stormwater wetland. For most of the species listed in Tables 20.1 and 20.2, the recommended planting density is one plant on 60-cm (24 in.) centers (or one plant per 4 sq ft), if the stormwater wetland is to be colonized in essentially 1 year. Planting herbaceous vegetation on 1-m centers (one plant per 1 sq m) will tend to have the wetland fully colonized after 2 years. It is not recommended to plant on densities less than 1 plant per 1 $m^2$.

Table 20.1 Stormwater Wetland Vegetation Recommended for North Carolina and Virginia. This Tier 1 List of Plants Have Been Found to Reliably Colonize Stormwater Wetlands. They Are Listed in Order from Most Water Tolerant to Least.

| Common name | Scientific name | Wetland zone(s) | Comments |
|---|---|---|---|
| Fragrant water lily[1] | *Nymphaea odorata* | I and II | Deepest fringe of Zone II only |
| Spatterdock | *Nuphar lutea* | I and II | Deepest fringe of Zone II only |
| Softstem bulrush | *Schoenoplectus tabernaemontani* | II and III | Former Scientific Name: *Scirpus validus* |
| Pickerelweed | *Pontedaria cordata* | II and III | Bright and showy purple/blue flower |
| Broadleaf arrowhead | *Sagittaria latifolia* | III | Broad leaves, white flowers in summer |
| Bulltongue arrowhead | *Sagittaria lancifolia* | III | White flowers in summer |
| Burreed or bur-reed | *Sparganium spamericanum* | III | Tolerates flowing water zones near inlets and outlets |
| Lizard's tail | *Saururus cernuus* | III and IV | Can dominate in drier years, distinctive thin white flower |
| Woolgrass | *Scirpus cyperinus* | III and IV | Tall, brown seed head in late summer; makes tall border. |
| Sedge | *Carex* spp. | III and IV | Many species available; good initial colonizer |
| Common rush | *Juncus* spp. | III and IV | Grows best at the water's edge; near evergreen[2] |

[1] While this species is listed as a native to North Carolina by USDA, some vegetation experts do not recommend its use.
[2] Near evergreen in Coastal Plain and eastern portions of Piedmont.

Table 20.2 Tier 2 Stormwater Wetland Vegetation in North Carolina and Virginia—List of Plants Found to Survive in Stormwater Wetlands, Often Adding Color. They Are Listed in Order from the Most Water Tolerant to the Least.

| Common name | Scientific name | Wetland zone(s) | Comments |
|---|---|---|---|
| Water lotus (American lotus) | *Nelumbo lutea* | I, edge II | Protrudes from deep pools; good for mountain wetlands; some concern that this plant is too aggressive |
| Arrow arum | *Peltandra virginica* | III | Similar appearance to Sagittaria |
| Swamp milk weed | *Asclepias incarnata* | III and IV | Orange flower in fall |
| Blue flag iris | *Iris virginica* or *versicolor* | III, IV edge | Showy blue (or other color) flower in late spring; grows at water's edge |
| Cardinal flower | *Lobelia cardinalis* | IV | Red flowers in late summer |
| Hibiscus (rose mallow) | *Hibiscus moscheutos and H. grandiflorus* | IV | Beautiful, showy white, and red flowers mid-late summer |
| Swamp rose | *Rosa palustris* | IV | Off-white bloom in spring |
| Joe pye weed | *Eupatorium purpureum* | IV and V | Purple-like bloom summer and fall |

Figure 20.6 Select Herbaceous Species That Survive in Stormwater Wetlands.

Many types of trees can survive in stormwater wetlands in temperate climates, including bald cypress (*Taxodium distichum*), River birch (*Betula nigra*), Sycamore (*Platanus occidentalis*), and Red maple (*Acer rubrum*). However, clusters of trees should be avoided due to their eventual harboring of mosquito larvae and pupae. A tree density of 3–4 trees per 1000 $m^2$ (10,000 $ft^2$) of wetland surface area is recommended. Because many trees will "volunteer" in a stormwater wetland, it is important to note to those responsible for wetland maintenance which trees are desirable and those to be removed.

If large enough, the wetland may have a dam face that is vegetated. The dam face should be completely free of trees and shrubs. The best vegetative cover for the dam face (and rear) is grass. If water is to flow over a grassed area at a velocity exceeding 1.2 m/s (4 ft/s), a turf reinforcement matting will be needed.

## 20.5 Wetland Soils and Vegetation Growth Media

As discussed earlier, excessive seepage rates from the wetland must be avoided. To prevent excessive exfiltration from wetlands, the in-situ soil is either compacted or a clay supplement is added and compacted into the wetland's base soil. Also important is the inclusion of collars on the outlet structure to prevent piping.

If unamended, the resulting compacted soil is a very difficult soil medium in which to establish plants; it is difficult for wetland plants to spread their roots through compacted soil. For that reason, it is strongly recommended that a layer of topsoil be

Figure 20.7 Top Soil Being Replaced on a Wetland Fringe of Wet Pond (Photo by Authors).

stored during construction (or be brought in) and placed over the compacted soil in the bottom of the wetland. The suggested thickness of the replaced topsoil layer is 8–16 cm (3–6 in.), with a 10-cm (4-in) minimum preferred (Figure 20.7). This topsoil is added back to the wetland to provide an easy path for root growth and a source of organic matter during the initial part of the life of the stormwater wetland. Increased performance of nitrate treatment has been noted when poor soils were amended with organic matter (Burchell et al. 2007). The topsoil is especially important in the shallow water (III) and temporary inundation (IV) zones.

Other organic amendments to wetland soils have been used, such as cotton gin wastes, with varying success. It is recommended that the emplaced topsoil (or growing soil) be limed and fertilized when the wetland is initially planted. The idea of fertilizing a stormwater wetland may be counterintuitive, particularly one that is supposed to remove nitrogen and phosphorus. However, during the initial year of growth, it is important that the plants survive and create a diverse vegetative ecosystem. This one-time application of fertilizer is simply protecting the investment. After the first year, in a healthy wetland, plants no longer require lime and fertilization. For proper agronomic fertilization rates, it is recommended that a soil test be taken.

## 20.6 Wetland Outlet Configuration

The stormwater wetland outlet has three functions: (1) detain the water quality volume for treatment within the wetland, (2) safely pass large events that exceed the water quality storm, and (3) allow for emergency or routine maintenance by lowering the pool elevation inside the wetland. Therefore, outlet design is of critical importance to proper wetland function. A multistage weir/orifice is commonly used as the wetland outlet.

The crest of the weir and the base of the inlet of the drawdown device are vertically separated by the wetland depth of storage. Assuming the depth of storage is 30 cm (12 in.), the distance from the crest of the weir to the orifice invert (bottom of the drawdown hole) of the drawdown device is also 30 cm. The drawdown device regulates normal pool depth and detains the runoff from the water quality storm for at least 2 days. To achieve maximum treatment of the runoff, the storage volume should be retained for as long as possible. However, water from one storm should be treated—and released—before the next storm arrives.

Wetland retention time is selected based on the average time between storms at the wetland locations. A maximum retention time should be set between the mean and median interval between rainfalls. To optimize runoff treatment, a minimum retention time must be established as well. For example, a minimum number of days to drain stored runoff can be set where the probability of a more frequent rainfall is one in three. The size and the number of drawdown holes (orifices) should be determined so that water stored from the water quality event is emptied between these minimum and maximum number of days.

The water quality pool is lowered by having the stored wetland water drain through small weep holes, or orifices. The orifices serve as the drawdown device and can be constructed by drilling holes through a wooden dam or having a riser pipe serve as the orifice. The orifice equation (Chapter 7) is used to calculate flow through submerged holes at the outlet structure:

$$Q = C_d \times A \times \sqrt{(2gh)}, \tag{20.3}$$

Assuming no inflow, the level of water in the wetland will drop as a function of the amount of water flowing through the orifice and therefore leaving the wetland. As the water elevation falls, the height of water above the orifice drops, decreasing the rate of flow through the orifice. Additionally, the area of the water surface will likely change with elevation. Therefore, an iterative process is generally the easiest way to calculate the time needed to draw down the level of the water. If the area is relatively unchanged, the integrated form of the orifice equation can be used (Chapter 7).

1. Set the height of the water above the drawdown device (assuming a water quality storm, this height should equal the weir crest elevation).
2. Calculate a flow rate ($Q$) at this height, using the orifice, or other governing flow, equation.
3. Determine the volume of water leaving the wetland by multiplying the flow rate by a preset time interval (such as 15 min).
4. Reducing the water in the wetland by the volume determined in step 3 lowers the height of the water level. To calculate a new water level, first establish a relationship between volume of water (storage) and water elevation (stage).
5. Use this new height to calculate a new flow and repeat the process.

**Example 20.3** The design of a wetland is such that the water surface area is 3000 $m^2$ when holding the water quality volume and 2700 $m^2$ at the designated low depth, 30 cm (at centerline) below that full mark. Find the time to drain this volume using a 8-cm diameter orifice. Use a 4-h increment.

***Solution*** The area of the orifice is calculated as

$$A = \pi(8\ \text{cm})^2/4 = 50.3\ \text{cm}^2$$

Therefore, orifice Equation (20.3) gives

$$Q = C_d \times A \times \sqrt{(2gh)} = 0.6(50.3/10{,}000\ \text{m}^2)\left[(2(9.81\ \text{m/s}^2)(0.30\ \text{m})\right]^{1/2}$$
$$= 7.32 \times 10^{-3}\ \text{m}^3/\text{s}$$

For a 4-h increment:

$$V = Qt = (7.32 \times 10^{-3}\ \text{m}^3/\text{s})(4\text{h})(3600\ \text{s/h}) = 105.4\ \text{m}^3\ \text{discharged}$$

The stage-discharge can be estimated from the change in surface area with depth. Assuming a linear relationship between surface area and depth:

$$\text{SA} = 2700 + D/H(3000 - 2700)\ \text{m}^2$$

Wetland volume is found from the average of the bottom and upper surface area:

$$V = D(2700 + \text{SA})/2\ \text{m}^3$$

Combining these two equations and solving for $D$, the depth of the wetland above the orifice centerline gives

$$V = 2700D + 500D^2 \text{ for } D \text{ in m and } V \text{ in m}^3$$

At $D$ = 0.3 m (full wetland), the volume is

$$V = 2700(0.3) + 500(0.3)^2 = 855\ \text{m}^3$$

After the first 4 h, the volume, $V_4$ = 855 − 105.4 = 749.6 m$^3$.
Solving for $D$ from the volume equation gives $D$ = 0.2647 m (26.47 cm).

With this new depth, the process is repeated in 4-h increments. The results are shown in the following table.

| $D$ (m) | $Q$ (m$^3$/s) | Cumulative $t$ (h) | $\Delta t$ (h) | $V$ drained (m$^3$) | $V$ (m$^3$) | New $D$ (m) |
|---|---|---|---|---|---|---|
| 0.300 | 0.007317 | 4 | 4 | 105.4 | 749.6 | 0.2647 |
| 0.2647 | 0.006873 | 8 | 4 | 99.0 | 650.7 | 0.2311 |
| 0.2311 | 0.006422 | 12 | 4 | 92.5 | 558.2 | 0.1994 |
| 0.1994 | 0.005965 | 16 | 4 | 85.9 | 472.3 | 0.1696 |
| 0.1696 | 0.005502 | 20 | 4 | 79.2 | 393.1 | 0.1419 |
| 0.1419 | 0.005032 | 24 | 4 | 72.5 | 320.6 | 0.1162 |
| 0.1162 | 0.004555 | 28 | 4 | 65.6 | 255.0 | 0.0929 |
| 0.0929 | 0.004071 | 32 | 4 | 58.6 | 196.4 | 0.0718 |
| 0.0718 | 0.003579 | 36 | 4 | 51.5 | 144.9 | 0.0531 |
| 0.0531 | 0.003079 | 40 | 4 | 44.3 | 100.5 | 0.0370 |
| 0.0370 | 0.002569 | 44 | 4 | 37.0 | 63.5 | 0.0234 |
| 0.0234 | 0.002045 | 48 | 4 | 29.4 | 34.1 | 0.0126 |

As given in the last cell, the water level has drained down to 1.26 cm at the end of 48 h.

The size and number of orifices govern how fast the water elevation drops. The designer can adjust these values until the total drawdown time is 2–4 days.

In lieu of routing stormwater through an orifice on a small time-step basis, an approximate time to dewatering the water quality volume can be calculated using the following equation:

$$T = \frac{V_{\mathrm{wq}}}{Q_{\frac{1}{3}\mathrm{Depth}}} \tag{20.5}$$

where $T$ is the time to dewater the water quality volume, $V_{\mathrm{wq}}$ is the water quality volume, and $Q_{1/3\ \mathrm{Depth}}$ is the flow rate associated with the height water above the centerline of orifice that is 1/3 that of the maximum depth at water quality pool. For example, a wetland with a water quality depth of 30 cm would have $Q_{1/3\ \mathrm{Depth}}$ calculated with 10 cm of head above the orifice centerline.

**Example 20.4** The design of a wetland is such that the water surface area is 3000 m$^2$ when holding the water quality volume and 2700 m$^2$ at the designated low depth, 30 cm (at centerline) below that full mark. Using Equation (20.5), estimate the time to drain this volume using a 8-cm diameter orifice.

***Solution*** From Example 20.3, the area of the orifice is 50.3 cm$^2$. The one-third depth is 10 cm. The orifice Equation (20.3) gives

$$Q_{1/3} = C_d \times A \times \sqrt{(2gh)} = 0.6(50.3/10{,}000\ \mathrm{m}^2)\left[2\left(9.81\ \mathrm{m/s}^2\right)(0.10\mathrm{m})\right]^{1/2}$$
$$= 4.23 \times 10^{-3}\ \mathrm{m}^3/\mathrm{s}$$

The total volume to be discharged is

$$V = 0.3\ \mathrm{m}(3000 + 2700)/2\ \mathrm{m}^2 = 855\ \mathrm{m}^3$$

Using Equation (20.5):

$$T = \frac{V_{\mathrm{wq}}}{Q_{\frac{1}{3}\mathrm{Depth}}} = \frac{855\ \mathrm{m}^3}{4.23 \times 10^{-3}\ \mathrm{m}^3/\mathrm{s}} = 2.02 \times 10^5\ \mathrm{s} = 56.2\ \mathrm{h}$$

The result is about 17% longer than the answer obtained from the iterative process.

For larger wetlands, the drawdown device could be a narrow weir. Calculating drawdown for this type of device uses the same procedure as that of an orifice, except with a new governing equation for the discharge device.

For rectangular weirs (Chapter 7):

$$Q = C_w \times H^{1.5} \times L \tag{20.4}$$

For stormwater wetlands serving smaller watersheds (10 ha (25 ac) and less), the typical orifice is quite small, with a diameter often measuring less than 5 cm (2 in.), leaving it prone to clogging. Several preventative measures can be taken to limit the potential for clogging. First and foremost, a trash rack surrounding the orifice should be included as part of all designs. Second, many designers are drawing water from lower portions of the deep pool by submerging the orifice inlet. This keeps floating debris from clogging the orifice. Photos of each are found in Figure 20.8.

The flashboard riser, a technology borrowed from controlled drainage in agriculture is an effective outlet for the stormwater wetland (Figure 20.9). The flashboard riser is flexible. It functions by placing tongue and groove boards (Figure 20.10) on-end to form an adjustable weir. The orifice is simply drilled through one of the boards to which a trash rack and/or downturn pipe is also attached. Flashboard riser variations include concrete and other metal outlet structures (Figure 20.11), all of which follow the same premise of adjustable water elevations. The type of lumber used should be an

Figure 20.8 Downturned Pipe on an Orifice and a Trash Rack in Place around an Orifice (Photos by Authors).

Figure 20.9 An Outlet Employing the Traditional Flashboard Riser (Photo by Authors).

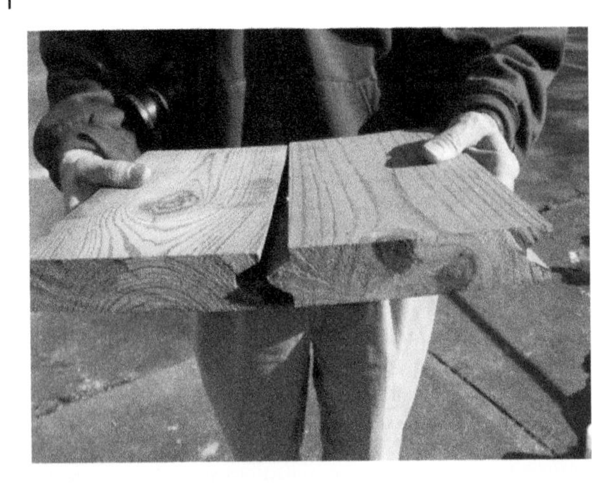

Figure 20.10 Small Flashboard Riser Boards Are Stacked Tongue-in-Groove (Photo by Authors).

Figure 20.11 Adjustable Outlet Structures Are Modifications of the Flashboard Riser (Photos by Authors).

environmentally friendly marine grade. Other variations of flashboard structures are available that employ materials other than wood; these materials are usually more expensive but may be easier to use or have a longer life. No matter the outlet type, it is advisable to install antiseep collars to prevent piping.

With a flashboard riser, the water level can be adjusted by adding or removing boards, or even by replacing a large board with a smaller one. The adjustable outlet allows a designer to compensate for potentially small (but important) grading mistakes inside the wetland. This is particularly important for the shallow water zone intended to be 5–10 cm (2–4 in.) in depth. Being a few centimeters higher or lower than the target elevation can significantly impact the mortality rate of the selected vegetation.

When the wetland needs to be drained for maintenance purposes (e.g., to plant the deep pools), a corresponding number of boards are removed. The flashboards need to be able to empty water from every wetland zone but the deep pools. High flows are able to overtop the highest board, which acts as a weir. In short, a flashboard riser is able to meet all three functions of an outlet.

Finally, the adjustable water level concept is particularly useful during plant establishment. Even if the outlet is designed to retain 30 cm (12 in.) of water during the water quality event, it has been observed that during the initial growing season a lower water elevation fluctuation promotes a higher plant survival rate. The maximum water elevation fluctuation should be kept at 10–15 cm (4–6 in.) during the initial growing season.

## 20.7 Wetland Construction

The construction and establishment of stormwater wetlands can be complex, demanding that stringent engineering and horticultural requirements be met. Water level control is critical and laser leveling will likely be required to establish pool and stormwater discharge depths. Wetland plants require the appropriate amounts of water, soil, nutrients, and sunlight, especially during establishment. It is important to reach a healthy stand of vegetation as quickly as possible; establishment is generally most successful when vegetation is planted in spring/early summer. Frequent control of water level may be necessary during establishment, through changing weir levels and even possibly irrigating. It may take several growing seasons for the wetland to become fully established and functional.

## 20.8 Wetland Variations

### 20.8.1 Wetland Design for Cold Water Species (Salmonids)

In certain environments and climates, especially where cold water is needed to protect fauna in receiving streams protect are a concern, wetland design features should be adjusted. Among them are plant selection and a requirement to draw water from the bottom of the deep pool located adjacent to the outlet structure. The latter is done to release the coldest water in the wetland to a receiving stream. In cold-water stormwater wetlands, it is important to select vegetation that does not float on the water, such as spatterdock and fragrant water lilies. The only deep pool plant that is suggested where cold water needs to be discharged is the American lotus, because the lotus does not float.

### 20.8.2 Off-line Stormwater Wetlands

Due to the regulatory challenges of constructing stormwater wetlands in streams, some stormwater wetlands are constructed "off-line." That is, water is diverted from a stream to the wetland and then released from the outlet of the wetland back to the stream. A relatively famous and well-functioning example of an off-line wetland has been constructed in Los Angeles County, CA, along the banks of the Los Angeles River (Figure 20.12). Sometimes off-line wetlands are the only possible practice from a retrofit perspective. Off-line wetlands can be more difficult to design and may perform more poorly than wetlands constructed in-line with an ephemeral stream.

Off-line wetlands tend to be sited in floodplains, which creates a challenge when flooding occurs. If the stream floods, it can force water to pond in the wetland for long periods of time. This is particularly true when the wetland is installed near a river. When a large river floods, flood waters could inundate the stormwater wetland for weeks, killing most of the wetland vegetation.

Figure 20.12 Off-line Stormwater Wetland Adjacent to the Los Angeles River in LA County, California (Photo by Authors).

Conversely, during dry periods when only small storms fall on a watershed, it is highly possible that too little (or no) water will enter the off-line stormwater wetland. Flow into off-line wetlands is often triggered by storms exceeding 1.3 cm (0.5 in.). If a stormwater wetland is built in-line with a channel, however, any storm that produces runoff will provide water to the wetland.

Off-line wetlands with a contributing drainage area of 2–4 ha (5–10 ac), smaller than a standard wetland, may be known as *pocket wetlands*. Pocket wetlands rely on intercepting the water table in order to maintain wetland conditions.

Off-line wetlands are designed to provide enough contact time within the SCM for pollutants to be removed. The amount of contact time is dictated by the pollutants of concern and the temperature. Characterizing flow rates, flow regimes, and residence times in off-line wetlands is essential. For wetlands intended to remove particulate matter, the particle size distribution of particulate matter, its specific gravity, and the range of operating temperatures must be known. Stokes Law and ideal sedimentation theory (Chapter 8) can be applied to determine the particulate matter removal efficiency by the wetland under storm or drain-down conditions.

### 20.8.3 Wetlands with High Flow Bypass

When a stormwater wetland is designed as a retrofit, not enough land is always available to properly size the wetland. In these situations, the wetland may need to have runoff bypass it at high flows, rather than flow directly through it. This, too, becomes an off-line system. When a substantially undersized wetland does not have a bypass, too much flow can enter the wetland, risking a "blow out" of vegetation. A good rule of thumb is that if the available area for a stormwater wetland is at least 50% of the required design surface area (Equation (20.2)), the wetland should be constructed in the ephemeral channel, without a bypass. If the available space is less than 50% of what is needed for a full-sized stormwater wetland, then a bypass should be constructed.

## 20.9 Water Quality Improvements in Stormwater Wetlands

The ability of a stormwater wetland to successfully treat stormwater is a function of storm intensity, runoff volume, and wetland size (area and volume; Barten 1987; Carleton et al. 2001; Meiorin 1989). These parameters affect retention time, which controls the removal efficiency of most pollutants, and flow velocity, which will affect the degree of bottom scouring and resuspension of settled solids.

Stormwater wetlands are successful in treating urban stormwater runoff as noted in multiple studies (Bulc and Slak 2003; Mitchell et al. 2002; Shutes et al. 1999, 1997). Generally speaking, stormwater wetlands are optimal for treating urban stormwater runoff because they can operate effectively under a wide range of hydraulic loads. Specifically, stormwater wetlands are capable of water storage and peak-flow attenuation (DeLaney 1995), nutrient cycling and burial (Reddy et al. 1993), metal sequestration (Odum et al. 2000), particulate matter removal (Kadlec and Knight 1996), and breakdown of organic compounds (Knight et al. 1999). Barten (1987), Carleton et al. (2001), and Meiorin (1989) all suggest that stormwater wetland performance is dependent upon hydraulic loading rate and detention time.

Because stormwater wetlands have flows that change with time, they do not operate under steady-state conditions. However, discharge flow, being orifice or weir controlled, may be relatively constant compared to the more dynamic inflow. The steady-state assumption allows simplified design and analysis of wetland systems.

Removal of particulate matter will occur during the storm event and as the wetland drains back to its baseline conditions. Simple sedimentation theory (Chapter 8) can be employed with wetland systems to evaluate removal of particulate matter. An average depth and average flow rate must be used for the steady-state analysis.

**Example 20.5** Following the wetland design from Example 20.4, a wetland fills to a point 30 cm above the base condition, draining at $4.23 \times 10^{-3}$ $m^3/s$. The water surface area is 3000 $m^2$ when holding the water quality volume and 2700 $m^2$ at the designated low depth. The size distribution of the incoming particulate matter is given in the first two columns of the following table (Sansalone et al. 1998). Determine the particulate matter removal in the wetland.

| Particle size distribution, $d_{x\_}$ | Particle size (μm) | Settling velocity (m/s) |
|---|---|---|
| 100 | 2200 | 4.22 |
| 80 | 1300 | 1.47 |
| 60 | 970 | 0.82 |
| 50 | 820 | 0.59 |
| 30 | 590 | 0.30 |
| 10 | 130 | 0.015 |
| 0 | 16 | 0.00022 |

*Solution* From ideal sedimentation theory, the critical particle settling velocity is given by the ratio of the flow rate per unit bottom surface area. In this case, assume that the flow rate remains constant. The surface area is estimated as the average area between high and low depth, 2850 $m^2$.

$$v_c = Q/A = \left(4.23\times10^{-3}\ m^3/s\right)/2850\ m^2 = 1.48\times10^{-6}\ m/s$$

The particle settling velocities in the table are calculated using Stoke's law ($T$ = 20°C, specific gravity = 2.6), given in column 3. Because of the size of the wetland, the critical settling velocity is much less than the velocity of the incoming particles. Therefore, in this case, ≈100% particulate matter removal is expected in this wetland.

The size of the critical particle for this wetland can be back-calculated using Stoke's law:

$$d = \sqrt{\frac{18\mu v_s}{\left(\rho_p - \rho_w\right)g}} =$$

$$\sqrt{\frac{18\left(1.00\times10^{-3}\ kg/m\,s\right)\left(1.48\times10^{-6}\ m/s\right)}{\left(2.6\left(1000\right) - 1000\ kg/m^3\right)9.81\ m/s^2}} = 1.3\times10^{-6}\ m = 1.3\ \mu m$$

The results of Example 20.5 reinforce the previous statements that wetlands are highly efficient in removal of sediment and particulate matter. Obviously simplifying assumptions are made in Example 20.5, but the predicted efficiency is very high. This primarily results from the large footprint required by the wetland, allowing adequate retention time in the SCM.

While low concentrations of particulate matter are commonly detected in wetlands discharge, their origin may lie not in unremoved stormwater sediment. Usually this particulate matter results from scoured materials, or even biomass of some form that has grown in the wetland.

The removal of nutrients in stormwater wetlands relies on shallow water, high primary productivity, the presence of aerobic and anaerobic sediments, and accumulation of natural litter (Mitsch and Gosselink 1993). Nitrogen is removed primarily by physical settlement of particulate nitrogen, denitrification, and plant/microbial uptake (Bulc and Slak 2003).

Carleton et al. (2001) explain the stormwater wetland performance as a function of wetland structure and hydrology, climate, soils, vegetation, percent watershed imperviousness, and other factors. This study modeled long-term pollutant removal using first-order steady flow design equations used for wastewater treatment wetlands, as a function of hydraulic loading rate and detention time. $NH_4$, $NO_2$-N, and $NO_3$-N removals are a function of hydraulic loading rate, while TP removal is primarily influenced by mean detention time. During intermittent high inflow rates, TP settled solids may resuspend. This offsets the influence of a low mean hydraulic loading rate and decreases the overall removal of TP.

Nitrogen transformation rates in wetlands may be based on typical first-order reaction kinetics, as described in Chapter 8. In the case of wetlands, it is also common to describe the rate in terms of mass per (wetland) bottom surface area per time, $J$; this rate can be derived by multiplying the standard reaction rate by the wetland depth, $H$, which, since $H$ is relatively constant among various wetlands, is then incorporated into an area-based rate constant, $k'$ (Kadlec and Knight 1996). For a first-order reaction:

$$J = rH = kHC = k'C \tag{20.5}$$

In wetland systems, a background or irreducible nitrogen concentration, $C^*$, may exist due to the activities of the vegetation. This changes the rate to (Kadlec and Knight 1996):

$$J = k'(C - C^*) \tag{20.6}$$

These expressions can be combined with the simple reactor models presented in Chapter 8. Integrated forms of these equations for various reactors are given in Table 20.3.

Table 20.3 Mass Balance Reactor Equations for First-order Area-based Reactions for Various Reactor Models under Steady-state Conditions. See Chapter 8 for Reactor Definitions.

| Reactor model | First-order reaction | First-order reaction with irreducible concentration, $C^*$ |
|---|---|---|
| Batch | $C = C_0 exp^{\left(-\frac{k't}{H}\right)}$ | $C = \left(C_0 - C^*\right) exp^{\left(-\frac{k't}{H}\right)} + C^*$ |
| CMFR | $C = C_0 \left(\frac{1}{1 + \frac{k'A}{Q}}\right)$ | $C = \frac{C_0 + \frac{k'AC^*}{Q}}{1 + \frac{k'A}{Q}}$ |
| PFR | $C = C_0 exp^{\left(-\frac{k'A}{Q}\right)}$ | $C = \left(C_0 - C^*\right) exp^{\left(-\frac{k'A}{Q}\right)} + C^*$ |
| n-CMFRs in series | $C = C_0 \left(\frac{1}{1 + \frac{k'A}{Qn}}\right)^n$ | No closed solution |

**Example 20.6** Following the wetland design from Example 20.5, nitrogen removal is modeled as a first-order reaction with $k' = 0.13$ m/day. The influent nitrogen event mean concentration (EMC) is 2.5 mg/L. Find the effluent nitrogen concentration in the wetland assuming average flow conditions:

(a) Using a plug-flow model
(b) Using a model of eight CMFRs in series.

*Solution* The flow rate is assumed constant at 4.23 × $10^{-3}$ $m^3$/s, equal to 365 $m^3$/day. The surface area is found as the average area between high and low depth, 2850 $m^2$.

a) For plug flow:

$$C = C_0 exp^{\left(-\frac{k'A}{Q}\right)} = 2.5\ \text{mg/L}\ exp^{\left(\frac{-(0.13\ \text{m/day})(2850\ \text{m}^2)}{365\text{m}^3/\text{day}}\right)} = 0.91\ \text{mg/L}$$

b) For 8 CMFRs in series:

$$C = C_0\left(\frac{1}{1+\frac{k'A}{Qn}}\right)^n = 2.5\ \text{mg/L}\left(\frac{1}{1+\frac{(0.13\ \text{m/day})(28503^2)}{(365\ \text{m}^3/\text{day})8}}\right)^8 = 0.96\ \text{mg/L}$$

**Example 20.7** Following the wetland design from Example 20.5, nitrogen removal is modeled as a first-order reaction with $k'$ = 0.13 m/day. The influent nitrogen EMC is 2.5 mg/L and the irreducible nitrogen concentration is 0.8 mg/L. Find the effluent nitrogen concentration in the wetland assuming average flow conditions using a plug-flow model.

*Solution* Using information derived in Example 20.6, for plug flow:

$$C = \left(C_0 - C^*\right) exp^{\left(-\frac{k'A}{Q}\right)} + C^* = \left(2.5 - 0.8\,\text{mg/L}\right) exp^{\left(\frac{-(0.13\ \text{m/day})(2850\ \text{m}^2)}{365\ \text{m}^3/\text{day}}\right)} + 0.8\,\text{mg/L} = 1.42\,\text{mg/L}$$

The irreducible concentration results in a higher nitrogen concentration leaving the wetland.

## 20.10 Other Stormwater Wetland Designs

### 20.10.1 Submerged Gravel Wetlands (SGWs)

Small-scale wetland systems can also be used as SCMs. An example of this is a submerged gravel wetland (SGW). It has the defining features of a wetland that include continuous saturated conditions in a crushed stone media, a permeable soil at the surface, and wetland vegetation. This allows the wetland unit processes for hydrologic and water quality improvement to be active. A small forebay is required for pretreatment.

Type C or D soils are ideal to maintain saturation in the SGW. An impervious liner may be used in soils where the hydraulic conductivity is too high to maintain saturated wetland conditions.

The SGW is divided into two treatment cells, each designed to hold 45% of the water treatment volume, with the forebay to hold the other 10% (UNHSC (University of New

Hampshire Stormwater Center) 2009). Each cell is filled with 60 cm of 1.9 mm (3/4 in.) crushed stone. About 7.5 cm of pea gravel is placed on top of the stone, with 20 cm of wetland soil (frequently similar to a bioretention media mix) placed on top of the pea gravel. The two cells are connected subsurface with a PVC pipe.

Each cell should have a minimum travel distance of 4.6 m. Within this constraint, the length:width ratio is 0.5, with the system designed to hold the water quality volume, as described above (UNHSC (University of New Hampshire Stormwater Center) 2009).

An orifice is constructed at the minimum water level as the outlet of the SGW. High-flow bypasses are included to allow surface flows from the first to the second cell and as an outlet from the wetland system.

**Example 20.8** Size the submerged zone for a SGW treating a water quality volume of 100 m$^3$.

*Solution* 10% of the water quality volume is stored in the wetland forebay and the remaining 90% is split between the two cells. Therefore:

$$V_w = 0.45\left(100\ \text{m}^3\right) = 45\ \text{m}^3$$

The porosity of the stone media is assumed as 30%, therefore the total volume is larger than the water volume.

$$V = 45/0.3 = 105\ \text{m}^3$$

The depth is 0.6 m, leaving the cross-sectional area as

$$A = 105\ \text{m}^3/0.6\ \text{m} = 250\ \text{m}$$

With a length to width ratio of 0.5, each cell has dimensions:

$$W = \left(250/0.5\right)^{1/2} = 22.4\ \text{m, and the length is 11.2 m.}$$

Consistent removal of nitrogen and phosphorus were noted in SGW mesocosm studies (Mangum et al. 2019). Nitrogen was removed via denitrification during the water storage between storm events, so discharge concentrations ranged from about 0.92 to 3 mg/L as N, from an input of approximately 4.0 to 4.5 mg/L as N. Effluent phosphorus ranged from 0.07 to 0.14 mg/L as P from an input of 0.40 mg/L as P; phosphorus appears to be adsorbed. Addition of compost to the media resulted in initial leaching of both N and P.

### 20.10.2 Ponds Transitioning to Wetlands

As noted, wetlands possess a number of unit processes that are beneficial to the management of stormwater. As a result, a number of techniques are available for converting other SCMs to wetlands, or to produce wetland conditions, in order to enhance stormwater performance.

**Example 20.9** Consider the SGW designed in Example 20.8. Nitrogen removal occurs during the stormwater storage between rainfall events and is modeled as first-order reaction with $k'$ = 0.13 m/day. The influent nitrogen EMC is 3.5 mg/L. Find the nitrogen concentration in the stored wetland water after 5 days of storage.

(a) Without an irreducible N concentration
(b) With an irreducible N concentration of 0.8 mg/L

*Solution* The depth is 0.6 m and the time is 5 days.
Using the batch equation model from Chapter 8:

$$C = C_0 exp^{\left(-\frac{k't}{H}\right)} = (3.5\ \text{mg/L})\, exp^{\left(\frac{-(0.13\ \text{m/day})(5\ \text{days})}{0.6\ \text{m}}\right)} = 1.18\ \text{mg/L}$$

With the irreducible concentration:

$$C = \left(C_0 - C^*\right) exp^{\left(-\frac{k't}{H}\right)} + C^* = \left(3.5 - 0.8\ \text{mg/L}\right) exp^{\left(\frac{-(0.13\ \text{m/day})(5\ \text{days})}{0.6\ \text{m}}\right)}$$
$$+0.8\ \text{mg/L} = 1.81\ \text{mg/L}$$

Stormwater retention/detention ponds have been used for many years for flood control and stormwater management. With new concerns for water quality, these ponds deliver deficient performance. Several modifications have been examined to promote wetland characteristics in ponds and provide greater water quality improvement. In some cases, this may be as simple as changing the outlet structure so that a shallow permanent pool is created, leading to wetland vegetation and resulting beneficial processes.

In a case study documented by Natarajan and Davis (2015a, 2015b, 2016), a large failed infiltration basin was storing water and was transitioning to a wetpond/wetland. The basin appeared to be clogged and always held ponded water. As a result, wetland vegetation was found prevalent in the SCM. The new SCM was able to impart some hydrologic management due to storage, limited infiltration, and ET. Water quality improvement was found due to sedimentation of suspended solids and the affiliated pollutants, including some heavy metals and phosphorus. Anoxic conditions were also documented and denitrification was indicated, resulting in reduced nitrogen concentrations (and loads) leaving the transitioned SCM.

### 20.10.3 Floating Wetlands

Floating wetlands are artificial wetlands that can be retrofitted to existing stormwater ponds or other SCMs with ponded water. Floating wetlands generally have a base material made of Styrofoam or some form of plastic mesh material. The floats are planted with wetland plants so that the root systems are below the water line. The plants will live hydroponically, taking up nutrients from the water. The floating wetlands are anchored in the ponds so they will not float far from the desired placement.

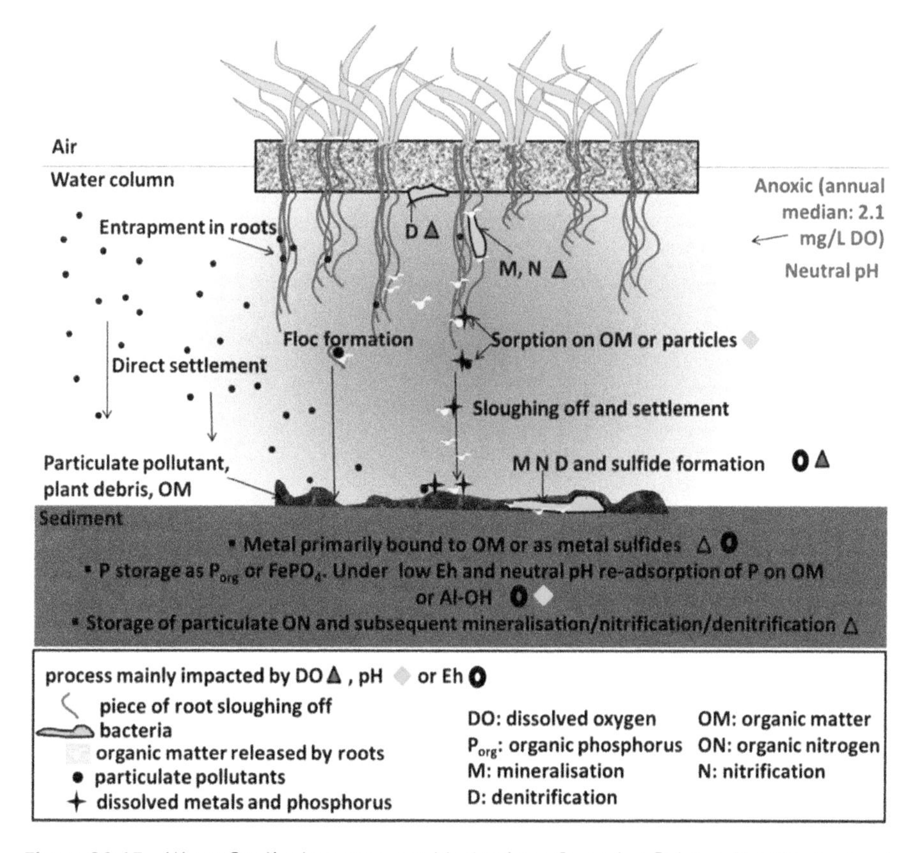


Figure 20.13 Water Quality Improvement Mechanisms Occurring Below a Floating Wetland (Borne et al. 2015, with Permission from ASCE).

Floating wetlands can increase the potential for sediment, nutrients, and metals removal in a ponded area. Water quality improvement mechanisms for floating wetlands are shown in Figure 20.13. Sediments can be filtered from the water via vegetation roots; particulate-affiliated pollutants are removed also (Borne et al. 2013a). Low dissolved oxygen concentrations may be promoted beneath the floating wetlands if they are present at high enough density (e.g., 50% surface area coverage). This can promote denitrification and may precipitate dissolved heavy metals as metals sulfides (Borne et al. 2013a, 2013b).

However, a number of challenges exist with floating wetland use. Plant growth must be ensured. If dissolved oxygen becomes excessively low for an extended period, plants can die. Invasive plants and other undesired volunteers may become established. The wetland may have to be retrieved from the open water and brought to a shoreline for maintenance. They will also shade any processes below the wetland. Overall, floating wetlands may offer several water quality improvements, but the costs and maintenance challenges must be considered. Table 20.4 presents information on floating wetlands design and maintenance.

Table 20.4 Design and Maintenance Information for Floating Wetlands (Basis is Borne et al. 2015, with Permission from ASCE).

| Parameter | Recommendations | Purpose |
|---|---|---|
| Floating treatment wetland size[1] | Design recommendations >50 $m^2$ to promote anoxic conditions underneath the FTW during the warmest months of the year | Anoxic conditions will promote nitrate removal by denitrification and stable storage of metals bound to organics/sulfides |
| Floating treatment wetland coverage ratio[1] | When feasible, promote large coverage ratio: ~50% of pond surface area. 20% may be sufficient winston et al. | Promote greater dissolved oxygen depletion in adjacent open water areas, increasing the extent of ponds' favorable conditions for denitrification/storage of metals |
| Inlet structure | If the aim is to provide water quality control rather than peak flow control, installing a high-flow bypass could be considered | Allow the diversion of flows in excess of the water quality design storm and promote maximum TSS and PZn treatment performance |
| Plants' characteristics[2] | Tolerant to periodic anoxic conditions and potentially high incoming pollutant loads. New Zealand site showed that dense fibrous root network occupying no more than 67% of permanent pool water depth, i.e., $R$ ratio < 0.67, and more than 55% of peak flow water depth, i.e., $R$ ratio > 0.55 was beneficial<br>Avoid tall plants and plants presenting large above-ground biomass loss during senescence | Promote perennial development of the plants<br>Promote particle entrapment<br>Prevent the plants from acting as a sail under high wind conditions which would allow the floating mat to drift or tip over. Limit the release of the accumulated pollutants in the plant shoots to the water column |
| Floating mat | Mat should allow the roots to reach the surface of the water or the media should be able to wick up water by capillary action | Sustain the plants' growth |
| Anchorage/ location | If a forebay is present, it is preferable to install the floating treatment wetland (FTW) immediately downstream, in the main body of the pond<br>Design an FTW extending over the width of the retention pond perpendicular to the incoming flow<br>Install the FTW close to the inlet and in a location easy to dredge; Khan et al. (2013) recommend a distance of 0.25 times the length of the pond for a rectangular shaped pond with an inlet centrally located over the width of the pond<br>Anchor the FTW and avoid the use of galvanized materials which can leach heavy metals into the water column. Allow for some movement from normal pool to the brink-of-overflow elevation | Avoid saturating the roots of the FTW with coarse particles and maximize the entrapment of the remaining fine particles by the FTW<br>Limit short circuiting around the FTW and force the incoming flow through the roots to increase particles entrapment<br>Increase the distribution of the inflow and thus the hydraulic efficiency and treatment performance. Facilitate removal of the most polluted sediments which accumulate below the FTW<br>Maintain the FTW at the desired location and prevent the FTW from moving violently around the pond during extreme weather, potentially leading to partial/complete clogging of the emergency overflow device |

| Parameter | Recommendations | Purpose |
|---|---|---|
| Retention pond | All invasive aquatic weeds should be removed prior to installation of FTWs | Prevent the FTWs from acting as new footholds for invasive aquatic weeds, which would promote their proliferation |
| Plants | Maintenance recommendations; install a netting or twine/plastic grids over the vegetation during the first six months | Prevent birds or turtles from feeding on the juvenile plants |
| Sediment | Identify an adequate disposal solution for the sediments before dredging: sediments could potentially be suitable for land application check with 40 CFR 503 thresholds | Maintain good water storage capacity |
| General | Regular monitoring of the site, e.g., at least quarterly, especially during the initial plant establishment period | Verify the healthy and perennial development of the plants and initiate remedial actions if needed, e.g., pulling out weeds, check location and anchorage of the FTWs, remove potential floating detritus, e.g., plastic bottles, branches, and so on trapped upstream of the FTW |

Note: Based on coupled New Zealand and North Carolina field observations. Parameters' list is not all-inclusive but from the combined observations from New Zealand and North Carolina, where feasible.
[1] Installing a mixing system like a cascade structure at the system outlet should be considered to promote reaeration during discharge and reduce potential toxicity downstream.
[2] Selecting several plant species meeting the mentioned criteria would limit the risk of loss of the plants due to disease or pest outbreak, as different species usually show different tolerances.

## 20.11 Inspection and Maintenance

Inspection of stormwater wetlands focuses on two interrelated components: water pathways and vegetation. The wetland was designed with a specific flow pattern in mind. If this pattern changes, the expected treatment may not be reached. Flow paths can be altered by deposition of sediment, scour of soil, unintended plant growth or death, objects emplaced by animals, and other conditions. Changes in flow patterns or depth can alter the type of the vegetation expected in each of the treatment zones. Lack of vegetation will reduce the pollutant treatment expected in the wetland.

Adjusting the water level in the wetland may be the only control available without resorting to extensive maintenance construction on the wetland. Reestablishment of vegetation can be costly and the desired vegetative conditions may take time to reach.

## References

Barten, J.M. (1987). Stormwater runoff treatment in a wetland filter: Effects on the water quality of Clear Lake. *Lake and Reservoir Management* 3 (1), 297–305.

Borne, K.E., Fassman, E.A., and Tanner, C.C. (2013a). Floating treatment wetland retrofit to improve stormwater pond performance for suspended solids, copper and zinc. *Ecological Engineering* 54: 173–182.

Borne, K.E., Fassman-Beck, E.A., Winston, R.J., Hunt, W.F., and Tanner, C.C. (2015). Implementation and maintenance of floating treatment wetlands for Urban stormwater management. *Journal of Environmental Engineering* 141 (11): 04015030.

Borne, K.E., Tanner, C.C., and Fassman-Beck, E.A. (2013b). Stormwater nitrogen removal performance of a floating treatment wetland. *Water Science and Technology* 68 (7): 1657–1664.

Bulc, T. and Slak, A.S. (2003). Performance of constructed wetland for highway runoff treatment. *Water Science and Technology* 48: 315–322.

Burchell, M.R., Skaggs, R.W., Lee, C.R., Broome, S., Chescheir, G.M., and Osborne, J. (2007). Substrate organic matter to improve nitrate removal in surface-flow constructed wetlands. *Journal of Environmental Quality* 36: 194–207.

Carleton, J.N., Grizzard, T.J., Godrej, A.N., and Post, H.E. (2001). Factors affecting the performance of stormwater treatment wetlands. *Water Research* 35: 1552–1562.

DeLaney, T.A. (1995). Benefits to downstream flood attenuation and water quality as a result of constructed wetlands in agricultural landscapes. *Journal of Soil and Water Conservation (JSWC) Science* 50: 620–626.

Imfeld, G., Braeckevelt, M., Kuschk, P., and Richnow, H.H. (2009). Monitoring and assessing processes of organic chemicals removal in constructed wetlands. *Chemosphere* 74: 349–362.

Kadlec, R.H. and Knight, R.L. (1996). *Treatment Wetlands*. Boca Raton, FL: CRC Press.

Knight, R.L., Kadlec, R.H., and Ohlendorf, H.M. (1999). The use of treatment wetlands for petroleum industry effluent. *Environmental Science & Technology* 33: 973–980.

Khan, S., Melville, B. W., and Shamseldin, A. (2013). Design of Storm-Water Retention Ponds with Floating Treatment Wetlands. *Journal of Environmental Engineering, American Society of Civil Engineers*, 139 (11): 1343–1349.

Knight, R.L., Walton, W.E., O'Meara, G.F., Reisen, W.K., and Wass, R. (2003). Strategies for effective mosquito control in constructed treatment wetlands. *Ecological Engineering* 21 (4): 211–232.

Malaviya, P. and Singh, A. (2012). Constructed wetlands for management of urban stormwater runoff. *Environmental Science & Technology* 42: 2153–2214.

Mangum, K.R., Yan, Q., Ostrom, T.K., and Davis, A.P. (2019). Nutrient Leaching from Green Waste Compost Addition to Stormwater Submerged Gravel Wetland Mesocosms. *Journal of Environmental Engineering* American Society of Civil Engineers, 146 (3): 04019128.

Meiorin, E.C. (1989). Urban runoff treatment in a fresh/brackish water marsh in Fremont, California. In: *Constructed Wetlands for Wastewater Treatment: Municipal, Industrial, and Agricultural* (ed. D.A. Hammer), 677–685. Chelsea, MI: Lewis.

Mitchell, G.F., Hunt, C.L., and Su, Y.M. (2002). Mitigating highway runoff constituents via a wetland. *Transportation Research Record* 1808: 127–133.

Mitsch, W.J. and Gosselink, J.G. (1993). *Wetlands*, 2nd e. New York, NY: Van Nostrand Reinhold.

Natarajan, P. and Davis, A.P. (2015a). Water quality performance of a "transitioned" infiltration basin Part 1: TSS, metals, and chloride removals. *Water Environment Research* 87 (9): 823–834.

Natarajan, P. and Davis, A.P. (2015b). Hydrologic performance of a "transitioned" infiltration basin managing highway runoff. *Journal of Sustainable Water in the Built Environment* 1 (3): 04015002.

Natarajan, P. and Davis, A.P. (2016). Performance of a "transitioned" infiltration basin in treating nutrients in highway runoff. *Water Environment Research* 88 (4): 291–305.
Odum, H.T., Woucik, W., and Pritchard, L. (2000). *Heavy Metals in the Environment*. Boca Raton, FL: CRC Press.
Reddy, K.R., DeLaune, R.D., DeBusk, W.F., and Koch, M.S. (1993). Long-term nutrient accumulation rates in the everglades. *Soil Science Society of America Journal* 57: 1147–1155.
Sansalone, J. J., Koran, J. M., Smithson, J. A., and Buchberger, S. G. (1998). Physical Characteristics of Urban Roadway Solids Transported during Rain Events. *Journal of Environmental Engineering, American Society of Civil Engineers*, 124 (5): 427–440.
Shutes, R.B.E., Revitt, D.M., Lagerberg, I.M., and Barraud, V.C.E. (1999). The design of vegetative constructed wetlands for the treatment of highway runoff. *The Science of the Total Environment* 235: 189–197.
Shutes, R.B.E., Revitt, D.M., Mungur, A.S., and Scholes, L.N.L. (1997). The design of wetland systems for the treatment of urban runoff. *Water Science and Technology* 35: 19–25.
UNHSC (University of New Hampshire Stormwater Center) (2009). *Subsurface Gravel Wetland Design Specifications* (R.M. Roseen, T.P. Ballestero, and J.J. Houle). Durham, NH.

## Problems

20.1 In stormwater wetlands (and other SCMs), the average ponding depth associated with the water quality volume is critical. Why? Illustrate (through mathematical expression) how much of a difference using average ponding depth vis-à-vis maximum ponding depth is to a property owner.

20.2 Find the area needed for a stormwater wetland to treat a 1400-$m^3$ water quality volume.

20.3 Find the area needed for a stormwater wetland to treat a 2000-$m^3$ water quality volume.

20.4 What water quality volume could be treated by a stormwater wetland with a surface area of 7500 $m^2$?

20.5 What water quality volume could be treated by a stormwater wetland with a surface area of 5500 $m^2$?

20.6 Size the deep-pool depth for a wetland based on the following design criteria: summer rainfall of 8 cm, with 30% entering the wetland. Watershed:wetland area ratio of 18%, ET loss estimated at 18 cm, with an infiltration rate of 0.02 cm/h; 15 cm of water reserve is required.

20.7 Size the deep-pool depth for a wetland based on the following design criteria: summer rainfall of 6 cm, with 28% entering the wetland. Watershed:wetland area ratio of 16%, ET loss estimated at 22 cm, with an infiltration rate of 0.015 cm/h; 15 cm of water reserve is required, as a factor of safety. Find the depth of the wetland deep pool.

20.8 The input nitrogen concentration to a wetland is 4.2 mg/L at a flow rate of 860 $m^3$/day. For a first-order (area based) rate constant of 0.60 m/day, find the wetland area necessary to reduce the nitrogen concentration to 1.2 mg/L. Use a plug-flow model.

**20.9** The input nitrogen concentration to a wetland is 4.2 mg/L at a flow rate of 860 $m^3$/day. For a first-order (area based) rate constant of 0.60 m/day, find the wetland area necessary to reduce the nitrogen concentration to 1.2 mg/L. Use a 6-CMFRs in series model.

**20.10** The input nitrogen concentration to a wetland is 4.2 mg/L at a flow rate of 860 $m^3$/day. For a first-order (area based) rate constant of 0.60 m/day, find the wetland area necessary to reduce the nitrogen concentration to 1.2 mg/L, with an irreducible nitrogen concentration of 0.6 mg/L. Use a plug-flow model.

**20.11** The input nitrogen concentration to a wetland is 4.2 mg/L at a flow rate of 860 $m^3$/day. The first-order (area based) rate constant is 0.67 m/day at 20°C. The wetland area is 1900 $m^2$. Find the minimum wetland temperature at which the discharge nitrogen concentration remains below 1.2 mg/L. Use a plug-flow model and a temperature adjustment $\theta$ of 1.012.

**20.12** For a water quality volume of 80 $m^3$, find the dimensions of the two cells of a submerged gravel wetland.

**20.13** For a water quality volume of 65 $m^3$, find the dimensions of the two cells of a submerged gravel wetland.

**20.14** Each cell (of two total) in a submerged gravel wetland has dimensions 5.0 m × 10.0 m × 0.6 m ($L \times W \times H$). Find the water quality volume treated by this SCM.

**20.15** Each cell (of two total) in a submerged gravel wetland has dimensions 6.0 m × 12.0 m × 0.5 m ($L \times W \times H$). Find the water quality volume treated by this SCM.

**20.16** A submerged gravel wetland has a depth of 0.9 m. Nitrogen removal occurs during the stormwater storage between rainfall events and is modeled as a first-order reaction with $k' = 0.15$ m/day. The influent nitrogen EMC is 2.8 mg/L. Find the nitrogen concentration in the stored wetland water after 5 days of storage.

**20.17** A submerged gravel wetland has a depth of 0.9 m. Nitrogen removal occurs during the stormwater storage between rainfall events and is modeled as a first-order reaction with $k' = 0.15$ m/day. The influent nitrogen EMC is 2.8 mg/L. Find the nitrogen concentration in the stored wetland water after 5 days of storage with an irreducible N concentration of 0.8 mg/L.

**20.18** A submerged gravel wetland has a depth of 4 ft. Nitrogen removal occurs during the stormwater storage between rainfall events and is modeled as a first-order reaction with $k' = 0.5$ m/day. The influent nitrogen EMC is 3.8 mg/L. Find the number of days of storage it would take for the nitrogen concentration in the stored wetland water to reduce to 2.0 mg/L.

**20.19** A submerged gravel wetland has a depth of 4 ft. Nitrogen removal occurs during the stormwater storage between rainfall events and is modeled as a first-order reaction with $k' = 0.5$ m/day. The influent nitrogen EMC is 3.8 mg/L. Find the number of days of storage it would take for the nitrogen concentration in the stored wetland water to reduce to 2.0 mg/L. The irreducible nitrogen concentration is 0.7 mg/L.

# 21

# Putting It All Together

## 21.1 Introduction

Selecting the "right" stormwater control measure (SCM) for a particular application will depend on many different factors and criteria. These include availability of land, characteristics of the site, stormwater control goals, other benefits expected from the SCM, and costs. The right SCM will allow the greatest benefits, while minimizing current and future costs, realizing that parameters like goals, costs, and climate may change in the future. In some areas, such as those with combined sewers, only hydrologic metrics may be important; in others, stormwater control may be dominated by water quality needs, and the water quality parameter may differ from site to site. More and more, multiple objectives must be met and designs that are able to do this will be the most successful.

## 21.2 SCM Hydrologic Performance Summary

A summary of the hydrologic performance of the various SCMs discussed in this book follows and is summarized in Table 21.1. Hydrologic performance considers the water loss pathways of infiltration, evapotranspiration (ET), and beneficial use. Storage is an important aspect to all SCMs that provide hydrologic benefit, as storage of rainfall/runoff must occur during a storm event so that it can then be infiltrated, evapotranspired, and/or harvested. Each of these are generally slow processes and will mostly occur interevent. Any design factor that increases the amount of storage, such as ponding or media depth or surface area, will increase the rate and volume of runoff eliminated.

Not every SCM enables alternate water pathways. For example, green roofs collect rainfall and minimize runoff. Infiltration is not possible from a green roof and all water loss must come from ET. Media storage is important to promote ET between storms and for vegetation survival. A raised underdrain is necessary to provide any substantial storage.

Beneficial use is the primary water pathway for water harvesting. Water harvesting design is driven by matching expected capture volume with use. Adequate storage

*Green Stormwater Infrastructure Fundamentals and Design*, First Edition. Allen P. Davis, William F. Hunt, and Robert G. Traver.

Companion Website: www.wiley.com/go/davis/greenstormwater

Table 21.1 Summary of Hydrologic Water Loss Pathway for the Green Stormwater Infrastructure SCMs.

| Green stormwater infrastructure SCM | Hydrologic water loss pathway | | | |
|---|---|---|---|---|
| | **Infiltration** | **ET** | **Beneficial use** | **Note** |
| **Green roof** | None | Primary | Minimal | Raised discharge can be helpful |
| **Water harvesting** | Minor[1] | None | Primary | Storage sizing is important |
| **Permeable pavement** | Primary | Minimal | Minimal | Raised discharge can be helpful |
| **Infiltration trenches** | Primary | Minimal | Minimal | Designed to infiltrate |
| **Sand filters** | Minimal | Minimal | Minimal | Not designed for volume reduction |
| **Bioretention** | Primary | Primary | Increasing | Infiltration and ET both important |
| **Swales and filter strips** | Primary | Minor | Minimal | Swales mostly conveyance |
| **Stormwater wetlands** | Minor | Minor | Minimal | Slow discharge. Large surface areas improve ET and infiltration |

[1]Some designs may allow infiltration.

must be available to allow to produce a reliable supply of harvested water to meet intended uses.

The vast majority of water loss provided by permeable pavements is due to infiltration. The pavement layer will prevent significant evaporation from occurring from pervious pavements. As with the green roofs, a raised underdrain creates more internal storage, encouraging infiltration up to several days after an event's completion.

Infiltration trenches are designed specifically to store and infiltrate stormwater. Sand filters are primarily used for improvement of water quality; they may not have any provision for stormwater volume reduction.

Infiltration and ET are both important water loss pathways for bioretention, making this GSI technology one of the most attractive for hydrologic management. Internal storage (or no underdrain) yields greater water storage and infiltration. Using the outlet structure to create a storage pool above the media will aid also.

Swales can sometimes infiltrate (depending upon their design configuration and underlying soil), but they are primarily conveyance SCMs. Filter strips rely on infiltration. In both, some surface water may be lost to ET.

Storage and slow release is the primary hydrologic function of stormwater wetlands. Infiltration is not a major pathway; wetlands are designed so that they rarely go dry. Wetlands are commonly designed to intersect with groundwater. Some water loss will occur via ET, the amount of which proportionally increases with surface area.

## 21.3 SCM Water Quality Performance Summary

A water quality performance summary of the various SCMs of the previous chapters is provided later. The performance summary is categorized by SCM. Only certain parameters are used to designate performance qualitatively based on available research. The removal of constituents is summarized on a low, medium, and high rating. These categories are based on measured data, supported by unit operation considerations. These designations also follow those used in a recent publication, *Pollutant Load Reductions for the Total Maximum Daily Loads for Highways* (NCHRP 2013). In many cases, inadequate information is available. The water quality constituents are as follows:

- Total suspended solids (TSS)
- Total P
- Dissolved phosphorus (DP)
- Particulate phosphorus (TP)
- Total N
- Total Kjeldahl nitrogen (TKN)
- Nitrate/Nitrite ($NO_3^-/NO_2^-$)
- Total zinc
- Total copper
- Dissolved copper (DC)
- Total lead
- Chloride
- Hydrocarbons
- Temperature

### 21.3.1 Green Roofs and Water Harvesting

In general, green roofs and water harvesting systems are not designed for the improvement of water quality; they are constructed as runoff reduction technologies. Rainfall (generally of good quality) falls upon the roof. Concerns of leaching from the green roof, especially nutrients from plant growth media and additives, must be addressed. Metal concentrations in green roof discharge have tended to be lower than those of conventional roof runoff.

### 21.3.2 Permeable Pavements

While multiple types of permeable pavement systems are available (permeable interlocking concrete pavers, permeable concrete, and permeable asphalt), water quality variation among the three is quite limited. In other words, pavement type selection is unlikely influenced by water quality performance. However, design alternatives, such as the creation of internal water storage, can impact how well permeable pavements improve water quality.

| TSS | TP | Chloride | TN | Total metals | Temperature |
|---|---|---|---|---|---|
| High | Medium | Low | Low (no IWS)<br>Medium (IWS) | High | High (no IWS)<br>Very high (IWS) |

For porous friction courses (PFCs):

| TSS | TP | PP | DP | Total copper | Dissolved copper | Total lead | TN | TKN | $NO_3^-/NO_2^-$ |
|---|---|---|---|---|---|---|---|---|---|
| High | Medium | High | Low | Medium | Low | Medium | Low | Low | Low |

### 21.3.3 Infiltration Basins

Infiltration basins provide treatment as the water exfiltrates into the surrounding soils. Particulate pollutants will exhibit high removal. Because infiltration basins tend to be used in sandy soil regions, runoff exfiltrating these basins undergoes filtration as in sand filters.

| TSS | PP | DP | TN | $NO_3^-/NO_2^-$ | Total metals | Temp. |
|---|---|---|---|---|---|---|
| High | High | Low | Medium | Low | High | High |

### 21.3.4 Sand Filters

Sand filters will provide excellent removal of particulate pollutants. Some removal of dissolved species has been documented as well.

| TSS | TP | PP | DP | Total zinc | Total copper | Total lead | TN | TKN | $NO_3^-/NO_2^-$ |
|---|---|---|---|---|---|---|---|---|---|
| High | Medium | High | Low | High | High | Medium | Medium | Medium | Low |

Different design variants, however, can improve pollutant removal. For example, enhancing sand filter media with 5% iron fillings (by weight) yields better performance.

| TSS | TP | PP | DP | Total zinc | Total lead | TN | TKN | $NO_3^-/NO_2^-$ |
|---|---|---|---|---|---|---|---|---|
| High | High | High | High | High | Medium | Medium | Medium | Low |

### 21.3.5 Bioretention

Bioretention contains media for filtration and adsorption. Moreover, it encourages infiltration. This practice provides several mechanisms for improving water quality.

| TSS | TP | PP | DP | Total zinc | Total copper | Total lead | TN | TKN | $NO_3^-/NO_2^-$ | Temp. |
|---|---|---|---|---|---|---|---|---|---|---|
| High | High | High | Medium | High | High | High | Medium | Medium | Low | Medium |

Designers can improve bioretention's treatment of nitrogen and temperature by incorporating internal water storage.

| TSS | TP | PP | DP | Total zinc | Total copper | Total lead | TN | TKN | $NO_3^-/NO_2^-$ | Temp. |
|---|---|---|---|---|---|---|---|---|---|---|
| High | High | High | Medium | High | High | High | High | High | High | High |

Addition of aluminum-based water treatment residual (or a similar aluminum amendment) can boost phosphorus removal without sacrificing water quality treatment of other pollutants.

| TSS | TP | PP | DP |
|---|---|---|---|
| High | Very high | High | High |

### 21.3.6 Vegetated Swales

Swales can provide for removal of particulate matter and affiliated pollutants through sedimentation when flow is laminar. If media or check dams are employed, a swale's pollutant removal improves. They provide little treatment for dissolved pollutants.

| TSS | TP | PP | DP | Total zinc | Total copper | Total lead | TN | TKN | $NO_3^-/NO_2^-$ | Chloride |
|---|---|---|---|---|---|---|---|---|---|---|
| Medium | Low | Medium | Low | High | Medium | Medium | Low | Low | Very low | Low |

A bioswale is a composite SCM, enhancing a typical swale's ability to remove pollutants because bioswales incorporate a filter medium, much like bioretention and sand filters. Limited study on bioswales show substantially improved TSS removal (such that it would be considered "high" in the table above), with likely benefits extending to many other pollutants.

### 21.3.7 Stormwater Wetlands

Nutrient removal by stormwater wetlands has been well documented, as has that for sediment and sediment-bound pollutants. Less information is available on treatment of toxic metals. Constructed stormwater wetlands create an environment where aerobic and anoxic zones are in near-proximity, which facilitates microbial transformations. Wet ponds are now being modified by floating wetlands to improve their water quality treatment. In temperature-sensitive receiving waters, wetlands are preferable to wet ponds because the former's effluent is cooler; however, wetland temperature mitigation is not as good as that provided by practices that promote infiltration.

| TSS | TP | PP | Metals | TN | TKN | $NO_3^-/NO_2^-$ | Temp. |
|---|---|---|---|---|---|---|---|
| High | High | High | Medium | Medium–high | Medium–high | Medium | Low |

## 21.4 Treatment Trains

Since multiple objectives frequently must be met with stormwater management infrastructure, a *treatment train* approach is commonly necessary. In a treatment train, a set of unit treatment or management processes are placed in series, each with a specific (set of) objective(s) to achieve. This is similar to how a water or wastewater treatment plant employs specific unit processes, each specifically optimized for removal of a specific pollutant. A pretreatment system, such as a forebay, is increasingly being employed in SCM design. The forebay will remove large sediment and gross solids that can readily clog the SCM. In some cases, a treatment train is incorporated into an SCM as part of its standard design. SCMs that include treatment of multiple pollutants, such as bioretention, generally operate in this manner. Additionally, traditional SCMs (e.g., wet ponds) can be modified to include an additional treatment component. Finally, the use of multiple SCMs in series to achieve several storm water management objectives is an obvious example of treatment trains.

## 21.5 SCM Treatment Train Examples

### 21.5.1 Treatment Trains within Individual SCMs

Several SCMs contain treatment trains, either as a default component in their design or as an add-on modification. For example, the bioretention SCM has at least two treatment unit processes inherent to the design, and typically more. During bioretention operation, the bowl storage will promote sedimentation, as well as a modest amount of evaporation. As water moves into the media, filtration of fine particulate matter will occur. Adsorption of several pollutants will take place onto chemically active media. Water stored in the media will be removed via ET. Finally, infiltration will likely occur as water exits to underlying soil.

Certain modifications to SCMs can create multiple unit processes in series (e.g., adding specific layers of different media). Various media layers can include (1) a layer of soil used for the vegetation, and (2) media amendments can be added as layers for enhanced removal of pollutants, such as phosphorus. Incorporating an internal water storage zone (and its associated temporarily saturated layer) for enhanced nitrogen removal within a bioretention cell is another example of an internal treatment train (Figure 21.1).

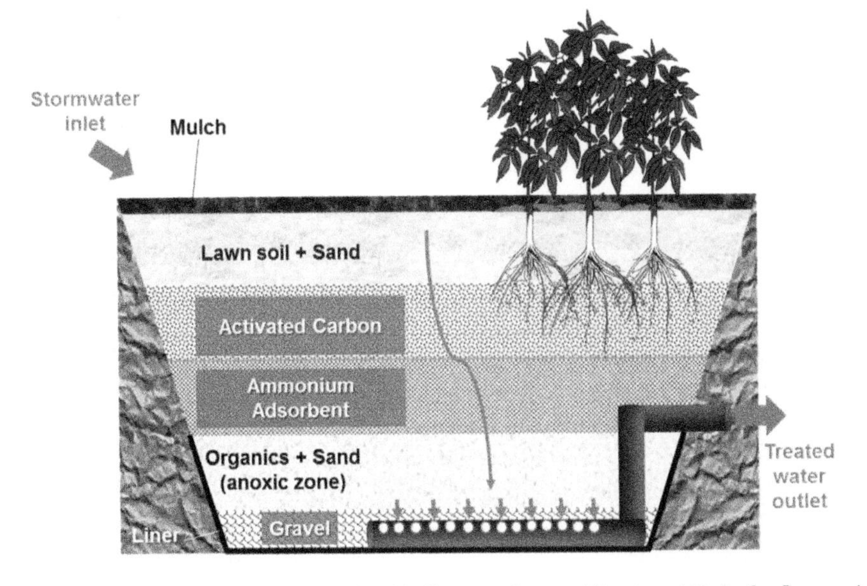


Figure 21.1 Layering of Bioretention Media, as an Internal Treatment Train, for Removal of N (Mohtadi et al. 2017).

These same approaches can be used to create treatment trains within other filtration-based practices such as green roofs and pervious pavements.

### 21.5.2 Incorporating Treatment Trains in Traditional SCMs

Because wet ponds are so commonplace across the suburban landscape, they provide opportunities for enhancement employing a retrofit treatment train approach. The most common opportunity is to add a filtration and sorption component. Examples of this retrofit (or design enhancement) include the addition of upflow filters to the outlet structure, permeable filter berms upstream of the outlet, and littoral shelf sand/media filters.

As the name implies, upflow filters force water discharged through an orifice to flow upward through a treatment medium. Some proof of concept field research has yielded generally positive results for sediment and phosphorus (Ryan et al. 2010; Winston et al. 2017), but applications of this treatment train component are likely restricted to outlets with good maintenance access. Permeable filter berms allow water to flow through them at low to moderate rates (and would be overtopped during larger storm events). Littoral shelf filters were first examined in Minnesota (Erickson and Gulliver 2010) and encourage a portion of pond water to flow through a phosphorus-adsorbing medium before being discharged to the outlet structure.

Floating wetlands were discussed in Chapter 20. A similar technology, floating treatment islands, has been employed in pond retrofit situations to incorporate additional pollutant removal mechanisms. They have been tested on multiple continents and have been shown to remove most pollutants if a sufficient number of islands are employed (Borne et al. 2013; Winston et al. 2013; Lynch et al. 2015). The addition of floating treatment islands likely improves (or creates mechanistic pathways for) nutrient uptake. Examples of several retrofits to wet ponds—and thus the creation of intra-SCM treatment trains are shown in Figure 21.2.

### 21.5.3 SCMs in Series

Placing two or more SCMs in series can lead to a number of advantages compared to only using a single SCM. Because different SCMs address different objectives (through

Figure 21.2 Pond Retrofits That Create Intra-SCM Treatment Trains: (Left) Littoral Shelf Media Filter in Rocky Mount, NC, and (Right) Floating Treatment Wetland in Auckland, New Zealand (Photos by Authors).

different treatment mechanisms), placing them in series can lead to efficient use of space.

#### 21.5.3.1 Bioretention to Infiltration Basin or Dry Detention

Placing a bioretention SCM "upstream" of an infiltration basin or dry detention basin can overcome some of the limitations of each (Figure 21.3). The bioretention cell will manage runoff from the majority of the smallest events, producing little or no runoff discharge. Only the largest events will lead to discharge to the infiltration basin (for large-scale infiltration and peak flow mitigation) or dry detention (for peak flow mitigation). Additionally, the bioretention cell will reduce the particulate matter concentrations that are loaded to the larger, downstream, basin, which prolongs the maintenance interval required of the second practice in series.

The reduction of water volumes and particulate matter loads to downstream basins allow a smaller footprint for the second practice, minimize maintenance requirements, and greatly prolong the active life of the infiltration basin. Dry detention, which provides limited pollutant removal mechanisms, and bioretention, which rarely mitigates peak flows from events with a greater than 2-year ARI, are a "natural" pair as part of a treatment train system. A bioretention facility upstream of other large detention devices such as wet ponds and constructed stormwater wetlands is occasionally implemented on sites required to meet both nutrient removal and peak flow needs.

#### 21.5.3.2 Treatment SCMs to Cisterns

A treatment train consisting of a series of bioretention cells draining to a cistern has been investigated for water harvesting (Doan and Davis 2017). The drainage area was a parking lot which drains to the bioretention cells (Figure 21.4). Water that percolated the bioretention media flowed through an underdrain, feeding a cistern. As noted in Chapter 18, bioretention is very effective in the removal of suspended solids, heavy metals, and some toxic organic compounds, allowing stormwater from a parking lot to be treated to a sufficient quality such that it can be used for basic outdoor uses. When water quality of the cistern was examined, tests showed low concentrations of heavy metals, but ample concentrations of N and P (Table 21.2) for fertigation. In fact, if fertigation is an intended use of this treatment train, bioretention media could be selected

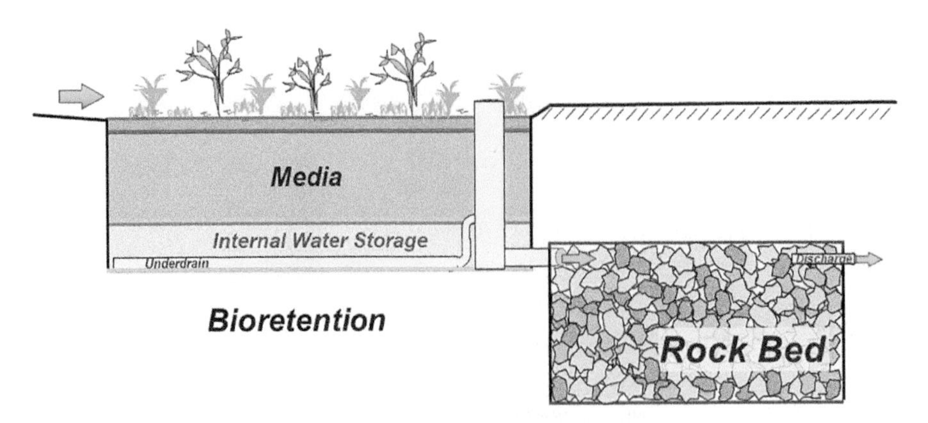


Figure 21.3 Diagram of Bioretention/Rock Bed in Series.

Figure 21.4 SCM Treatment Train Consisting of a Bioretention Cell Accepting Runoff from a Parking Lot as the First Step. Underdrain Flow from the Bioretention Cell Enters a Cistern, Where a Solar Pump Provides Water for Vegetable Garden Irrigation from Rain Barrels (Photos by Authors).

Table 21.2 Runoff Water Quality and Water Quality in Cistern (Figure 21.4) after Bioretention Treatment. Tap Water Quality Is Included for Comparison (Doan and Davis 2017, with Permission from ASCE).

| Parameters | Average inflow | Average cistern | Tap water ($n$ = 8) |
|---|---|---|---|
| TSS (mg/L) | 164 ± 104, $n = 27$ | 14 ± 10, $n = 27$ | Not measured |
| TP (mg/L) | 0.53 ± 0.32, $n = 27$ | 0.17 ± 0.04, $n = 27$ | 0.28 ± 0.01 |
| TN (mg/L) | 2.23 ± 1.33, $n = 25$ | 1.08 ± 0.53, $n = 25$ | 1.85 ± 0.13 |
| EC (mS/cm) | 0.51 ± 0.83, $n = 18$ | 0.11 ± 0.02, $n = 18$ | 0.23 ± 0.01 |
| pH | 7.25 ± 0.42, $n = 18$ | 6.98 ± 0.49, $n = 18$ | 7.11 ± 0.12 |
| Cu (μg/L) | 19.3 ± 14.7, $n = 22$ | 15.7 ± 17.6, $n = 22$ | 166 ± 6.5 |
| Pb (μg/L) | 5.3 ± 3.1, $n = 14$ | 0.87 ± 1.3, $n = 14$ | 0.88 ± 0.36 |
| Zn (μg/L) | 58 ± 56.2, $n = 19$ | 12.2 ± 6.3, $n = 19$ | 81 ± 38.1 |

that had minimal nutrient removal or even purposely "leached" N and P for this use. The net result of the bioretention/cistern treatment train results in both minimal runoff from the parking lot and a water supply apt for fertigation.

A similar approach to treating stormwater runoff for cistern use has been employed with permeable pavement as the initial SCM (in lieu of bioretention). This concept was studied in Ohio (Figure 21.5) and while metals concentrations leaving the permeable pavement were low (as typical), nutrient concentrations were moderately high, again making for a "good" fertigation source (Winston et al. 2020).

#### 21.5.3.3 Treatment Trains Employing Swales

Because swales (1) are often used to convey water and (2) provide typically modest amounts of pollutant treatment, they are often a part of multi-SCM treatment trains. Usually, swales are the first component of the treatment train as a water collection SCM before being delivered to a second practice downstream. This example is illustrated in Figure 21.6 which shows a treatment train consisting of a vegetated swale feeding a bioswale, which discharges into an infiltration trench. This system manages and treats runoff from a large parking garage.

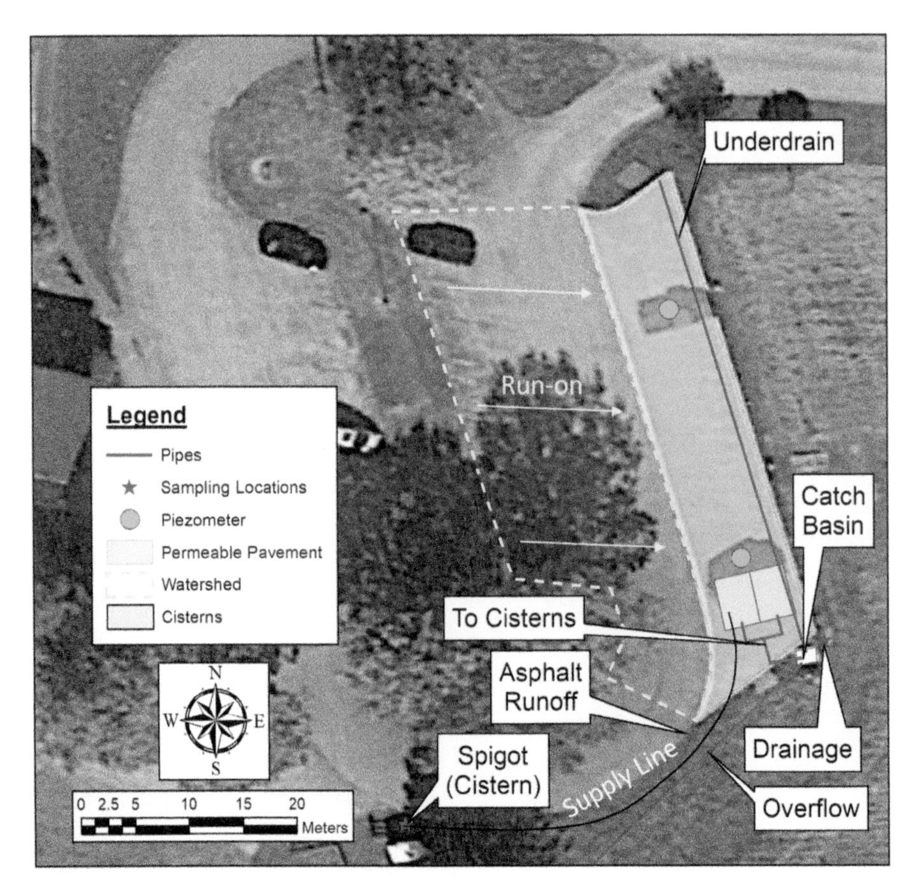


Figure 21.5 A Permeable Pavement-cistern Treatment Train Installed and Tested in Parma, Ohio (Winston et al. 2020).

Figure 21.6 Grass Swale–bioswale-infiltration Trench Treatment Train at Villanova University (Photo by Authors).

Figure 21.7 A Typical Stormwater Treatment Train along a Divided Highway in Eastern North Carolina: Water from Pervious Friction Course Asphalt Flows through a Filter Strip En Route to a Wetland Swale (Photo by Authors).

Another common treatment train involving swales is associated with highway runoff treatment, where stormwater flows from pavement, through a filter strip (also part of the roadway shoulder), and then into a swale. This "classic" treatment train is illustrated in Figure 21.7 and has been shown to be quite effective at trapping sediment and other pollutants (Barrett et al. 1998; Winston et al. 2012). More details on these SCMs are provided in Chapter 19.

## 21.6 Quantifying Performance in SCM Treatment Trains

Quantifying hydrologic and water quality performance of SCMs in treatment trains is not consistent, and the most common currently employed method is erroneous in its load-reduction calculation. The percent removal metric (while typically employed) likely results in a major overestimation of treatment. Consider an example of a first SCM that is assumed to remove 80% of a pollutant (say TSS) and this SCM is then followed by a second SCM that also is assigned 80% removal. Would this pairing truly reduce TSS load by an overall 96% [1–(1–0.8)(1–0.8)]? Likely not.

As discussed in several previous chapters, the efficiency of the latter unit processes in series will depend on the pollutant concentration discharged into them. Each unit process or SCM should be evaluated individually based on the condition to which it is exposed. The discharge flow and pollutant concentrations from the preceding SCM become the input concentrations to the second (Figure 21.8). If the inflow concentration is low, additional pollutant reduction may not occur unless pollutants are exposed to new (to the pollutant) unit processes.

One metric employed to evaluate treatment trains is based upon average effluent concentrations discharged by SCMs. Studies have demonstrated that vegetation-and-soil-based SCMs tend to discharge toward median effluent concentrations (McNett et al. 2011; Li and Davis 2016), also discussed in Chapter 11. When this concept is coupled with the idea that individual SCMs reduce stormwater volume but eventually have minor impact on concentration, it is possible to calculate a pollutant load using the following equation:

$$Load = Vol_{INF} \times DVol_{SCM1} \times DVol_{SCM2} \times C_{SCM2} \times f \tag{21.1}$$

where load is the discharge pollutant load (in kg or lbs); $Vol_{INF}$ is the volume of water entering the first SCM of the treatment train; $DVol_{SCM1}$ is the fraction of volume leaving the first SCM (%); $DVol_{SCM2}$ is the fraction of volume leaving the second SCM (%); $C_{SCM2}$ is the concentration of water leaving the second SCM (mg/L); and $f$ is the appropriate unit conversion factor.

Inherent in this analysis is that the final SCM of the treatment is the "gatekeeper" concentration for the system. The median effluent concentration of the last SCM will dictate the concentration part of the load discharged. While this calculation is not fully rigorous, it is considered a more accurate estimation of treatment train performance than simple repetition of the pollutant load per cent removal efficiency.

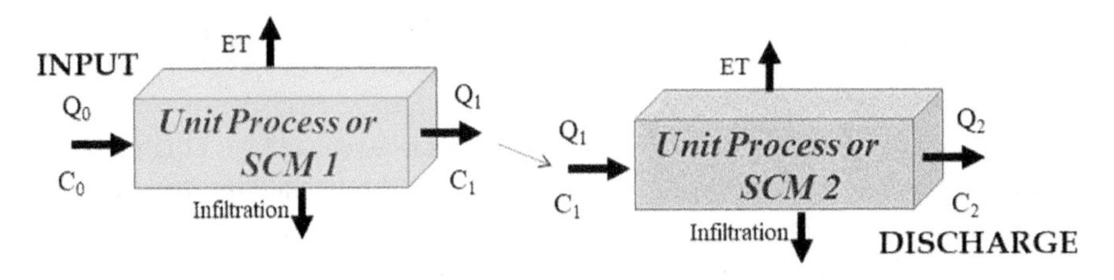


Figure 21.8 Series Approach to Evaluating Hydrology and Water Quality Improvements from Two Unit Processes or SCMs in Series.

## 21.7 Real Time Controls

Real-time control (RTC) of SCMs appears to be an upcoming design and operation feature (Kerkez et al. 2016). The premise behind RTC is that SCMs will be equipped with sensors that collect data in real time allowing for preprogrammed operation of the practice. One example is an SCM that communicates with the National Weather Service (NWS). When the NWS forecasts runoff-producing precipitation (rainfall) within a 36-hour window the SCM will "check" to see if sufficient volume is present to capture the storm event (via a depth measurement). If storage capacity is limited, then the RTC will dictate prestorm discharge so that volumetric capacity in the SCM is created and available to capture the entirety of the storm (assuming the event is not large). By the time the storm event arrives, the SCM will ideally be sufficiently dewatered and thus no discharge will take place intra-event. Conversely, if dry weather is expected, water left in storage is available for beneficial use, to infiltrate, and/or to promote denitrification or some other water quality benefit. The benefits of eliminating intra-event discharge are multiple including (1) a reduction in peak flows released to stream banks (thus limiting scour), (2) a reduction in the change of surcharge of wastewater treatment plants in combined sewer overflows (CSOs), and (3) an increase in retention time of stormwater so that pollutants such as various nitrogen species are provided ample time to eventually be denitrified. An example of an RTC rainwater harvesting system is illustrated in Figure 21.9.

Figure 21.9 A Real-time Control (Gray Box) Manages Water Stored in a Series of Black Tanks (Top Left) That Is Used to Irrigate a Garden in New Bern, NC (Top Right). In Advance of Large Rainfall Events (Bottom Left), Stored Water Was Often Discharged into an Infiltration Zone (Bottom Right) (Photos by Authors).

New developments in soil moisture and water quality sensors can provide dynamic information so that SCMs can respond to changing conditions and deliver the best performance possible. Early research (e.g., Gee and Hunt 2016; Schmitt et al. 2020) of this management tool provides evidence of its effectiveness; however, its ultimate application will likely be focused on larger systems and with property owners with ample financial resources.

## 21.8 Designing for Climate Change

During the lifetime of a SCM, it is likely that climate change will impact both the characteristics of the runoff input and the SCM performance. Temperatures and precipitation patterns will change with climate change. Many climate models predict somewhat longer dry periods punctuated by more intense rainfall. Larger storm volumes are also indicated. Projected seasonal changes in precipitation in the United States are shown in Figure 21.10. In general, the winter is wetter and the summer is drier, but this depends on the geographical area.

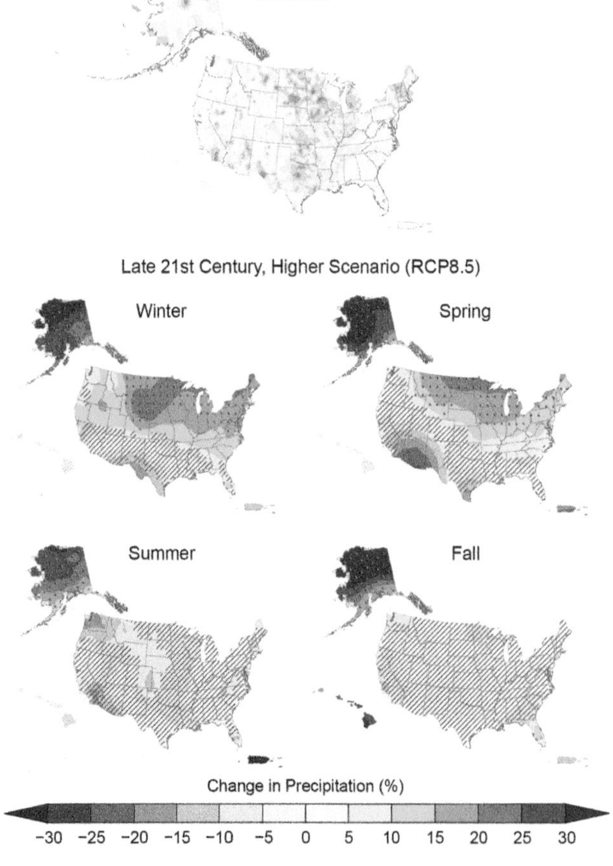


Figure 21.10 Projected Changes in Seasonal Precipitation in North America. http://www.globalchange.gov.

A change in long-term weather patterns will influence SCM performance. Obviously, the degree of change impacts whether an SCM can fulfill its design intent. Other factors such as increased urbanization (and the heat island associated with more impervious surfaces), SCM construction tolerances, and (lack of) SCM maintenance may even be more important issues for which designers must account. Some SCMs such as green roofs and other vegetation-based technologies may become more attractive in urban areas under climate change scenarios due to their evaporative cooling impacts.

Figure 21.11 shows predicted changes to IDF curves in Auburn AL based on three different downscaled climate change models (Mirhosseini et al. 2013). While the details are different for the three models, some overall trends are apparent. For rainfall events less than 4 h duration, lower intensity events are predicted for both 10-year and 100-year events. For larger durations, some models predicted higher intensities, while others predicted lower. Zhu et al. (2012) note that such analyses are highly specific to the region under investigation.

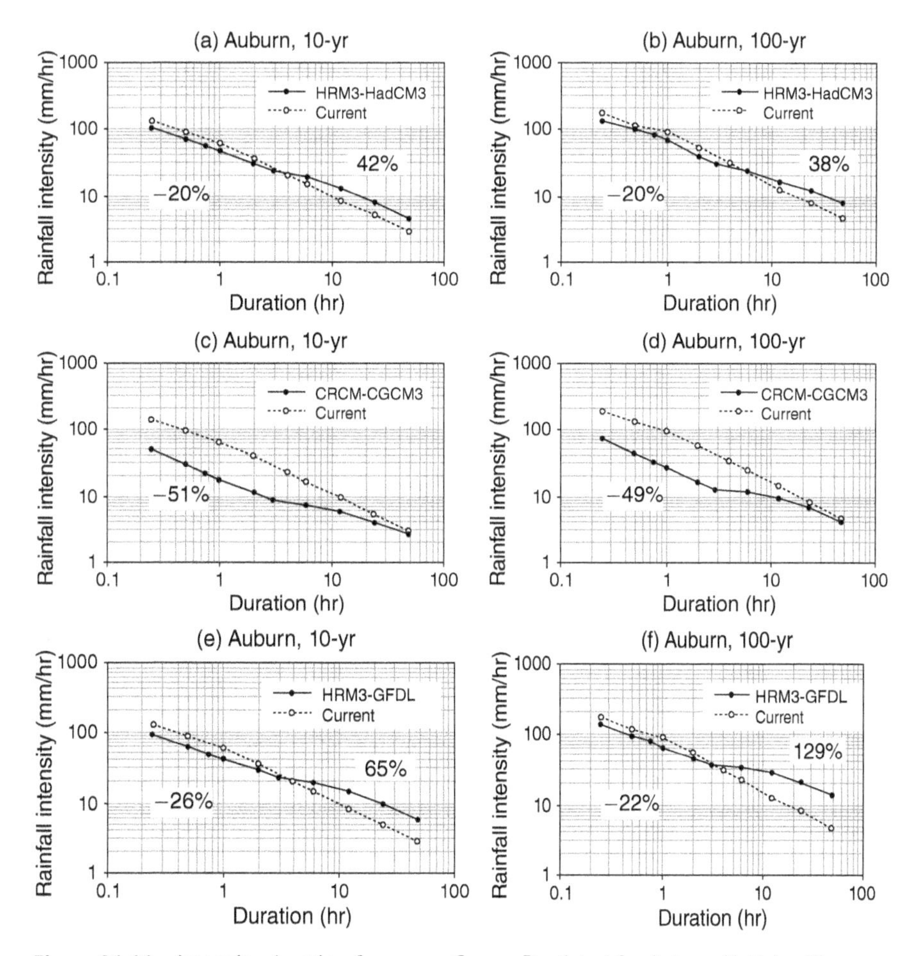


Figure 21.11 Intensity-duration-frequency Curves Predicted for Auburn AL Using Three Climate Change Models, Compared to Existing IDF Curves (Mirhosseini et al. 2013).

Increased rainfall/runoff volume and higher flow rates will lead to greater instances of bypass/overflow in green infrastructure SCMs. This produces a greater fraction of untreated water discharged to the local environment (Hathaway et al. 2014). To address the issue of greater runoff volume and flows, the SCMs will need to contain greater storage volumes, or more will have to be constructed throughout the watershed. Hathaway et al. (2014), for example, suggest that the bowls of bioretention cells might need to be 2–4 in (5–10 cm) deeper to account for changing rainfall patterns in North Carolina, if the annual capture volume provided in 2050 is to be the same as 2010. Other research (Smolek 2016) suggests that little additional gravel depth would be necessary in Northeast Ohio to achieve similar hydrologic capture in 2010 and 2050. The contrasting guidance provided in this research illustrates how variable—and therefore site specific—climate change may be for SCM design.

In addition to the obvious impact on the size and number of practices, many other factors that control GI SCM design and performance will have to be considered under various climate change scenarios. Increasing temperature will modestly increase infiltration rates due to the relationship of water viscosity with temperature (greater than 4°C). Chemical reactions and biological processes that occur during, and especially between, storm events will be also faster at higher temperatures. Vegetation survival differs at different temperature ranges and planting palettes may need to be adjusted. Nature is demonstrating how local plant palettes will likely change as certain species' migrations are observed. Vegetation long-regarded as coastal (and therefore warmer temperature tolerant) have migrated inland to the NC Piedmont (e.g., coastal dog hobble, *Leucothoe axillaris*). Microbial populations may shift due to differing temperatures, moisture contents, and chemical conditions. Watershed-scale processes that shape the hydrologic and water quality characteristics of incoming runoff may be altered. Overall, SCM performance as we know it now will likely be different in the future.

An additional impact of climate change is rise in sea level. This directly impacts existing urban stormwater infrastructure in coastal regions as water levels rise in receiving streams, thus elevating groundwater. Backwater retards SCM drainage and higher groundwater levels reduce the depth available for installing subsurface and/or infiltration-based SCMs.

Finally, increasing global temperatures will increase urban demand for water. This will produce added stresses on local surface and groundwater supplies. Stormwater will likely become a more valuable resource. Perhaps we should expect direct capture, on-site treatment, and local harvesting technologies to become more prevalent. Large-scale examples of this already exist in semi-arid locales like Santa Monica, California, or long-termed drought-stricken places like Queensland, Australia.

The magnitude of impact of a changing rainfall pattern on SCMs to that of incorrect construction oversight (Hunt et al. 2011) or misguided maintenance (Wardynski and Hunt 2012) is very similar. In fact, studies argue that poor maintenance has a more pronounced impact on long-term SCM effectiveness than our changing climate (Blecken et al. 2017; Hathaway et al. 2014).

## 21.9 Greener Infrastructure: What Does the Future Hold?

The field of green stormwater infrastructure is changing rapidly. Many jurisdictions are seeing GI as an effective way to address CSO and TMDL challenges. Recent research, monitoring, and modeling studies have provided strong evidence that GI SCMs are effective when designed, installed, and maintained appropriately. Research has also provided fundamental mechanistic information about SCM performance.

Better predictions and understanding of site-specific conditions will allow for novel designs to meet specific, targeted goals. Research will continue to provide greater understanding of operation and to guide improved design and performance. Advanced modeling will allow for optimization of SCM installations by evaluating various design scenarios. Because this is an evolving field, many areas of GI are still very poorly understood, including the role of vegetation, long-term fates of pollutants, optimization of maintenance, and dynamic performance evaluation.

Advances in technology may allow for flexible operations of SCMs, depending on existing and expected weather conditions (Kerkez et al. 2016). This will potentially help maintenance providers who can inspect "suspect" practices based on erratic behavior as observed by RTCs; RTC should make SCMs more efficient.

The public and decision makers need to be informed on the benefits of GI stormwater infrastructure and how it is integrated into our urban lifestyle. Information indicates that decision makers remain cautious on integrating GI throughout the built environment due to concerns of inconsistent performance and high-maintenance requirements (Meng et al. 2017). Continued research, education, and experience with well-designed and maintained systems will facilitate increasing overall confidence in GI stormwater infrastructure.

Finally, as noted early in this text, and as demonstrated in many areas of the United States and world, urban infrastructure inequity is being recognized by many jurisdictions. Efforts are being made in many cases to address aspects of this inequity. It has become clear that infrastructure decisions, including green infrastructure, must be made using not just engineering, economic, planning, and similar traditional metrics but must also consider the historic and social needs of the neighborhood. The neighborhood should be a partner in the decision-making process. The most effective stormwater control measure is one that will integrate within the neighborhood and be embraced by the residents. Such an SCM may have the greatest probability of being effective and attractive for a long time.

## References


Barrett, M.E., Walsh, P.M., Malina, J.F., and Charbeneau, R.J. (1998). Performance of vegetative controls for treating highway runoff. *Journal of Environmental Engineering, American Society of Civil Engineers* 124 (11): 1121–1128.

Blecken, G.-T., Hunt, W.F., Al-Rubaei, A.M., Viklander, M., and Lord, W.G. (2017). Stormwater control measure (SCM) maintenance considerations to ensure designed functionality. *Urban Water Journal, Taylor & Francis* 14 (3): 278–290.

Borne, K.E., Fassman, E.A., and Tanner, C.C. (2013). Floating treatment wetland retrofit to improve stormwater pond performance for suspended solids, copper and zinc. *Ecological Engineering* 54: 173–182.

Doan, L. and Davis, A.P. (2017). Bioretention/cistern/irrigation treatment train to minimize stormwater runoff. *J. Sustainable Water Built Environ.* 3 (2): 04017003.

Erickson, A.J. and Gulliver, J.S. (2010). *Performance Assessment of an Iron-Enhanced Sand Filtration Trench for Capturing Dissolved Phosphorus*. Minneapolis, MN: Report, St. Anthony Falls Laboratory, University of Minnesota.

Gee, K.D. and Hunt, W.F. (2016). Enhancing stormwater management benefits of rainwater harvesting via innovative technologies. *Journal of Environmental Engineering, American Society of Civil Engineers* 142 (8): 04016039.

Hathaway, J.M., Brown, R.A., Fu, J.S., and Hunt, W.F. (2014). Bioretention function under climate change scenarios in North Carolina, USA. *Journal of Hydrology* 519: 503–511.

Hunt, W.F., Greenway, M., Moore, T.C., Brown, R.A., Kennedy, S.G., Line, D.E., and Lord, W.G. (2011). Constructed storm-water wetland installation and maintenance: Are we getting it right? *Journal of Irrigation and Drainage Engineering* 137 (8): 469–474.

Kerkez, B., Gruden, C., Lewis, M., Montestruque, L., Quigley, M., Wong, B., Bedig, A., Kertesz, R., Braun, T., Cadwalader, O., Poresky, A., and Pak, C. (2016). Smarter stormwater systems. *Environmental Science & Technology* 50 (14): 7267–7273.

Li, J. and Davis, A.P. (2016). A unified look at phosphorus treatment using bioretention. *Water Research* 90: 141–155.

Lynch, J., Fox, L. J., Owen, J. S., Jr., and Sample, D. J. (2015). Evaluation of commercial floating treatment wetland technologies for nutrient remediation of stormwater. *Ecological Engineering*, 75: 61–69.

McNett, J. K., Hunt, W. F., and Davis, A. P. (2011). Influent Pollutant Concentrations as Predictors of Effluent Pollutant Concentrations for Mid-Atlantic Bioretention. *Journal of Environmental Engineering-ASCE*, 137 (9): 790–799.

Meng, T., Hsu, D., and Wadzuk, B. (2017). Green and smart: perspectives of city and water agency officials in Pennsylvania toward adopting new infrastructure technologies for stormwater management. *Journal of Sustainable Water in the Built Environment* 3 (2): 05017001.

Mirhosseini, G., Srivastava, P., and Stefanova, L. (2013). The impact of climate change on rainfall Intensity–Duration–Frequency (IDF) curves in Alabama. *Regional Environmental Change* 13 (1): 25–33.

Mohtadi, M., James, B.R., and Davis, A.P. (2017). Adsorption of compounds that mimic urban stormwater dissolved organic nitrogen. *Water Environment Research* 89 (2): 105–116.

National Cooperative Highway Research Program (NCHRP) (2013). *Pollutant Load Reductions for the Total Maximum Daily Loads for Highways*. Synthesis 444 (NCHRP 2013). (S.A. Abbasi and A. Koskelo), Washington DC: Consultants, Transportation Research Board.

Ryan, P., Wanielista, M., and Chang, N.-B. (2010). Nutrient reduction in stormwater pond discharge using a Chamber Upflow Filter And Skimmer (CUFS). *Water, Air, and Soil Pollution* 208 (1): 385–399.

Schmitt, Z. K., Hodges, C. C., and Dymond, R. L. (2020). Simulation and assessment of long-term stormwater basin performance under real-time control retrofits. *Urban Water Journal*, 17 (5), 467–480.

Smolek, A.P. (2016). *Monitoring and Modeling the Performance of Ultra-urban Stormwater Control Measures in North Carolina and Ohio*. Ph.D. Dissertation. Raleigh, NC.: North Carolina State University. https://catalog.lib.ncsu.edu/catalog/NCSU3574323.

Wardynski, B.J. and Hunt, W.F. (2012). Are bioretention cells being installed per design standards in North Carolina? A field study. *Journal of Environmental Engineering* 138 (12): 1210–1217.

Winston, R.J., Arend, K., Dorsey, J.D., and Hunt, W.F. (2020). Water quality performance of a permeable pavement and stormwater harvesting treatment train stormwater control measure. *Blue-Green Systems, IWA Publishing* 2 (1): 91–111.

Winston, R.J., Hunt, W.F., Kennedy, S.G., Merriman, L.S., Chandler, J., and Brown, D. (2013). Evaluation of floating treatment wetlands as retrofits to existing stormwater retention ponds. *Ecological Engineering* 54: 254–265.

Winston, R.J., Hunt, W.F., Kennedy, S.G., Wright, J.D., and Lauffer, M.S. (2012). Field evaluation of storm-water control measures for highway runoff treatment. *Journal of Environmental Engineering* 138 (1): 101–111.

Winston, R.J., Hunt, W.F., and Pluer, W.T. (2017). Nutrient and sediment reduction through upflow filtration of stormwater retention pond effluent. *Journal of Environmental Engineering, American Society of Civil Engineers* 143 (5): 06017002.

Zhu, J., Stone, M.C., and Forsee, W. (2012). Analysis of potential impacts of climate change on intensity–duration–frequency (IDF) relationships for six regions in the United States. *Journal of Water and Climate Change, IWA Publishing* 3 (3): 185–196.

## Problems

**21.1** List five different SCMs and provide four considerations for their selection. Example:
Stormwater wetland. Consideration: water table near the surface is best.

# Appendix A

Physical Properties of Water in SI Units.

| Temp. (°C) | Specific weight (kN/m$^3$) | Density (kg/m$^3$) | Absolute viscosity × $10^{-3}$ (N s/m$^2$) | Kinematic viscosity × $10^{-6}$ (m$^2$/s) |
|---|---|---|---|---|
| 0 | 9.805 | 999.8 | 1.781 | 1.785 |
| 5 | 9.807 | 1000.0 | 1.518 | 1.518 |
| 10 | 9.804 | 999.7 | 1.307 | 1.306 |
| 15 | 9.798 | 999.1 | 1.139 | 1.139 |
| 20 | 9.789 | 998.2 | 1.002 | 1.003 |
| 25 | 9.777 | 997.0 | 0.890 | 0.893 |
| 30 | 9.764 | 995.7 | 0.798 | 0.800 |
| 40 | 9.730 | 992.2 | 0.653 | 0.658 |
| 50 | 9.689 | 988.0 | 0.547 | 0.553 |
| 60 | 9.642 | 983.2 | 0.466 | 0.474 |
| 70 | 9.589 | 977.8 | 0.404 | 0.413 |
| 80 | 9.530 | 971.8 | 0.354 | 0.364 |
| 90 | 9.466 | 965.3 | 0.315 | 0.326 |
| 100 | 9.399 | 958.4 | 0.282 | 0.294 |

Physical Properties of Water in English Units.

| Temp. (°F) | Specific weight (lb/ft$^3$) | Density (lb s$^2$/ft$^4$) | Absolute viscosity × $10^{-5}$ (lb s/ft$^2$) | Kinematic viscosity × $10^{-5}$ (ft$^2$/s) |
|---|---|---|---|---|
| 32 | 62.42 | 1.940 | 1.931 | 1.931 |
| 40 | 62.43 | 1.938 | 1.664 | 1.664 |
| 50 | 62.41 | 1.936 | 1.410 | 1.410 |
| 60 | 62.37 | 1.934 | 1.217 | 1.217 |
| 70 | 62.30 | 1.931 | 1.059 | 1.059 |
| 80 | 62.22 | 1.927 | 0.930 | 0.930 |
| 90 | 62.11 | 1.923 | 0.826 | 0.826 |
| 100 | 62.00 | 1.918 | 0.739 | 0.739 |
| 120 | 61.71 | 1.908 | 0.609 | 0.609 |
| 140 | 61.38 | 1.896 | 0.514 | 0.514 |
| 160 | 61.00 | 1.896 | 0.442 | 0.442 |
| 180 | 60.58 | 1.883 | 0.385 | 0.385 |
| 200 | 60.12 | 1.868 | 0.341 | 0.341 |
| 212 | 59.83 | 1.860 | 0.319 | 0.319 |

*Green Stormwater Infrastructure Fundamentals and Design*, First Edition. Allen P. Davis, William F. Hunt, and Robert G. Traver.

Companion Website: www.wiley.com/go/davis/greenstormwater

# Index

Tables shown in **bold** Figures shown in *italic*



*Green Stormwater Infrastructure Fundamentals and Design*, First Edition. Allen P. Davis, William F. Hunt, and Robert G. Traver.

Companion Website: www.wiley.com/go/davis/greenstormwater